ISBN 978-1-5284-4308-1
PIBN 10917699

1 MONTH OF
FREE
READING

at

www.ForgottenBooks.com

By purchasing this book you are eligible for one month membership to ForgottenBooks.com, giving you unlimited access to our entire collection of over 1,000,000 titles via our web site and mobile apps.

To claim your free month visit:

www.forgottenbooks.com/free917699

English
Français
Deutsche
Italiano
Español
Português

www.forgottenbooks.com

Mythology Photography **Fiction**
Fishing Christianity **Art** Cooking
Essays Buddhism Freemasonry
Medicine **Biology** Music **Ancient
Egypt** Evolution Carpentry Physics
Dance Geology **Mathematics** Fitness
Shakespeare **Folklore** Yoga Marketing
Confidence Immortality Biographies
Poetry **Psychology** Witchcraft
Electronics Chemistry History **Law**
Accounting **Philosophy** Anthropology
Alchemy Drama Quantum Mechanics
Atheism Sexual Health **Ancient History**
Entrepreneurship Languages Sport
Paleontology Needlework Islam
Metaphysics Investment Archaeology
Parenting Statistics Criminology
Motivational

UNITED STATES REPORTS

VOLUME 182

CASES ADJUDGED

IN

THE SUPREME COURT

AT

OCTOBER TERM, 1900

J. C. BANCROFT DAVIS

REPORTER

THE BANKS LAW PUBLISHING CO.
21 MURRAY STREET, NEW YORK
1901

REPRINTED IN TAIWAN

JUSTICES

SUPREME COURT

DURING THE TIME OF THESE REPORTS.

MELVILLE WESTON FULLER, CHIEF JUSTICE.
JOHN MARSHALL HARLAN, ASSOCIATE JUSTICE.
HORACE GRAY, ASSOCIATE JUSTICE.
DAVID JOSIAH BREWER, ASSOCIATE JUSTICE.
HENRY BILLINGS BROWN, ASSOCIATE JUSTICE.
GEORGE SHIRAS, JR., ASSOCIATE JUSTICE.
EDWARD DOUGLASS WHITE, ASSOCIATE JUSTICE.
RUFUS W. PECKHAM, ASSOCIATE JUSTICE.
JOSEPH McKENNA, ASSOCIATE JUSTICE.

PHILANDER CHASE KNOX, ATTORNEY GENERAL.
JOHN KELVEY RICHARDS, SOLICITOR GENERAL.
JAMES HALL McKENNEY, CLERK.
JOHN MONTGOMERY WRIGHT, MARSHAL.

TABLE OF CONTENTS.

TABLE OF CASES REPORTED.

v

Table of Cases Reported.

Table of Cases Reported.

TABLE OF CASES

CITED IN OPINIONS.

ix

TABLE OF CASES CITED.

TABLE OF STATUTES

CITED IN OPINIONS.

(A.) STATUTES OF THE UNITED STATES.

CASES ADJUDGED

SUPREME COURT OF THE UNITED STATES,

AT

OCTOBER TERM, 1900.

DE LIMA *v.* BIDWELL.

ERROR TO THE CIRCUIT COURT OF THE UNITED STATES FOR THE
SOUTHERN DISTRICT OF NEW YORK.

No. 456. Argued January 8, 9, 10, 11, 1901.—Decided May 27, 1901.

By the Customs Administrative Act of 1890 an appeal is given from the decision of the collector "as to the rate and amount of the duties chargeable upon imported merchandise," to the Board of General Appraisers,
who are authorized to decide "as to the construction of the law and the
facts respecting the classification of such merchandise; and the rate of
duties imposed thereon under such classification;" but where the merchandise is alleged not to have been imported at all, but to have been
brought from one domestic port to another, the Board of General Appraisers has no jurisdiction of the case, and an action for money had and
received will lie against the collector to recover back duties assessed by
him upon such property, and paid under protest.

With the ratification of the treaty of peace between the United States and
Spain, April 11, 1899, the island of Porto Rico ceased to be a "foreign
country" within the meaning of the tariff laws.

Whatever effect be given to the act of March 24, 1900, applying for the benefit of Porto Rico the duties received on importations from that island
after the evacuation by the Spanish forces, it has no application to an action brought before the act was passed.

THIS was an action originally instituted in the Supreme Court
of the State of New York by the firm of D. A. De Lima & Co.

against the collector of the port of New York, to recover back duties alleged to have been illegally exacted and paid under protest, upon certain importations of sugar from San Juan in the island of Porto Rico, during the autumn of 1899, and subsequent to the cession of the island to the United States.

Upon the petition of the collector, and pursuant to Rev. Stat. sec. 643, the case was removed by certiorari to the Circuit Court of the United States, in which the defendant appeared and demurred to the complaint upon the ground that it did not state a cause of action, and also that the court had no jurisdiction of the case. The demurrer was sustained upon both grounds, and the action dismissed. Hence this writ of error.

In this and the following cases, which may be collectively designated as the "Insular Tariff Cases," the dates here given become material:

In July, 1898, Porto Rico was invaded by the military forces of the United States under General Miles.

On August 12, 1898, during the progress of the campaign, a protocol was entered into between the Secretary of State and the French Ambassador on the part of Spain, providing for a suspension of hostilities, the cession of the island and the conclusion of a treaty of peace. 30 Stat. 1742.

On October 18, Porto Rico was evacuated by the Spanish forces.

On December 10, 1898, such treaty was signed at Paris, (under which Spain ceded to the United States the island of Porto Rico,) was ratified by the President and Senate, February 6, 1899, and by the Queen Regent of Spain, March 19, 1899. 30 Stat. 1754.

On March 2, 1899, an act was passed making an appropriation to carry out the obligations of the treaty.

On April 11, 1899, the ratifications were exchanged, and the treaty proclaimed at Washington.

On April 12, 1900, an act was passed, commonly called the Foraker Act, to provide temporary revenues and a civil government for Porto Rico, which took effect May 1, 1900.

This case was argued with No. 507, *Downes* v. *Bidwell;* No. 501, *Dooley* v. *United States;* No. 502, *Dooley* v. *United*

States; No. 509, *Armstrong* v. *United States.* The briefs and the arguments were reported at length in a book entitled " The Insular Cases," compiled and published pursuant to a resolution of the House of Representatives passed in the Second Session of the 56th Congress, and containing both the briefs of counsel and their oral arguments. They amounted to 1075 pages. Of course it is impossible to reproduce all here, even if it were desirable.

Mr. Frederick R. Coudert, Jr., for plaintiff in error. *Mr. Charles Frederick Adams* and *Mr. Paul Fuller* were on his brief.

The questions of law involved are: *First,* whether the said circuit court " *had jurisdiction* of the cause of action alleged in the complaint against the defendant." *Second,* whether " the complaint states facts sufficient to constitute a cause of action against the defendant."

1. The questions are raised under the following circumstances: " On or about the 6th day of November, 1899, the defendant " (being at the time " the duly appointed and commissioned collector of customs of the United States at the port of New York, in the actual and unrestricted exercise of his functions as such collector, and fully vested with all the powers and authority of his said office ") " did under color of his said office and through the . . . exercise of the powers and authority in him vested for the purposes of the performance of his duties as such collector, . . . demand and by duress of goods collect from the plaintiffs' said firm of D. A. De Lima & Co., as alleged duties upon certain sugars, the product of the island of Puerto Rico, consigned to (said) plaintiffs at the port of New York, and brought thither from the port of San Juan in the said island during the month of July, 1899, by steamer *Salamanca* (the said sugars being those mentioned and described in warehouse entry No. 117,587, bond No. 1224, liquidated September 11, 1899), the sum of two thousand four hundred and fifty dollars and fifty-eight cents ($2450.58), which sum the plaintiffs were . . . against their will and in spite of their formal protest duly made, com-

pelled to pay, and did pay, in order to obtain possession of the
said sugars, ... which the said defendant, enabled so to
do by the power and authority of his said office, had detained,
was detaining, and threatened to continue to detain from them,
exacting as a condition to the delivery thereof such payment of
said alleged duties. . . .

". . . On or about the 14th day of September, 1899, the
defendant, being such collector as aforesaid, did, under color of
his said office, and through the . . . exercise of the powers
and authority in him vested for the purposes of the performance
of his duties as such collector, . . . demand and by duress
of goods collect from the plaintiffs' said firm of D. A. De Lima
& Co., as alleged duties upon certain sugars, the product of the
island of Puerto Rico, consigned to the plaintiffs at the port of
New York, and brought thither from the port of San Juan in
the said island during the month of June, 1899, by steamer
Evelyn (the said sugars being those mentioned and described in
consumption entry No. 95,684, liquidated Sept. 11, 1899), the sum
of five thousand four hundred and fifty-two dollars and sixty-
one cents ($5,452.61), which sum (the) plaintiffs were . . .
against their will and in spite of their formal protest duly made,
compelled to pay, and did pay, in order to obtain possession of
said sugars, . . . which the said defendant, enabled so to do
by the power and authority of his said office, had detained, was
detaining, and threatened to continue to detain from them, exacting as a condition to the delivery thereof such payment of such
alleged duties. . . .

". . . On or about the 1st day of September, 1899, the
defendant being such collector as aforesaid, did, under color of
his said office and through the . . . exercise of the powers
and authority in him vested for the purpose of the performance
of his duties as such collector . . . demand and by duress
of goods collect from the plaintiffs' said firm of D. A. De Lima
& Co., as alleged duties upon certain sugars, the product of the
island of Puerto Rico, consigned to (the) plaintiffs at the port of
New York, and brought thither from the port of San Juan, in
the said island, during the month of —————, 1899, by steamer
Catania (the said sugars being those mentioned and described

in consumption entry No. 89,319, liquidated September 1st, 1899), the sum of five thousand two hundred and forty-two dollars and seventeen cents ($5242.17), which sum (the) plaintiffs were . . . against their will and in spite of their formal protest duly made, compelled to pay, and did pay, in order to obtain possession of their said sugars, . . . which the said defendant, enabled so to do by the power and authority of his said office, had detained, was detaining, and threatened to continue to detain from them, exacting as a condition to the delivery thereof, such payment of such alleged duties. . . ." (Facts stated in the complaint and admitted by the demurrer, Record pp. 3, 4 and 5.)

Having thus, under protest, paid the said alleged duties exacted from them as a condition to the delivery to them of the sugars in question, the plaintiffs in error brought this suit to recover back the same, in the Supreme Court of the State of New York.

By writ of *certiorari*, dated March 22, 1900, and sued out by the defendant Bidwell, through Henry L. Burnett, Esq., United States attorney, acting as attorney for said defendant, the said suit was removed into the said Circuit Court of the United States for the Southern District of New York, in the Second Circuit.

Thereupon the said United States attorney, acting as attorney for said defendant, interposed a demurrer to the complaint upon the following grounds:

"*First.* Upon the ground that it does not state facts sufficient to constitute a cause of action against the defendant.

"*Second.* Upon the ground that this court has no jurisdiction of the cause of action alleged in said complaint against said defendant."

By its decree, filed October 17, 1900, the said Circuit Court "ordered, adjudged, and decreed that the said demurrer . . . be sustained, both on the ground that the complaint does not state facts sufficient to constitute a cause of action against the defendant, and on the further ground that this court has no jurisdiction of the cause of action alleged in the complaint against the defendant;" and on the same day judgment was

signed and filed, "that the complaint be dismissed" with
costs.

.To review the said judgment this writ of error has been
brought.

I. It is not true "that (the) court has no jurisdiction of the
cause of action alleged in said complaint against (the) defendant." The action being one against a Federal official for acts
done by color of his office, and the remedy provided by the
customs administrative act not being available (inasmuch as the
plaintiff does not "concede that the [sugar] is imported merchandise"), the jurisdiction of the court to entertain this action
is entirely beyond question.

II. It is not true that the complaint "does not state facts
sufficient to constitute a cause of action:"

1. Puerto Rico was not, in June or September, 1899, a "foreign country" within the meaning of that term as used in the
tariff act of 1897 (under authority of which, and of which alone,
the defendant claimed the right to collect as duties the sums
mentioned in the complaint).

2. Even if—in denial of the foregoing contention—the tariff
act of 1897 had to be construed as in fact purporting to authorize the collection of duties on goods brought from Puerto Rico
into New York in June or September, 1899, then, in that aspect
of it, and to that extent, the act in question must be held unconstitutional and ineffectual to justify the exaction complained
of in this case.

a. Congress cannot "lay and collect" any "duties" save
such as are "uniform throughout the United States;"

b. "Duties" collectible "on goods brought from Puerto Rico
into New York in June or September, 1899," would have been
duties not "uniform throughout the United States," Puerto
Rico having been, ever since the ratification of the treaty with
Spain (antedating the period in question), a part of "the United
States:"

(1) Treaties "ceding" territory to the United States make
the territory so "ceded" a part of the United States within
the meaning of the provision of the Constitution as to the uniformity of duties throughout the United States.

(2) The treaty with Spain "ceded" Puerto Rico to the United States as of the date when such treaty became effective (a date antedating the period here in question). There was nothing to postpone or suspend the operation of the treaty as a present cession of the island, in the circumstance—the only one which has been suggested to that effect—that it (the treaty) provides that the Congress shall determine the civil rights and political status of the native inhabitants of the ceded islands and that the Spanish-born inhabitants may have one year in which to choose whether to preserve or abandon their allegiance to Spain.

It is not true "that the court has no jurisdiction of the cause of action alleged in said complaint against (the) defendant." The action being one against a Federal official for acts done by color of his office, and the remedy provided by the customs administrative act not being available (inasmuch as plaintiff does not "concede that the sugar is imported merchandise"), the jurisdiction of the court to entertain this action is entirely beyond question.

The defendant's claim (in his "second ground" of demurrer) that the court has no jurisdiction of this action, is based, as appears by his brief in the court below, on the view "that the entire and only existing remedy for all claimants for duties alleged to be illegally exacted is to be found in the customs administrative act of June 10, 1890, which has provided for a new course of procedure on behalf of such claimants, repealed the preëxisting rights of action in such cases, and relieved the collector from liability for his decisions or actions as to customs duties."

In other words, the argument is that "the act of 1890" on the one hand provided a "remedy" (distinct from an action, such as the present one, against the collector), of which special remedy the plaintiffs here might have availed themselves to secure a decision of the issue they have sought to present in this suit; while on the other hand, the said act in effect prevented the valid bringing of such an action as the present by repealing (in sec. 29) "sections 2931 and 3011" of the Revised

Statutes ("providing for an exclusive statutory right of action"), and expressly (in sec. 25) "relieving the collector from liability," etc.

To the contrary of this, we respectfully submit:

First. That the "remedy" and the "procedure" provided by the customs administrative act of June 10, 1890, have no application whatever to, and are not available in, cases which (like the present one) are not "customs" cases at all (the merchandise not having been "imported"); and,

Second. That the act of 1890 has not prevented the valid bringing of such an action as the present, in a case such as that set up by the complaint herein, by its repeal of sections 2931 and 3011 of the Revised Statutes and its provision that collectors should not be liable for or on account of any of the matters mentioned in that connection in section 25 of the act.

I. That the "remedy" and the procedure provided by the customs administrative act of 1890 are not available in cases which (like the present one) are not "customs" cases at all, has been distinctly laid down by this court in its unanimous opinion in the *Fassett* case, as the following quotations show:

"It is contended on behalf of Fassett that when he, as collector, took possession of the yacht and decided that she was dutiable, the only remedy open to her owner was to pay under protest the duties assessed upon her, and in that way secure possession of her, with the right thereafter, as provided in sections 14 and 15 of the customs administrative act, of June 10th, 1890, 26 Stat. 131, 137, 138, to obtain a refund of those duties by taking an appeal from the decision of the collector to the Board of General Appraisers, and appealing, if necessary, from that board to the Circuit Court of the United States."

"The idea embodied in the libel is, that if the yacht was not an imported article, the act of the collector in forcibly taking possession of her was tortious, and, as that act was committed on the navigable waters of the United States, the District Court, as a court of admiralty, had jurisdiction, in a cause of possession, to compel the restitution of her. The libel presents for the determination of the District Court, as the subject-matter of the suit, the question whether the yacht is an imported article,

within the meaning of the customs revenue laws." p. 483.
" The libellant had no other remedy than the filing of this libel.
He has none under the customs administrative act, of June, 1890.
By § 14 of that act, the decision of the collector as to 'the rate
and amount' of duties chargeable upon imported merchandise
is made final and conclusive, unless the owner, etc., . . .
The appeal provided for in § 15 brings up for review in court
only the decision of the Board of General Appraisers as to the
construction of the law, and the facts respecting the classification
of imported merchandise, and the rate of duty imposed thereon
under such classification. It does not bring up for review the
question of whether an article is imported merchandise or not;
nor, under § 15, is the ascertainment of that fact such a 'decision' as is provided for. The decisions of the collector from
which appeals are provided for by § 14 are only decisions as to
'the rate and amount' of duties charged upon imported merchandise, and decisions as to dutiable costs and charges, and decisions
as to fees and exactions of whatever character. Nor can the
court of review pass upon any question which the collector had not
original authority to determine. The collector has no authority
to make any determination regarding any article which is not
imported merchandise; and if the vessel in question here is not
imported merchandise the court of review would have no jurisdiction to determine any matter regarding that question, and could
not determine the very fact which is in issue under the libel in the
district court, on which the rights of the libellant depended (*i. e.,*
the question whether the yacht was 'imported merchandise').

" Under the customs administrative act, the libellant, in order
to have the benefit of proceedings thereunder, must concede that
the vessel is imported merchandise, which is the very question
put in contention under the libel, and must make entry of her
as imported merchandise, with an invoice and a consular certificate to that effect, and thus estop himself from maintaining the
fact which he alleges in his libel, that she is not imported merchandise." *In re Fassett,* 142 U. S. pp. 486-7.

The principle of the case has never been repudiated or qualified by this court, and the only supposed authority against it
which the learned district attorney was able to cite in the Cir-

cuit Court is the decision in *Lascelles* v. *Bidwell*, 102 Fed.
Rep. 1004, the entire report of which reads as follows: "*Lascelles* v. *Bidwell*, (Circuit Court, S. D., New York, March 19,
1900). Motion for preliminary injunction. Charles Henry
Butler, for the motion. Henry L. Burnett, U. S. Atty., opposed.
Lacombe, circuit judge. Motion denied on authority of *Cruikshank* v. *Bidwell*, 176 U. S. 73. Complainant has an adequate,
summary, and expeditious remedy at law under the customs administrative act."

As the existence of "an adequate remedy at law," even
though not "under the customs administrative act," afforded
ample ground for the denial of the motion for an injunction,
the specification of the customs act as affording the remedy at
law, was clearly not of the essence of the ruling, but in the
nature of a merely incidental *dictum.* It is hardly to be supposed that, had the learned Circuit Judge had distinctly in mind
at the time of writing the reasoning and doctrine above quoted
from the unanimous opinion of this court in the *Fassett* case,
he would have announced his impression that Lascelles had a
remedy "under the customs administrative act," without giving
his reasons for thinking so, notwithstanding that the very essence
of Lascelles' contention was that his Puerto Rico sugar "is not
imported merchandise," and that this court has held in the *Fassett*
case that that is a contention which is not raised, but surrendered,
by proceedings under the act mentioned, since, "in order to
have the benefit of proceedings thereunder, (one) must concede
that the (article) is imported merchandise . . . and . . .
estop himself from maintaining the fact which he alleges, that
(it) is not imported merchandise."

As the memorandum itself shows, Judge Lacombe denied
the Lascelles motion for an injunction "on authority of *Cruikshank* v. *Bidwell*, 176 U. S. 73." The report of that case
shows that an injunction was there denied on the ground that
the "remedy at law" was adequate; but so far from there being any intimation or implication that such remedy at law was
to be had "under the customs administrative act," the opinion
distinctly points to a suit against the collector as constituting
the remedy referred to (the gravamen of the complaint there,

as here, being the absolute lack of authority on the part of the collector, instead of a merely erroneous exercise of authority vested in him):

"The sole ground of equity jurisdiction put forward," declares the opinion, "is the inadequacy of remedy at law in that the injury threatened is not susceptible of complete compensation in damages. The mere assertion that the apprehended acts will inflict irreparable injury is not enough. Facts must be alleged from which the court can reasonably infer that such would be the result, and in this particular we think the bill fatally defective. The matter in dispute was averred to be 'the value of the said teas and the right to import teas.' Confessedly the value of these teas was known, and their destruction capable of being compensated by recovery at law. The official character of the collector, the provisions of the act, and the regulations of the Secretary of the Treasury in execution thereof would not constitute a defense if the act were unconstitutional" (which was what was alleged). There was no intimation that the collector would be unable to respond in judgment, and, moreover, section 989 of the Revised Statutes provides that when a recovery is had in any suit or proceeding against a collector for any act done by him, probable cause being certified, 'the amount recovered shall, upon final judgment, be provided for and paid out of the proper appropriation from the Treasury.' *The Conqueror*, 166 U. S. 110, 124." *Cruikshank* v. *Bidwell*, 176 U. S. 81, 82.

There can be no question that action by an administrative officer, in a case other than that in which action by him is contemplated by the statutes conferring his official authority, is as completely unauthorized and unofficial as would be action under a statute which was itself void as unconstitutional. The statute would not protect him from personal liability any more in the one case than in the other. By analogy, therefore, the *Cruikshank* decision is an authority against instead of for the idea that the remedy provided by the customs administrative act is "the entire and only existing remedy" (or is one available at all) for those whose cause of action against the collector is not that he erred as to details of a customs case, but that as a mere trespasser he assumed to act officially in a case in which, inasmuch as there has been no importation of merchandise, he has no authority whatsoever to act at all.

It being thus apparent that the remark in the *Lascelles* case was only *obiter dictum*, and, moreover, a *dictum* inconsistent with the principle of both the *Fassett* and the *Cruikshank* rulings of the Supreme Court (rulings which that court has never repudiated, doubted, or qualified) we beg leave to submit, with all respect to Judge Townsend, that he was mistaken in declaring in his opinion in *Goetze* v. *United States*, 103 Fed. Rep. 74, that the "preliminary question had been disposed of in the suit of *Lascelles* v. *Bidwell*," at least in the sense of establishing the availability of the remedy provided in the act of 1890 in cases in which, as in those of *Lascelles*, *Goetze*, and the present one, "the very question put in contention," namely, whether or not the merchandise had been "imported," would be "conceded" by proceedings under the act.

On the contrary the clear effect of the authorities cited, as well as of the principles of the subject, is undoubtedly that the "remedy" and the "procedure" provided by "the customs administrative act of June 10, 1890," have no application whatever to cases which (like the present one) are not "customs" cases at all (the merchandise not having been "imported"), and in which accordingly, the "illegality" complained of is not an erroneous exercise of the collector's authority in a case in which he was authorized to act as collector, but the radical "illegality" involved in his having, as a mere trespasser, assumed to act as collector in a case not one of the kind of case in which alone the statutes contemplated and authorized his acting officially.

2. Having thus seen that the defendant is in error in the first part of his theory as to our remedy—in his notion, namely, that the procedure provided by the act might have been available to us for securing a decision of the issue raised by the complaint—we beg now to submit that he is equally in error in the second part of that theory, since in point of fact (his argument to the contrary notwithstanding) the act of 1890 has not prevented the valid bringing of such an action as the present (in a case such as that set up by the complaint herein) by

its repeal of sections 2931 and 3011 of the Revised Statutes and its provision that collectors should not be liable for or on account of any of the matters mentioned in that connection in section 25 of the act.

The defendant's inference from the repeal of the sections named and the declaration of "exemption from liability" in § 25—the inference, namely, that the right to sue the collector in a case such as the present no longer exists—is based upon the assumption that "the right to sue the collector in a case such as the present" existed only by virtue of sections 2931 and 3011, and upon the further assumption that the matters in respect of which § 25 declares the collector to be exempt from liability, include a "matter" such as that which constitutes the gravamen of our complaint. Both assumptions ignore the essential distinction (recognized by this court in the *Fassett* and *Cruikshank* cases) between matters which are really "customs" matters and those which are not really such at all. Owing to their thus ignoring that distinction, both assumptions are erroneous, making fallacious the inference based upon them.

Consider, first, the repeal of sections 2931 and 3011. What does that "repeal" amount to? Simply the substitution of a new procedure in "customs" cases for the old procedure in "customs" cases. Those "sections" were portions of the old "customs administrative act" embodied in Title XXXIV of the Revised Statutes, the official heading of which is "collection of duties upon imports." The act of 1890 is simply a revision of that system. Both the original and the revision assume as a fact that merchandise is to have been "imported." Neither had any bearing upon or reference to the remedy available to one whose grievance is that the man who happens to be collector has assumed to act as such in a case in which, no "importation" having been made, he was not really authorized to act officially at all. The "repeal" of the sections regulating the old procedure in customs cases, to make way for the revised procedure in customs cases, did not destroy the right of action in non-customs cases, for the simple reason that such last-named right of action (against a mere trespasser) was not created by and did not depend upon or have any connection with the "sections" mentioned.

What now about the provision of § 25 of the act of 1890, "relieving the collector from liability for his decisions or actions as to customs duties?" In the light of the *Fassett* distinction, this difficulty proves as unsubstantial as that of the repeal of the two irrelevant sections. Obviously, the "relieving of the collector from liability for his decisions or actions as to customs duties" cannot mean the exemption of Mr. George R. Bidwell, the individual, from liability for "decisions or actions" having nothing to do with "customs duties" and made or performed in a case in which, inasmuch as there has been no "importation," he did not and could not decide or act as, or in any sense be, the "collector" at all.

Indeed, the text of § 25 on its face shows that the exemption from liability thereby secured to the "collector" is strictly restricted to customs matters, and by no means extends protection to the individual who, in customs cases, is collector, in respect of "any determination regarding any article which is not imported merchandise," which kind of "determination" this court in so many words declares that "the collector has no authority to make." 142 U. S. 487. The section reads as follows:

"SECTION 25. From and after the taking effect of this act no collector or other officer of the customs shall be in any way liable to any owner, importer, consignee, or agent of any merchandise, or any other person, for or on account of any rulings or decisions as to the classification of said merchandise, or of duties charged thereon, or the collection of any dues, charges, or duties on or on account of said merchandise, or any other matter or things as to which said importer, consignee, or agent of such merchandise might under this act be entitled to appeal from the decision of said collector or other officer, or from any board of appraisers provided for in this act." 26 Stat. 141.

This language clearly restricts the collector's exemption from liability to matters as to which an appeal can be had under the act from the decision of the collector. This court has held that, under the act—

"The court of review cannot pass upon any question which the collector had not original authority to determine. The collector had no authority to make any determination regarding any article which is not imported merchandise." *In re Fassett*, 142 U. S. 479, 486.

In other words, no appeal can be had under the act from "any determination (by the collector) regarding any article which is not imported merchandise." Consequently in a case in which the decision complained of is one "regarding (an) article which is not imported merchandise" the collector is not "relieved from liability" by § 25.

It thus becomes plain that neither the repeal of sections 2931 and 3011, nor the exemption provision in § 25 of the act of 1890, really prevent the valid bringing of an action against the defendant Bidwell, notwithstanding his collectorship, in a case where the determination complained of was one which, because it regarded an article which was not imported merchandise, he "had no authority to make."

It is true, indeed, that in his brief in the Circuit Court the learned district attorney categorically imputes to this court a decision inconsistent with this conclusion; but we respectfully insist that in this he was demonstrably mistaken.

His citation reads: " In the case of *Schoenfeld* v. *Hendricks* (152 U. S. 691, affirming 57 Fed. Rep. 568, in this circuit), the Supreme Court also held that ' the right to maintain an action at law against the collector to recover duties paid, whether existing by virtue of the statutory or common law' (*sic*); ' was taken away by sections 25 and 29 of the customs administrative act of June 10, 1890.'"

As a matter of fact the Supreme Court "held" nothing of the sort. It certainly did not hold that the "common law" "right to maintain an action at law against the collector" "was taken away by section 25 and section 29 of the customs administrative act." Though appearing in the brief between quotation marks (precisely as it is above repeated), the language given as embodying the supposed "holding" nowhere appears in the report of the case in this court, either in the "headnotes" or in the opinion. On the contrary, the opinion affirmatively shows that what was "held" to have been "taken away" by the act of 1890 was simply the statutory right of action against a collector (in customs cases) until then existing under sec-

tions 3011 and 2931 of the Revised Statutes (152 U. S. 693);
while the reason for holding the "common law" right of action
unavailable in such a case as Schoenfeld's (which is of course all
that was "held" or even intimated, *obiter*, in the *Schoenfeld*
decision) is that indicated in the following statement in the
opinion (p. 695): "We are of opinion that this action would not
lie at common law, the money being required by section 3010
to be paid into the Treasury." In the light of the reason thus
given, and on the principle *cessante ratione legis cessat ipsa lex*,
it is clear that the principle of the *Schoenfeld* decision holds
only in cases to which the requirement of "section 3010," that
"the money . . . be paid into the Treasury," can itself be
held to apply. Can that requirement be sanely held to apply
to any but "customs" cases? Look at the text of the enactment in question: "SECTION 3010. All money paid to any collector of the customs, or to any person acting as such, for
unascertained duties or for duties paid under protest against the
rate or amount of duties charged, shall be placed to the credit
of the Treasury of the United States, and shall not be held by
the collector or person acting as such, to await any ascertainment of duties, or the result of any litigation in relation to the
rate or amount of duty legally chargeable and collectible in any
case where money is so paid."

This section, being a part of Article XXXIV, on the "Collection of duties upon imports," would be presumed to apply
only to cases in which merchandise had been in fact "imported."
Furthermore, the very wording of the provision affirmatively
shows that it is only money which the collector gets in "customs"
cases proper that he is directed to "place to the credit of the
Treasurer." The direction for immediate payment into the
Treasury is in so many words explained to be made in order to
prevent the money being "held by the collector to await"—
what?—"any ascertainment of duties, or, the result of any
litigation in relation to the rate or amount of duty legally chargeable," etc.

Now, in a case in which there has been, in fact, an "importation" of merchandise, the collector has statutory authority,
for the purposes of the performance of his functions, to decide

officially, in the first instance, all questions involved in the "ascertainment" of duties and the determination of their "rate and amount;" and he is in such cases authorized to receive "duties" paid before definitive "ascertainment," or paid "under protest against the rate or amount of duties charged." Such "duties," and such duties only—"duties" the amount of which has either not been "ascertained" at all, or not conclusively ascertained as against the objection of the importer—are, under § 3010, to be at once on receipt "placed to the credit of the Treasurer." Where the essential "jurisdictional fact" exists, of an actual importation from a foreign country, the collector's errors as to details do not make his acts unauthorized or unofficial, and therefore his collections, though subject to revision, are deemed provisionally valid and as having been made by authority of the Government, and they may therefore well be the subject of such a provision as that of § 3010, as to the paying of the money into the Treasury. But in a case in which there has been in fact no importation at all the individual who holds the office of collector has simply "no authority" at all, and his erroneously holding that there has been an importation does not give him authority, or convert an exaction of money by him upon that theory into an official or authorized collection of "duties" such as can be deemed to be either the "unascertained" duties or the "duties paid under protest against the rate or amount of duties charged," which (and which alone) the statute directs the collector to deposit in the Treasury. As this court has said in the *Fassett* case, "The collector has no authority to make any determination regarding any article which is not imported merchandise." 142 U. S. 487. In such a case, therefore, he is a mere trespasser if he exacts money as if for "duties," and the law cannot be supposed to have contemplated any such trespass by him, nor, therefore, to have provided for the "paying into the Treasury" of the proceeds thereof.

The *Schoenfeld* case, 152 U. S. 691, was in fact a "customs" case, there having been an importation of merchandise. The money sued for there had been paid "for duties paid under protest as to rate or amount of duty charged," etc. To the money paid to the collector in that case, consequently, the provisions of sec. 3010 literally applied. It was entirely appropriate, therefore, for the court to say, as it did: " We are of opinion that this action would not lie at common law, the money being required by sec. 3010 to be paid into the Treasury." To read this as intended to apply to a case materially different from the *Schoenfeld* case itself (as not being a " customs " case at all) would be to give it a sense in which it would be clearly *obiter dictum.*

Nor can it validly be urged against the maintenance of this action, that, whether compelled thereto by sec. 3010 or not, the defendant, supposing as he did that this was a customs case, did in fact deposit the money here in question, and his having done so should have the same effect toward exempting him from liability as it would have had in a case to which sec. 3010 applied. In the first place, this supposed " actual," though voluntary, payment into the Treasury does not appear by the record, and is not to be presumed, it being, *ex hypothesi*, not required by law. Secondly, the reason why a deposit of the moneys required by sec. 3010 to be deposited exempts the collector from personal liability is simply this, that by that very requirement the United States adopts the collection as its own act, and takes its agent's place in any litigation as to the propriety of such collection (as respects " rate and amount "): This reason obviously does not hold where the collector's act is one which is wholly unofficial and unauthorized, as being one concerning " an article which is not imported merchandise." That his having acted in good faith, and in fact deposited the money in the Treasury, is not in law a bar to a " judgment " against him (as distinguished from an execution) is made entirely clear by the explicit provisions of section 989 of the Revised Statutes, which was not " repealed " by the law of 1890, but, on the contrary, has been distinctly recognized by this court in cases much later than the *Schoenfeld* case, *The Conqueror*, 166 U. S. 124; *Cruikshank* v. *Bidwell*, 176 U. S. 81, as being in full force and operation. It reads as follows : SEC. 989. Whenever a recovery is had in any suit or proceeding against a collector or other officer of the revenue for any act done by him, or for the recovery of any money exacted by or paid to him and by him paid into the Treasury, in the performance of his official duty, and the court certifies that there was probable cause for the act done by the collector or other officer, or that he acted under the directions of the Secretary, or other proper officer of the Government, no execution shall issue against such collector or other officer, but the amount so recovered shall, upon final judgment, be provided for and paid out of the proper appropriation for the Treasury."

It would seem to be beyond contradiction that this section—which is quite as clearly in force as "section 3010," or the "act of 1890"—distinctly proves the policy of the law to be to permit, in some cases, "a recovery" (*i. e.*, a judgment) "against a collector . . . for the recovery of any money exacted by or paid to him and by him paid into the Treasury" (though execution is not to issue against the official, and the "final judgment" against him is to be paid out of the Treasury, if the court certifies to "probable cause, etc."). In what sort of a case could this provision find scope and application if not in a case such as the present, in which, the article not being imported merchandise, the intervention of the collector was wholly unauthorized and therefore unofficial, instead of being simply erroneous as to details? In the teeth of this statute, declared in the *Cruikshank* case to be in force, it seems impossible to insist that the collector's having paid the money into the Treasury is in any way incompatible with the "recovery" of a "final judgment" against him therefor.

II. It is not true that the complaint "does not state facts sufficient to constitute a cause of action":

a. Porto Rico was not, in June or September, 1899, a "foreign country" within the meaning of that term as used in the Tariff Act of 1897 (under authority of which, and of which alone, the defendant claimed the right to collect as duties the sums mentioned in the complaint).

b. Even if—in denial of the foregoing contention—the Tariff Act of 1897 had to be construed as in fact purporting to authorize the collection of duties on goods brought from Porto Rico into New York in June or September, 1899, then, in that aspect of it, and to that extent, the act in question must be held unconstitutional and ineffectual to justify the exaction complained of in this case.

c. Congress cannot "lay and collect" any "duties" save such as are "uniform throughout the United States."

d. "Duties" collectible "on goods brought from Porto Rico into New York in June or September, 1899," would have been duties not "uniform throughout the United States"; Porto Rico having been,. ever since the ratification of the treaty with Spain (antedating the period in question), a part of the United States."

Treaties "ceding" territory to the United States make the territory so "ceded" a part of the United States within the meaning of the provisions of the Constitution as to the uniformity of duties throughout the United States.

The treaty with Spain "ceded" Porto Rico to the United States as of the date when such treaty became effective (a date antedating the period here in question). There was nothing to postpone or suspend the operation of the treaty as a present cession of the island, in the circumstance—the only one which has been suggested to that effect—that it (the treaty) provides that the Congress shall determine the civil rights and political status of the native inhabitants of the ceded islands and that the Spanish-born inhabitants may have one year in which to choose whether to preserve or abandon their allegiance to Spain.

These cases present the question whether under the Constitution the Government is authorized to impose a tax upon merchandise brought into the port of New York from the island of Puerto Rico after the cession of that island to the United States by formal treaty, duly ratified and proclaimed.

Such a tax has been here imposed on the supposed authority of the Customs Revenue Act of 1897 (Dingley Act).

The Dingley Act provides for the imposition of a customs duty on sugars imported from foreign countries, and notwithstanding the acquisition by the United States of the island of Puerto Rico under the treaty with Spain of December 10, 1898, ratifications of which were exchanged on the 11th day of April, 1899, the collector of the port of New York exacted the payment of customs duties on sugar brought into said port from Puerto Rico in the months of June and July, 1899, as though it had been imported from a foreign country.

As a basis for the examination of this question we submit the following propositions :

A. A treaty duly entered into is law, and has the force of a statute until superseded by subsequent enactment.

B. The treaty of Paris ceded Puerto Rico to the United States. Puerto Rico then came completely under the sovereignty and dominion of the United States. The political map of the world was changed and Puerto Rico became geographically a part of the United States, or of what Marshall called the "American Empire," under the statutory name of Porto Rico.

C. The clause of the treaty leaving the determination of the "civil rights and political status" of the native inhabitants to Congress was merely declaratory of the power given by the Constitution to withhold political rights and franchises and to establish civil government and enact municipal law in all places where no state government exists.

D. All territory lawfully acquired and taken under sovereign jurisdiction is a part of the United States.

E. The Constitution is a charter or grant of powers conferred upon the Federal Government by the people of the United States. The Federal Government has no existence outside of this Constitution. Hence it is a confusion of terms to speak of territory to which the United States has acquired title as not being within our "constitutional boundaries" or incorporated into the United States. It is a misapprehension of the nature of our institutions and of the function of the organic law of our national existence, known as the "Constitution," to speak of any part of the nation being beyond its boundaries, or to speak of its "extension" over portions or over all of the national territory. There is no boundary to the Constitution other than the whole sphere of the activity of the Federal Government. Outside of that sphere, beyond that boundary, the Federal Government can only act by usurpation—a government of force—not of law, and officials assuming to act for the United States outside of the prescriptions of the Constitution are, however well intentioned, outside of the law.

F. This is the elementary rule of constitutional functions. But it does not follow that, because all government finds its sole authority in the constitutional grant, every prescription of the Constitution, its delegations, limitations, and prohibitions can always and at all places be made applicable to all governmental action in all circumstances. These are applicable according to varying place and circumstance.

The unquestioned proposition that the government is powerless to act outside of the charter of its existence does not of necessity imply that the Bill of Rights—the prohibition against cruel and unusual punishments—operates at once throughout any territory over which the Government of the United States exercises jurisdiction—military, transitory, or permanent.

G. Territory held by military occupation during hostilities or as an incident thereto is subject to the rule of the President as Commander in Chief under the Constitution. No limitations are placed upon his power as Commander in Chief, save such as must be implied—*i. e.*, to wage only civilized warfare. But the freedom from limitation does not arise from the inapplicability of the restraints of the Constitution; on the contrary, it is a freedom granted by the Constitution, which gives him, in case of war, the usual powers of military commanders recognized by international law.

H. Territory acquired by the law or treaty-making power, and hence coming under the sovereign jurisdiction of the United States, may be governed by the Executive until Congress undertakes to govern it.

As long as war lasts the Executive continues his military rule as Commander in Chief. Upon ratification of the treaty of peace he continues his rule under his general duty and power to execute the laws, but as a *de facto* civil government, pending any action of Congress for the government of the new territory. This doctrine was followed by the political authorities in the case of California and was defined and upheld in *Cross* v. *Harrison* (*vide* the opinion of Judge Magoon, legal adviser to the War Dept., Sen. Doc. No. 594, 56th Cong., 1st Sess.).

But in any event the new territory is part of the United States pending its definite organization under the powers given to Congress.

I. As soon as the military status ceases and a *de facto* civil government is carried on, even by army officers, the civil rule being reëstablished, it is subject to the constitutional requirements. These territories are but " political subdivisions of the outlying dominion of the United States." Congress is supreme in legislating for them; it has all the powers of the people of the United States, except such as have been expressly or by implication denied and prohibited by the terms of the Constitution.

J. Within those prohibitions or limits Congress has supreme power. These limitations and prohibitions, however, are its " constitutional boundaries," outside of which it may not go.

The only question, therefore, is:

Has Congress ignored these prohibitions and gone beyond these limits in its government of Porto Rico; in other words, violated the constitutional restrictions which lie at the center and foundation of the Federal powers?

K. The Dingley Act in terms imposed a duty on goods imported from foreign countries. It could have no application to goods from Porto Rico, which ceased to be a foreign country upon the ratification of the treaty ceding it to the United States. To apply it to Porto Rico would make it obnoxious to the constitutional prohibition (Art. I, Section VIII) which prescribes that "all duties, taxes, and imposts shall be uniform throughout the United States."

The tax was levied at the port of New York on sugar from Porto Rico. No tax was leviable upon like merchandise from any other part of the United States. This is not the uniform taxation required by the Constitution.

This legislation was enacted by Congress as the lawmaking body for the whole United States and affected every port in the United States. It was not a local tax or excise for the benefit of a particular locality.

L. The precedents adduced from our former acquisitions of territory do not militate against this view.

M. The inhabitants of the ceded territories are citizens of the United States in the sense in which all persons owing immedi-ate and complete allegiance to the United States and inhabit-ing its territory are citizens. In that sense the word citizen is the equivalent of "national" or subject. See Senator Fora-ker's report on Porto Rico, No. 249, Feb. 5, 1900, p. 12.

N. While the argument from the consequences is not always the best argument, it is perhaps in this case, owing to its im-portance, relevant.

The only consequence of the construction contended for to which the Government objects is the uniformity of duties clause. All the other rights secured to citizens and others within the United States by the prohibitions granted by the Constitution are admittedly applicable. The danger feared is the possibility of the American markets being thrown open to the products of the ceded territories. It is not an objection which the court can take into consideration in deciding this case.

I. Under the Constitution of the United States a treaty
is the supreme law of the land.

That this proposition has never been dissented from or doubted in this court is well known. It would not be necessary to dis-cuss it now were it not for the fact that it has recently been challenged and a novel doctrine has been broadly asserted. It has been claimed that a treaty is a mere contract between two nations of no effect as law within the United States until given such effect by act of Congress; in other words, the treaty is of no legal force save as an international obligation.

This suggestion cannot appeal to this court. The question was settled positively, clearly, and without possibility of equiv-ocation by the "Great Chief Justice:"

"A treaty is in its nature a contract between two nations, not a legislative act. . . . In the United States a different principle is established. Our Constitution declares a treaty to be the law of the land. It is, consequently, to be regarded in courts of justice as equivalent to an act of the legislature wher-ever it appears, of itself, without the aid of any legislative pro-visions." Marshall, C. J., *Foster* v. *Neilson*, 2 Peters, 253.

" . . . A treaty, it is true, is in its nature a contract between two nations, and is often merely promissory in its character, requiring legislation to carry its stipulations into effect. . . . If the treaty operates of its own force and relates to a subject within the power of Congress, it can be deemed in that particular only the equivalent of a legislative act to be repealed or modified at the pleasure of Congress. In either case, the last expression of the sovereign will must control." *Chinese Exclusion Case,* 130 U. S. 600.

"A treaty is primarily a compact between independent nations. It depends for the enforcement of its provisions on the interest and honor of the governments which are parties to it. . . . But a treaty may also contain other provisions which confer certain rights upon the citizens or subjects of the nations residing in the territorial limits of the other which partake of the nature of the municipal law."

" . . . The Constitution gives it (the treaty) no superiority over an act of Congress in this respect, which may be repealed or modified by an act of a later date." *Head Money Cases,* 112 U. S. 597. See also *Geoffrey* v. *Riggs,* 133 U. S. 258–271.

II. BY THE TREATY OF CESSION PORTO RICO BECAME A PART OF THE UNITED STATES.

By the treaty of Paris Spain ceded Porto Rico to the United States, and by such cession, we submit, Porto Rico became a part of the political entity known as the United States.

It is now claimed, and as we believe for the first time by a court of the United States, that territory may come under the complete and absolute sovereignty and dominion of the United States and yet remain foreign.

Judge Townsend has held in the case of *Goetze* v. *The United States* that although the title to the soil of Porto Rico is in the United States and no other country has any rights there of any character, yet Porto Rico was, subsequent to the treaty, a foreign country within the meaning of the statutes of the United States, imposing duties upon goods coming from foreign countries. The reasoning by which this conclusion is reached is so novel and important that it will justify a close examination.

He says: "By cession the title to the soil became *de jure*, but in the status of the islanders as foreigners, and so in the status of Porto Rico as a foreign country no change was to be made until Congress should determine its character." 103 Fed. Rep. 17. "Thus we see that in all previous cessions of territory there has been a special provision in the treaty for incorporating the inhabitants within the United States. Whether a treaty stipulation would be sufficient to incorporate the territory into the Union is not clearly established. . . . There is (in the treaty of 1898) no provision for the incorporation of the inhabitants within the Union as there has always been in prior treaties." 103 Fed. Rep. 76. "There has been found, then, no reason either on principle or authority why the United States should not accept sovereignty over territory without admitting it as an integral part of the Union or making it bear the burden of the taxation uniform throughout our nation. To deny this power is to deny to the nation an important attribute of sovereignty," etc. 103 Fed. Rep. 86.

The sentences quoted contain the reasoning of the Government, and, as we believe, the fallacy upon which their position is based. These fallacies are endorsed by the Attorney General, who says in his *Goetze* brief, p. 4:

"That the treaty-making power—the President and the Senate—as evidenced by the language of the treaty of Paris, did not intend to make Porto Rico and the Philippine Islands integral parts of the United States, but intended in several particulars to reserve their final status for adjustment by Congress." And at page 8: "There is no doubt that it was the intention of the treaty of Paris not to make the ceded islands a part of the United States."

The Government of the United States may sustain as to any given territory three relations: (1) Sovereign jurisdiction. (2) Temporary occupation of foreign soil. (3) Foreign territory over which it has no jurisdiction.

In the last case it has no relations with the inhabitants; in the second it is merely the *de facto* sovereign over certain territory; this sovereignty cannot under the Constitution affect the political status of the inhabitants since the allegiance which they owe to the United States is but temporary and only as an incident of war, their former allegiance reverting with the return of the former sovereign. *The Castine Case, United States* v. *Rice,* 4 Wheat. 246, and *Fleming* v. *Page,* 9 How. 615.

In the first case, and that is the position of Porto Rico, the power of Congress over the political status is plenary. Political rights are franchises which may be given or taken away by Congress in the territories, *i. e.,* the places over which it has exclusive local jurisdiction. *Murphy* v. *Ramsey,* 114 U. S. 15.

The treaties to which Judge Townsend referred endeavored to settle the political status of the countries ceded by provisions that they should be admitted into the Union as soon as possible, and the Attorney General (*Goetze* brief, p. 66) emphasizes this position.

Granting that by the treaty the inhabitants of Porto Rico acquired neither civil nor political rights, yet that did not make Porto Rico a foreign country.

A foreign country is a country under a sovereignty other than that of the United States. "By a foreign port may be understood a port within the dominions of a foreign sovereign and without the dominions of the United States." Mr. Justice Story in *United States* v. *Heyward,* 2 Gall. 501. See also Chief Justice Spencer in *King* v. *Parks,* 19 Johns. 375. Also Treasury Regulation 835, approved in *Stairs* v. *Paislee,* 18 How. 526. This Porto Rico admittedly was not.

What Judge Townsend meant, then, was simply that until Congress had legislated, the inhabitants had no political rights, and their private or civil rights remained unchanged.

Incorporation of the inhabitants within the United States means, if anything, that the inhabitants shall be made part of the body politic, *i. e.,* enter the union as a State, as was intended in the case of Louisiana, which we shall hereafter examine.

This is very different from making territory a part of the United States, which is all the present case contends for.

The fact that the inhabitants of a country ceded by treaty to the United States are still under the military authority of the Government awaiting the action of Congress organizing a local government is entirely apart from the question as to whether the territory, regardless of the status, race, or color of its inhabitants, is a part of the United States.

Let us assume that Porto Rico was inhabited by roving Indian tribes and had no other inhabitants, could it be contended that although we had acquired title to the soil, the Indians being tribes which were not, while maintaining their tribal relations, citizens of the United States, therefore the territory in question was a foreign country? Certainly not.

As the Attorney General says, *Goetze* brief, p. 6: "The basis of the custom laws is not ownership, but (1) the geographical origin of the shipment, and (2) the nature of the goods."

The learned judge and the Attorney General confuse the idea of acquiring territory, and thus enlarging the boundaries of the United States, with the withholding of political rights. They make the one depend upon the other. This is clear from the expression (in the *Goetze* case) that the United States "may accept sovereignty without admitting it (the territory) as an integral part of the Union." If by an integral part of the Union he means a political part, *i. e.*, a State, we assent to the proposition.

The political power of the Union is in the inhabitants of the States—those of the Territories have none.

The incorporation of new territory into our body politic would and must mean the incorporation of the inhabitants into our political people *i. e.*, into people of the States.

This we do not contend for.

Had nothing been said in the treaty as to the inhabitants, their political status and within certain limitations their civil rights would have been entirely within the power of Congress. In previous treaties acquiring territory the United States had usually promised the ceding country that its inhabitants should have admission to statehood.

This had been the usual course.

In the present instance the American Government, desiring that the disposition of the question should be left entirely to the Congress, was not satisfied to negatively refrain from promises to Spain, but, in order that no misunderstanding should occur in the future, expressly stipulated with Spain that Congress should determine these questions. It would have been proper for Spain to have asked that her subjects in Porto Rico should be admitted to and incorporated in the Union of States. She did not do so, but left the matter absolutely to the United States.

This clause in the treaty then left the United States free to deal with the inhabitants as she chose—subject always to the prohibitions of the Constitution.

Its sovereignty over the territory is thus emphasized, not diminished.

III. EFFECTS OF ANNEXATION. The fallacy underlying all the reasoning of the learned court below, and of the counsel for the Government, seems to be based upon the following reasoning:

"We have the authority of *Fleming* v. *Page*, that acquiring title to the soil of the territory making it part of the United States as regards other nations does not bring it within the sphere of the Constitution. If, then, it is not acquisition of soil which extends our constitutional boundaries, what does accomplish this result? In order to extend the boundaries recognized by other nations, the extension of dominion by acquisition is sufficient."

To speak of soil coming within the sphere of the Constitution seems to us to be a misuse of language. It was held, and rightly held, in *Fleming* v. *Page*, that where the armies of the United States had overrun, conquered, and held an extent of territory, other nations would recognize that the United States was a *de facto* Government in and over such territory.

This is an elementary rule of international law which we do not question.

This was occupation, not acquisition.

How the country over which the authorities of the United States had established a *de facto* government was to be organized and governed, is a question with which international law has no concern. Under the Constitution of the United States

the Government has power to wage war and to carry out all
the duties necessary and incident to the waging of such war.
When it occupies foreign territory it is doing so in pursuance
of a power delegated to it by the Constitution, and while the
Constitution as such does not affect the territory or soil over
which the United States troops exercise jurisdiction, it is by
reason of the grant of power contained in the Constitution
that the United States troops are there carrying on legitimate
warfare, and are not mere adventurers or revolutionists.

What the learned judge means by bringing the territory
within the sphere of the Constitution we do not exactly understand.

If he means that our jurisdiction there is not exercised in
pursuance of the Constitution, we claim that he is incorrect in
his postulate of constitutional law. If, however, he means that
the jurisdiction is only temporary military jurisdiction, and that
the clauses of the Constitution in regard to the bill of rights
and uniformity of taxation do not and cannot apply, we accede
to his view entirely.

The confusion in his reasoning seems to arise from want of
appreciation of the fact that the Constitution applies both to
peace and to war. That there is, so to speak, a Constitution for
peace and one for war.

This is no new theory, but was clearly and ably expressed by
John Quincy Adams in the House of Representatives in 1836.
He said:

" There are, then, in the authority of Congress and in the Executive, two classes of powers altogether different in their nature
and often incompatible with each other—war power and peace
power. The peace power is limited by regulations and restricted
by provisions in the Constitution itself. The war power is only
limited by the usage of nations. This power is tremendous. It
is strictly constitutional, but it breaks down every barrier so
anxiously erected for the protection of liberty and of life."

This war power is, then, unlimited, except by the limitation
which may fairly be implied from the Constitution that the
war allowed to be waged shall be civilized warfare; that is to
say, warfare according to the rules and regulations recognized

by civilized nations, not warfare as known to and practiced by the Apaches and Zulus. That in carrying on such warfare in accordance with the public law of the world the Government of the United States has the right to exercise a temporary jurisdiction over territory belonging to another nation is unquestioned. That jurisdiction, however, is and must remain temporary, until either the treaty-making or law-making power of the Government has acted.

The President, as Judge Taney said, cannot enlarge the territorial boundaries of the United States. The nation whose soil we are occupying and whose jurisdiction we have temporarily ousted has what might be termed in private law a right of reverter, and when the United States withdraws its troops the world recognizes that the sovereignty belongs to the nation temporarily dispossessed. The boundaries could not " be enlarged or diminished as the armies on either side advanced or retreated." *Fleming* v. *Page*, 9 How. 615.

But, and here we think the learned court in the *Goetze* case failed to appreciate the distinction, if the law or treaty-making power enacts that the territory over which the military arm of the Government has extended shall come under the permanent absolute sovereign jurisdiction of the United States, then, and then only, a new and different status arises. "The United States, it is true, may extend its boundaries by conquest or treaty . . . but that can be done only by the treaty-making power or the legislative authority, and is not a part of the power conferred upon the President by the declaration of war." *Fleming* v. *Page*, 9 How. 614. The former sovereign then loses all right of reverter and the territorial limits of the United States are in so far enlarged. See *Cross* v. *Harrison*, 19 How. It is, therefore, erroneous to say that "it is not acquisition of soil which extends our constitutional boundaries." What was meant is probably that occupation of soil did not extend our boundaries.

"Constitutional boundaries," we submit, is a misleading if not meaningless term. The Constitution is the life of the Government of the United States. Wherever that Government goes it goes by virtue of that Constitution or grant from the

sovereign people which made the Government and which gave
it as a government certain powers and withheld from it others.
When, therefore, the Government of the United States was in
Porto Rico, in Cuba, and in the Philippines during the war
with Spain, it was because the Constitution gave it the right
to wage war. The constitutional boundaries, therefore, if the
phrase be claimed to have any meaning, we must again insist
can only mean the entire sphere of activity within which the
Government moves.

We repeat here the contention of the Government:

"In order to extend boundaries recognized by other nations
the extension of dominion by conquest is sufficient. To extend
constitutional boundaries there must be some extension of organic law to the inhabitants or of institutions over the territory.
The sphere of application of the Constitution is determined not
by considerations of title to land, but by recognition of the political status of its inhabitants" (opinion in the *Goetze* case); or,
as the Attorney General phrases the same contention, "acquired
territory as [is] neither bound nor privileged by that instrument
until brought within its operation either by express compact in
the treaty or by act of Congress." p. 10, Brief.

The difference between our position and the reasoning of the
learned judge and the Attorney General is fundamental and
admits of no compromise.

If they be correct, we were in Porto Rico from the treaty of
peace down to the recent act for the government of that island
without any constitutional authority.

If this be so, our Government and officers had no warrant
for their acts in the Constitution, and, however well-meaning
they might have been, they were in law mere usurpers; they
were acting without the law and without the authority of the
sovereign creating the law.

Granting that the title to the soil came rightfully to the
United States; that the island was completely under its dominion and jurisdiction, all its agencies in that island and all its
actions there were in pursuance of the Constitution.

By this we mean—and we desire to make this point very
clear, as it seems to us that misconception of its force has led to

the fallacy underlying the decision under consideration—that the agencies of the Government acting in Porto Rico had the powers and only the powers conferred by the Constitution, and that in their actions there they were subjected to all its applicable limitations, restrictions, prohibitions, or delegations.

IV. THE EXTENSION THEORY. The clear effect of annexation as shown in the point above is sought to be avoided by a theory that the Constitution extends to certain places and not to others.

The so-called extension of the Constitution has been a premise upon which much reasoning has been based. This reasoning we believe to be fallacious, because the premise is a misleading one.

Our claim is that the Constitution as such cannot be extended by the legislature. This use of the term " extension " is a misnomer.

The cases relating to the application of the constitutional provision in regard to jury trials in the District of Columbia and in Utah have been fully discussed in the other cases now pending before this court, and to do so here would thus involve endless repetition. The cases to which we refer are the following: *Reynolds* v. *United States*, 98 U. S. 145; *Callan* v. *Wilson*, 127 U. S. 540; *Springville* v. *Thomas*, 166 U. S. 707; *Bauman* v. *Ross*, 167 U. S. 548; *Thompson* v. *Utah*, 170 U. S. 343; *Capital Traction Co.* v. *Hoff*, 174 U. S. 1; *American Publishing Co.* v. *Fischer*, 166 U. S. 464; *Black* v. *Jackson*, 177 U. S. 363. These cases decide that the Congress cannot make any law in violation of the prohibitions of the Constitution.

In order, however, to avoid the conclusion that these cases authoritatively settle the proposition that Congress in legislating for the Territories is bound by the limitations expressly contained in the Constitution, the learned counsel for the Government claim that in all the cases cited Congress had legislated that Constitution into the Territories, *i. e.*, extended the Constitution. Hence it was there in force by Congressional action, and the cases referred to were properly decided.

It is true that with one possible exception this theory is no-

where foreshadowed in these decisions.   They are all based on
the Constitution itself, not the Constitution by act of Congress.

Passing this objection, however, the position criticised is unsound for the following reasons:

(1) If the Constitution is in the Territories as an act of Congress, it is a mere law, and can be recalled in whole or in part
by the same power that projected it.

(2) The Constitution is a constitution or creation of a government, not a system of laws applicable to any particular territory.
The Government created thereby has jurisdiction over certain
territory, but the Constitution only affects the territory indirectly because of its operation upon the Government.   To extend the Constitution to a territory does not establish a government for the territory.   It can be changed, modified, abrogated
—it cannot be extended.   The Government which it has ordained
may, in the march of time, rule all the peoples of the earth, but
the Constitution would not be thereby extended—the same Government would have extended its dominions, but the Constitution would be the same instrument operating in the same way,
viz., upon the Government.

(3) The organic acts for the Territories and the Revised Statutes enact that no law shall be passed for the Territories "inconsistent with the Constitution."   Assuming for the argument
that the contentions of counsel for the Government are correct,
to the effect that the Constitution was only made for and can
only apply to the States of the United States, we must then
read into all the general prohibitory clauses of the Constitution
the word States, e. g., Congress shall make no law respecting
an establishment of a religion within the States.

No person shall be held to answer for a capital or otherwise
infamous crime " within the States " unless on a presentment, etc.

Thus read there would have been nothing inconsistent with
the Constitution in the laws held unconstitutional in *Callan*
v. *Wilson*, or *Springville* v. *Thomas*.   If the Congress was allowed by the Constitution to enact laws for the trial of capital
cases without jury in the Territories, then such laws are not inconsistent with the Constitution.   The truth is that the legislation of Congress on this point was merely declaratory of its

own powers. It knew that laws violating the prohibitions were " inconsistent with the Constitution " wherever civil government prevailed. If the views of the learned Attorney General are correct, the legislation of Congress was the merest nullity because none of the laws declared unconstitutional are inconsistent with the Constitution read in the light of his novel theory.

It is respectfully submitted that Congress in establishing a government in the Territories and enacting an organic act defining the powers of the local legislature used out of abundant caution the language cited as part of one complete scheme, a portion of which was merely declaratory.

The Constitution is not a physical substance. It is in the nature of a grant or power, or what would be termed, in private law, a power of attorney. A real Constitution is a grant of rights or powers by a sovereign. The sovereign cannot be limited, for he is the source of all law. Judge Matthews in *Yick Wo* v. *Hopkins*, 118 U. S. 370.

If the sovereign, so called, is limited by some external power, then he is not the real sovereign; it is the power imposing the limitation that possesses sovereignty. This is so because sovereignty is something which cannot be limited. It is the ultimate power. The sovereignty in the United States is in the people of the States.

It was contended during a long period of our history, and the contention finds adherents in our day, that the sovereignty of the United States was in the States of the Union, and that they, as States, and not the people, created the Constitution.

In the great case of *McCulloch* v. *Maryland*, 4 Wheat. 416, it was argued by one of the ablest advocates of that theory that the Constitution was created by the acts of the sovereign and independent States. Chief Justice Marshall met the proposition and answered it.

He said : " To the formation of a league such as was the Confederation the state sovereignties were certainly competent. But when, 'in order to form a more perfect union,' it was deemed necessary to change this alliance into an effective government, possessing great and sovereign powers, and acting directly on the people, the necessity of referring to the people,

and of deriving its powers directly from them, was felt and acknowledged by all.

"The Government of the Union, then (whatever might be
the influence of this fact on the case), is emphatically and truly
a government of the people.　In form and in substance it emanates from them.　Its powers are granted by them and are to
be exercised directly on them and for their benefit.

"This Government is acknowledged by all to be one of enumerated powers."

The limitations of the Constitution upon the Federal Government are not limitations upon the American nation.

The American nation is sovereign.

It can go where it wishes, can act where it wishes, acquire
territory where it wishes, treat the inhabitants as it wishes, and
its powers are only limited by the physical force which may be
brought to bear against it by other sovereigns.

But the Government is not sovereign.

Again we desire to respectfully submit that a great deal of
the reasoning upon which our opponents rely is based upon the
inability to distinguish this salient fact: That the people of the
United States are sovereign, and that the Government is not, is
the great fact which distinguishes our constitutional law from
that of most of the civilized nations of Europe.

It was a great departure from and a great improvement upon
the political science and upon the law and institutions which
had preceded it.　It did not make us a crippled nation, as the
Attorney General suggests, but a nation that has permanently
protected itself against usurpations by its own agents.

The court below said: "If the United States is to be denied
this common attribute of sovereignty, it must be admitted that
the treaty of Paris is so far unconstitutional; but if our nation
has this power in common with other nations, then the treaty
is valid."

Here, again, we find this precise fallacy, the confounding of
the nation with the Government.　If the Government possesses
all the powers of the nation, then there is no question before
the court for decision.

To state that because other nations or states possess certain powers the Government of the United States must possess them, or the nation be a crippled one, is an absurdity.

The difference between the other nations referred to and the United States is that in those nations the body of officials constituting the government are endowed by the people with all the powers of the state or of the sovereignty. They can take property without due process of law; they can try in any mode which they may desire; they can abridge the freedom of the press; they can violate all those rights which we are accustomed to call sacred and inalienable.

The United States as a sovereign people can do all these things, but they were unwilling to allow their officials to do them, and until their ideas shall have changed, so that they no longer believe certain rights important or fundamental, these limitations placed upon the Government will doubtless remain there.

But to argue from this that the sovereign nation called the United States is any less powerful than other nations, or cannot pursue any course or policy which it may desire, is, we submit, due to a failure to appreciate the basic elements of our constitutional law.

"The *de jure* title to the soil," says the learned judge, "was in the United States, but its inhabitants were foreigners to the Union, and the provision for the uniformity of duties had no application there."

If by foreigners to the Union he means persons without political rights, then we acquiesce in the proposition. Citizens of the United States residing in the District of Columbia or in the Territory of Oklahoma, or residing abroad and having lost their residence within the States of the Union, are then foreigners to the Union.

Many, if not most, of the provisions of the Constitution may be inapplicable to the inhabitants of Porto Rico, but this is true of many inhabitants of the United States. Aliens of all races, whether Aryan or Mongolian, inhabiting the United States, may in this sense be foreigners to the Union, yet they possess certain rights which the Government cannot infringe (*Yick Wo* v. *Hopkins*, 118 U. S. 370), "not because those provisions were enacted for them, but because they are essential limitations inherent in the very existence of the American Government." **Secretary** Root's report of 1899.

The Constitution does not act directly upon the people of Porto Rico or the United States.

It is upon the Government that the Constitution acts directly.

The officers of the Government cannot take property within the District of Columbia without due process of law. They cannot try a man in the Territory of Oklahoma without indictment by a grand jury and trial before a petit jury and with the other safeguards known to the common law, and yet the individual whose property is so protected in the District of Columbia, or whose life is so safeguarded in the Territory of Oklahoma, may not be, and often is not, a citizen either in the general or political sense and has no direct relations to the Constitution; he is an inhabitant of the United States, and as such (temporarily subject to its jurisdiction) he is entitled to certain rights because the people of the United States have chosen to place certain limitations on the Government.

The people may take his property without due process of law and they may try him without a jury, if they so desire. They have elected to do otherwise, and until they terminate that election he possesses immunities against the action of the Government.

In other words, there are certain spheres within which the Government, at least under normal circumstances—that is to say, peace—cannot tread, by reason of those inhibitions in the Constitution. The inhabitant has rights, or what may be better called, viewed at least from the Government standpoint, immunities.

While the military status lasted the prohibitions and limitations of the Constitution did not apply.

Martial law is the will of the commander—that is to say, it is no law—and, therefore, while martial law existed by virtue of the Constitution, the Porto Ricans had no rights thereunder because the Constitution granted them none.

It is this absence of immunities on the part of the inhabitants of territory under the sovereign dominion of the United States during the existence of military government which the learned judge apparently had in mind when he speaks of the Porto Ricans being foreigners to the Constitution. The Constitution did not spread about them its protecting ægis; because during the military period the usual limitations did not apply.

The Constitution does not apply as a whole to every action of the Government in every particular locality. Wherever and however acting, it is acting under some clauses or provisions of organic law and may not be affected by others.

There are, in other words, in our system, broadly speaking, two kinds of government permitted by the Constitution. (1) Military government, which means the suspension of all immunities, and (2) the normal or peace Constitution. This may in turn be properly divided into two portions, viz.: (I) The Federal, in which the powers of government are divided between the local or state governments and the General Government (and the greater portion of the Constitution applies to the Government in this Federal capacity); (II) the local, or that government in Territories or places in which no state government exists.

As a territorial government Congress has all the powers which it possesses as a Federal Government, and together therewith all the powers which the state governments possess, save such powers as may be expressly inhibited to both governments by the Constitution and reserved to the people. *Nat. Bank* v. *Yankton*, 101 U. S. 129.

But a curious sophism has recently been advanced. It is contended that land may be within the sovereign jurisdiction of the United States, the Government may exercise unlimited jurisdiction over it, and yet that such land or territory is not territory of the United States. It is difficult to combat this assertion, because it is a mere assertion, resting upon no logical basis whatever. "All territory within the jurisdiction of the United States not included in any State must necessarily be governed by or under the authority of Congress. The Territories are but political subdivisions of the outlying dominion of the United States." *Nat. Bank* v. *Yankton, supra.*

The Territories are nothing more than outlying dominion of the United States.

The learned court below in the *Goetze* case says:

" New territory is not brought under the Constitution by acquisition of the soil, otherwise *Fleming* v. *Page* could not have been decided as it was. This is done either by an incorporation of the inhabitants into the Union, or by an extension of our laws and institutions throughout the territory. This cannot be done by conquest, but only by legislation or treaty. *Fleming* v. *Page*. Here the treaty recognizes and makes complete the *de facto* title gained by conquest. The island is not thus brought under the Constitution unless the treaty supplements the confirmation of title by an incorporation of the inhabitants into the Union under the Constitution or by the extension of our institutions. . . ."

That the incorporation of the inhabitants into the political body constituting the sovereign people of the United States has nothing to do with the immunities of persons within the territory seems thus abundantly established.

V. THE QUESTION OF THE STATUS OF PORTO RICO UNDER THE PRESIDENTIAL GOVERNMENT CANNOT AFFECT THIS CASE.

Again it has been urged that the imposition of duties here complained of is made valid by reason of the President's prerogatives over conquered territory held under military sway.

But the tax having been imposed at New York upon goods of a New York merchant under the general tariff law, it is immaterial that Porto Rico may have been under a military form of government.

The Executive claimed that war existed, and the military status continued, and we believe the courts cannot view the matter in any other light, but must follow the coördinate branch of the Government.

Assuming, however, the truth of this, it does not follow therefrom that territory ceded to the United States is not a part thereof. The fact that the Executive still continues a *de facto* government originated under the law of belligerent rights does not affect the question. It has never been judicially determined, however, that when war has notoriously ceased and peace reigns triumphant the Executive in such a *de facto* government can still exercise full war rights without the restraints and restrictions imposed upon government by the Constitution.

The Government and the courts have established a contrary doctrine in the case of California *Cross* v. *Harrison* and *Leitendorfer* v. *Webb, supra,* where it was held that the presidential government originated in belligerent rights and remained the *de facto* government until Congress chose to legislate otherwise.

It is respectfully submitted that while war actually continues the executive power is there as Commander in Chief, and that when war has ceased, and it is so recognized by the Executive, the Executive · remains as the government in a civil capacity for the purpose of executing the laws of the United States.

International law, or that great body of usage prevailing among nations, is, of course, only a part of our law and binding upon our Government, as far as it has been recognized, acted upon, and adopted by our tribunals. But even viewed from this precise standpoint, our courts have recognized and adopted the rule of international law, that when territory is ceded the law of the old government remains in force. The sanction, of course, is in the acquiring government, and the old laws are laws because sanctioned by the new sovereign of the ceded territory. As the laws of the former sovereign, they are without force; as the usages of the inhabitants sanctioned by the new sovereign, they obtain the dignity of law, and this law remains unchanged until Congress chooses to act. And as the laws of such new sovereign they cannot prevail if they are in conflict with the fundamental principles of the new sovereign's constitution.

"Every nation acquiring territory by treaty or otherwise must hold it subject to the constitution and laws of its own government." *Pollard's Lessee* v. *Hagan,* 3 How. 212–225.

"Every nation which acquires territory by treaty or conquest holds it according to its own institutions and laws." *Fleming* v. *Page,* 9 How. 615.

"By this substitution of the new supremacy, although the former political relations of the inhabitants were dissolved, their private relations, their rights vested under the government of their former allegiance or those arising from contract or usage remained in full force and unchanged, except in so far as they were in their nature and character found to be in conflict with the Constitution and laws of the United States." *Leitendorfer* v. *Webb*, 20 How. 177.

It is therefore clear that such law is only good when not in contravention of the Constitution or laws of the United States, which might possibly apply to new territories.

In enforcing such law the Executive is merely enforcing the law of the United States, and, we respectfully submit, is acting in a civil capacity.

In such capacity he has not the rights which he would have as Commander in Chief during hostilities, and, therefore, the immunities of the Constitution for the protection of life, liberty and property operate in favor of the individual in the ceded territory; that is to say, they operate as restraints upon the Government there because it has ceased to be military and become civil.

This theory was the one adopted by the court in *Cross* v. *Harrison,* and was tersely summed up as follows: "This government *de facto* will, of course, exercise no power inconsistent with the powers of the Constitution of the United States, which is the supreme law of the land."

We submit, however, that the determination of this question is not necessary to the decision of this case.

The duties were levied upon a merchant at the port of New York, a place within the Southern District of New York, a portion of the territory constituting the United States under the civil government of a State.

The form of government in territory belonging to the United States may be military or civil, but the territory is for that reason none the less a part of the United States and, therefore, according to the Constitution, duties must be uniform throughout.

To claim that because a part of the United States may temporarily be under military government goods coming therefrom must be taxed as goods coming from foreign countries seems to us the result of great confusion of thought.

Judge Taney's illustration in *Fleming* v. *Page*, to the effect that ports remained foreign to the revenue laws until these laws had erected the machinery of custom-houses, collection districts, inspectors, and collectors, was clearly not necessary to the decision, and as a *dictum* was in itself incorrect as the historical precedents invoked were mistakenly stated and have been ignored by this court in the later case of *Cross* v. *Harrison,* 16 How. 164.

Duties must be uniform throughout the United States, and it is a matter of indifference under what particular form of government any portion of the United States be.

Were the State of New York declared to be under the military government of the United States, we respectfully submit that during the time of such military occupation goods coming from New York into New Jersey or into the District of Columbia belonging to merchants there could not be taxed on the theory that New York was not a part of the United States.

VI. *The meaning of the United States.* This brings us to a consideration of the Government plea that in the uniformity clause the term " United States " does not mean what it plainly implies.

It may be admitted, as Judge Townsend says, that other nations may take territory under their sovereignty, which they do not annex and make part of themselves.

That the people of the United States could do this and could declare that the inhabitants of territory annexed in future should have no rights recognized by the Constitution is clearly demonstrated.

That the present officials of the United States can do this we deny.

The analogy to other countries is misleading. The Constitution of the United States is a peculiar one. In the European states the government is also the state or nation. The same power which legislates also makes the constitution. Many of the European nations have a so-called constitution, but that instrument is not a constitution strictly, but merely a *charte constitutionelle* or charter, an instrument by which the government gives to the people certain rights. The government possessing all the rights of the sovereign nation and being itself sovereign can, when acting in territory not covered by this *charte*, govern as it wishes.

This is the absolute reverse of the United States. In the United States the people endow the Government. And the people of the United States, in addition to other inhibitions which they have placed upon their Government, have declared that duties must be uniform throughout the United States.

References, therefore, to the constitutional history of other nations can have no bearing whatever.

The one question, and the sole question, for decision is, whether Porto Rico, within the meaning of this clause, is a part of the United States.

While in one sense this is a political fact, it is also a fact affecting a property right protected by the Constitution, and as such a fact the court will, of course, feel bound to decide it.

The advocates of the position taken by the collector must claim broadly and without reservation, in order to maintain their contention, that the clause of the Constitution requiring uniformity of duties throughout the United States refers only to the thirteen original States and the States to be formed in the future, "because the term United States as there used (in the uniformity clause) means only territory comprised within the several States of the Union." Brief of Attorney-General in *Goetze* case, p. 5.

The claim in substance is that the term United States as used in the Constitution can have only two meanings: (1) The collective name of the States which were united together under the Constitution and mentioned in the Declaration of Independence and in the Articles of Confederation; this is the original and literal meaning of the word. (2) The corporate name of the nation.

That as used in the United States Constitution the term "United States" frequently refers to the States united does not admit of question. It is, however, admitted that it is used with great frequency in another sense as the political entity exercising governmental power.

In the Pinckney draft of the Constitution, evidently with a view to make clear one of the meanings of the term "United

States" as used in the Constitution, appears the following: "The United States shall be forever considered as one body corporate in law, and entitled to all the rights and privileges which to bodies corporate do, or ought to, appertain."

That it has, however, a third meaning, is also evident. · It means not only the States united and the body corporate or governmental power which represents them, but it means— and this is its ordinary meaning in the language of the day— that whole portion of the earth's surface over which the flag of the United States flies in sovereign dominion.

It is clear, therefore, that we are not restricted to the meaning of the term as it appears in the Articles of Confederation. It is argued with more emphasis than plausibility, that because it meant in that instrument the States united, it can mean nothing more in the Constitution of the United States, and that the phrase "United States" in the tax clause of the Constitution is equivalent to "The United States in Congress assembled."

We submit that this is a misinterpretation of history.

During the confederate period the thirteen States were thirteen distinct political sovereignties united together by a compact which was strictly an agreement in the nature of a treaty. They were not a nation.

The creation of the Constitution, however, wrought a fundamental change; a pouring of new wine into old bottles. Some of the form remained, but the spirit was gone.

A people practically homogeneous·in law and language had chosen to organize itself into a political governmental unity; an idea which had existed only when the consciousness of the people had become by the organization of the Constitution an objective reality.

A nation did not spring into being as the poets have it, because the nation existed. But the nation established for itself a government and by the Constitution gave it the necessary organization. This change was so radical that it is absurd to say that the term "United States" as used in the Constitution was used in the same sense as it had been used in the old Confederation.

The United States, indeed, sometimes might mean the States
of the United States.   But it meant something more besides.

Since the treaty with England of September 3, 1783, a vast
tract of unorganized land had come into the possession of the
people inhabiting the thirteen States, formerly the British colonies of North America.   Whether this tract of land belonged
to the individual States or to the people was long a mooted
question, and the dispute arising therefrom was the main cause
leading to the formation of a more perfect union and the adoption of the Constitution.   But, from the time that this vast
tract of territory came within the sovereign dominion and jurisdiction of the United States, that term ceased to mean only
the States united.

As was said by Madison in the Federalist (No. 38): "We
may conclude that the Northwest Territory will soon become
a national tract, and Congress having assumed the government of this Territory, has attempted to do more.   They have
appointed officers and have prescribed the conditions upon which
States may be admitted into the Union.   All this has been done,
and done without the least color of constitutional authority."

It was clear that the Government of the Confederation had
never had any constitutional right to govern this Territory.
The people of the United States, even under that imperfect
organization, took upon themselves the task which resulted in
the celebrated ordinance of 1787, by which the inhabitants of
this Territory were accorded not only the ordinary civil rights,
which in that primitive age were considered so important as to
be inalienable, but also certain political rights.

Of course until the cession of these lands to the General Government by the people of the United States, the latter term
could have but one meaning.   It would have been perfectly
possible, and even proper, for the people to have used another
word to designate the entire domain made up of the original
States and the new land, and which John Marshall called the
American Empire.

But the draughtsman of the Constitution chose to use the
same word to designate three things: The States, the corporate
name of the nation, and the whole territory over which the

people of the United States through either their general or
state governments had jurisdiction.

It is admitted by an able advocate of the view under criticism
that, as far as the United States has been concerned, "At all
events no such new term has been adopted and hence United
States is the only term which we have had to designate either
individually or collectively the States and Territories, and accordingly, while it has always been used for the former of these
purposes, it has sometimes been used for the latter." Professor
Langdell, Harvard Law Review, Feb. 1900.

Or, as the learned Attorney General says, the word has the
third meaning in "an international sense designating the extent of our dominion as a sovereign nation," and explains the
admission by stating that the term in this sense is one of common usage—that is to say, conventional, and that it has no constitutional or legal meaning, and that, therefore, the Constitution
cannot be supposed to have intended it for that purpose.

So far from its being probable that the framers did not mean
to use the word in its so-called international sense, the history
of that time demonstrates quite conclusively that the exact
opposite was their intention.

The great ordinance for the government of the Northwest
Territory, drawn originally by Jefferson, and somewhat modified before it passed through Congress, was in some respects a
prototype of the Constitution itself. It embodied the ideas
which led up to the foundation of the Constitution, based upon
the political philosophy adhered to by most of the framers of
the Constitution. It gave to the hardy and self-reliant pioneers
in that Territory political rights of self-government and secured
to them the guarantees of personal freedom in accordance with
the most enlightened rules of the common law. That this ordinance was regarded as sacred and as unchangeable as the law
of the Medes and Persians, appears from its language, which
declares it to be a compact between the people of the Territories
and the people of the States, unchangeable except by consent.
Almost the first act of the first Congress, in which many of the
framers of the Constitution sat, was to reënact the Northwest
ordinance in its entirety. It is idle to say that their doing this

involved the notion that the people therein were not sufficiently protected by the Constitution, as the learned Attorney-General assumes. The Constitution gave them no right of local self-government. It was necessary to enact some law conferring upon them political rights, and therefore the ordinance was re-enacted by Congress, the original ordinance having been adopted prior to the adoption of the Constitution.

The fact that the ordinance contained many of the provisions of the subsequent Constitution in no manner supports the theory of the learned Attorney General that "the accepted doctrine was that such guarantees and rights must be conferred by Congress." p. 102 of *Goetze* brief.

Unnecessary provisions are sometimes inserted in statutes out of abundant caution. *McAllister* v. *U. S.*, 141 U. S. 174, 187.

VII. THE UNIFORMITY CLAUSE IS NOT IN THE NATURE OF A LAW ITSELF, BUT PROHIBITS THE CONGRESS FROM PASSING CERTAIN LAWS.

In further considering the reach of this uniformity clause or the consequent breadth to be assigned to the term "United States," it is proper to recall the difference between the rule of interpretation to be given to a statute, and that to be given to an organic act whose object was to restrict the statute-making power, and prohibit the enactment of a certain class of obnoxious legislation.

As the ordinance was framed before the Constitution it seems strange to claim that "the history of the ordinance for the government of the Northwest Territory also proves that the statesmen of that day did not accept the doctrine that the guarantees enjoyed by the inhabitants of the States were possessed by the inhabitants of the Northwest Territory neither by virtue of the Articles of Confederation nor the fact that they had theretofore been within the jurisdiction of one of the States." Atty. Gen. *Goetze* brief, p. 102.

The provisions of the Constitution relating to the States have often been put in the statutes creating the machinery necessary to carry them out. Without this machinery many of these enactments are lifeless.

This is true of the original judiciary act drawn by Mr. Ellsworth and of many of the early statutes.

The prohibitions in the Constitution against direct taxation, unless in proportion to representation, uniformity in duties, and the bill of rights, are, however, all of a negative nature.

They forbid the Government to do certain things and it does not require legislation to carry out the prohibition. In other words, the Government cannot legislate in contravention of them.

The Constitution intended that all the inhabitants of the States and Territories under the sovereign dominion of the United States should have the equal protection of the laws and the Constitution.

As was said by Judge Bradley in *Boyd* v. *United States*, 116 U. S. 616, regarding the Fourth Amendment: "As every American statesman during our revolutionary and formative period as a nation was familiar with this monument of English freedom (referring to Lord Camden's decision in *Entic* v. *Carrington* and three other king's messengers, which was the *Wilkes* case) and considered it as the true and ultimate expression of constitutional law, it may be confidently asserted that its propositions were in the minds of those who framed the Fourth Amendment to the Constitution," etc. In was therefore true, historically and legally, "That the District of Columbia or the territory west of the Missouri is not less within the United States than Maryland or Pennsylvania; and it is not less necessary on the principles of our Constitution that uniformity in the imposition of imposts, duties, and excises should be observed in the one than in the other. Since then the power to lay and collect taxes, which includes direct taxation, is obviously coextensive with the power to lay and collect duties, imposts, and excises, and since the latter extends throughout the United States, it follows that the power to impose direct taxes also extends throughout the United States." *Loughborough* v. *Blake*.

Admitting that the Constitution uses the term "United States" in several senses, it would then follow that we must seek for the meaning of the term in the context.

It is not reasonable to suppose, however, that different senses would be given to the word in the same clause. "That Congress shall have power to lay and collect taxes . . . to provide for the common and general welfare of the United States," etc., but all duties must be uniform throughout the "United States."

The United States for whose debts and general welfare the proceeds of the taxes are to be devoted must mean the same United States throughout which they are to be uniform.

It is respectfully submitted that it can hardly be seriously contended that Congress cannot apply the proceeds of the general taxation to the general defence and welfare of the Territories as parts of the United States.

If our opponents are logical they must deny this and Congress would, therefore, not have the power to apply the proceeds of general taxation to the welfare of the people of Oklahoma or New Mexico, or to defend them in case of invasion.

The Constitution also provides that Congress shall have power to pass a uniform rule of naturalization. It has been recognized by the Supreme Court that the early laws passed by the Congress in which sat many of the members of the convention are contemporaneous interpretations of the highest value.

An examination of the naturalization law will show that that statute was intended to include the Territories as well as the States.

The act of January 29, 1795, c. 20, 1 U. S. 414, which was an act to provide a uniform rule of naturalization, includes the Territories of the United States within the term "United States." It declares that any alien may become a citizen of the United States upon complying with certain requisites.

He shall declare before one of the courts that he has resided in one of the States aforesaid or within the Territory within which such court is held at least one year. His time of residence within the Territory is evidently included within the five years within which he shall reside within the United States, and it is evident that the statute uses the term "United States" in the same sense that Chief Justice Marshall used it as the Great American Empire. That it also used it in the sense of States united is evident from the first article, "That any alien being a free white person may be admitted to become a citizen of the United States or any of them." Certainly in the naturalization law the word was used in both senses.

Elk v. *Wilkins*, 112 U. S. 102, virtually takes the same view.

It is unsound to argue that because in some contexts the word is used meaning individual States it may not in others mean to apply to all the dominions over which the Government exercises jurisdiction.

The flexibility with which the word "State" may be used, and the underlying principle that when used in some legal enactment or document the context must be considered and the word may be understood in its conventional and ordinary meaning as well as in the legal or historical meaning, is well illustrated in the case of *Geofroy* v. *Riggs*, 133 U. S. 258.

It is there held that the word "States" or "Union" may include the District of Columbia, although strictly speaking the District of Columbia is not a State. It is a political entity possessing the right to local self-government and may properly fall within the designation of State as understood generally in the language of diplomacy and international law. "To insure reciprocity in the meaning of the treaty it would be necessary to hold that by the term United States or Union is meant all the political States in the country. . . . It is not only those political communities called the States, but also those which constitute the political bodies called the Territories and the District of Columbia." *Geofroy* v. *Riggs, supra.*

The question of the meaning of this term arises very clearly under the Fourteenth Amendment in the phrase, "All persons born or naturalized in the United States." This phrase has been interpreted by the Supreme Court in the famous case of *Wong Kim Ark* v. *United States*, 169 U. S. 649, as follows: "These provisions are useful in their application to all persons within the territorial jurisdiction. It is accordingly enacted by section 1997 of the statutes that all persons within the jurisdiction of the United States shall have the same rights in every State or Territory."

As was said in the *Slaughter House Cases*, 16 Wall. 36, 74: "Not only may a man be a citizen of the United States without

being a citizen of a State, but an important element is necessary
to convert the former into the latter. He must reside within
the State to make him a citizen of it, but it is only necessary
that he should be born or naturalized in the United States to be
a citizen of the Union."

Which is the " United States " as here distinguished from the
several " States ? "

It has been demonstrated —

That the term "United States " was meant by the framers of
the Constitution to include States and Territories or the outlying dominion under the jurisdiction of the United States;

That the Constitution itself shows that it was used in this
sense in the uniformity taxation clause ;

That the early laws of the United States carrying out the
Constitution so interpreted it and that the meaning given to it
by Chief Justice Marshall as the equivalent of the Great American Empire was the meaning intended by the Fourteenth
Amendment.

This meaning is the ordinary general meaning in which it is
understood, not only by American citizens, but by people throughout the world. The historical, legal, and constitutional uses of
the term are therefore in accord.

The fact that it sometimes means the States of the United
States and at other times the Government is immaterial, as each
time that it occurs in the Constitution its meaning must be determined by the context.

It is true that an eminent statesman, Mr. Webster, at times
contended, both before the Supreme Court and in Congress, that
the Constitution did not apply to the Government of the United
States when it was acting in the Territories, and that it had
there no limitation.

" Congress," he says, " has full legislative powers in the Territories without any grants from the State. What is Florida?
It is no part of the United States."

In the great debate with Calhoun, however, Mr. Webster admitted that the laws of Congress governing Territories were
based upon the power granted in the Constitution to make all
necessary rules and regulations for the Territories of the United

States, thereby admitting that a portion of the Constitution, at least, applied to the Territories. He was also forced to admit that the constitutional inhibitions on the General Government were everywhere in force. 20 Cong. Globe, 252, Feb. 1849.

An examination of the position taken by Mr. Webster shows that he had in mind political rights, and that when he asked about Florida and said it was no part of the United States because not represented in Congress, he had in mind the political rights of the people of the States recognized by the Constitution, which extend as well to the inhabitants of Territories.

The position, however, which he was forced to take resolves itself simply into the assumption that under the rules and regulations clause Congress can do what it wishes in the Territories. This position has been so frequently overruled by the courts that it is scarcely necessary now to argue it.

It is certainly inapplicable to the uniformity clause, because even if the Bill of Rights by any strained construction of the Constitution be held not to go with Congress into the Territories, certainly the uniformity clause, as has been shown, applies to the whole United States, and therefore limits Congress when legislating for the Territories.

The learned counsel for the Government has set forth the debate in the Senate on the Walker amendment proposing the extension of the Constitution to California. He considers Calhoun the father of the theory that the Constitution can have effect in the Territories, and believes the doctrine to have had its origin in the desire of the advocates of slavery to carry that institution into the Territories.

If Webster adopted a position which even his ability and ingenuity failed to sustain, this position is fairly attributable to his fear that the doctrine that the Constitution extended to the States would involve the proposition (not as we believe a necessary consequence) that slavery should also be allowed to exist in the Territories free from the power of Congress to interfere with it.

He was, therefore, looking at the question from a partisan standpoint, and his opinion on the question as a legal proposition was as much influenced thereby as that of Calhoun.

The debate referred to has been well described by Von Holst, a historian whose hostility to slavery and its advocates is a most marked characteristic of his able and exhaustive work.

He says: " The amendment in this modified form gave rise to an interesting and important constitutional.debate. Webster objected to it on the ground that it gave the President unlimited authority over the district, but he also maintained that it was impossible to extend the Constitution in so general a way to a Territory. It was indeed the moral duty of Congress in its legislation for' the Territory to preserve the. principles of the Constitution, but it was not absolutely necessary. The Territories were not a part, but a possession of the United States.

" Calhoun, on the contrary, maintained that the Constitution, which was of itself the supreme law of the land, extended *pro-prio vigore* and *eo ipso* also to the Territories, even though its provisions were not all applicable there. If the Constitution does not extend to the Territories, whence did Congress get the authority, which existed only by virtue of the Constitution, to exercise any government over the Territories?

" Calhoun was evidently right, although Webster had good grounds for astonishment that the radical upholder of State rights should support' this view. The courts of the United States have decided that the Constitution has a legal existence. The relation of the Union to the Territories is, therefore, a legal relation in and under the Constitution which is wholly independent of the legislation of Congress, of which it is in fact the basis. The fact that the legislative action of Congress is required in order to make this legal relation effective is by no means, as Webster seemed to think, in contradiction of this relation, for, as Calhoun rightly said, the legislative action of Congress is equally necessary in order to put into operation the provision of the Constitution relating to the States. Unquestionably there is an essential difference between the nature of the legal relations of the States to the Union and that of the Territory. The distinction following Webster's line of thought is closely followed by Cooley's saying, ' the Constitution is made for States, not for Territories.' . . .

*　　*　　*　　*　　*　　*　　*　　*

" And equally incontestable is its (the Supreme Court's) further declaration that the powers of the Federal Government in regard to the persons and property in Territories cannot be greater than those guaranteed to the citizens of the State. Calhoun had asked whether Congress could create a nobility and an established church in the Territories." Von Holst's Constitutional History, vol. 3, p. 444.

Exemption from the uniformity clause has been sought in the fact that Congress, acting as the local legislature, may impose special taxes for the use of a special locality, as the States may do in the territory over which they have legislative power.

When Congress is acting as the local legislature in the Territories, and taxing there, it is contended that it is not bound by the uniformity clause.

Such taxes are not for the common welfare of the United States, but are to defray the expense of the government of the locality, and in the dual position which Congress occupies in our system, as Federal Government and as local government for the territory of the United States not erected into States, it has the power to tax for local purposes.

Taxes, therefore, levied in Porto Rico, the proceeds of which are applied for the benefit or maintenance of the government of the island, may, perhaps, be defended upon the ground that they are imposed in the exercise of the right which Congress has in the Territories.

But no question of this kind can arise in this case. The tax was imposed under the Dingley Act, a law for the taxation of all goods coming into the United States of America and for the benefit of the Treasury of the United States. Congress in passing this law was acting as the General Government, and no question of its power as the local legislature can possibly be raised. The tax was levied on the goods of a New York merchant at the port of New York and is unaffected by the status of Porto Rico, it being once admitted that Porto Rico was a part of the United States.

VIII. PRECEDENTS DRAWN FROM OUR HISTORY DO NOT SUSTAIN THE POSITION OF THE GOVERNMENT.

The precedents attempted to be drawn by the learned coun-

sel for the Government from the history of the Louisiana and
Mexican annexations under the treaties with France and Mexico, respectively, are not in point.

The learned Attorney General states (page 31 of his *Goetze*
brief):

"It is a common error, long disseminated and many times
repeated, to assert that Jefferson was under the belief that the
United States had no constitutional power to acquire foreign
territory. . . .

"An examination, however, of his writings and of his whole
course of action with reference to the Louisiana purchase, especially with reference to the constitutional question, shows
conclusively that Mr. Jefferson's doubt was not with reference
to the power of the United States to acquire foreign territory,
but rather as to the right to annex it to and make it a part of
the United States."

The learned counsel thinks this point of very great importance.

As a matter of history his view is perhaps correct, although
even as to this there is considerable doubt.

What Jefferson did certainly doubt, and the history of the
time and the debates in Congress tend to show it, was the
power of Congress to admit new States to the Union from the
ceded territory without even a Constitutional Amendment or
the consent of all the States.

Mr. Jefferson had instructed Mr. Livingston, then American
minister in Paris, that in no event should a provision be inserted
in the treaty with the French Government providing that
States should be erected in the new territory, as he evidently
did not believe that this could be legally done, and was therefore unwilling that the Government should take upon itself an
obligation which it could not carry out.

Mr. Jefferson knew the jealousy which the States felt of each
other and the sectional feeling which prevailed. He felt that
an attempt to form States out of this vast territory would give
rise to controversy, and with this in view he so instructed Mr.
Livingston. Mr. Livingston, however, for reasons which doubtless justified the wisdom of his act, disobeyed the instructions
of Mr. Jefferson.

The First Consul desired to insert a provision in the Louisiana treaty to the effect " that the inhabitants be incorporated into the Union of the United States," etc. It was necessary to conclude the treaty with great rapidity, as France was verging upon a war with England, and the opportunity presented by the proposition of the First Consul to cede the whole Louisiana territory seemed so favorable to Mr. Livingston that he thought no obstacles should be interposed to its immediate execution. It was for this reason that the clause was inserted in the treaty, contrary to the express instructions of the President. This clause, as appears from its wording, can mean only one thing. The new territory was to be admitted among the States of the Union, and its inhabitants to be citizens of such States as soon as possible.

The history of the time, as outlined in the foregoing, proves this beyond question. Adams' History of United States, vol. II, chap. II to V.

As was said by Mr. Jefferson and quoted by the Attorney General: It is most necessary [to convene Congress] because they will be obliged to ask from the people an amendment of the Constitution authorizing their receiving the province into the Union."

" The Constitution has made no provision for our holding foreign territory, still less for incorporating foreign nations into our Union."

" I think it would be safer not to permit the enlargement of the Union but by amendment of the Constitution."

" I am aware of the force of the observations you make on the power given by the Constitution to Congress to admit new States into the Union without restraining the subject to the territory then constituting the United States. But when I consider that the limits of the United States are precisely fixed by the Treaty of 1783, that the Constitution expressly declares itself to be made for the United States, I cannot help believing that the intention was to permit Congress to admit into the Union new States which should be formed out of the territory for which and under whose authority alone they were then acting. I do not believe it was meant that they might receive

England, Ireland, Holland, etc., into it, which would be the case under your construction." pp. 33 to 36, Attorney General's brief.

The learned Attorney General, however, seems to assume that the expression of the treaty that the territory shall be admitted into the Union, etc., means something different from the union of the States. He says: "This correspondence demonstrates conclusively that whatever doubt Jefferson had as to the constitutional authority for the Louisiana Treaty related, not to acquiring territory, but to the right either of the treaty-making power or of Congress to annex it to or incorporate it into the Union."

If by this the Attorney General means to incorporate it into the Union *as a State*, we agree with his assertion. We cannot see what other meaning it can possibly have, and yet the Attorney General finds in this history of the Louisiana acquisition precedent for the proposition that territory may be acquired and held by our Government as a colony or province, not a part of the United States.

The meaning which the learned Attorney General seems to have in mind is that the Union included not only the actual States, but that portion of the States which had been ceded to the General Government and which was usually known as the Northwest Territory. He seems to think, further, that as the framers had intended that that territory should be erected into States, it stood upon a different basis from territory thereafter acquired, and that incorporation into the Union did not necessarily mean as a State, but meant to place the new territory in the same position as that formerly held by the Northwest Territory.

We fail utterly to appreciate the force of this argument.

Even assuming it to be true that Jefferson and his advisers, as well as the framers of the Constitution, contemplated that this territory was held in trust for the purpose of erecting States out of it, nevertheless there was nothing in the Constitution to show that this territory should be held differently and governed differently from territory thereafter acquired.

Admitting, as it is claimed, that Jefferson did assume that

the United States Government had the power of acquiring territory, it would then, according to the contention of the learned counsel for the Government, come under the rules and regulations clause.

This is the clause, however, of the Constitution in which he finds warrant for the government of the Northwest Territory.

Therefore, the Northwest Territory and the new acquisitions must have stood on a precisely similar footing. As is said by the learned Attorney General (page 102 of his brief): "The history of the ordinance for the government of the Northwest Territory also proves that the statesmen of that period did not accept the doctrine that the guarantees enjoyed by the inhabitants of the States were possessed by the inhabitants of the Northwest Territory, neither by virtue of the Article of Confederation nor by the fact that they had theretofore been within the jurisdiction of one of the States."

The phrase "union" therefore meant the union of States, and when Mr. Jefferson and his advisors doubted the propriety of admitting the territory into the Union, they did not mean the union of States and Territories, but the union of the States. Besides, it is respectfully submitted that that is the undoubted meaning of the word "union."

The debates cited at so much length clearly show that the only question was as to the constitutionality and propriety of the stipulation of the treaty admitting the new territory into the Union.

Many in Congress shared Mr. Jefferson's doubts, at least as far as the question of admitting the new territory to statehood was concerned.

It did not seem to be clearly understood at that time whether the treaty was of itself operative so to admit the inhabitants, or whether an act of Congress was necessary, or whether both together without an amendment of the Constitution or the consent of all the States could accomplish the object.

That Art. III of the Treaty of 1803 was considered by Congress and by the Louisiana inhabitants as intending to provide for an admission of their territory as a State is evidenced by the remonstrance and the Congressional reply, which we ex-

cerpt: " Your honorable body seems to have adopted a construction of this article which would suspend its performance until
some period fixed by the principles of the Constitution and to
have read the article thus: 'The inhabitants shall be incorporated into the Union and admitted to the enjoyment of all
the rights, etc., as soon as the principles of the Federal Constitution will permit.' We, on the contrary, contend that the
words 'according to the principles of the Federal Constitution,'
as they are placed in the sentence form no limitation, that they
were intended as a description of the kind of rights we were to
enjoy, or, at most, relate to the mode in which they were to be
conferred, and that the article contemplates no other delay to
our reception than will be required to pass the necessary laws
and ascertain the representation to which we are entitled." To
this remonstrance the Committee of Congress replied: "We
consider, in the first place, that the clause, which is the ground
of our claim, is a stipulation made expressly in favor of the inhabitants of Louisiana then existing, because the French Government had no right to stipulate the incorporation of the future
citizens of Louisiana. We think that the words 'as soon as possible, according to the principles of the Constitution,' evidently
express that this incorporation is to be executed without any
unnecessary delay, and that it is to take place on the same principles by which the Constitution has regulated the rights of the
individual States, and of the citizens of the United States, in
relation to the Federal compact. We humbly think that any
interpretation tending to procrastinate the incorporation of the
present inhabitants of Louisiana into the Union is directly opposite to the spirit of the third article of cession of our country,
the object of which is unquestionably to secure that advantage
to the inhabitants who are annexed to the United States by that
treaty; that, consequently, any condition depending on future
circumstances ought to be inadmissible, because it would expose
the inhabitants who existed in Louisiana when the treaty was
made to be kept out of the enjoyment of rights which have been
stipulated for them."

The only difference of opinion was as to the time when such
statehood should be conferred.

These doubts and difficulties were evidently borne in mind by the Government when it concluded the treaty with Mexico, and they were avoided, as appears from the clause of that treaty.

MEXICO TREATY.

Should be incorporated into the Union and be admitted at the proper time (not immediately *proprio vigore* of the treaty), but by act of the Congress of the United States to the enjoyment of all rights of citizens of the United States. p. 66, Attorney General's brief in *Goetze* case.

LOUISIANA TREATY.

The inhabitants of the ceded territory shall be incorporated into the Union of the United States and admitted as soon as possible according to the principles of the Federal Constitution to the enjoyment of all rights, advantages, and immunities of citizens of the United States.

We respectfully submit that Jefferson believed the Government could annex territory, though he doubted whether such territory could be admitted into the Union, the question in this case is not affected.

The contention of the Government is, that this territory and all territories, save the original States and the States subsequently admitted, are not affected by the inhibitions placed by the Constitution on the action of the Government.

This they claim to be true of all territory owned by the United States from the earliest time to the present; that is, their argument applies equally to the Northwest Territory and to the island of Porto Rico.

There is here failure to distinguish between political rights on the one hand and the immunities against the actions of the Government which the people of the United States have created by the Constitution on the other hand.

The statesmen of Jefferson's day were, many of them, unwilling that Louisiana should be admitted into the Union, have two Senators and Representatives in Congress, and disturb what they believed to be a very nice adjustment of interests.

What they doubtless feared was political power. They had no desire to oppress Louisiana, establish an order of nobility or a religion, to take property without due process of law, or to tax the inhabitants for their own benefit. That none of these things were in their minds was evident from the course pursued. She was given all the guarantees of liberty and the machinery to carry them out. Our customs laws and tariff were extended to her and her inhabitants were not cut off from our markets. No debates, no struggles, no doubts can be found in the history of the time as to the right of her people to have all these things. It is, therefore, manifest that the statesmen who opposed the treaty opposed it, not because they feared to grant those things which were given so freely and so unanimously, but because they feared the subsequent admission of States from Louisiana and the injection of new political forces and interests into the Union of the United States. The act of March 26, 1804, for the government of Louisiana enacted a full bill of rights in entire accord with the Constitution.

It is true that in October, 1803, the House hurriedly enacted a bill providing for immediate temporary government, by the President, transferring to him all the powers held by the former Spanish officials. That this was a temporary measure appears upon its face and the bill above referred to for the government of the territory was passed within a year, yet even in this haste the safeguard was inserted that these powers should be exercised for maintaining and protecting the inhabitants of Louisiana in the full enjoyment of their liberty, property, and religion.

While this latter bill was under consideration, Dr. Eustace, of Massachusetts, made a speech largely relied upon by the learned Attorney General in support of his point that the Constitution had no effect in Louisiana. Dr. Eustace said: "The people, in my opinion, are at present unprepared for and undesirous of exercising the elective franchise. The first object of the Government is to hold the country. How? By protecting the people in all their rights and by administering the government in such a manner as to prevent any disagreement among them—to use no other term. . . . When they should be better acquainted with the principles of our Government, and shall have become desirous of participating in our privileges, it will be full time to extend to them the elective franchise. Have not the House been informed from an authentic source since the cession that the provisions of our institutions are inapplicable to them?"

And yet this speech was made in support of a bill which guaranteed full civil rights.

The view was then held by Congress, and probably rightly held, that the people should remain under what we term territorial government for some time before they should be admitted as States.

In this connection the learned Attorney General seems to believe that Gouverneur Morris's statement as to what he intended by the rules and regulations clause of the Constitution should have some weight. While we scarcely believe that a communication contained in a private letter as to what one member of a convention desired that the law should mean, can be considered as a factor by this tribunal, nevertheless we respectfully submit that if the letter is to be given any weight at all it goes to show that the convention took a view opposite from that advocated by the Government.

Gouverneur Morris says: "I always thought that when we should acquire Canada and Louisiana, it would be proper to govern them as provinces and allow them no voice in our councils. In wording the third section of the fourth article, I went as far as circumstances would permit to establish the exclusion. Candor obliges me to add my belief, that had it been more pointedly expressed, a strong opposition would have been made."

Mr. Morris's idea seems to have been that newly acquired territory should not, under the Constitution, be admitted to statehood. If he meant that it should be denied the ordinary common-law rights guaranteed by the Constitution, he did not say so. However, even assuming, as the Government seems to do, that this was his intention, he apparently shrank from announcing it to the convention. He, as a member of that convention, and a prominent participant in its debates, doubtless understood the views of all those present, and so sure was he that no scheme of colonial government, such as he apparently had in mind,

could be engrafted upon the Constitution, that he endeavored
by means of a subterfuge to inject into the Constitution something which might be twisted into granting a power which the
other members of the convention did not wish to confer upon
the Government.

The rules and regulations clause, viewed in the light of history, referred to granting titles to land in the Northwest Territory and otherwise disposing of and regulating it. It was a
substitute for the Pinckney draft "to appropriate the unappropriated lands of the United States."

Gouverneur Morris's redraft of this clause was passed without
opposition, and it was evident that the framers of the Constitution saw no other meaning in it than that in the Pinckney draft.
That so able a man as Morris should have been compelled to
attempt to confer upon the Government by the Constitution
such a power in such a way is very clear evidence of the inten-
tions of the majority who framed the Constitution. As he himself admits, had his intention been expressed, a strong opposition would have been made. That this opposition would have
been strong enough to override his views would seem not im-
probable from his fear and failure openly to express them.

The cases relating to the territorial courts have been so fully
discussed in the briefs already presented to the learned court,
that further comment is not required. We may only say that
they do not affect the question as to whether territory newly
acquired by treaty, and as yet unorganized, is within the limits
of the United States and subject to the uniformity clause of the
Constitution.

The regular judicial courts of the United States were clearly
established by the Constitution for the purpose of exercising
jurisdiction in reference to certain specified matters and within
the States of the United States. The language of the Constitution makes this clear in itself. They were adapted to carry
out the Federal system of government.

They are, this court has said, " parts of the Federal system,
invested with the judicial power of the United States, expressly
conferred by the Constitution and to be exercised in correlation
with the presence and jurisdiction of the several state courts

and governments." *Hornbuckle* v. *Toombs*, 18 Wall. 648, 655. Cited with approval in *McAllister* v. *U. S.*, 141 U. S. 174, 183.

On the other hand, the courts established by the Congress within the Territories have jurisdiction not only over matters which the Constitution specially reserves to the Government as a Federal Government, but general jurisdiction over all cases arising between man and man, and which in the States are within the jurisdiction of the state courts.

In other words, they are not Federal courts, but municipal courts.

The most, then, that these cases decide is that the territorial courts are not the courts mentioned in the Constitution. The ultimate ground upon which these decisions do and must rest, is the fact that the Territories are not States, and therefore the constitutional courts would be inapplicable to them. "The distinctions between the Federal and state jurisdictions, under the Constitution of the United States, has no foundation in these territorial governments, and consequently no such distinction exists either in respect to the jurisdiction of their courts or the subjects submitted to their cognizance. They are legislative governments, and their courts legislative courts, Congress in the exercise of its powers in the organization and government of the Territories combining the powers of both the state and Federal authorities. There is but one system of government or of laws operating within their limits, as neither is subject to the constitutional provisions in respect to state and Federal jurisdiction." *Benner* v. *Porter*, 9 How. 235.

This question is entirely different from the question at bar. The inhibitions placed upon the central Government are general in their language and are applicable to the Government and not to any particular territory or any particular circumstances. That Chief Justice Marshall so understood it is very clear from the expressions used by him in the *Canter* case. He admitted that by the treaty at least the citizens of Florida were citizens of the United States. If that was so, it is very clear that Mr. Webster's contention that Florida was not part of the United States was considered unsound. But having held Florida to be a part of the United States, the court then proceeds to show territorial courts to be local courts under the act of Congress, and not courts of the United States. It is little less than absurd to say that Chief Justice Marshall, though considering citizens of Florida to be citizens of the United States, considered Florida to be a foreign country.

Some concern has been expressed with reference to the effect of the nationalization of the uncivilized tribes that may inhabit the invaluable possessions acquired under the Treaty of Paris—a dread of the sufferage wielded by hordes of untamed Malays.

The Attorney General has, we believe, dissipated this fear by the position which he assumes for the Government at page 60 of the *Goetze* brief: "The political status of the native Indian tribes within territory acquired by the United States by treaty has been uniformly regarded as unaffected by the cession. A long line of special treaties with such tribes and numerous acts of legislation by Congress on the subject of Indians and Indian rights show that these people have always been regarded as *quasi* foreign."

This position is sustained by precedent at once abundant and illustrious, from *Worcester v. Georgia,* 5 Pet. 1, 17 (1826), down to the most recent date, through *Kagama* v. *United States,* 118 U. S. 375; *Talton* v. *Mayes,* 163 U. S. 376; *Elk* v. *Wilkins,* etc.

The Indians have from the beginning been considered and held as distinct political communities, owing a primary allegiance to their tribal authorities, and not subject to the complete jurisdiction of the United States.

For this reason their birth within the United States does not confer upon them the citizenship which the Constitution attaches to such birth in one subject to the jurisdiction.

While the Indians, however, have occupied under the law the anomalous position of independent though subservient nationality, the territory they occupy has never ceased to be territory of the United States, within the geographical boundaries of and subject to the sovereignty and dominion of the nation; in every sense a part of the United States, to the extent that birth within such territory was enough to endow the person so born with citizenship—unless he owed immediate allegiance to some tribe.

The only question that remains is: Whether uncivilized tribes in our new Asiatic or Caribbean possessions may be assimilated to these Indian tribes, if the Government or Congress should choose so to treat them.

That the existence of tribal relations and the savage state should have a like effect in either case goes without saying, unless there is some constitutional inhibition; unless, in other words, the relation of Indian tribes, which has prevailed since the Constitution was adopted, can be shown to have been limited by such Constitution to North American Indians.

We submit with confidence that no such limitation can be found.

This court in *United States* v. *Kagama*, 118 U. S. 374, says:

"The Constitution of the United States is almost silent in regard to the relations of the Government which was established by it to the numerous tribes of Indians within its borders." p. 278.

The court then proceeds to point out that the only clauses relevant are the power to regulate commerce with the Indian tribes, and the apportionment of direct taxation excluding Indians not taxed.

It was the ownership of the territory, and the right of exclusive sovereignty over the same, which was lodged in the Federal Government that gave that Government the right of controlling the actions of the Indians, excluding from any such privilege even the state government within whose borders an Indian reservation was located.

In *Elk* v. *Wilkins*, the relation of the Indian born within the United States and subject to tribal government was determined not under any specific Indian clause in the Constitution, but by the clause relating to citizenship by birth, and under the XIVth amendment.

In other words, there are virtually no Indian clauses in the Constitution, certainly nothing to confine the regulation of our intercourse with uncivilized tribes within our borders to North American Indians.

In the acquisition of Mexican territory, additional uncivilized tribes were brought in and dealt with.

On the acquisition of Alaska, the uncivilized tribes in that Territory, whose racial characteristics are as distinct from the North American Indian as both are from the Malay and the Tagal, were dealt with on the same basis. The treaty provides:

"The inhabitants . . . with the exception of uncivilized native tribes shall be admitted to the enjoyment of all the rights, advantages, and immunities of citizenship, etc." Art. III, March 30, 1867.

"The uncivilized tribes will be subject to such laws and regulations as the United States from time to time adopt in regard to aboriginal tribes."

On the acquisition of the Philippines, the general commanding entered into a treaty with the head of the Sulu tribes, who there enjoys the title and certain attributes of a sultan.

We can see no difficulty, as indeed we see little relevancy, in the relation of the Indian question to the sovereignty of the United States over all territory within its borders, and the obligation which the Constitution establishes of uniform imposts throughout those borders.

IX. The Effects of Cession upon the Question of Citizenship.

It is contended by the Government that: "The conceded power to acquire territory by treaty or by conquest includes the right to prescribe what terms the United States will agree to as fixing the status of its inhabitants."

We have elsewhere shown that the status of the inhabitants is a matter apart from and outside the principles governing this case.

The question of customs duties has nothing to do with citizenship or nationality.

The brief of the learned Attorney General expresses our views admirably.

"The right to bring merchandise into the United States is a right entirely within the regulation of Congress; such a right in no wise differs as to either citizens or aliens. Citizenship carries with it no special or peculiar privileges at the custom-house. The American, the Spaniard, the Porto Rican are treated alike. The basis of the customs laws is not ownership, but (1) the geographical origin of the shipment, and (2) the nature of the goods." Brief Atty. Gen. in *Goetze* case, p. 6.

Under these circumstances, and with this concession, it might seem superfluous to discuss the vexed question of citizenship had not the learned Attorney General deemed it important, if not relevant, and discussed the question at some length in his brief.

In support of his proposition he cites two precedents.

(1) The status of the free negro prior to the civil war, and the amendments to the Constitution.

(2) The history of our relations with the Indians.

It may be a cause for surprise that he should have adduced in support of such an important proposition the two least cred-itable instances in our history.

His argument seems to sum itself up as follows:

The inhabitants of Porto Rico are not citizens of the United States because (1) the power to confer citizenship is one which the Government has not in this instance chosen to exercise; (2) such citizenship is not expressly conferred either by the Constitution, the laws, or the treaty; this appears from the fact (as shown by the *Dred Scott* case) that free negroes were not citizens and that the members of the Indian tribes have always been held to be not citizens but *quasi* foreigners who could only acquire citizenship by naturalization.

These questions seem so important as to require somewhat full examination.

"The law knows nations only as political communities and as sovereign States. The nationality, therefore, as a legal attri-bute of persons, is connection with a certain body politic, mem-bership in a particular State. The members of a State are called its subjects or citizens. The former term if properly construed is applicable to the people of any nation without re-gard to the form of government, for every State is based upon the relation of its members to its sovereign." Encyclopædia

Political Science and United States History, article Nationality, by Munroe Smith.

The question of citizenship in the United States has always been confused because of the use of that word in a dual sense. The word "citizen" has two meanings.

It means in the first sense, primarily and' properly, the persons exercising political rights and members of the ruling body politic.

In the second sense, it is applicable to the whole people of any nation without regard to the form of government. Citizenship in the latter sense means simply subject to the allegiance of a particular State or nation. In this sense it has precisely the same meaning as the term "subject." Story on the Constitution, Cooley's edition and notes, §§ 1932-33-34, cited at length by Sen. Foraker, p. 12, Rep. No. 249, 5th Feb. 1900, 56th Cong. 1st Sess.

All the members of a nation, subject to its jurisdiction, or, as the common law has it, " born under the actual obedience " are subjects or citizens in this sense. The word "subject" has been somewhat discredited by reason of its usual reference to feudal or absolute monarchies where none or few of the subjects are citizens in the sense of possessing political rights. The learned Attorney General is in error in supposing that "the term does not imply anything as to the nature or form of the government of which one is a subject." p. 72, *Goetze* brief. By reason of the disfavor that this term has thus fallen into, it is now found in no constitutionally governed nation save England.

The rule of the common law upon this subject is plain and well settled both in England and America. Except in the case of children of ambassadors, who are in theory born upon the soil of the sovereign whom the parent represents, a child born in the allegiance of the king is born his subject without reference to the political *status* or condition of its parents. Birth and allegiance go together. 1 Blackstone, 366 ; 2 Kent's Com. 39, 43 ; *Ingles* v. *The Sailor's Snug Harbor*, 3 Pet. 120 ; *U. S.* v. *Rhodes*, 1 Abb. U. S. Rep. 40 ; *Lynch* v. *Clarke*, and authorities there cited ; 1 Sandf. Ch. 630.

This is nothing more than declaratory of the rule of the common law as above stated. To be a citizen of the United States by reason of his birth, a person must not only be born within its territorial limits, but he must also be born subject to its jurisdiction—that is, in its power and obedience. *McKay* v. *Campbell*, 3 U. S. Courts Rep. Ninth Circuit, 118, p. 129. See also *Elk* v. *Wilkins*, 112 U. S. 99.

In order, however, to avoid the ambiguity due to this dual sense, the Germans and the French make use of the word "nationals" to denote all persons subject to the allegiance of the state, *i. e.*, forming a part of the nationality, including both holders and nonholders of political rights. Generally speaking, therefore, nationals and aliens would include every person within a given territory and would indicate the legal relations which they hold to the public authority of such territory.

Nationals are again divided into two classes, those possessing political rights and those who do not possess them. The latter class would include women, minors, and persons who, for a variety of reasons other than alienage, do not possess the political franchise. *Minor* v. *Happersett*, 21 Wall. 162.

The Fourteenth Amendment, declaring that all persons born or naturalized in the United States and subject to their allegiance are citizens, uses the word in the sense of national or subject.

Before the Fourteenth Amendment the only apparent exception was due to the peculiar incidents of our history which made the negro something different from the ordinary human being—half man, half beast—something partly within the domain of natural history and partly within that of politics.

" The citizenship of the negro had been denied in the *Dred Scott* case on the assumption that citizenship and subjection were not indentical ideas; that a person might be a subject without being a citizen. In declaring that citizenship is acquired in the same manner in which subjection is established at common law, the Fourteenth Amendment has placed the equivalency of these terms and established the citizenship of a negro beyond the possibility of a doubt." Encyclopædia Pol. Sc. Article Nationality.

In the recent leading case on the question of nationality and

citizenship (*Wong Kim Ark*, 169 U. S.) Justice Gray, writing
for the court, says: "In *Dred Scott* v. *Sandford* (1857), 19
How. 393, Mr. Justice Curtis said:

"'The first section of the second article of the Constitution
uses the language—a natural-born citizen. It thus assumes
that citizenship may be acquired by birth. Undoubtedly this
language of the Constitution was used in reference to that
principle of public law well understood in this country at the
time of the adoption of the Constitution, which referred citizenship to the place of birth. 19 How. 576.

"'Allegiance is nothing more than the tie or duty of obedience of a subject to the sovereign under whose protection he
is; and allegiance by birth is that which arises from being born
within the dominions and under the protection of a particular
sovereign. Two things usually concur to create citizenship;
first, birth locally within the dominions of the sovereign; and
secondly, birth within the protection and obedience, or in other
words, within the ligeance of the sovereign—that is, the party
must be born within a place where the sovereign is at the time
in full possession and exercise of his power, and the party must
also at his birth derive protection from and consequently owe
obedience or allegiance to the sovereign as such *de facto*. . . .

"'Subject and citizen are in a degree convertible terms as applied to natives; and though the term citizen seems to be appropriate to republican freemen, yet we are equally with the inhabitants of all other countries subjects, or we are equally bound
by allegiance and subjection to the Government and law of the
land.' 2 Kent. Com. 258, note.

"Passing by questions once earnestly controverted, but finally
put at rest by the Fourteenth Amendment of the Constitution, it
is beyond doubt that, before the enactment of the civil rights
act of 1866, or the adoption of the constitutional amendment,
all white persons at least and born within the sovereignty of
the United States, whether children of citizens or of foreigners,
excepting only children of ambassadors or public ministers of a
foreign government, were native-born citizens of the United
States.

"The fundamental principle of the common law with regard

to English nationality was birth within the allegiance, also called 'ligealty,' 'obedience,' 'faith,' or 'power,' of the King. The principle embraced all persons born within the King's allegiance and subject to his protection. Such allegiance and protection were mutual—as expressed in the maxim, *protectio trahit subjectionem, et subjectio protectionem*—and were not restricted to natural-born subjects and naturalized subjects, or to those who had taken an oath of allegiance; but were predicable of aliens in amity, so long as they were within the Kingdom. Children born in England of such aliens were, therefore, natural-born subjects. But the children born within the realm, of foreign ambassadors, or the children of alien enemies, born during and within their hostile occupation of part of the King's dominions, were not natural-born subjects, because not born within the allegiance, obedience, or the power, or, as would be said at this day, within the jurisdiction of the King."

Proceeding from these unquestioned principles, it naturally follows that the natives of Porto Rico and the other ceded islands are United States nationals, or, as the learned Attorney General prefers to term them, American subjects.

They are subjects or nationals in the same sense that women, minors, inhabitants of Oklahoma and Arizona are subjects or nationals.

Persons in States requiring an educational qualification for voting, who cannot attain to this qualification, are also in this position.

And, certainly, if the learned counsel means no more than this, he is right when he says that, "To be called an American subject is no disgrace."

That the treaty carries out this idea is very clear, for it declares that natives of the peninsula of Spain who have not elected to remain Spanish subjects shall be deemed to possess the *nationality* of the territory in which they reside. Of course, the nationality of the territories depends upon the nation under whose jurisdiction the territories are, and as this jurisdiction is the United States, the phrase is equivalent to saying that citizens of the territory who do not elect to remain Spanish citizens become American nationals, or, again, as the learned Attorney General prefers to style them, American subjects.

It is very clear that Porto Rico and the Island of Guam have no nationality of their own, nor can any territory which does not possess sovereignty or autonomy be said to have any nationality. The inhabitants of these possessions of the United States are subject to its obedience and are, therefore, its nationals or subjects.

The negotiators of the treaty with Spain undoubtedly understood the treaty—as making all the inhabitants who did not elect to remain Spaniards, American citizens or nationals.

The Spanish commissioners claimed that—

" The American commission refuses to acknowledge the right of the inhabitants of the countries ceded or relinquished by Spain to choose the citizenship with which, up to the present, they have been clothed. And, nevertheless, this right of choosing, which is one of the most sacred rights of human beings, has been constantly sacred since the day when man was emancipated from serfdom. This sacred right has been respected in treaties of territorial cession concluded in modern times." Annex to Protocol No. 21, treaty of peace between United States and Spain of Dec. 10, 1898.

To this the following reply was made:

" The American commissioners do not so understand the article upon the subject of citizenship submitted by them as a substitute for the article proposed by the Spanish commissioners. An analysis of the article will show that *Spanish subjects, natives of Spain* are allowed a year's time in which, by the simple process of stating in a court of record their intention so to do, they may preserve their allegiance to Spain.

" Such persons have the fullest right to dispose of their property and remove from the territory, or, remaining, to continue to be Spanish subjects or elect the nationality of the new territory.

" As to natives, their status and civil rights are left to Congress, which will enact laws to govern the ceded territory. This is no more than the assertion of the right of the governing power to control these important relations to the new government. The Congress of a country which never has enacted laws to oppress or abridge the rights of residents within its domain, and whose laws permit the largest liberty consistent with the preservation of order and the protection of property, may safely be trusted not to depart from its well-settled practice in dealing with the inhabitants of these islands." Annex 1 to Protocol No. 22, treaty of peace between United States and Spain of Dec. 10, 1898.

In view of these assertions of the treaty makers, is it reasonable to claim that this treaty was intended to empower Congress for the first time in its history to govern " dependencies" without regard to Constitutional immunities ?

But the learned counsel for the Government, if we understand him correctly, claims that annexation of territory by mere treaty cession which makes no provision for conferring citizenship upon the inhabitants leaves them aliens until Congress chooses to enact otherwise.

The Louisiana, Florida, Mexican, and Alaskan treaties provided that the inhabitants shall be admitted to the enjoyment of the rights and privileges of citizens of the United States, and from this he infers that without such stipulation they would not have been citizens.

As to Louisiana, Florida, and Alaska, the stipulation evidently refers to the full citizenship incident to statehood ; not to " naked citizenship," to borrow Justice Curtis's phrase, or, as we have termed it, " nationals."

The Alaskan treaty is peculiar in that it excepts uncivilized tribes.

" The inhabitants of the ceded territories, according to their choice, reserving their natural allegiance, may return to Russia within three years; but if they prefer remaining in the ceded territory they, with the exception of uncivilized native tribes, shall be admitted to the enjoyment of all the rights, advantages, and immunities of citizens of the United States."

In that treaty remaining three years was considered equivalent to renouncing the Russian allegiance.

The Attorney General considers this privilege of election a suspension of citizenship by the United States, and finds in

this proof that the Constitution did not affect the question.
Brief, p. 58. It is respectfully submitted that the inhabitants
of Alaska had been Russian citizens or subjects; that it is usual
under the general postulates of international law to allow persons to retain the allegiance to their former masters, if they so
desire. The provision in the Alaskan treaty simply gave the
inhabitants three years to decide whether they would retain
their former allegiance. Their citizenship was not suspended;
they were Russian citizens until they chose to become American citizens.

This treaty is analogous to the treaty with Spain. The
Spanish-born inhabitants of the ceded islands are allowed one
year in which to decide whether they wish to retain their former citizenship. In case they should retain it, their allegiance
was due to Spain and their reliance for protection was upon her.
Should they not retain it, they then became United States nationals. Treaty of Paris, Art. IX.

The other inhabitants of the islands have not been accorded
this privilege for reasons fully set out in the documents of the
Peace Commission. Senate Doc. 64, 1898.

As far as the United States was concerned, the latter people
could not remain like natives of the peninsula, Spanish subjects,
but became at once United States nationals. That the United
States might have given them power to remain Spanish subjects
is doubtless true, but it did not choose to do so.

Pothier thus lays down the principle, says Mr. Lawrence, in
reference to the acquisitions which had been made by France
before the French Revolution: "When a province is united to
the Crown, its inhabitants must be regarded as Frenchmen
whether they were born before or after the union."

Pothier carries the principle so far as to say: "There is every
reason to think that the foreigners who are established in these
provinces, and who have there obtained, according to the laws
in force, the rights of citizenship, must, after the annexation,
be considered citizens equally with the native inhabitants of
those provinces, or, at least, with foreigners naturalized in
France."

And applying the same principle in the cases of loss and

restoration of territory, he says: "When a province is dismembered from the Crown, when a conquered country is restored by the treaty of peace, the sovereignty over the inhabitants is changed. Citizens at the time of the conquest or since the conquest, or if born since the union, citizens by their birth till the dismemberment of the province, become foreigners." Traité des Personnes, Part I, tit. 2, sec. 1, cited by Lawrence, Appendix to Wheaton, 897.

The treaty of April 26, 1798, for the incorporation of the Republic of Geneva with the French Republic, declared that the Genevese who inhabited the city and territory of Geneva, as well as those who were in France or elsewhere, became and were native-born Frenchmen (français nés), and the treaty for the annexation of Mulhausen also declared that the citizens and inhabitants of Mulhausen and its dependencies became and were native-born citizens (français nés). Referring to these treaties, Mr. Lawrence says: "It is not, however, understood that these special declarations varied the conditions of the inhabitants of these small republics from that of the numerous countries and provinces which were incorporated with France between 1789 and 1814.

"These relations established as to Geneva and Mulhausen were applicable to all the annexations.

"They were the 'immediate consequences,' says Fœlix (Revue de Droit Français et Etranger, Tom. II, page 328, Naturalization Collective), 'of every union of territory, according to the existing law of nations, and since it is no longer the custom, even after the conquest of a country, to reduce its inhabitants to a condition inferior to that of the conquering country.'"

This custom which, as Fœlix says, has fallen into honorable disuse, is apparently what the Attorney General desires to revive by placing Porto Ricans on the footing of the "1135 free people of color in New Orleans in 1803," that is, at the time of its cession to the United States.

The dismemberment of populated territory from a State on the one hand, and its incorporation into a new nationality on the other, operate as a collective naturalization *ipso facto.*

"Annexation of territory, either by peaceful cession or as a result of war, invariably carries with it a change of nationality. This is what is called collective naturalization." Pradier Fœdéré, Droit International Public, ed. 1885, vol. III, p. 721.

"Treaties of annexation generally give an option to individuals owing allegiance to the State whose territory is annexed. This option may be manifested either by emigration simply, or by a declaration of intention accompanied by emigration; sometimes a simple declaration is made without resorting to emigration. In any case inaction or silence imports adhesion to the new order of things—tacit acceptance of the nationality newly imposed." *Ibid.* 1, p. 723. "It is a doctrine of natural law that conquest or peaceful cession relieves the inhabitants from all bonds of allegiance towards the sovereign of the passing territory and enjoins fidelity on their part to the new régime. In fact, the inhabitants having had the choice of leaving the country or continuing their residence therein, it is but just that their permanent sojourn in the annexed territory should be construed as a tacit declaration of their fidelity to the conqueror." Calvo, Droit International Theorique et Pratique, ed. 1896, vol. IV, p. 394.

Fœlix, cited by Lawrence, *supra*, says that "change of nationality results either by mere operation of law or from the act of the individual." Of the former he says, "cession of territory furnishes another example." "There can be little or no doubt," says Halleck, "that the inhabitants of Florida, as intimated by Chief Justice Marshall, were entitled *without* the treaty stipulation, to the 'privileges, rights, and immunities' of citizens in this more extended sense of the term; but their right to be incorporated in the Union, and participate in political power, was derived from the treaty and not a necessary consequence, under the law of nations, of the transfer of their country and allegiance." Halleck's Int. Law, § 13, p. 824.

"A collective naturalization of all the inhabitants is effected when a country or province becomes incorporated in another country by conquest, cession, or free gift." Phillimore, vol. I, p. 449, ed. 1879.

The treaty thus confers upon the inhabitants the "nationality of the territory to which they belong."

As to Porto Rico, that nationality is of course the United States. *Boyd* v. *Thayer*, 143 U. S. 162.

Porto Rico is not a country in the political sense, and hence can have no independent nationality of its own.

International law knows no State or nation of Porto Rico. It is not a member of the family of nations. Its inhabitants can only be either aliens, *i. e.*, persons owing allegiance to a sovereignty other than the United States, or nationals, *i. e.*, (passive) citizens of the United States.

Congress may of course naturalize, by annexing territory, the inhabitants, and, as we have demonstrated, mere cession and transfer of territory has this effect without special stipulation in the treaty.

A treaty provision to that effect is therefore merely declaratory of the rule of international law. As was said by the present learned Chief Justice: "Persons not thus subject to the jurisdiction of the United States at the time of birth cannot become so afterward except by being naturalized either individually as by proceedings under the naturalization acts, or collectively, as by the force of a treaty by which foreign territory is acquired." *Wong Kim Ark*, 169 U. S.

"A person born out of the jurisdiction of the United States can only become a citizen by being naturalized either by treaty, as in the case of the annexation of foreign territory, or by authority of Congress, exercised either by declaring certain classes of persons to be citizens, as in the enactments conferring citizenship upon foreign-born children of citizens, or by enabling foreigners individually to become citizens by proceedings in the judicial tribunals as in the ordinary provisions of the naturalization acts." *Wong Kim Ark*, 169 U. S. 649, 702.

As to persons born subsequent to the acquisition, the question is even clearer. The Fourteenth Amendment has enacted a rule of law into the Constitution which overrules treaties and legislation.

Prior to such amendment had the Government desired to violate the common-law rule adopted by the United States, it could have declared in a case like that of Porto Rico that all of the inhabitants should remain citizens of Spain. The territory would none the less have been a part of the United States, but its inhabitants would have been aliens and subjects to a foreign jurisdiction. Such an incongruous result would, in the absence of Constitutional restriction, have been possible. The inhabitants of such territory would then have owed temporary allegiance to the United States such as aliens within its jurisdiction now owe it; but because a part of the territory is populated by aliens that territory is none the less within the geographical boundaries of the United States. The question could only arise as to inhabitants born before the cession, but as this treaty has provided otherwise, the question is academic.

The Fourteenth Amendment enacting the common-law rule of citizenship into the dignity of constitutional provision, settles the status of persons born since the cession. "The Fourteenth Amendment of the Constitution, in the declaration that 'all persons born or naturalized in the United States and subject to the jurisdiction thereof are citizens of the United States and of the States wherein they reside,' contemplates two sources of citizenship, and two only—birth and naturalization. Citizenship by naturalization can only be acquired by naturalization under the authority and in the forms of law. But citizenship by birth is established by the mere fact of birth under the circumstances defined in the Constitution. Every person born in the United States, and subject to the jurisdiction thereof, becomes at once a citizen of the United States and needs no naturalization." *Wong Kim Ark*, 169 U. S. 649.

The main precedent, however, upon which the learned Attorney General seems to rely is that of the position of the free negroes before the civil war; because he says:

"Suppose a cession of a small island with half a dozen inhabitants—must the United States agree to permit them to remain and accept them as citizens? It might be the purpose of the Government to use the island solely as a fort or military reservation. . . . And if such restriction on its right to acquire exists, how does it resist the rights of uncivilized tribes in Alaska and in the Mississippi and New Mexican regions to be counted also as citizens? Or the 1135 'free people of color' in New Orleans in 1803, to say nothing of the slaves."

The learned Attorney General then proceeds to show from the *Dred Scott* decision that free negroes were not citizens. We may admit that the free negroes before the war and during the civil war occupied an anomalous position.

The case of *Dred Scott* simply held that the negro was so low in the scale of humanity that the States could not, by conferring freedom upon him, make him capable of becoming a citizen of the United States in the broad or passive sense. He was, therefore, neither citizen nor subject, but a being who, under the Constitution, was something different and apart from the rest of humanity.

His anomalous position was thus described by Chief Justice Taney: "In the opinion of the court the legislation and the histories of the times and the language used in the Declaration of Independence show that neither the class of persons who had been imported as slaves, nor their descendants, whether they had become free or not, were then acknowledged as a part of the people, nor intended to be included in the general words used in that memorable instrument.

"It is difficult at this day to realize the state of public opinion in relation to that unfortunate race which prevailed in the civilized and enlightened portion of the world at the time of the Declaration of Independence, and when the Constitution of the United States was framed and adopted. But the public history of every European nation displays it in a manner too plain to be mistaken.

"They had for more than a century been regarded as beings of an inferior order and altogether unfit to associate with the white race either in social or political relations; and so far inferior, that they had no rights which the white man was bound to respect; and that the negro might lawfully and justly be reduced to slavery for his benefit. He was bought and sold, as an ordinary article of merchandise and traffic, whenever a profit could be made by it. This opinion was at that time fixed and universal in the civilized portion of the white race. It was regarded as an axiom in morals as well as in politics, which no one thought of disputing or supposed to be open to dispute; and men in every grade and position in society daily and habitually acted upon it in their private pursuits as well as in matters of public concern, without doubting for a moment the correctness of this opinion.

" And in no nation was this opinion more firmly fixed or more uniformly acted upon than by the English Government and the English people. They not only seized them on the coast of Africa and sold them or held them in slavery for their own use, but they took them as ordinary articles of merchandise to every country where they could make a profit upon them and were far more extensively engaged in this commerce than any other nation in the world.

" The opinion thus entertained and acted upon in England was naturally impressed upon the colonies they founded on this side of the Atlantic. And, accordingly, a negro of the African race was regarded by them as an article of property and held, and bought and sold as such, in every one of the thirteen colonies which united in the Declaration of Independence and afterwards formed the Constitution of the United States. The slaves were more or less numerous in the different colonies, as slave labor was found more or less profitable. But no one seems to have doubted the correctness of the prevailing opinion of the time." pp. 407–408, 19 How.

* * * * * * * *

" The question with which we are now dealing is, whether a person of the African race can be a citizen of the United States and become thereby entitled to a special privilege by virtue of his title to that character and which, under the Constitution, no one but a citizen can claim.

* * * * * * * *

" The only two provisions which point to them and include them treat them as property, and make it the duty of the Government to protect it; no other power in relation to this race is to be found in the Constitution, and as it is a Government of special delegated powers, no authority beyond these two provisions can be constitutionally exercised."

Mr. Justice Curtis in his dissenting opinion uses the following apposite language (p. 583): " And my opinion is that, under the Constitution of the United States, every free person born on the soil of a State, who is a citizen of that State by force of its constitution or laws, is also a citizen of the United States.

"I will proceed to state the grounds of that opinion.

"The first section of the second article of the Constitution uses the language 'a natural born citizen.' It thus assumes that citizenship may be acquired by birth. Undoubtedly this language of the Constitution was used in reference to that principle of public law well understood in this country at the time of the adoption of the Constitution which referred citizenship to the place of birth. At the Declaration of Independence and ever since the received general doctrine has been, in conformity with the common law, that free persons born within either of the colonies were subjects of the King; that by the Declaration of Independence and the consequent acquisition of sovereignty by the several States all such persons ceased to be subjects and became citizens of the several States, except so far as some of them were disfranchised by the legislative power of the States, or availed themselves seasonably of the right to adhere to the British Crown in the civil contest and thus to continue British subjects. *McIlvaine* v. *Coxe's Lessee*, 4 Cranch, 209; *Inglas* v. *Sailors' Snug Harbor*, 3 Pet. 90; *Shanks* v. *Dupont*, 3 Pet. 42."

  * * * * * * * *

"A naturalized citizen cannot be President of the United States, nor a Senator till after the lapse of nine years, nor a Representative until after the lapse of seven from his naturalization. Yet, as soon as he is naturalized, he is certainly a citizen of the United States. Nor is any inhabitant of the District of Columbia or of either of the Territories eligible to the office of Senator or Representative in Congress though they may be citizens of the United States. So in all the States numerous persons, though citizens, cannot vote or cannot hold office either on account of their age or sex, or the want of necessary legal qualifications. The truth is, that citizenship under the Constitution of the United States is not dependent on the possession of any particular political or even of all civil rights; and any attempt so to define it must lead to error. To what citizens the elective franchise shall be confided is a question to be deter-

mined by each State, in accordance with its own views of the necessities or expediencies of its condition. What civil rights shall
be enjoyed by its citizens, and whether all shall enjoy the same,
or how they may be gained or lost, are to be determined in the
same way."

* * * * * * * *

" It rests with the States themselves so to frame their constitutions and laws as not to attach a particular privilege or immunity to mere naked citizenship."

It thus appears the condition of the negro was such that he
was not in the legal sense a person. Whether free or slave, he
was something capable of being reduced to property, and, therefore, he did not fall within any category which would fit the
genus man.

But assuming that it is necessary to classify him at all, it may
be said that he belongs to the class of " nationals," and further
was placed in a subclass by himself (under the Constitution of
the United States as interpreted by the court in the *Dred Scott*
case), and that as member of that subclass he owed allegiance
to the United States, but was incapable of possessing constitutional rights such as the right to sue in the Federal courts,
which was expressly guaranteed to the citizens of the United
States.

Thus political rights were accorded to some citizens and civil
rights to all save the negro.

It was for the purpose of removing from our Constitution
this disability that the Fourteenth Amendment was enacted.
By it the negro stepped from the domain of zoölogy into that of
history.

What rights human beings owing direct and immediate obedience to the sovereign in whose jurisdiction they may reside are
to possess is a question for that sovereign to determine in a constitution or by legislation, but subjection or nationality merely
express a relation of fact, to wit, allegiance and protection.

The inhabitants of Porto Rico who were born subsequent to
the cession and who do not owe any direct, immediate allegiance
to any foreign nation are citizens or subjects of the United
States.

As such citizens or subjects they possess whatever rights are generally conferred upon that class by law.

Neither by the laws of the United States nor the Constitution does any subdivision of that class exist incapable by nature of possessing any rights of any character. This anomalous position was confined to the free negro before the Fourteenth Amendment.

If the learned Attorney General dissents from this proposition, as to the inhabitants of the ceded territory, he can only do so upon the ground upon which Judge Taney held negroes not citizens, namely, that they were persons capable of being considered as property, and therefore too degraded to come within that category. If the learned counsel means anything else than this his argument is irrelevant.

If he means this, we can only say that his views have been repudiated by the American people in the civil war, by three amendments to the Constitution of the United States by this court, and by forty years of advancing civilization.

X. It is erroneous to assume that the decision in this case can or will involve the right of the United States to own, possess, or govern colonies.

The only question involved is as to how the United States shall govern its colonies. From the beginning it has possessed colonies or dependencies.

Morris, in his work on colonization, volume II, at page 292, speaking of Russia, says:

"These recent efforts of Russia recall, if the digression be here permitted, that in this sense the United States have likewise, throughout their actual career, been engaged in the real work of colonization, although the extension of the Republic may not generally be recognized as such a manifestation. The casual observer is prone to attach to this idea the idea of distance, to believe that for the application of the term colony to a dependency, the latter must necessarily be remote from the metropolis. The fact is, that the relation is based on certain peculiar mutuality of rights. What difference can it possibly make that the possession be isolated by the depths of the sea, by a voyage over the seas occupying a month, or by a journey on the land of a similar interval.   If the thought of separation by water be disassociated, cannot the settlement of Louisiana, California, and the Northwest Territory well be claimed as some of the greatest episodes of history?   In Alaska the inhabitants are still occupied in the work of colonization."

That those territories have not usually been thought of as colonies, because they were not separated from the United States by large bodies of water, does not make them any less colonies.   They are colonies just as much as Canada, New Zealand, or Australia are colonies of Great Britain.   They had, however, been governed better and more liberally than the colonies of any power in the world, and the history of colonization shows that the methods of the Government of the United States are being imitated by the other nations of the world.

Therefore in acquiring and governing new territories, dependencies, or colonies, we have continuous precedents extending back to the formation of the Constitution, but in governing these territories without according them as of right certain immunities which have always been deemed by the American people fundamental rights, we should be equally reversing the precedents of one hundred years.

It is idle to say that Congress will give them these rights independent of the Constitution.   The question is not what Congress will do, but what it can do.   Congress has heretofore passed laws which were unwise, and has passed laws which have been declared by this court to be unconstitutional.   There is no guaranty that they will not pass such laws again.

The government of Great Britain not many years ago passed a law (Ashburton Act) practically confiscating property in England and Ireland, and allowing the courts to fix the rents which the tenants should pay the landlords.   Such an act would be utterly impossible under our system of government, as long as our present Constitution endures.

The learned Attorney General in his brief, page 12, says: " No one pretends that Congress, irrespective of any limitation of the Constitution, could properly make and enforce a law to take without cause property of one person and vest it in another."

We ask why Congress cannot do this. If there is no legal limitation upon Congress, what limitation is there? If it be said a moral obligation, the answer is that it is no limitation whatever. Is there any reason to suppose that the Congress of the United States might not be willing to do as was done by the Parliament of England, which did the thing the Attorney General claims that Congress could not do irrespective of the limitations of the Constitution, namely, to take property of one individual and vest it in another?

But the Attorney General adds: "These are all despotic powers which no Congress would claim, much less attempt to exercise." History scarcely teaches the lesson that a body of men will not exercise all the power which they possess. Rather the contrary is true, and the instances which we have in our own history in the attempt to make laws inconsistent with the Constitution would hardly lead us to assent to the proposition of the learned Attorney General, that while Congress had the power to ignore these rights, it is certain that they would not do so. Colonies frequently, if not usually, suffer from too much government, rarely, if ever, from too little.

XI. We have now considered every one of the strictly legal arguments advanced in support of the Government's position. But arguments of another class have been presented, and a word must be said in reference to them.

It has been said that the due regard for constitutional limitations would make us a "crippled nation," and that, like "humpbacked Richard," we would be the laughing stock of nations as we "halt by them."

The argument from the consequences which may attend upon the interpretation of a constitutional provision is not always the best argument, but is one which may sometimes be considered. As was said by this court in *Maxwell* v. *Dow*, 176 U. S. 590: "The argument, we admit, is not always the most conclusive which is drawn from the consequences urged against the adoption of a particular construction of an instrument. But when, as in the case before us, these consequences are so serious, so far reaching and pervading, so great a departure from the structure and spirit of our institutions; when the effect is to fetter and degrade the state governments by subjecting them to the control of Congress in the exercise of powers heretofore universally conceded to them of the most ordinary and fundamental character; when, in fact, it radically changes the whole theory of the relations of the state and Federal governments to each other and of both these governments to the people, the argument has a force that is irresistible in the absence of language which expresses such a purpose too clearly to admit of doubt. We are convinced that no such results were intended by the Congress which proposed these amendments, nor by the legislatures of the States which ratified them."

We assert that the only consequence which the Government of the United States fears from an adverse decision is the necessity for free trade between the new possessions, or colonies, and the States of the United States.

It has been assumed that the guarantees of certain civil rights conferred by the Constitution might be incompatible with the government of newly acquired possessions. This assumption, however, is entirely negatived: 1. By the history of our past acquisitions, and 2. By the attitude already assumed by the Government in regard to our new territories.

In all our past acquisitions, not only those of the Northwest Territory, inhabited by English speaking people, but in territory acquired from Mexico, Spain, and France, inhabited by people different in language, law, and religion, we have not feared, but, on the contrary, we have hastened to confer on the inhabitants all the rights and liberties which centuries of conflict led our ancestors to believe essential, if not sacred.

That these concessions have not retarded the development of our former colonies or territories is matter of public history.

The first act for the government of Louisiana conceded trial by jury and provided for the other guaranties of the bill of rights.

The military governments in the territories wrested from

Mexico by conquest and confirmed by treaty gave the inhabitants these rights.

Not only have they been accorded to the civilized inhabitants of former acquisitions, but by recent legislation provision is made for the trial of Indians by the United States courts, and they are tried by the methods known to the common law and sanctioned by the amendments to the Constitution.

There is nothing in the Constitution incompatible with the proper administration of such territory. It is not probable that we will find it necessary to establish an order of nobility or to prohibit the free exercise of religion or of the right of the people peaceably to assemble.

We will scarcely find it useful to quarter soldiers in the houses of the inhabitants in time of peace or to establish torture in order to compel witnesses to tell the truth, or burning at the stake as a means of capital punishment.

And why, if we can try Indians by the ordinary methods of petit and grand jury, should we deny this right to the people of Porto Rico and the Philippines, who have been accustomed to Spanish criminal and civil law, which, whatever may be its deficiencies, is certainly preferable to the Indian tribal customs! Why should we desire to require excessive bail or prescribe cruel punishment?

The Government evidently desires to do none of these things. As the Secretary of War has said, they do not mean to interfere with what he terms: "The underlying principles of justice and freedom which we have declared into our Constitution, and which are the essential safeguards of every individual against the powers of government, not because these provisions were enacted for them, but because they are essential limitations inherent in the very existence of the American Government. To illustrate: The people of Porto Rico have not the right to demand that duties should be uniform as between Porto Rico and the United States because the provision of the Constitution was not made for them."

We quote this to show exactly the effects which the Government fears from a decision that Porto Rico is within the United

States; not the granting of common-law rights to the people, but the opening of our markets to the colonial products.

In the *Goetze* case involving this question, " certain industries" have filed a brief in which they state that their interests are equally important with those of the Government of the United States. Those interests are commercial interests, whose desires, according to their brief, is that the American Government shall have the power to impose a tariff upon the products of these islands, which would shut them out from competition.

It is therefore evident,

(1) From the nature of the rights guaranteed by the Constitution.

(2) From the views of our Government as outlined by the Secretary of War.

(3) From the attitude of the industrial interests here represented, and

(4) From the history of the Government of our former acquisitions, that the only effect of a decision by this court in favor of the Government would be to allow the shutting out of the products of these places from our markets, or, in other words, the taxation of the inhabitants of the new possessions for the benefit of some inhabitants of the States of the United States.

That any danger so great in its extent and dire in its nature would follow from the impossibility of imposing such commercial restrictions is hardly so evident or certain a factor as to influence the decision in this case.

But, even assuming that the people of the United States are unwilling to consume tobacco from Porto Rico and the Philippine Islands, or to allow it to be sold in their markets, they can prevent this by constitutional amendment.

The learned Attorney General stated in his argument in the *Goetze* case that England taxed the products of her colonies at her custom-house. While the fact that the English Government follows a certain policy may not prove absolutely that such policy is a wise and beneficent one, yet we admit that the acts of the British Government, as the acts of a wise and prudent Government in matters of finance, are entitled to respect.

We would, however, call attention to the fact that the

English colonies also exclude English goods from their markets, and as they possess a local self-government, with which the English Government interferes in no respect, can care for their own interests as well as the mother country, and, therefore, the case is scarcely analogous to that at bar.

There is probably no provision in the Constitution which demonstrates more conclusively the wisdom of the framers of that instrument than the uniformity clause.

It is argued that the insertion of this clause was due to the desire on the part of some States that a majority of the House of Representatives should not build up industries in some States to the loss of others, or, in other words, that the industries of the smaller States should not be at the mercy of the larger ones. While this reason may have been applicable to the States, it is inapplicable to the Territories according to the contention of the learned Attorney General.

But we submit that it is more applicable to the Territories because, as they possess no representation, they are defenceless, and should the States impose burdens upon them by taxing their products they would thus have complete and absolute power to do so. Having no representation, the Territories could not defend themselves as might even a minority of the States from this form of oppression. It is true that taxation without representation is only a political right.

We have no right to assume that the framers of the Constitution, realizing this, were willing that their "posterity" should have no protection against this taxation. As the people of the Territories could not, under the Constitution, have representation, there was only one principle that could protect them, and that was uniformity. It is good for the governed and the governing, the rulers and the ruled, to feel the pressure of the same law. "If it be said that the principle of uniformity established in the Constitution secures the district from oppression in the imposition of indirect taxes, it is not less true that the principle of apportionment, also established in the Constitution, secures the district from any oppressive exercise of the power to lay and collect any direct taxes." C. J. Marshall in *Loughborough* v. *Blake*, 5 Wheat. 324.

Is it not possible that the framers of the Constitution may have foreseen the possibility that interested persons in the thirteen States might desire to build themselves up at the expense of or free from competition of the Northwest Territory? Certainly no provision more admirably adapted to prevent this could be framed than that requiring that all duties should be uniform throughout the United States.

We contend, therefore, that far from the effects which would follow from the adoption of our interpretation of this clause being incompatible with good government of the newly acquired Territories the direct opposite would be the result. They would have the inherent rights which the Government does not wish to deny them. They would possess, besides, freedom from that danger which English-speaking men have always held most important—unjust taxation. They cannot be represented in fixing their taxes; their only safeguard, therefore, is that their rulers cannot tax them without equally taxing themselves, and upon the best known principles of human nature it would be difficult to find an incident so calculated to guarantee the inhabitants of the newly acquired Territories from being unequally taxed in the disposition of their products and properties for the benefit of certain industrial interests.

XII. It is inevitable that upon questions of the breadth and far-reaching importance of those here presented all concerned in their settlement should look with anxiety on the result of their deliberations and seek to avoid in reaching a conclusion any disastrous effect upon the nation.

The Attorney General has told us that the maintenance of the limitations invoked by the plaintiffs in error would make us a crippled nation, incapable of meeting the requirements which duty and destiny impose upon this Republic.

That we have a government of limited powers he has frankly conceded. We desire to impress upon him that the dangers of such limitation and their possible inadaptability to circumstances that might arise in a distant future were quite present to the master minds from whose contact sprang the instrument which has held together this mighty nation through nullification, secession, and civil war for more than a hundred years.

They provided for the emergency.

Article V of the Constitution prescribes a method for its amendment, which may be readily applied whenever the people agree with the contentions of the Government in the present case, and determine that the limitations imposed upon their agents have ceased to be a benefit or have become an obstacle to the good government for which they sought.

We will not enlarge upon this theme, but will leave the case in the hands of the court with citations from two of its justices, and from one whom it is not yet considered sentimental to call the Father of his Country.

Mr. Chief Justice Fuller says: "Differences have often occurred in this court—differences exist now—but there has never been a time in its history when there has been a difference of opinion as to its duty to announce its deliberate conclusions unaffected by considerations not pertaining to the case in hand." *Pollock* v. *Farmers' Loan & Trust Co.*, 158 U. S. 634, 635. "Still less can we recognize the doctrine that because the Constitution has been found in the march of time sufficiently comprehensive to be applicable to conditions not within the minds of its framers, and not arising in their time, it may therefore be wrenched from the subjects expressly embraced within it, and amended by judicial decision without action by the designated organs in the mode by which alone amendments can be made." *McPherson* v. *Blackie*, 146 U. S. 36. Mr. Justice Harlan says:

"If some of the guarantees of life, liberty, and property which at the time of the adoption of the national Constitution were regarded as fundamental and as absolutely essential to the enjoyment of freedom, have, in the judgment of some, ceased to be of practical value, it is for the people of the United States so to declare by the amendment of that instrument." *Maxwell* v. *Dow*, 176 U. S. 617.

Washington, in his farewell words to his fellow-countrymen, says:

"If in the opinion of the people, the distribution or modification of the constitutional powers be in any particular wrong, let it be corrected by an amendment in the way in which the

Constitution designates. But let there be no change by usurpation; for though this in one instance may be the instrument of good, it is the customary weapon by which free governments are destroyed. The precedent must always greatly overbalance in permanent evil any partial or transient benefit which the use can at any time yield."

We confidently submit that no authority has been shown to justify the exaction of duties complained of by the plaintiff in error, and that the judgment of the court below should be set aside and the case remanded with instructions to give judgment for the plaintiff.

New York, January 2, 1901.

Mr. Attorney General. for the United States.

I. INDIRECT TAXES NEED NOT BE UNIFORM THROUGHOUT " THE TERRITORY " OF THE UNITED STATES.

The Constitution has not provided for absolute uniformity of duties under all circumstances.

States may still impose imposts and duties on imports and exports and duty of tonnage, provided Congress consent thereto. Constitution, Art. I, sec. 10, pars. 2 and 3.

The uniformity clause of the Constitution refers to the States and not to Territories.

(*a*) The historical reasons for its insertion into the Constitution prove this.

(*b*) The phrase "throughout the United States" elsewhere used in the Constitution refers only to the States.

Article I, section 8, paragraph 4; Article II, section 1, paragraph 3. See *Sturges* v. *Crowninshield*, 4 Wheat. 122.

Similar meanings should be attached to the same phrase wherever it occurs unless some different meaning is clearly indicated by the context.

(*c*) The power to tax within the limits of territory is not derived from article 1, section 8, paragraph 1, but from the general power to make all needful rules and regulations respecting the territory belonging to the United States.

(*d*) The States, by the compact of submission to the Government organized under the Constitution, were to stand on a perfect equality with each other. The Congress was forbidden to exercise any discrimination between the States or their several ports.

As to the Territories no such compact was made. The full power of taxation was conferred on Congress along with the power to govern them; and in the exercise of the power Congress possesses unrestricted discretion both as to the subjects of taxation and the places where it shall be levied and those where it shall not be levied.

As between the different Territories there is no compact in favor of uniformity. Such uniformity is not essential for the protection of the States as between each other, because the Territories are the common property of all of the States, and whatever is done as to territorial taxation is done by the authority of the States and for their equal benefit.

(e) There are obvious reasons of prudence and policy for not requiring the revenue laws, which must be·uniform throughout the States, to be uniform also throughout the Territories.

This is expressly decided to be so as to direct taxes. *Loughborough* v. *Blake*. For the same reasons, and for other reasons as well, the same is true as to indirect taxes.

The internal revenue, or tariff, or license laws, made for the States, may be inconvenient, oppressive, unprofitable, impolitic, for the Territories. Those laws may be wise and politic for some and very unwise and impracticable for other Territories.

Congress ought to possess, and we contend does possess, the power to vary its system of taxation according to the location, conditions, and circumstances of the different Territories.

Otherwise not only will the Government be embarrassed and hampered, but actual injustice will be done to some sections of our possessions.

(f) It is conceded that Congress has such power to vary the system of taxation for local purposes.

But in principle and in reality there is no difference between local taxation and general taxation upon territorial property.

There were reasons why the limits of taxation upon the States should be fixed by the Constitution.

Equality between the States was the prime and most evident object to be attained. To that end were established the rule of apportionment as to direct taxes and the rule of conformity throughout all the States as to duties, imposts, and excises.

Story, answering the question why "duties, imposts, and excises" are required to be uniform throughout the United States, says :

"The answer to the latter may be given in a few words. It was to cut off all undue preferences of one State over another in the regulation of subjects affecting their common interests. Unless duties, imposts, and excises were uniform, the grossest and most oppressive inequalities, vitally affecting the pursuits and employments of the people of different States, might exist. The agriculture, commerce, or manufactures of one State might be built up on the ruins of those of another ; and a combination of a few States in Congress might secure a monopoly of certain branches of trade and business to themselves, to the injury, if not to the destruction, of their less-favored neighbors. The Constitution, throughout all its provisions, is an instrument of checks and restraints, as well as of powers. It does not rely on confidence in the General Government to preserve the interests of all the States. It is founded in a wholesome and strenuous jealousy, which, foreseeing the possibility of mischief, guards with solicitude against any exercise of power which may endanger the States, as far as it is practicable. If this provision as to uniformity of duties had been omitted, although the power might never have been abused to the injury of the feebler States of the Union (a presumption which history does not justify us in deeming quite safe or certain), yet it would, of itself, have been sufficient to demolish, in a practical sense, the value of most of the other restrictive clauses in the Constitution. New York and Pennsylvania might, by an easy combination with the Southern States, have destroyed the whole navigation of New England. A combination of a different character, between the New England and the Western States, might have borne down the agriculture of the South ; and a combination of a yet different character might have struck at the vital interests of manufacturers." 2 Story on Constitution, sec. 957.

He discusses the cognate clauses of the Constitution relating to taxation by the States, showing that all of those clauses are a part of one and the same system and have the same object, viz., the regulation of taxes within the States and by the States. "No State shall, without the consent of Congress, lay any imposts or duties on imports or exports, except what may be absolutely necessary for executing its inspection laws; and the net produce of all duties and imposts laid by any State on imports and exports shall be for the use of the Treasury of the United States; and all such laws shall be subject to the revision and control of Congress. No State shall, without the consent of Congress, lay any tonnage duty." In the first draft of the Constitution the clause stood: "No State, without the consent," etc., "shall lay imposts or duties on imports." The clause was then amended by adding "or exports," not, however, without opposition, six States voting in the affirmative and five in the negative; and again, by adding "nor with such consent, but for the use of the Treasury of the United States," by a vote of nine States against two. In the revised draft the clause was reported as thus amended. The clause was then altered to its present shape by a vote of ten States against one; and the clause which respects the duty on tonnage was then added by a vote of six States against four, one being divided. So that it seems that a struggle for state powers was constantly maintained with zeal and pertinacity throughout the whole discussion. If there is wisdom and sound policy in restraining the United States from exercising the power of taxation unequally in the States, there is, at least, equal wisdom and policy in restraining the States themselves from the exercise of the same power injuriously to the interests of each other. A petty warfare of regulation is thus prevented, which would rouse resentments and create dissensions, to the ruin of the harmony and amity of the States. The power to enforce their inspection laws is still retained, subject to the revision and control of Congress; so that sufficient provision is made for the convenient arrangement of their domestic and internal trade, whenever it is not injurious to the general interests. Idem, sec. 1016.

"No tax or duty shall be laid on articles exported from any

State. (*a*) No preference shall be given by any regulation of commerce or revenue to the ports of one State over those of another; (*b*) Nor shall vessels bound to or from one State be obliged to enter, clear, or pay duties in another."

The obvious object of these provisions is to prevent any possibility of applying the power to lay taxes or regulate commerce injuriously to the interests of any one State so as to favor or aid another. If Congress were allowed to lay a duty on exports from any one State, it might unreasonably injure, or even destroy the staple productions or common articles of that State. The inequality of such a tax would be extreme. In some of the States the whole of their means result from agricultural exports. In others a great portion is derived from other sources; from external fisheries, from freights, and from the profits of commerce in its largest extent. The burden of such a tax would, of course, be very unequally distributed. The power is, therefore, wholly taken away to intermeddle with the subject of exports. On the other hand, preferences might be given to the ports of one State, by regulations either of commerce or revenue, which might confer on them local facilities or privileges in regard to commerce or revenue. And such preferences might be equally fatal, if indirectly given under the milder form of requiring an entry, clearance, or payment of duties, in the ports of any State other than the ports of the State to or from which the vessel was bound. Idem, sec. 1013, 1014.

It is not necessary to rely upon "inherent powers" in order to sustain the authority of Congress to govern territory. The power to govern territory is expressly conferred.

It is given without limitation. Subject it may be to general restrictions contained in the Constitution, but these restrictions are not strictly upon the power to govern territory; they are not local in any sense.

Such is the prohibition against creating titles of nobility or the passage of bills of attainder.

There is no reservation of power in the people of the Territories. This reservation is to the people of the States.

The States granted to the Federal Government some powers; others they reserved to themselves or to their people.

On the subject of the government of the territory of the United States the States reserved nothing; they granted the power to make all needful rules and regulations respecting it.

The power of Congress over the States is the exact converse of its power over territory.

In legislating for the States, Congress has only the powers expressly or impliedly granted. The States have reserved all other powers.

But in legislating for territory the States have no power. Congress has it all. There is no residuary power left anywhere.

There can be no government in a Territory, except by the will of Congress. But States are organized governments which Congress cannot destroy, or even interfere with, except in special matters where the States have delegated the power by means of the Constitution.

Congress alone is the judge of what laws for territory are needful. There are no residuary powers.

II. POWER OF TAXATION.

This power to govern territory, which is so absolutely conferred on Congress, includes the power to tax, either by direct or indirect methods.

Taxation is necessary for the purpose of government.

Revenue must be provided, either by appropriation out of the general treasury or by local assessments, in order to govern any territory.

Therefore tax laws are "needful" as to Porto Rico and the Hawaiian Islands.

The subject we are now inquiring about is taxation. Congress has the power of taxation. That is conceded. Is the power of taxation absolute or limited?

It is absolute in some respects; limited in others.

What are the limitations?

Direct taxes must be apportioned among the several States which compose this Union.

Duties, imposts, and excises must be uniform throughout the United States.

Perhaps the objects for which taxes, etc., may be levied are also limited to the payment of the debts, the common defence, and the general welfare.

Are there any other limitations?

I know of none. In all other respects, then, the power of taxation is unrestricted. We might properly say absolute, unlimited, arbitrary.

It would be highly inaccurate, however, to say that it is despotic.

Congress can fix absolutely the subjects of taxation, the rates, the methods of assessment and collection. Its power in these respects is absolute—or equivalent to what counsel for appellants improperly call "despotic;" but it is not unconstitutional.

The question here is whether duties on merchandise, imported into the United States from the insular possessions, or into the islands from the United States, may be constitutionally laid. That they may be so laid is undoubted, unless such a system violates the clause of the Constitution which requires duties to be uniform throughout the United States.

There is no principle of inherent justice or personal rights at stake. It is a pure matter of geographical equality. There is no question of arbitrary or despotic power such as counsel have imagined.

We concede the power must be one exercised under the Constitution. We find the power in the Constitution, not outside it, nor beyond it, nor contrary to it.

It is not, in any correct way of speaking, a question of whether the Constitution "extends," or "follows," or "goes;" it is a question of the extent of the power conferred on Congress by the Constitution and how far it is limited by that instrument.

Federal taxation is either general or local.

Local taxes are levied under Article I, section 8, paragraph I.

Local taxes are for the support of territorial or non-state governments.

The Porto Rican tariff is of the local kind.

General taxes are of two kinds, direct; and what, for brevity, may be called indirect, meaning thereby duties, imposts, and excises.

Direct taxes must be laid on all the States alike; none may

be exempted. They may be, but they need not be, laid on the territorial possessions.

In the same way we contend that indirect taxes must be uniformly imposed throughout the States which compose the Union; that they may be extended, but do not need to be extended, to the territorial possessions.

The power of Congress to tax the Territories for local purposes is not limited or restricted. Counsel for appellants would, probably, though improperly, call that a claim of despotic power.

The taxes (using the term in its general significance) authorized to be imposed within the States are Federal taxes. They are imposed within the jurisdiction of the States, which themselves constitute separate and, in some senses, independent and sovereign governments, possessing and required to exercise, for their own internal needs to carry on their own administration, the power of taxation. The power of the States to raise money by taxation for domestic uses does not depend in any way upon the Federal Government. The right of the United States to tax for national purposes property within the limits of the States is a concession made by the States to the General Government.

The concession thus given was guarded by some limitations, and those limitations naturally were only such as the States demanded for their own protection to prevent inequality.

No such condition exists as to territory. The government of territory, whether denominated local or general, is all Federal. All the officers of a Territory are but agents of the General Government. The legislatures of the Territories exercise only delegated powers—are nothing but legislative agents.

In *Gibbons* v. *District of Columbia*, 116 U. S. 404, it was declared that the power of Congress, legislating as a local legislature for the District, to levy taxes for District purposes only, in like manner as the legislature of a State may tax the people of a State for state purposes, was expressly admitted in *Loughborough* v. *Blake*, and has never since been doubted.

" In the exercise of this power Congress, like any state legislature unrestricted by constitutional provisions, may at its discretion wholly exempt certain classes of property from taxation, or may tax them at a lower rate than other property." Per Gray, J.

If the unrestricted right of local taxation for the support of territorial governments be, as we submit it is, conceded, then it may be useful to follow the subject further, in order to show how useless such a distinction would be as a corollary to the doctrine which denies the right to lay special taxes for general purposes on territorial property by means of port duties when carried back and forth between the ports of the United States and the islands.

The laws, the administration, and the revenues of the Territories are subject to the absolute control of Congress.

Congress may repeal the whole form of government existing in a Territory; may destroy the legislature, vacate all the offices, and take over all the public funds and absorb them into the common Treasury. It may appropriate out of the Federal Treasury all the money necessary to carry on a territorial government, omitting all local taxation. We must not forget that "territory belonging to the United States" is the common property of the United States and is to be administered at the common expense and for the common benefit of the States united, who jointly, as a governing entity, own it.

Porto Rico and the Philippines were not won by arms and taken over by treaty through the efforts or influence or at the expense of the inhabitants, but through the might of the United States, upon their demand and upon their contribution of $20,000,000 to Spain, and upon the assumption by treaty of solemn national obligations which the United States, not the islands or their inhabitants, are bound to observe and keep.

The inhabitants of the islands are not joint partners with the States in their transaction.

The islands are "territory belonging to the United States," not a part of the United States. The islands were the things acquired by the treaty; the United States were the party who acquired them, and to whom they belong. The owner and the thing owned are not the same.

It is not a very sensible construction of the Constitution which will forbid Congress to do directly what it will permit it to do indirectly, and yet, if the contention of the appellants is correct, it follows that Congress cannot levy import duties on goods taken from the United States into Porto Rico, or *vice versa*, provided those duties are levied for general purposes, although Congress may levy any kind of a tax it chooses on the merchandise after it has been admitted into Porto Rico, provided it be for local purposes, and may then by legislative act take over such taxes into the General Treasury, to be paid out at the pleasure of Congress for general purposes.

Taking into consideration the relation of the Federal Government to territory, the fact that it is the common property of all the States.; that the General Government through Congress must support and administer the government of the territory, if it is to have any government; that Congress alone has the power, and the discretion as well, to say whether there shall be any organized government in any particular territory, and what such government, if allowed, shall be, it is necessary to concede the broadest discretion to Congress in determining the means by which and the sources from which the revenue to carry on the government of such territory shall be raised.

In so far as the question of taxation for local territorial purposes is concerned, therefore, it is clear that the express tax clauses of the Constitution are not applicable, and neither the law of apportionment nor of uniformity exists.

The right to tax merchandise in Porto Rico for local purposes existing in Congress, it is immaterial what the merchandise consists of, or where it originated, or who is its owner.

So also it is not perceived that to withhold the levy of the tax until the merchandise is brought into a port of the United States modifies or destroys the power to tax for such local purposes. The tax may be imposed in Porto Rico or held in abeyance until the merchandise reaches a port in the United States when, as a preliminary to its admission (not after its admission), it may be taxed for the support of the government of the islands.

One Territory may be taxed for the support of its local government, while another may be supported wholly from the Gen-

eral Treasury of the United States. In this there would be a
technical inequality, but the practical wisdom and justice of it
might be universally conceded.

The question would be one of governmental discretion vested
in Congress, which neither the States nor the courts of justice
are entitled to review.

The legality of the collection of duties on imports from Porto
Rico between the date of the evacuation and the date on which
the Porto Rico act took effect has been expressly recognized
and confirmed by Congress in the act entitled "An act appropriating, for the benefit and government of Porto Rico, revenues collected on importations therefrom since its evacuation by
Spain and revenues hereafter collected on such importations
under existing law," approved March 24, 1900. Acts of Fifty-
sixth Congress, first session, page 51.

This act directs that the amount of customs revenue received
on importations by the United States from Porto Rico since the
evacuation of Porto Rico by the Spanish forces on the 18th of
October, 1898, to the 1st of January, 1900, together with any
further customs revenue collected on importations from Porto
Rico since the 1st of January, 1900, or that shall hereafter be
collected under existing law, shall be placed at the disposal of
the President, to be used for the government now existing and
which may hereafter be established in Porto Rico, and for the
aid and relief of the people thereof, and for public education,
public works, and other governmental and public purposes
therein until otherwise provided by law.

Every provision of the Porto Rico act is for the peculiar and
local benefit of the insular government. The revenue is all paid
into the insular treasury to be used to support the local establishment created by the act.

A special protective duty on coffee, a product of Porto Rico,
is laid for the benefit and encouragement of the coffee growers
of the Territory.

Upon the point that laws of Congress do not extend in operation to territory unless such extension be expressed in the
statute, I desire, in addition to what was said in my brief in
the *Goetze* case, to add the following additional remarks:

Many instances of legislation show that Congress has always considered something more than the term "United States" to be necessary when it designed a statute to extend to territory.

The internal revenue laws are one instance.

See especially the act of 1868, 15 Stat. 125, where the word "State" is specifically defined to include a Territory. Sec. 104.

Also section 107, where the phrase "the exterior boundaries of the United States" is used in order to include all territory within the geographical limits of this country.

See also section 1891, Revised Statutes. It is to be remarked that this section refers only to organized Territories, and not to one organized territory.

Section 2145, Revised Statutes, extends criminal statutes to the Indian country, which would not be necessary if criminal statutes extended there of their own force.

Thomas H. Benton, for thirty years a senator from Missouri, was as able and distinguished a statesman as the territory included in the Louisiana purchase has ever produced. His views on the general question under discussion were strong and positive; his long service in the National Legislature, his familiarity with the course of public events, increased to an unusual degree by his practice of recording for publication the incidents of political and legislative discussion, his great ability as a constitutional lawyer, and his patriotic devotion to the best interests of our country, render his opinions and statements of superlative value. Some of his expressions concerning the subject of the extension of the Constitution to the Territories were quoted in the brief of the United States in the *Goetze* and *Pepke* cases. Still more remarkable passages, evincing the same views which the Government's counsel have maintained in these arguments, are found in a little book put forth by Mr. Benton in 1857, entitled "Historical and Legal Examination of the Dred Scott Case."

Referring to the history of the formation of the Constitution, he says:

"Who were the parties to it? The States alone. Their delegates framed it in the Federal convention; their citizens adopted it in the state conventions. The Northwest Territory was then in existence, and had been for three years ; yet it had no voice, either in the framing or adopting of the instrument—no delegate at Philadelphia, no submission of it to their will for adoption. The preamble shows it was made by States and for States. Territories are not alluded to in it. The body of the instrument shows the same thing, every clause, except one, being for States ; and Territories, as political entities, never mentioned once ; and the word 'territory,' occurring but once, and that as property, assimilated to other property—as land, in fact, and as a thing to be disposed of—to be sold. Now, you never sell a territorial government, but you sell property ; and in that sense alone does the word 'territory' occur, and that but once in the whole instrument. Tried by the practice under it, and the Territory is a subject, without a political right—no right to vote for President or Vice-President, or Senator, or Representative in Congress ; nor even to vote through their Delegate on any question in Congress—all their officers appointable and removable by the Federal authority, even their judges—their territory to be cut up as Congress pleases ; even parts of it to be given to Indians ; no political rights under it, except as specially granted by Congress ; no benefit from any act of Congress, except specially named in it, or the act specially extended to them, like the subject colonies and dependencies of Great Britain. How can the Constitution go to them of itself, when no act of Congress under it can go to them·unless specially extended? Far from embracing these Territories, the Constitution ignores them, and even refuses to recognize their existence where it would seem to be necessary—as in the case of fugitive from service and from labor. ˙Look at the clause. It only applies to States—Fugitives from States to States.[1] Why? Because the ordinance of 1787, the organic law of the Territories, made that provision for the Territories, and about in the same

_____

[1] "No person held to service or labor in one *State*, under the laws thereof, and escaping into *another*, shall, in consequence of any law or regulation therein, be discharged from such service or labor, but shall be delivered up on the claim of the party to whom such service or labor may be due." Article 4, sec. 2.

words, and before it was put in the Constitution.[1] In both places it is an organic provision, barren of execution until a law should be passed under it to give it effect—which was done in the fugitive-slave and criminal act of 1793, that act applying to Territories as well as to States, and so carrying both the Constitution and the ordinance into effect (p. 26).

"The whole Constitution was carried out upon the principle of ignoring the existence of Territories. I speak of Territories, implying political existence and organization, in contradistinction to territory signifying land, and repeat that, as political entities, the Constitution ignores them. This may be seen in every clause—strongly in the two instances just given and in those previously given, and still more strongly in the article which relates to the establishment of courts. If there is one branch of the Government which, above all others and more than all others, concerns the whole body of the community, it is the judicial department. The administration of justice, civilly and criminally, may reach every individual of a country. No age or sex, no rank, no condition of rich or poor, no conduct— not even that of virtue and merit itself—is secure from litigious involvement. The first care of the organic legislating power is to give a judiciary to the people; and this is what our Constitution has carefully done, as far as our system of government required its action. It has provided for the trial of all cases which could invoke the Federal authority—all between citizens of different States, and between citizens and foreigners, and for all cases arising under the Federal laws—all cases, in short, which were not left to the state courts—so that between the two systems the citizens should have a remedy for every wrong. Did this extend to the Territories? Not at all! The Federal judiciary system does not reach them, nor the state systems either. What then? Are they without courts? By no means. Congress supplies them, and in a way to show that they do not

---

[1] "Provided, always, that any person escaping into the same (the Northwest Territory) from whom labor or service is lawfully claimed in any one of the original *States*, such fugitive may be lawfully reclaimed, and conveyed to the person claiming his or her labor or service, as aforesaid." Ordinance of 1787, art. 6.

do it under the Federal Constitution, or in conformity to any
state constitution known in our America.   They made judges
to hold office for a term of years, subject to be removed by the
President, like any common officeholder, and several have been
so removed; and they gave codes of law, both civil and criminal, not only over the organized Territories reduced to our possession, but over the wild territory still in the hands of the
Indians.   By the decision of the Supreme Court this would
seem to be unconstitutional and void, a consequence which
seemed to set hard on one of the brother justices who had acted
under these laws, and who, while agreeing in the decision upon
the Missouri compromise act, did it for a different reason from
that which would have condemned his own action.[1] Certainly
all this legislation was incompatible with the Constitution, but
no violation of it, because the Constitution did not reach these
territories, either civilized or savage."

Referring to the act for the government of the Louisiana
territory, Mr. Benton declared : " The bill thus passed received
the approbation of the President the same day it was laid before him; and to those who are acquainted with the working
of the legislative machinery, it may well be believed that the
whole proceeding was in concert with the Administration; that
Mr. Jefferson picked out Mr. Breckenridge to bring in the bill;
that its principles were settled in Cabinet meeting; that Mr.
Madison drew it, and that every question in relation to it was
duly considered before it was submitted to final action.   And
thus, this first instance of Congress legislation upon newly acquired territory was as high an instance of disregard of the
Constitution as the imagination could conceive, being nothing
less than the continuation of the Spanish regal despotism; the
President taking the place of the King of Spain; Governor

---

[1] " It is due to myself to say that it is asking much of a judge, who has
for nearly twenty years been exercising jurisdiction, from the western Missouri line to the Rocky Mountains, and, on this understanding of the Constitution, inflicting the extreme penalty of death for crimes committed where
the direct legislation of Congress was the only rule, to agree that he has
been all the while acting in mistake, and as an usurper."   Mr. Justice
Catron.

Claiborne the place of the intendant-general, Morales; the laws of Spain remaining in force and administered by American judges, and the whole provincial administration going on as if no change of government had taken place. It was a royal despotic government, and everybody knew it, and no one thought of testing it by the Constitution (some few new members in the House excepted) than by the Koran " (p. 60).

" And now for the men who passed these acts—who established these governments—so incompatible with the Constitution and so fully asserting absolute power over this new territory. Who were they ? They were the men of the Revolution—of the ordinance of 1787—of the Constitution of that year—of the first administration of the Federal Government in its early age —and the authors of the acquisition of Louisiana. Mr. Jefferson was President, Mr. Madison Secretary of State, and the two Houses of Congress filled with men who had acted their good part in founding and putting into operation the new Federal Government. These were the men who did these things and who ought to be allowed to know something of their own work ; and, if they did not, somebody existing at the time ought to have known of their dreadful usurpations and proclaimed them to the world. No such discovery was made " (p. 69).

Speaking of the doctrine of the *proprio vigore* extension of the Constitution to territory, Mr. Benton said: " Mr. Calhoun declared its effect when he proclaimed it, saying:

" ' I deny that the laws of Mexico can have the effect attributed to them (that of keeping slavery out of New Mexico and California).   As soon as the treaty between the two countries is ratified, the sovereignty and authority of Mexico in the territory acquired by it becomes extinct, and that of the United States is substituted in its place, conveying the Constitution with its overriding control over all the laws and institutions of Mexico inconsistent with it.'   Oregon Debate, 1848.

" This is the declared effect of the transmigration of the Constitution to free territory by the author of the doctrine; and great is the extent of country, either acquired or to be acquired, in which the doctrine is to have application. All New Mexico and California at the time it was broached ; all the Territories now held, wherever situated, and as much as can be added to them—these additions have already been considerable, and vast and varied accessions are still expected. Arizona has been acquired; fifty millions were offered to Mexico for her northern half, to include Monterey and Saltillo; a vast sum is now offered for Sonora and Sinaloa, down to Guaymas; Tehuantepec, Nicaragua, Panama, Darien, the Spanish part of Santo Domingo, Cuba, with islands on both sides of the tropical continent. Nor do we stop at the two Americas, their coasts, and islands, extensive as they are, but circumvolving the terraqueous globe, we look wistfully at the Sandwich Islands, and on some gem in the Polynesian group, and plunging to the antipodes pounce down upon Formosa in the China Sea. Such were the schemes of the last administration, and must continue, if its policy should continue. Over all these provinces, isthmuses, islands, and ports, now free, our Constitution must spread (if we acquire them, and the decision of the Supreme Court stands), overriding and overruling all anti-slavery law in their respective limits, and planting African slavery in its place, beyond the power of Congress or the people there to prevent it" (p. 29).

III. THE INTERNAL REVENUE LAWS HAVE NOT BEEN HERETOFORE UNIVERSAL IN APPLICATION.

Internal duties: Under the Constitution the internal revenue laws should be as universal and uniform in application as the tariff laws. Were they framed for universal application, and have they been so applied?

The first internal revenue tax was on spirits distilled in the United States, and was levied by the act of March 3, 1791, which, for purposes of collection, provided "that the United States shall be divided into fourteen districts, each consisting of one State." 1 U. S. Stat. sec. 4. pp. 199, 200. That act provided (secs. 14 and 15) that duties should be paid upon all spirits distilled "within the United States," but no provision was made for the collection of the tax in the territory not included in the boundaries of the existing fourteen States. Other instructive phrases of that act are as follows (secs. 53, 55): "Without the limits of the United States;" "relanded in any other part of the same;" "within the limits of any part of the United States."

The following acts, June 5, 1794, as to carriages, and of the same date as to retail liquor licenses (id. 373, 376), although imposing a tax upon "all carriages for the conveyance of persons" and upon "every person who shall deal in the selling of wines," respectively, provided for collection only in the districts created by the act of 1791. Another act of the same date (id. 378) expressly extends the tax on distilled spirits and stills to "the territories northwest and south of the river Ohio" by authorizing the President to erect new districts and appoint the necessary officers in that region; and still another act of that date laid a duty upon snuff and refined sugar "manufactured or made in the United States," without any indication that the extension of the distilled-spirits tax over the Northwest Territory should also cover these additional articles. Similarly the act of June 9, 1794 (id. 397), imposing duties on property sold at auction, refers seemingly to the "several supervisors of the revenue" and the "respective districts" in the fourteen States only. These duties, along with the tax on stamped paper, act of July 6, 1797, 1 Stat. 527, in which the phrase "throughout the United States" is used, were altered, amended, or repealed, and later reënacted by various acts, in the interval before the whole body of internal revenue laws was repealed in 1817; but the entire course of legislation shows that the taxes were not applied outside the States included in the original act or those subsequently admitted, unless the tax laws were expressly extended to the Territories of the United States. Thus, the act of July 11, 1798, 1 Stat. 591, fixed the compensation of officers employed in collecting the internal revenues, but mentions no districts except those in the sixteen States then forming the Union. But, consistently with the extension of the distilled spirits tax to the Northwest Territory, it appears that there was a supervisor of revenue in that district whose compensation was fixed by section 4 of the act of April 6, 1802, 2 Stat. 148.

By the act of July 22, 1813, 3 Stat. 22, the collection of direct taxes and internal duties was jointly regulated, and no provision was made for the collection of either species of tax outside the eighteen States at that time. The last section of that act, requiring separate accounts of the direct taxes and internal duties to be kept, indicates that the only sums received were those received from "each State" as enumerated in the beginning of the act.

The act of August 2, 1813, with the previous acts therein referred to, 3 Stat. 82, and note *a*, reënacted the various internal duties which had previously been abolished, and charged the collectors appointed under the acts, *supra*, erecting the various collection districts in the States, with the collection of the duties imposed and by section 2 expressly authorized the President "to divide respectively the several Territories of the United States and the District of Columbia" into convenient districts for the purpose of collecting the internal duties specified and to appoint collectors, thus for the first time extending these laws generally and comprehensively to the territory of the United States outside the limits of the States; and that act provided (sec. 3) that the several duties "shall be laid and collected in the several Territories of the United States and in the District of Columbia in the same manner and under the same penalties" as in the "districts" of the old and reënacted laws, that is, in the States; and extended the existing acts to the "several Territories of the United States and to the District of Columbia." In other sections of that act and throughout the later acts such phrases as "within the several Territories of the United States and the District of Columbia" and "within the United States or Territories thereof" constantly appear. *Vide* act December 21, 1814, 3 Stat. 152; act January 18, 1815, id. 180.

The internal revenue law of July 1, 1862, 12 Stat. 432, which was the basis of all the succeeding laws amending its provisions or supplying new provisions, provided "that the States and Territories of the United States and the District of Columbia" should be divided into convenient collection districts, and "within the United States or Territories thereof" and "of the United States or Territories" (e. g., secs. 75, 82) are the phrases used to describe or locate the persons or property subject to tax.

The most important subsequent acts are those of March 3,

1863, 12 Stat. 713; March 7, 1864, 13 Stat. 14; June 30, 1864, id. 223, which was a new general act supplanting the act of 1862, of which section 46 provided for the execution of the law "in a State or Territory of the United States or any part thereof, or within the District of Columbia," as soon as the authority of the United States therein shall be reëstablished, if, for any cause, the laws could not be executed therein; and the last section, carried into the Revised Statutes as section 3140, provided that wherever the word "State" is used in the act, it shall be construed to include the Territories and the District of Columbia where such construction is necessary to carry out the provisions of the act. The act of July 13, 1866, 14 Stat. 98, was also a general law and largely reduced the duties; and the act of July 20, 1868, 15 Stat. 125, made new provisions for the taxation of distilled spirits and tobacco. Section 104 of this act also construed the word "State" as including a Territory and the District of Columbia, and section 107 provided that the internal revenue laws imposing taxes upon distilled spirits, fermented liquors, tobacco, snuff, and cigars shall be held and construed to extend to such articles produced within the exterior boundaries of the United States, whether within a collection district or not. The latter section was construed by the court in the *Cherokee Tobacco Case*, 11 Wall. 616, which determined that the section "extends the revenue laws over the Indian Territories only as to liquors and tobacco. In all other respects the Indians in those Territories are exempt." The dissenting opinion held that "it was not the intention of Congress to extend the internal revenue law to the Indian Territory; that Territory is an exempt jurisdiction," partly on the ground that the express and special privilege given to the Cherokees by the treaty of 1866 was not repealed by the subsequent general law, and partly on the ground that the language of section 107 could be applied to territory within the exterior boundaries of the United States without embracing the Indian Territory, to wit, to the Territory of Alaska. It was not suggested in the court below (*United States* v. *Tobacco Factory*, 1 Dill. 264; Fed. Cas. No. 16,528), nor in the Supreme Court that without express provision by Congress these laws would extend to the Territories, because, being taxes or duties, they must constitutionally "be uniform throughout the United States." The debates in Congress show that section 107 was offered as an amendment or addition to the act in the Senate by Mr. Sherman, and was adopted without explanation or debate. Cong. Globe, part 4, 2d session, 40th Cong. 1867–68, p. 3779.

It is to be noticed that the act of 1868, *supra*, section 55, and the act of June 6, 1872, 18 Stat. 230, amending the same (sec. 12), in making it a misdemeanor intentionally to reland distilled spirits shipped for exportation used the phrase " within the jurisdiction of the United States."

Beginning with the passage of the act of July 14, 1870, 16 Stat. 256, the large list of internal revenue taxes was gradually reduced until the enactment of the war revenue act of 1898, the language of which adds no specially significant phrase to the former legislation, although the construction given to it in *Knowlton* v. *Moore, post,* confirms our contention as to the purpose and scope of the rule of uniformity.

Direct taxes: Direct taxes do not, perhaps, present a close analogy, being imposed by the rule of apportionment "among the several States according to their respective numbers." The States and their respective quotas were necessarily specified in such laws, and there were no general expressions to render doubtful the divisions of territory in which direct taxes were intended to be laid. It is worthy of remark, however, that if the duties which are to be "uniform · throughout the United States" must also apply universally throughout acquired and dependent territory, then quite as clearly must be applied universally the direct taxes which are to be " apportioned among the several States." The uniform duties clause and the direct tax provisions both show the scrupulous care of the framers of the Constitution for equality among the States, and neither rule looks beyond the States to apply a fixed and self-acting ordinance to regions which the future might annex, but were not then in the States, nor under the definite compact recognized by the Constitution as to the Northwest Territory.

How has Congress construed their power and function relative to direct taxes? The act of 1798, 1 Stat. 580, provided for valuations in the States, but not in the Northwest Territory, although this coterminous region was recognized as intimately connected territory of the United States, or even as a portion thereof, under the Constitution. So also the similar act of 1813 provided (3 Stat. 22), and the direct taxes of 1813 and 1815 (id. 53, 164) were so laid. On the other hand, a direct tax was expressly imposed by Congress in the District of Columbia by the act of 1815, 3 Stat. 216, which was before the court in *Loughborough* v. *Blake, infra.* The enactment itself is proof that the interposition of Congress was conceived to be necessary, not only to provide the collecting machinery in the District, but also to carry the constitutional provisions beyond the limits of the States in "laying a direct tax upon the United States." And the decision simply determined that Congress had this power.

The direct tax of 1861, 12 Stat. 292, was specifically apportioned among the existing States, Territories, and the District of Columbia, and although the income tax imposed by the forty-ninth section of that act was levied upon the annual income of "every person residing in the United States," Congress was careful in subsequent sections to provide the machinery for the assessment and collection of that tax, not by any general phrase such as "throughout the United States," but "in each of the States and Territories of the United States, and in the District of Columbia."

The census acts, upon which the direct tax laws are based, show that in 1790 (1 Stat. 101) the marshals of the several districts (each one of the fourteen States then constituting a judicial district) were directed to take the enumeration. In 1800 (2 Stat. 11) the direction was given to the marshals of the several districts and the secretaries of the Northwest Territory and the Mississippi Territory. In 1810 (id. 564) the marshal of the District of Columbia and the secretaries of the additional Territories were added. In 1820 (3 Stat. 548) marshals having then been provided for the Territories, those officers alone were specified, and so the law continued for several decades, the act of 1850 (9 Stat. 428) providing the machinery under which the subsequent censuses were taken until the establishment of the

Census Office by the act of March 3, 1879, 20 Stat. 473, which regulated the taking of the census "within each State or Territory."

Thus, while Congress has provided throughout the United States and its Territories for the enumeration upon which direct taxes have been apportioned,.except in 1790, when, indeed, the Northwest Territory was in large part wild and unoccupied, but little more so than some of the States, it never seems to have been supposed that such taxes must be levied beyond the States or apportioned to the Territories unless Congress saw fit so to provide.

IV. ALASKA.

In 1868 the customs, commerce, and navigation laws were extended over Alaska (Rev. Stat. sec. 1954), but not the internal revenue laws, except so far as section 107 of the act of July 20, 1868, had that effect. Nor was any further change made in this respect by the act of 1884, 23 Stat. 24, which provided a civil government for Alaska but not fully organized territorial government, under which civil status the act of March 3, 1899, 30 Stat. 1253, gives to the "district of Alaska" (meaning, doubtless, the judicial district) a code of criminal procedure. Section 477 of this act recognizes the wide application of the taxes on intoxicating liquors and impliedly directs their enforcement in Alaska as follows: "That nothing in this act shall in any way repeal, conflict, or interfere with the public general laws of the United States imposing taxes on the manufacture and sale of intoxicating liquors, for the purpose of revenue, and known as the 'internal revenue laws.'" In practice, internal revenue duties have been collected in Alaska upon liquors and tobacco since December, 1872, when the Territory was added to the internal revenue district of Oregon by Executive order under the authority of sections 103 and 107 of the act of 1868, *supra*, for which action the laws embodied in section 3141, Revised Statutes, would also give authority.

So far as the internal revenue records show, it seems that internal revenue duties have in the past been collected in Alaska and the Indian Territory only upon the articles subjected to tax by section 107 of the act of 1868; for which, perhaps, a practical reason also might be given, namely, that in the conditions prevailing in those Territories for a long period after 1868 no articles were produced subject to tax except those named in section 107, other such articles entering those districts tax paid.

V. OKLAHOMA.

A point was made on the former argument as to the sources from which the Government has acquired this Territory.

The "Indian country," defined in the act of June 30, 1834, 4 Stat. 1729, and described in *United States* v. *43 Gallons of Whisky*, 93 U. S. 188, and *Bates* v. *Clark*, 95 U. S. 204, was narrowed under advancing civilization and by successive treaties which extinguished the Indian titles, and at the time of the adoption of the Revised Statutes, comprised the region known as the Indian Territory, the somewhat indefinite boundaries of which had been gradually defined as new States and Territories were erected. The previous laws, preserved in chapter 4 of Title XXVIII of the Revised Statutes, were applied to its government, and it was itself definitely bounded by the act of March 1, 1889, 25 Stat. 783; 1 Supp. R. S. 670, and note. By the act of May 2, 1890, the Territory of Oklahoma was erected and organized, and the limits defined to include a certain portion of the Indian Territory and the "Public Land Strip," with a provision for incorporating into the Territory the unoccupied portion of the "Cherokee Outlet," and lands remaining in the Indian Territory, whenever the respective Indian tribal owners should assent.

A question of long standing between the United States and Texas as to the title to what was known as "Greer County" being involved in that act, the dispute was settled by the decision in *United States* v. *Texas*, 162 U. S. 1, which held that the title to that portion of Oklahoma Territory was in the United States, and that the tract had been acquired by the United States under the treaty with Spain of 1819. It is manifest from the reasoning of the opinion in that case and the authorities cited, and especially from the compromise act of September 9, 1850, 9 Stat. 446, by which the northern and western boundaries

of Texas were defined, and all territory claimed by her exterior
to said boundaries was relinquished, that all the land now included in the Territory of Oklahoma had been claimed by the
United States against Spain and her successors in title and sovereignty, Mexico and Texas, as under the Louisiana purchase,
and that the Territory of Oklahoma as now constituted was
necessarily embraced either in the Louisiana purchase or under
the treaty of 1819 with Spain, or under the cession of territory
by Texas in 1850.   All of this country lies far east of the cession
by Mexico in 1848, and there seems to be no doubt that the entire territory was included in the Louisiana purchase, excepting
the portion decided by *United States* v. *Texas*, to have been
acquired under the treaty with Spain, and excepting the " Public Land Strip," which apparently was part of the territory
claimed by Texas exterior to her boundaries as settled, which
she surrendered in 1850.   It is evident that in one of these
ways all of the Indian Territory and Oklahoma must have
been acquired, since the three acquisitions in question (whatever
may have been the variations in boundary lines and surveys)
taken together covered the whole of that country.   See " The
Louisiana Purchase" by the present Commissioner of the General Land Office, pages 36, 39.   The passage on page 36 says
that the Louisiana purchase proper embraces   . . .   " all of
the Indian Territory and *part* of Oklahoma Territory."   It is
learned from the Land Office that the only parts of Oklahoma
Territory not included in the Louisiana purchase are those here
stated to have been acquired under the treaty with Spain or
through the cession by Texas.

The Indian Territory was added to the internal revenue district of Kansas August 8, 1881, in the same way as Alaska was
added to the district of Oregon ; and Oklahoma, since its separation from the Indian Territory, remains in the Kansas district.

We have seen that the one hundred and seventh section of
the internal revenue act of July 20, 1868, was construed in the
Cherokee Tobacco case to carry the internal revenue laws as to
distilled spirits, fermented liquors, tobacco, snuff, and cigars to
the Indian Territory as then constituted.   The Oklahoma act
of 1890 contains (sec. 28) a provision generally applied in ex-

press terms to all the Territories as they are organized, namely, "That the Constitution and all the laws of the United States not locally inapplicable, except so far as modified by this act, have the same force and effect as elsewhere within the United States." This is in accordance with section 1891 of the Revised Statutes, which applies this provision to "all the organized Territories and in every Territory hereafter organized as elsewhere within the United States." It is evident that the internal revenue laws are not inapplicable in an organized Territory, and such provisions taken in connection with the authority conferred upon the President by section 3141, Revised Statutes, are the ground upon which all the internal revenue laws are executed in the organized Territories; while section 107 of the act of 1868, *supra*, is the original basis for collecting the tax upon distilled spirits, fermented liquors, tobacco, snuff, and cigars in the Territory of Alaska, thus far not fully organized in the legal sense, to which, however, the act of 1899 extended the taxes on intoxicating liquors.

It seems that in practice at the present time taxes accruing under the war revenue act, as well as all internal revenue taxes, are collected in Alaska and the Indian Territory. This practice is based partly on section 107 (*ante*), reënacted as section 3448, Revised Statutes, and, as to Alaska, under the act of 1899 (*supra*), and partly on the ruling of the internal revenue authorities that these laws operate with respect to Alaska so as to subject to stamp tax articles not produced in the Territory but destined for consumption there. The practice means no more than that now, as in former years, the growth and manufacture of tobacco and production of spirits in the Indian Territory and the sale of these articles in Alaska are properly made to bear their burdens under the law; and substantially that other articles subject to tax before the war revenue act of 1898 are not produced in those Territories, but must enter them correctly stamped or tax paid. And taxes are levied there under the latter act because its language, construed in the light of its evident purpose and spirit, has been held by the Treasury Department to carry its provisions over those two Territories.

VI. HAWAII, PORTO RICO.

On this review of the status of the Territories in respect to the internal revenue laws, and of the varying action by Congress under different circumstances—always in strict conformity to the doctrine that these laws do not, without special provision, of themselves or by force of the Constitution, apply to the territorial possessions or dominion of the United States —it is logical and consistent to find Congress recognizing in the Alaska act of 1899, as above shown, the validity of the internal duties of most general importance, as previously extended there and established in practice; providing that the Constitution and, with certain exceptions, the laws of the United States shall have equal force and effect in Hawaii, and that the Territory shall constitute an internal revenue district (secs. 5, 87, act of April 30, 1900, 31 Stat. 141); and recognizing in section 3 of the act of April 12, 1900 (id. 77), the internal revenue taxes at Porto Rico, and in section 14 excepting our internal revenue laws from those statutes of the United States which are to have the same force and effect in Porto Rico as in the United States.

Thus, finally, in the case of the internal revenue laws to a striking degree, and also in the case of the direct tax laws—a somewhat analogous instance—Congress has uniformly and specifically legislated for the Territory or Territories of the United States whenever it was their intention to execute those laws beyond the limits of the States; and the only case in which their action has been challenged or questioned was the *Cherokee Tobacco* case, wherein the legislation was resisted, not on grounds which drew in question the constitutional authority of Congress as now presented, but simply because the Indian treaty established a lawful· exemption which, it was claimed, had not been repealed by Congress. Administrative practice, dealing through a long period of time with many novel, different, and peculiar conditions, has followed this view of the matter with substantial consistency; and no decisions on these laws can be found in which the soundness of the Government view is doubted or controverted, much less overthrown.

Counsel for Armstrong contend that the term "United

States" means the United States Government, composed of States and outlying Territories and embracing the people residing in both the States and outlying Territories (p. 33).

In this contention they entirely disregard the fact that the term "United States" is used sometimes in a geographical sense, sometimes in a sense describing the governing entity, and sometimes as describing the States of the Union.

It is also asserted that the theory of our Government is that duties are to be levied and collected upon the products of foreign countries. "Until now whoever dreamed that we could collect duties upon our own people."

The States, in the days of the Confederation, levied duties upon goods brought from one State into another. The States are still denominated foreign so far as the judgments of their courts are concerned.

The quality of "foreign," in connection with tariff laws, is one inserted only by the statute. Great Britain always imposed duties on merchandise brought into her home ports from the colonies, and does so now. Many of her colonies impose duties *ad libitum* upon imports from the home country.

The question is not one of domestic and of foreign ports, but one relating to the States and to Territories, the former being the constituent parts of the Union and the latter being territory belonging to the United States.

On page 47 of their brief counsel make the astonishing statement that for nearly one hundred years no distinction has been made between that part of the national domain which was States and that which was Territories. The direct opposite of this is the truth, as shown by the history of our Government, its legislation, and its judicial decisions.

Counsel assert that the President of the United States has no right to exercise legislative function. If by this is meant that he is not a legislative branch of the Government within the meaning of the Constitution, no fault can be found with the doctrine. It is too elementary to be even alluded to. But when as commander in chief he exercises government over conquered territory, he has, by the undoubted law of nations, the right not only to govern but to make laws for the territory so occupied. The legislative functions thus exercised are not a part of the legislative power conferred by the Constitution upon Congress, and have no relation to it. They are merely incidents under the public law of belligerent right, vested by the Constitution in the President as commander in chief of the army and navy. They are not unconstitutional, but are exercised by virtue of the Constitution, not by any express clause which confers them, but are implied in the functions and duties of the commander in chief. Such legislative functions are not national, but local and peculiar, and relate only to the particular extent of country occupied by the military forces. This doctrine is so well understood and has been so frequently asserted, both by the executive and by the courts, that citation seems hardly necessary. I refer especially to what was said on this subject in my brief in the *Goetze* case (p. 117, etc.), and what was said by this court in the case of *Cross* v. *Harrison*, 16 Howard, 164.

In that case, speaking of the continuance of the temporary government of California and New Mexico, the court said:

"It had been instituted during the war by the command of the President of the United States. It was the government when the territory was ceded as a conquest and it did not cease, as a matter of course or as a necessary consequence of the restoration of peace. The President might have dissolved it by withdrawing the army and navy officers who administered it, but he did not do so. Congress might have put an end to it, but that was not done. The right inference from the inaction of both is that it was meant to be continued until it had been legislatively changed. No presumption of a contrary intention can be made. Whatever may have been the causes of delay, it must be presumed that the delay was consistent with the true policy of the Government."

This claim of counsel for the appellant would be subversive and destructive of every vestige of organized government set up and sustained in the Philippine Islands from the time of our occupation of Manila until the present time, notwithstanding Congress has permitted the executive department to continue in the administration of the government of those islands without interference or action on its part.

VII. TARIFF AND REVENUE LAWS OF THE UNITED STATES DO NOT TAKE EFFECT IN CEDED TERRITORY IMMEDIATELY UPON THE RATIFICATION OF THE TREATY OF CESSION.

Counsel for appellants contend that immediately on the ratification of the treaty with Spain, and immediately upon the approval of the resolution annexing the Hawaiian Islands as territory of the United States, the tariff laws of the United States, became operative in the territories thus acquired.

Such a construction of the law and Constitution could not be made without grave prejudice to the United States, and ought not to be made unless the Constitution clearly and unmistakably requires it. Such a construction would overrule the direct provisions of Congress in the Hawaiian act, and the manifest purpose of the President and the Senate in negotiating the Paris treaty.

It is not to be credited that the founders of the Government intended the Constitution and laws of the United States to have such absolute and inconvenient application.

There must be in the nature of things a time between the deed of acquisition and the assumption by the United States of the full government of acquired territory when the relation between the Federal Government and the acquired territory will be inchoate. In these particular cases, when the treaty was ratified and the Hawaiian resolution approved, there were no collection districts, no revenue officers, no provision for turning over the proceeds of the revenue to the General Treasury, no means of enforcing the criminal laws passed to punish frauds upon the revenue, or anything, in fact, to enforce to the slightest extent the rights of the Government, or the provisions of the law, which, it is contended, nevertheless extended to the new possessions. Neither Porto Rico nor the Philippine Islands were possessed at the time of their acquisition of any autonomous government of their own after the Spanish sovereignty was eliminated. They were incapable of levying or collecting taxes for their own support.

There might be cases of the acquisition of territory which possess no organized form of government whatsoever, not even of a local or municipal kind. Whether such territory should have any local government would depend entirely upon the will of Congress; the contention of the appellants would create the absurd necessity of having acts of Congress as to revenue and other matters extended in theory through tracts of country in which they were utterly incapable of enforcement, all the agencies of government being absent.

It could never have been contended that such a condition of theoretical law and practical anarchy should arise.

Cross v. *Harrison* is authority against the position of appellants on this point.

Mr. Solicitor General for the United States.

If the court please: Before entering upon a discussion of the grave questions raised in these five cases, I desire very briefly to refer to some matters of jurisdiction. I do this, not for the purpose of securing a disposition of the cases other than upon the merits, but because counsel have adopted in these different suits different and inconsistent methods of testing the constitutionality of revenue exactions, and the Government does not desire to be taken as acquiescing in what it considers an improper course of procedure.

In the *Goetze* case, already argued fully before the court, the method taken of raising the question whether duties could lawfully and constitutionally be levied upon goods imported from Porto Rico after the treaty of peace and before the act of Congress, was by a protest under the customs administrative act, which was passed upon first by the collector and then by the board of appraisers, and then came through the regular judicial channel to this court. We believe that that method was the proper one of raising the questions sought to be raised, but that method has not been pursued in these cases.

In the *De Lima* and the *Downes* cases, the goods coming from Porto Rico to New York were entered under the customs laws and the duties were paid. It is said they were paid under protest and for the purpose of securing the possession of the goods. But they were paid. Having paid the duties, we submit that the importer could not bring a common-law action against the collector to recover them back. The case of *In re Fassett*, 142 U. S. 479, does not apply. That was a case where Mr. Vanderbilt brought a pleasure yacht into the port of New York. He did not enter it and pay the duties upon it. He declined to do so, and when the collector seized the yacht he brought the proper action in a United States court to recover possession of the vessel. Now, if counsel desired to stand upon the proposition that no articles had been imported into the United States within the meaning of the revenue law, they should have refused to enter the goods, and then have taken the proper steps to secure possession of them. But they entered them, and they paid the duties upon them, and now they seek to bring an action against the collector to recover back the money paid, although the law required the collector to pay that money into the Treasury of the United States, and has expressly provided that he shall not be subject to a suit of this kind.

We also make the point that in one of these cases, the *Downes* case, there is not involved a sufficient sum of money to give the United States court jurisdiction, our claim being that there must have been involved the sum of $2,000, when it appears in the record that only six hundred odd dollars was involved.

MR. JUSTICE HARLAN. Does that apply to revenue cases?

THE SOLICITOR GENERAL. This is not a revenue case, so opposing counsel insist. They don't concede it is a revenue case; they insist it is a common-law action to recover back money unlawfully exacted by an officer outside his authority and without authority.

In the *Dooley* cases and in the *Armstrong* case, suits have been brought against the Government of the United States. In the *Armstrong* case the suit was brought in the Court of Claims; in the *Dooley* cases under the concurrent jurisdiction act, in the United States Circuit Court. Now, if these cases are revenue cases, the suits do not lie. Suits cannot be brought against the United States either in the Court of Claims or in the Circuit Court to recover back revenue collected by officers

of the United States. That jurisdiction has not been given to
those courts, nor such a privilege accorded to those who pay
money into the Treasury of the United States. And the reason is obvious. If such suits lie, there is no statute of limitations, and the Government could never know the amount of
claims outstanding against it resulting from the collection of
revenue through its agents. The Government has, therefore,
provided exclusive methods of determining whether revenue
was rightfully collected or not. When those methods are pursued, the officers of the Government are able to tell right along
what claims exist against it, and Congress can provide for
them. On the other hand, if these cases are not revenue cases,
then they sound in tort, and neither court, as I understand,
takes jurisdiction of cases of that sort. And so for these reasons, which are supported, as we think, by the authorities, we
claim that the courts below had no jurisdiction of any of these
cases.

Now I come to a consideration of the very serious questions
raised in these cases. And in order that the court may understand how the questions arise, and the order in which I shall
discuss them, I desire to state categorically the specific duties
which were collected, the validity of which is contested.

In the first place, there were duties collected on goods imported into Porto Rico from the United States, during the
military occupation of the island, after the signing of the protocol and before the ratification of the treaty of Paris. Such
were some of the duties collected in the *Armstrong* case. I
had supposed that similar duties were exacted in the first *Dooley* case, but I find I am mistaken.

In the second place, there were duties collected on goods imported into Porto Rico from the United States during the military occupation, but after the cession of Porto Rico by the
ratification of the treaty and before the passage of the Porto
Rican act. Such duties were collected in the *Armstrong* case
and in the first *Dooley* case.

In the third place, there were duties collected on goods imported from Porto Rico into the United States after the ratification of the treaty of Paris and before the taking effect of the

Porto Rican act.   Such were the duties exacted in the *De Lima* case.

In the fourth place, there were duties collected on goods coming into the United States from Porto Rico after the Porto Rican act took effect.   The validity of these exactions is brought in question in the *Downes* case.

Finally, there were duties collected on goods coming into Porto Rico from the United States after the taking effect of the Porto Rican act.   Such were the duties exacted in the second *Dooley* case.

I shall first consider the validity of the duties exacted in Porto Rico by the President prior to the treaty.

These duties, we claim, were imposed in Porto Rico by Executive order during the military occupation of the island prior to the ratification of the treaty of peace, and were rightfully levied by the President, as commander in chief, acting under belligerent right, at a time when hostilities between the United States and Spain had only been suspended, not terminated, and when Porto Rico had not been ceded to the United States, and when the right and obligation of conducting a civil government by the military authority was imposed upon the President.   I am at a loss to perceive any reasonable grounds for opposing the validity of these exactions.   It appears from the brief in the *Armstrong* case that the authority of the President in promulgating those executive orders and providing a civil government for the island is attacked as being an exercise of a legislative power in a time of peace, and also—they say—when Porto Rico had been ceded to the United States and had become a part of the United States.   Apparently, from a reading of their brief, the position of counsel in the *Armstrong* case is logically this :

First. By the protocol Porto Rico was ceded to and became a part of the United States.

Second. That the suspension of hostilities which followed the signing of the protocol ended the war and brought about peace.

Third. That consequently an end was put to the authority of the President to govern Porto Rico under the war power.

Now, these propositions seem to me so absurd that to state them is to refute them. I really feel as if I ought to beg the pardon of the court for calling attention to the provisions of the protocol. The protocol says, in the second article, "Spain will cede to the United States the island of Porto Rico." That is not a cession; that is a promise to cede in the event a treaty of peace should be concluded and ratified. The protocol also provides in the sixth article, "Upon the conclusion and signing of this protocol hostilities between the two countries shall be suspended," not terminated. And it further provides in the fifth article, that the United States and Spain "will each appoint not more than five commissioners to treat of peace." There was no peace then. There was a suspension of hostilities and a promise to cede, and a provision that commissioners should be appointed to treat of peace; but there was no peace, and no termination of hostilities, and no cession of Porto Rico; and if the two countries had failed to conclude a treaty of peace, or that treaty had failed of·ratification, the suspension of hostilities would have terminated and the war would have been resumed. So our claim is that during this entire period, until peace had been concluded, the President was in the legitimate exercise of the war power; and that brings me to another suggestion.

Counsel talk about peace, about there being no war in Porto Rico, about the protocol placing a limitation upon the power of the President acting under belligerent right. They assume that under the war power all the President can do is to fight. It is true the President makes war in order to win a peace, and to that end he fights, as commander in chief, and he invades the enemy's territory and subjugates it if he can, and he holds and occupies it. After he has conquered the enemy's territory, he stops fighting there because there is no one there to fight, but his power does not therefore cease under belligerent right. It then becomes his duty to occupy and hold this subjugated territory until disposed of by the treaty of peace, and in exercising that duty he should put in operation a government there that will cover the entire field of civil life, that will preserve order and protect life and property, and collect revenues sufficient to

pay the expenses of the provisional government he thus institutes. He has a right to provide courts; he has a right to provide courts, not to pass upon purely military questions, but on
all questions that arise between man and man, within the occupied territory. These propositions are so elementary it seems
to me hardly necessary to refer to the authorities. I may do
so later.

Now, I desire for but a moment to refer to the necessity in
this case of the President providing a new system of customs
regulations in Porto Rico. At the time the war began the commerce of Porto Rico was largely with Spain and with Cuba.
Necessarily, the customs regulations were framed so as to meet
that condition. When the war came and we occupied Porto
Rico, naturally this trade was cut off. It was an impossibility
then, having proper regard for the interests of the people of
Porto Rico, to continue in force, unmodified, the Spanish customs laws. The President therefore put in force new customs
regulations, and he changed them as developing circumstances
showed they ought to be changed in the interests of Porto Rico
and of the United States. He placed on the free list many
articles brought into Porto Rico from the United States. For
instance, all food supplies, implements of industry, machinery,
etc., and in every way he endeavored to put in operation there
a system of customs laws, enforced by the military authority,
which might, if necessary, be continued in force after the conclusion and ratification of a treaty of peace, and until Congress
should legislate for the island.

I refer in my brief to the cases of *Cross* v. *Harrison*, 16 How.
164, *Leitensdorfer* v. *Webb*, 20 How. 176, *The Grapeshot*, 9
Wall. 129, the *Mechanics' Bank* v. *The Union Bank*, 22 Wall.
276, and the *United States* v. *Rice*, 4 Wheat. 246, in support of
what the President did in Porto Rico with reference to customs
and revenues, both before and after the treaty of Paris was
made. In the case of *Cross* v. *Harrison*, the customs laws and
regulations for the conquered territory of California were first
put in operation by the President through the military commander. It was a war tariff, and that war tariff continued to
be enforced in California after the ratification of the treaty of

peace which, according to the contention of opposing counsel, made California a part of the United States. The war tariff, which was not the tariff then in force under the laws of the United States in the ports of the United States, was enforced until, I think, in August, 1848, when word was brought to California of the ratification of the treaty. Then there was substituted for that war tariff, by the order of the military commander, a tariff that was based upon and I suppose faithfully reproduced the provisions of the customs law then in force throughout the United States, and duties continued to be collected under that tariff until the arrival of agents of the Government authorized to put in force there the laws of the United States with reference to customs. But the court sustained the validity of the duties collected under all of these circumstances, even after the ratification of the treaty of peace. It held that the government which was rightfully instituted by the President under the law of belligerent right, continued in force necessarily and properly until another government should be substituted by Congress, and all the things done by the provisional government under authority of the President were sustained by the court in that case. The court said (p. 193): "The territory had been ceded as a conquest, and was to be preserved and governed as such until the sovereignty to which it had passed had legislated for it. That sovereignty was the United States, under the Constitution, by which power had been given to Congress to dispose of and make all needful rules and regulations respecting the territory or other property belonging to the United States, with the power also to admit new States into this Union, with only such limitations as are expressed in the section in which this power is given. The government, of which Colonel Mason was the executive, had its origin in the lawful exercise of a belligerent right over a conquered territory. It had been instituted during the war by the command of the President of the United States. It was the government when the territory was ceded as a conquest, and it did not cease as a matter of course or as a necessary consequence of the restoration of peace. The President might have dissolved it by withdrawing the army and navy officers who administered it, but he did not do so. Congress could have put an end to it, but that was not done. The right inference from the inactivity of both is, that it was meant to be continued until it had been legislatively changed. No presumption of a contrary intention can be made. Whatever may have been the causes of delay, it must be presumed that the delay was consistent with the true policy of the Government. And the more so as it was continued until the people of the Territory met in convention to form a state government, which was subsequently recognized by Congress under its power to admit new States into the Union."

And here is the conclusion of the case: " Our conclusion, from what has been said, is, that the civil government of California, organized as it was from a right of conquest, did not cease or become defunct in consequence of the signature of the treaty or from its ratification. We think it was continued over a ceded conquest, without any violation of the Constitution or laws of the United States, and that until Congress legislated for it, the duties upon foreign goods imported into San Francisco were legally demanded and lawfully received by Mr. Harrison, the collector of the port, who received his appointment, according to instructions from Washington, from Governor Mason."

In the argument so far I have briefly treated of the questions that arise from the importation into Porto Rico of goods from the United States, both before and after the treaty of peace, and before the taking effect of the Porto Rican act. Now, of course, there may be said to be involved in the collection of duties in Porto Rico on goods brought from the United States under the treaty of Paris, and before the Porto Rican act went into effect, a question similar to that which arises with regard to the exaction of duties on goods shipped into Porto Rico from the United States under the Porto Rican act. But I do not care to consider or discuss that question at this time. I prefer to take up and discuss the question which has been raised, and which in some respects is the vital question, as to the effect of the ratification of the treaty upon the relation of Porto Rico— and of course the Philippines—to the United States, because that is the primary question in these cases.

Counsel contend that upon the ratification of the treaty, and

upon the cession of Porto Rico to the United States, that territory became a part of the United States within the meaning of
the general grant of taxing power to the Federal Government,
subject to the limitation contained in that provision which requires " all duties, imposts, and excises to be uniform throughout the United States." In discussing the effect of the treaty,
I shall not repeat the historical argument so fully and elab-
orately presented by the Government in the discussion of the
Goetze case. I shall rather attempt, after going over the terms
of the treaty, to analyze the pertinent provisions of the Constitution of the United States, with a view of determining what
was the real meaning intended by the framers of the Constitution to be given to the words " the United States " used in that
connection.

Reduced to a legal proposition, the denial of the power which
has been exercised and is being exercised by the President and
by Congress in the new possessions, amounts to this : Ceded
territory becomes, by the act of cession, an integral part of the
United States, to which the Constitution of its own force at
once applies, placing its people, its products, and its ports on an
immediate equality with ours, and conferring upon them all the
rights, privileges, and immunities enjoyed under the Constitution by the people, the products, and the ports of the several
States. Moreover, the limitations of the Constitution apply
there as here, requiring the same taxes, duties, imposts, and excises to be collected, and the same Anglo-Saxon system of trial
by jury to be used. Their people become at once our people, citizens of the United States, our ports become their ports,
and our markets their markets. They are free to come here or
to sell their products here, while our taxes and our laws, however
unsuitable, must go there.

There is nothing obscure about this doctrine. It is plain and
unmistakable. The act of cession is all powerful ; its effect im-
mutable. As soon as the title passes, the territory is incorporated within the United States, and the Constitution *ex proprio
vigore* does the rest. The proposition is true as stated, or not
true at all. Either the mere act of cession, irrespective of the
terms of the treaty (which I shall consider later) and regardless

of the action of Congress, makes acquired territory a part of the United States in the constitutional sense, or it does not. If it does, the treaty-making power, in acquiring territory, so far as the status of that territory is concerned, is necessarily limited to providing for the mere act of cession. It can make no terms. It cannot take temporarily or provisionally, or for this purpose or that. It can give no pledges; it can grant no privileges; it can reserve no questions for future disposition; in short, although called the treaty-making power, and granted without limitation, it is stripped of its proper functions; it cannot treat; it is lame, impotent, impossible, ridiculous.

On the other hand, if the territory does not, by the mere act of cession, become immediately an integral part of the United States in the constitutional sense, of necessity the provisions of the treaty and the action of Congress must determine whether it shall or shall not become or be deemed a part of the United States, and, if ever, when. In other words, the acquired territory becomes not a part but a possession of the United States —territory, to use the language of the Constitution, belonging to the United States—and its disposition and government rest, under the Constitution, with the treaty-making power and with Congress.

MR. JUSTICE BROWN. If it be territory belonging to the United States, then does it fall within the provisions of the Dingley act, which requires duties to be assessed upon goods from foreign countries, or does it not cease to be a foreign country?

MR. SOLICITOR GENERAL. I think not; not within the meaning of the customs law. The Dingley law treated as foreign all territory outside of the limits of the United States, meaning the States and Territories then treated for customs purposes as the United States, and that condition remained until Congress saw fit to change it.

In the noted case of *Fleming* v. *Page*, 9 How. 614, Mr. Justice Taney says that "the United States may demand the cession of territory as the condition of peace, in order to indemnify its citizens for the injuries they have suffered or to reimburse the Government for the expenses of the war." And in this connection I might also refer to the language of Chief Justice

Marshall in the famous *Canter Case*, 1 Peters, 541, in which he says that acquired territory "becomes a part of the nation to which it is annexed either on the terms stipulated in the treaty of cession, or on such as the new master shall impose." And in the case of *Cross* v. *Harrison*, 16 How. 164, Mr. Justice Wayne uses this language (p. 197): "By the ratification of the treaty California became a part of the United States." So it did, in the international sense—in the legislative sense—subject to the dominion of the United States, to be ruled and regulated by Congress, under the power granted to make all needful rules and regulations respecting the territory belonging to the United States. And he continues: "And as there is nothing differently stipulated in the treaty with respect to commerce, it became instantly bound and privileged by the laws Congress had passed to raise a revenue from duties on imports and tonnage."

MR. JUSTICE BROWN. That case did not involve the question involved here of an importation from California to New York.

MR. SOLICITOR GENERAL. That is true.

MR. JUSTICE BROWN. It involved quite a different question. That involved a case of importation from an admittedly foreign country into the United States.

MR. SOLICITOR GENERAL. Yes, although the court did say, if I remember correctly, that if these goods had been allowed by the military authorities to enter California free of duty, then duty would have been exacted on them in the ports of the United States if taken there. Here is what the court says on page 192: "The best test of the correctness of what has just been said is this: That if such goods had been landed there duty free, they could not have been shipped to any other port in the United States without being liable to pay duty." Of course, California was contiguous territory, and it was very much better, as a matter of policy, to bring it as soon as possible within the operation of the customs laws of the United States, and that was what was done. But that does not apply to Porto Rico or the Philippines.

Now if territory may be acquired for the purposes, or any of the purposes, mentioned by Chief Justice Taney, it certainly may be taken and held upon such conditions as may be proper and necessary to carry the purpose into effect. Territory acquired to indemnify and reimburse may be taken and held as a pledge, or as a possession, provisionally, temporarily, or indefinitely, with the reserved power of disposition and control suitable to accomplish the desired end. To incorporate such territory into the Union and make it a part of the United States would defeat the very object of the acquisition. Once there it would have to stay, for no power exists within the Union to dismember it.

If Chief Justice Taney was wrong, and we cannot take territory *sub modo* to indemnify or reimburse us, but only to make it a part of the United States, then, before the President carries a war into the enemy's country, he should send ahead his advance agents—a commission to ascertain and report whether the territory he proposes to invade and subjugate is fit to be made a part of the United States. For observe, neither the treaty-making power nor Congress can, according to the contention of the other side, prevent that result if a cession follows conquest. Before the President sent Dewey to Manila he should have satisfied himself that the Philippines were suitable for incorporation into the Union, for we could destroy the Spanish power there only at the risk of having to assume the burdens of sovereignty ourselves.

The Constitution, while vesting in the President and Senate the treaty-making power, provides that: "This Constitution, and the laws of the United States which shall be made in pursuance thereof; and all treaties made, or which shall be made, under the authority of the United States, shall be the supreme law of the land." The treaty of Paris was made under the authority of the United States, and contains the terms upon which we acquired these territories. It is unique in this, that while former treaties of cession all provided that the civilized inhabitants of the ceded territories should ultimately — not immediately, but ultimately—become citizens of the United States, and be incorporated in the United States, this treaty left the determination of their civil rights and political status to Congress.

Mr. Justice Harlan. State that proposition again.

Mr. Solicitor General. I say that the treaty of Paris is unique in this, that while former treaties, such as the Florida treaty, the Louisiana treaty, and others, provided that the civilized inhabitants of the ceded territories should ultimately—not immediately, but ultimately, in the course of time—become citizens of the United States, this treaty, the treaty of Paris, left the determination of their civil rights and political status to Congress.

Mr. Justice Harlan. What treaty has used the word "civilized"?

Mr. Solicitor General. I do not assume to quote the precise language of the particular treaties, but simply state the effect of them.

Mr. Justice Shiras. The treaty with Russia used that term.

Mr. Solicitor General. Yes, the Alaskan treaty does use it. It distinguishes the uncivilized tribes there.

Let me refer to some of the provisions of the treaty of Paris. Spain ceded to the United States the island of Porto Rico, the island of Guam, and the archipelago known as the Philippine Islands. Spanish subjects, natives of the Peninsula, residing in such territories, were given one year from the exchange of the ratifications—that is, until April 11, 1900—to preserve their allegiance to Spain by making a declaration in a court of record. In default of this they were to be held to have renounced it and to have adopted the nationality of the territory in which they may reside—not to have adopted the nationality of the United States, to which the treaty ceded the islands, but to have adopted the nationality of the territory in which they may reside. Then directly after that comes this provision : " The civil rights and political status of the native inhabitants of the territories hereby ceded to the United States shall be determined by the Congress."

Spaniards residing in the territories were to be subject, under Article XI, to the jurisdiction of the courts of the country—not the courts of the United States—pursuant to the ordinary laws governing the same—presumably the Spanish or civil law—and were to have the right to appear and pursue the same course

therein "as citizens of the country to which the courts belong"
—not as citizens of the United States. Article IV reads as follows: "The United States will, for the term of ten years from
the date of the exchange of the ratifications of the present
treaty, admit Spanish ships and merchandise to the ports of the
Philippine Islands on the same terms as ships and merchandise
of the United States."

With regard to this, it is obvious that, unless a separate system of customs regulations is adopted for the Philippines, which
applies to goods shipped into the Philippines from the United
States, then the treaty, if observed, throws open the ports of
the Philippines absolutely to Spanish ships and Spanish importations, and provides an open door into the Philippines, and
thence into this country, for whatever goods Spain sees fit to
send there. I do not intend to pursue an argument of policy
based upon this provision, but simply call attention to the fact
that the treaty itself negatives the view that these islands were
to become a part of the United States within the meaning of
our customs laws.

Again, for ten years Spanish scientific, literary, and artistic
works were to be admitted free of duty into all the ceded territories, and that provision, as counsel has stated, has been incorporated into the Porto Rican act, for the purpose of carrying out the pledge of the treaty. In short, neither of these
provisions can be carried out if the Constitution requires our
customs regulations to apply in those islands as here in the
United States.

The purpose of these provisions is plain. Although under
the power and protection of the United States, the territories
are to have their own laws, their own courts, their own ports,
their own commerce, their own citizenship, their own system
of revenue. A separate and distinct existence under, but without, the United States, in the purely constitutional sense, as
used in the general grant of taxing power, is contemplated.
The parties to the treaty both knew that the location and condition of these islands would not permit their incorporation into
the United States and the application to them of those laws of
commerce, of revenue, and of civil and criminal procedure which

the Constitution, according to the contention of opposing counsel, requires to be uniform throughout the United States. They
provided, therefore, for a system of government which should
be adapted to local conditions and needs.

Now, are we free to disregard the plain provisions of the
treaty, which the Constitution says shall be the supreme law of
the land? If so, what becomes of the consent of the treaty-
making power to the acquisition? Would the President and
the Senate have consented to take the territories upon any other
terms? Would Spain have consented to cede them? Certainly
the treaty never intended to make these tropical islands, with
their savage and half-civilized and civilized people, a part of
the United States in the constitutional sense, and just as certainly did make them a part of the United States in the international sense.

Mr. Justice Harlan. What do you mean by the international
sense?

Mr. Solicitor General. I am just going to explain. The term
"the United States" may mean the territory which governs,
or the territory over which the Government extends. The
former is the constitutional, the latter the international, or, it
may be, the legislative sense. In the latter sense, in the international or legislative sense, States and Territories, all places
subject to the jurisdiction of the national power, combine to
constitute what Chief Justice Marshall in *Loughborough* v.
Blake, 5 Wheaton, 319, termed "The American Empire," "Our
Great Republic." "Does this term," said he, referring to "the
United States," "designate the whole or any particular portion
of the American empire? It is the name given to our great
republic, which is composed of States and Territories." The
great Chief Justice was clearly correct in holding that the taxing power extends throughout the United States in the international or legislative sense, although the limitation of the Constitution on the taxing power for Federal purposes applies, as we
contend, only throughout the United States in the constitutional
sense. What we are concerned with is, of course, the constitutional sense. For the vital question is whether the constitutional limitation upon the Federal taxing power which applies

"throughout the United States" operates in the new territories. As stated in the preamble—

MR. JUSTICE PECKHAM. Do you find any case where any such distinction has been drawn as you make now—between the United States in the constitutional sense and the United States in the international sense ?

MR. SOLICITOR GENERAL. I think I could if it were desirable. I am going on to show what these words "the United States" mean in the constitutional sense. I think it perfectly apparent that the phrase "the United States" in the international sense comprehends all territory which is subject to our dominion.

MR. JUSTICE PECKHAM. Yes; I understand what you state, but my question was whether you have in mind, or had come across in your research, any case in which such a distinction was drawn, between the United States in the constitutional sense and the United States in the international sense.

MR. SOLICITOR GENERAL. The distinction has been clearly drawn in a decision of this court between the word "State" as used in the Constitution and the word "State" as used in a treaty, in the international sense. Thus, it was held in *Geofrey* v. *Riggs*, 133 U. S. 258, that the District of Columbia is a "State" in the international sense, but certainly it is not a State within the meaning of the Constitution. That has been expressly held in *Hepburn* v. *Ellzey*, 2 Cranch, 445.

As stated in its preamble, the Constitution of the United States was ordained and established by "the people of the United States" "for the United States of America." There is no ambiguity about the meaning of the words "United States of America," as here used. They mean the States united under the Constitution, and are named individually in the second section of the first article, relating to the apportionment of representatives among the then existing United States.

MR. JUSTICE HARLAN. The existing United States—those constituting the existing United States ?

MR. SOLICITOR GENERAL. No, I did not say that. I said that the United States which framed and adopted the Constitution are named specifically in the Constitution at the place stated. They were the thirteen colonies which had first become the

United States in the Declaration and under the Confederation, and which, through their people, framed the present Constitution, in order, among other things, "to form a more perfect Union." There never was any doubt in those days as to what that term meant. This conclusively appears from the sixth article, which provides that all debts contracted before the adoption of the Constitution "shall be as valid against the United States under the Constitution as under the Confederation."

MR. JUSTICE HARLAN. And that would include the States, of course, which afterwards came into the Union before the debts were paid?

MR. SOLICITOR GENERAL. You could hardly say that they were "under the Confederation." They were not "United States under the Confederation." Undoubtedly the debts would be valid against the United States, including the States which were subsequently admitted.

MR. JUSTICE WHITE. Do you make a distinction in your mind or is there any distinction, from the consideration which you have given to this case, between the States and the Territories of the United States, and the States and the territory of the United States? Does not "the territories" in these cases which you have quoted from refer to territories in which Congress has organized a government, thus making them impliedly a part of the United States? Does not the article of the Constitution giving power to dispose of the "territory" suggest a distinction between the Territories which have been organized, and "territory" belonging to the United States as such?

MR. SOLICITOR GENERAL. Does your honor mean to ask me whether territories subsequently acquired came within the power thus granted to Congress to make all needful rules and regulations for the government of the territory of the United States, or is it confined simply to the territory which existed at the time of the adoption of the Constitution, outside of the thirteen States?

MR. JUSTICE WHITE. You quoted the language of Chief Justice Marshall in *Loughborough* v. *Blake*, and then you speak of the United States in the constitutional and the international sense of the words "United States." But that language of

Chief Justice Marshall, in which he spoke of "Our Great Republic," "The American Empire," was used with reference to the exercise of the taxing power.

MR. SOLICITOR GENERAL. I know it was. He was correct, as I take it, in his conclusion that the taxing power of the United States extends over all the territory belonging to the United States; that it extends over all the States and Territories if Congress sees fit to exercise it. But I think what he says—which is the basis of the claim that the limitation that duties, excises, and imposts shall be uniform throughout the United States, applies to the Territories as well as the States—was not requisite to the decision of the case before him, and I am endeavoring to argue was incorrect.

MR. JUSTICE WHITE. That is my question. My question was to ascertain whether you were challenging the statement of Chief Justice Marshall in that case or whether you were concurring in it.

MR. SOLICITOR GENERAL. I have to challenge it.

MR. JUSTICE BROWN. The general expression, you mean?

MR. SOLICITOR GENERAL. I say looked at from the point of view of the decision he was correct, because in a geographical sense "the United States," throughout which Congress may exercise the taxing power for Federal purposes, includes necessarily all territory subject to the dominion of the United States. Now, that is the international or legislative sense. But I submit the constitutional sense covers only the States, and was so intended by the framers of the Constitution.

The primary source of the sovereign power was the people of the thirteen original States. These men believed they were forming a government which would endure for ages, and would dominate a continent, and probably territory outside—islands beyond the seas. In the treaty of alliance which Benjamin Franklin concluded with France, in 1778, there was this provision in the fifth section:

"If the United States should think fit to attempt the reduction of the British power remaining in the northern parts of America, or the islands of Bermudas, those countries or islands in case of success, shall be confederated with, or dependent upon the said United States."

So from that we can see how far-reaching was the vision of the stalwart men of the early days. Now, notwithstanding this expansive outlook, it does not appear that the fathers of the Constitution worried themselves about "the consent of the governed" outside of the States they lived in, which alone were to participate in political power. They formed a government in which the people of the States were alone represented and adopted a Constitution which, in its distribution and limitation of powers, applied almost wholly to the States, united or several.

In the early case of *Hepburn* v. *Ellzey*, 2 Cranch, 445, the question came before the Supreme Court whether a citizen of the District of Columbia could maintain an action against a citizen of Virginia. In support of the jurisdiction Mr. Lee insisted that to give the term "State" a limited construction would deprive the citizens of the District of the general rights of citizens of the United States and put them in a worse condition than aliens; and he put the pertinent question whether, in the face of the provision that "no tax or duty shall be laid on any articles imported from any State," Congress could levy a tax or duty on articles exported from the District of Columbia. But the court properly held that a citizen of the District is not a citizen of a State and cannot use the United States courts as such, Chief Justice Marshall saying: "The members of the American confederacy only are the States contemplated in the Constitution."

Yesterday, in connection with a quotation which I made from the case of *Loughborough* v. *Blake*, Mr. Justice White put to me a question in which he desired my opinion as to whether I recognized any difference between the words "the Territories" as used by Chief Justice Marshall and "the territory" which the Constitution places under the disposition of Congress. I did not hear the question distinctly nor comprehend the full purport of it. I do not recognize that the power of Congress over territory belonging to the United States ceases when such territory is organized and brought under the operation of the laws of the United States; but I do recognize a distinction between unorganized territory and the territories to which Chief

Justice Marshall may possibly have referred. If I gave the court the impression that I intended to say that, in using those words, Chief Justice Marshall referred to the States and Territories, meaning thereby to cover all territory under the dominion of the United States, which I had defined, whether correctly or incorrectly, as the international meaning, I think I was wrong. I am inclined to think that what Chief Justice Marshall had in mind was " the United States" in the legislative sense, meaning thereby the States of the Union, the District of Columbia, and the organized Territories, to which Congress had applied the revenue laws of the United States, thus including all that territory within the phrase " the United States," as designating the territory to which Congress had applied the revenue laws of the United States. So, really, there are four meanings which may be conveyed by the phrase " the United States."

In the first place, it may mean the sovereignty itself, what Chief Justice Marshall called " that grand corporation."

In the second place, it may mean, geographically, what Chief Justice Marshall calls " the American Confederacy," composed of the members of the Union, the States inhabited by the people who participate in the Government of the United States; and this is what I have termed the constitutional sense.

In the third place, in a geographical and legislative sense, it may mean the States and the District of Columbia and the Territories, which Congress has seen fit to treat as the United States for legislative purposes; over which Congress has extended, and to which it has applied, the laws of the United States which are applicable.

And in the fourth place, it may mean something broader, which is the international sense, as I take it; that is, all territory, wherever situated, under the dominion of the United States, whether organized or not, and whether ever brought within the operation of the specific laws of the United States. And our claim is that newly acquired territory does not become a part of the United States in the legislative sense until Congress shall so determine.

In the case of *Hepburn* v: *Ellzey*, 2 Cranch, 452, in which Marshall, C. J., defined the " American Confederacy," he said:

" The members of the American Confederacy only are the States contemplated in the Constitution. The House of Representatives is to be composed of members chosen by the people of the several States; and each State shall have at least one Representative. The Senate of the United States shall be composed of two Senators from each State. Each State shall appoint, for the election of the Executive, a number of electors equal to the whole number of Senators and Representatives. These clauses show that the word ' State' is used in the Constitution as designating a member of the Union."

The States alone are the members of the American Confederacy. They constitute the Union, and the Union and 'the United States are equivalent terms in the Constitution. Thus the Constitution and " the laws of the United States " are made the supreme law of the land; yet Congress is to provide for calling forth the militia to execute " the laws of the Union." All legislative powers granted are vested in the Congress " of the United States," but the President is required from time to time to give to the Congress information of the state " of the Union."

In the first article, defining the legislative powers, it is provided that Representatives and direct taxes shall be apportioned "among the several States which may be included within this Union." This does not include the Territories, but does operate, evidently, throughout the United States.

Duties, imposts, and excises shall be uniform "throughout the United States." This, as we claim, is a geographical limitation, requiring indirect taxes to operate generally throughout the United States— that is, among the several States composing the Union. The history of the adoption of this provision will be found in interesting form in the learned opinion of Mr. Justice White in the case of *Knowlton* v. *Moore*, 178 U. S. 41, sustaining the constitutionality of the Federal tax on legacies. In the original draft the provision prohibiting any preference to the ports of one State over those of another, and that conferring and limiting the taxing power, were placed together. They really mean the same thing, that the States of the Union shall be treated alike in the regulation of commerce and the imposition of taxes. The uniformity required in each case was a uniformity among the several States of the Union, and this is shown by the decision in the *Cherokee Tobacco Case,* 11 Wallace, 616, affirming the constitutionality of the act of 1868 extending the excise tax on liquors and tobacco alone to the Indian Territory. A minority of the court held that, in view of the treaty provisions, it was not the intention of Congress to extend even the tax on liquor and tobacco to the Indian Territory. Obviously, the court was unanimous in the opinion that, although the Indian Territory was within the exterior boundaries of the United States, the provision of the Constitution requiring excises to be uniform throughout the United States did not apply within the Indian Territory.

The Constitution gives Congress power to regulate commerce " among the several States," and to establish a uniform rule of naturalization and uniform laws on the subject of bankruptcy " throughout the United States." Now, we submit that this latter was to remedy the mischief resulting from the diverse and conflicting legislation of the several States upon these subjects by securing uniform provisions throughout the States of the Union. I refer to No. 41 of the Federalist, written by Mr. Madison, upon that point, in which he says such was the object of that provision. The early laws of this character applied only within the States. The recent acts have properly been extended to the Territories, which Congress in its discretion has seen fit to include within the limits of the United States, legislatively treated.

It is provided that " no tax or duty shall be laid on articles exported from any State;" but nothing is said about any Territory. And that "no preference shall be given by any regulation of commerce or revenue to the ports of one State over those of another; nor shall vessels bound to or from one State be obliged to enter, clear, or pay duties in another;" but nothing is said about the ports of any Territory.

The prohibitions of the tenth section of the first article apply only to the States. "No State shall pass any bill of attainder or *ex post facto* law, or law impairing the obligation of contracts, or grant any title of nobility. No State shall, without the consent of Congress, lay any imposts or duties on imports or exports," etc. All these limitations apply only to the States of the Union.

In the second article, which grants and defines the Executive power, it is provided that Congress may determine the date on which the electors shall give their votes, which day shall be the same " throughout the United States." Necessarily, the United States here means the States of the Union which alone take part in electing the President. Later, it is provided, that during his term of office the President shall not receive, in addition to his stated compensation, any other emolument from " the United States or any of them," showing that the States united were alone in mind.

MR. JUSTICE BREWER. Do you think in that connection that the various Territories can add to the President's salary ; in view of that, can the various Territories add to the emoluments of the President?

MR. SOLICITOR GENERAL. No, I think the spirit of this would prevent that. I think there is no direct application to the Territories, but I dare say the spirit of it would forbid what you suggest. Territorial action might, in a certain sense, be treated as the action of the United States, seeing that a Territory could not act outside of the authority of the United States, being un- der the complete control of Congress. It might, in a certain sense, be treated as the action of the United States, if a Territory attempted to do that. However, I prefer to say that the general spirit of this provision applies and would prevent what is suggested by your honor.

The third article applies to the judicial power of the United States. It has been repeatedly held that the territorial courts are not organized under this article, and are, therefore, not courts of the United States. The article constantly keeps in mind the relation of the United States to the several States, and of those States and their citizens to one another. No mention is made of the Territories or their citizens.

The fourth article guards the rights of each State and its citizens with respect to every other State. The public acts of each shall have full faith and credit in all others. The citizens of each shall be entitled to the privileges and immunities of the citizens in the several States. Fugitives from justice shall be surrendered; new States may be admitted into " this Union ;" and a republican form of government to every State in the Union is guaranteed. But there is no safeguard or guarantee whatever in the case of a Territory and its citizens. No republican form of government for the Territories is guaranteed. On the contrary, just preceding the guarantee to the States, and following the provision for the admission of new States, the following grant of plenary power is made:

" Congress shall have power to dispose of and make all needful rules and regulations respecting the territory or other property belonging to the United States."

Notice the phraseology. Territory is treated as property, as something distinct from the United States—something belonging to the United States, a subject to be ruled and disposed of by Congress in its discretion as conditions may require, without being hampered by the restrictions which were framed for the States.

MR. JUSTICE BREWER. Right there, do you understand that Congress has absolute power over territory acquired, to do as it pleases with it?

MR. SOLICITOR GENERAL. No; I deny that utterly, as I shall show to your honor.

MR. JUSTICE BREWER. What limitations?

MR. SOLICITOR GENERAL. I shall point out specifically the limitations later. I say that Congress is subject to all applicable limitations, and I shall point out later what I mean by applicable limitations, in view of the decisions of this court.

In the case of *McCulloch* v. *Maryland*, 4 Wheaton, 442, in which the supremacy of the United States within the sphere of its action was sustained, Chief Justice Marshall, emphasizing the authority conferred on Congress to select the means for carrying into execution the powers vested by the Constitution, said: " The power to make all needful rules and regulations respecting the territory or other property belonging to the United States is not more comprehensive than the power to make all laws which shall be necessary and proper for carrying into execution the powers of the Government."

Apparently, he took the territorial grant as the test and standard of plenary power, as the maximum of comprehensiveness.

The Thirteenth Amendment contains an explicit recognition of the fact that a place subject to the jurisdiction of the United States is not necessarily a part of the United States, for it provides: "Neither slavery nor involuntary servitude, except as a punishment for crime whereof the party shall have been duly convicted, shall exist within the United States, or any place subject to their jurisdiction."

In this connection, in addition to the many instances cited by the Attorney General where Congress has drawn a distinction between the United States and the Territories, let me refer to the act of March 2, 1807, 2 Stat. 426, prohibiting the importation of slaves into this country. That act provided that it should be unlawful for any person to import or bring from any foreign country any slaves—now, I am quoting—"into the United States or the Territories thereof." And in the subsequent act of 1818, 3 Stat. 450, which supplemented this act, the same phraseology was used, the first section providing that it should be unlawful to import any negroes "into the United States or Territories thereof."

And as illustrating the fact that this court has drawn a distinction between the rights before this court of Territories and territorial legislation, as distinguished from States and state legislation, I wish to refer the court to the case of *Miner's Bank* v. *Iowa*, 12 Howard, 1, in which the court held that the validity of a territorial act repealing the charter of a bank granted by a Territory, could not be brought before the Supreme Court, under the twenty-fifth section of the judiciary act, either on the ground that there was drawn in question the validity of a statute of, or an authority exercised under, any State, or on the ground that there was drawn in question the validity of a statute or authority exercised under the authority of the United States. In holding that there was not drawn in question the validity of an act passed by a State, Mr. Justice Daniel, speaking for the court, said (p. 7): "In order to give this court jurisdiction, the statute, the validity of which is drawn in question, must be passed by a State, a member of the Union, and a public body owing obedience and conformity to its Constitution and laws. That if public bodies, not duly admitted into the Union, undertake as States, to pass laws which might encroach on the Union or its granted powers, such conduct would have to be reached either by the power of the Government to put down insurrection or by the ordinary penal laws of the States and Territories within which these bodies are situated and acting; but their measures are not examinable by this court upon a writ of error. They are not States, and cannot pass statutes within the meaning of the judiciary acts.

"Other cases cited by the court, in the opinion just quoted [referring to the case of *Scott* v. *Jones*, in the 5th Howard], might be adduced to show the difference ever taken by the court in reference to its relation to the States as States, and as contradistinguished from the Territories of the United States. It seems to us, that the control of these territorial governments properly appertains to that branch of the Government which creates and can change or modify them to meet its views of public policy, viz., the Congress of the United States. That control certainly has not been vested in this court, either in mode or substance, by the twenty-fifth section of the judiciary act."

In holding that the territorial charter could not be regarded as an act of Congress, the court said: "The charter of the Bank of Dubuque enacted in all its details and powers ever possessed by it (and according to which it was in fact organized) by the legislature of Wisconsin, must be looked upon as the creature of that legislature. To regard it as we are urged to do by the argument of the plaintiff in error, would constitute it rather a bank of the United States, situated without the United States, and operating within the Territory of Wisconsin."

And I think in the opinion the court will find the word "without" italicized—"*without* the United States."

I believe that a careful examination of the Constitution leads but to one conclusion, that the power of Congress over the Territories is plenary and absolute. Whether it follows from the power to acquire and hold territory, or is conferred by the clause of the Constitution which declares that "Congress shall have power to dispose of and make all needful rules and regulations respecting the territory or other property belonging to the United States," it is full and complete, and is unhampered by those limitations and restrictions which were intended to apply only within the States of the Union.

There is a line of decisions of the Supreme Court running back to the early days which sustains this view. Some years after the decision in *Loughborough* v. *Blake*, the case of *Insurance Company* v. *Canter*, 1 Pet. 511, came before the Supreme Court, over which Chief Justice Marshall still presided. A court of the Territory of Florida, composed of a notary and five jurors, had sold a wrecked cargo of cotton on a salvage claim and transferred the title to Canter, the purchaser. It was insisted that upon the acquisition of Florida it became a part of the United States over which the Constitution extended, and that under the Constitution admiralty jurisdiction could be exercised only by the courts of the United States. It had to be conceded that the territorial court was not organized in accordance with the Constitution, which requires judges to be appointed for service during good behavior. The opinion of Chief Justice Marshall is worthy of careful study. Its logic is unanswerable. While the power of Congress to govern ceded territory was declared to be inevitable and absolute, the limitations of the Constitution upon the exercise of the judicial power of the United States was expressly held to be confined to the States, the Chief Justice saying (p. 545): "Although admiralty jurisdiction can be exercised in the States in those courts, only, which are established in pursuance of the third article of the Constitution, the same restriction does not extend to the Territories. In legislating for them, Congress exercises the combined powers of the General and of a state Government."

The doctrine thus enunciated by the great Chief Justice has been approved and followed by his successors in a long line of cases, I think all of which were cited by the Attorney General. Note the language used. Chief Justice Waite speaks of the Territories as "the outlying dominion of the United States"

101 U. S. 129, 133—an apt phrase. " The outlying dominion ! "
Lying outside of what ? Outside of the governing body—the
United States. The " outlying dominion of the United States,"
not a part of the United States. He says that Congress "may
do for the Territories what the people, under the Constitution
of the United States may do for the States," the fullest and
clearest expression of Constitutional power without limitation.

MR. JUSTICE HARLAN. Please read that again.

MR. SOLICITOR GENERAL. That Congress " may do for the Territories what the people, under the Constitution of the United
States, may do for the States." Can there be any fuller expression of plenary power than that ? Mr. Justice Matthews says
that " the people of the United States, as sovereign owners of
the National Territories, have supreme power over them and
their inhabitants." " It rests with Congress to say whether, in
a given case, any of the people, resident of the Territory, shall
participate in the election of its officers, or the making of its
laws." 114 U. S. 15, 44. In other words, Congress can at any
time repeal an act giving local government to a Territory, and
take the authority to itself. Mr. Justice Bradley says that " It
would be absurd to hold that the United States has power to
acquire territory and no power to govern it when acquired."
136 U. S. 1, 42. And Mr. Justice Harlan says that " The whole
subject of the organization of the territorial courts, etc., was
left by the Constitution with Congress, under this plenary power
over the Territories of the United States." 141 U. S. 174, 188.
And then he inquires, " Has Congress, under ' the general right
of sovereignty ' existing in the Government of the United States
as to all matters submitted to its exclusive control, including
the making of needful rules and regulations respecting the Territories of the United States, any less power over the judges of
the Territories than a State, if unrestrained by its organic law,
might exercise over the judges of its own creation ? " 141 U. S.
174, 1890. And Mr. Justice Gray says that, " By the Constitution, as is now well settled, the United States, having rightfully
acquired the Territories, and being the only Government which
can impose laws upon them, has the entire dominion and sover-

eignty, national and municipal, Federal and state, over all the Territories, so long as they remain in a territorial condition."

And now I come to the subject of limitations. Are there no limitations on this plenary power of Congress to govern the Territories? I believe there are. If there are any who believe that the President or Congress can govern the new possessions outside of the Constitution, and wholly irrespective of all its limitations, I am not of them. Neither the executive, nor the legislative, nor the judicial branches of the Federal Government can act except through a power conferred by the Constitution. Wherever a particular power is exercised the limitation placed upon it by the Constitution must be observed. The Constitution was formed by the people of the thirteen original States. They provided the Government, conferred upon it certain powers, and subjected it in the exercise of some of these powers to certain limitations. It expressly prohibited the exercise of certain powers under any circumstances, and wholly irrespective of the place where exercised. Moreover, since certain powers were reserved to the States composing the Union, certain limitations and prohibitions were laid upon the States. In any case involving the exercise of a power claimed under the Constitution, the first question is, Was the power granted? and the next is, What are the limitations?

The difficulty of a clear conception of the important question in these cases has been increased by the use of campaign catchwords, of political phrases. "The Constitution follows the flag" is one of these. It is made use of to induce people to believe that the Government is contending that the President and Congress, in dealing with the new possessions, avowedly act outside of the Constitution; that the Government claims that the Constitution stays here, within the United States, leaving the President and Congress power unlimited and despotic with respect to the new possessions. This claim is designed and calculated to put both the President and Congress in a position obnoxious to a liberty-loving people. The position is one they have never taken and do not now occupy. Both the President and Congress concede, as I understand it, that they have no power except under the Constitution, and that they are subject in the exercise of their powers to every limitation properly applicable. The Constitution and the flag go together. Wherever the flag flies as the symbol of the sovereignty of this country it is raised by an authority created and existing under the Constitution. The flag now floats in the Philippines by virtue of the war and treaty-making power through which we have acquired that territory. It was raised in Porto Rico under the same authority. It waves there now as the symbol of the sovereignty of the Republic over rightfully acquired territory, which the Constitution expressly intrusts the regulation and disposition of to Congress. The Constitution is in force in the Philippines and is in force in Porto Rico, but not all of its provisions. Only those provisions operate there, or operate on Congress in legislating for the new possessions, which the framers of the Constitution intended should apply. Opposing counsel speak of the Constitution as if all of its provisions apply everywhere throughout the scope of the authority of the government it creates. This is not true. The United States, in the broadest sense, is composed of States and Territories, organized and unorganized. There are certain prohibitions and limitations which clearly apply only to the States as bodies politic. They were not intended to and do not apply to the Federal Government at all. There are other limitations which apply to the General Government when acting within the States united under the Constitution. There are other limitations which apply both throughout the States and the Territories, organized and unorganized. There are other limitations which apply everywhere, both within and without the United States in the broadest sense. So, after all, it is a question of the scope and application of specific limitations. Because an inapplicable limitation is not in force in the new possessions, it does not follow that applicable prohibitions and limitations can or would be ignored.

To repeat, the United States of America—which Chief Justice Marshall, in *Dixon* v. *The United States*, said is "the true name of that grand corporation which the American people have formed, and the charter will, I trust, long remain in full force and vigor"—is a body politic, of which the States alone are integral constituent parts, they only, as the same Chief Justice said in *Hepburn* v. *Ellzey*, being "the members of the American Confederacy," and this governing entity exercises sovereignty over "the American Empire," "our Great Republic," which is composed of States and Territories—and, in the broadest sense, if he does not mean by this, territory unorganized, then over that too. The Territories are not integral parts but possessions of this "grand corporation." The governing unit, composed of the States, possesses and exercises dominion over the Territories, subject only to the applicable restrictions and limitations of the Constitution. All the provisions of the Constitution do not and cannot have uniform operation both within the States and Territories whose political *status* and relation to the governing body are so widely different. It is true that every part of the national domain is within the jurisdiction of the Constitution, but it does not follow that every part is subject to all of its provisions. Each part is subject to some one or more of them, but all parts are not subject to all of them.

The Territories, not being parts, but possessions, of the governing body, are not within the scope or purpose of those limitations and restrictions which were designed to preserve and protect the rights of the States composing the Union. In legislating for the Territories Congress is not limited to jealously guarded national powers, but exercises the combined powers of the General and of a state Government.

Mr. Justice Harlan. Where is the *Dixon* case you referred to?

Mr. Solicitor General. In 1 Brockenbrough, 177. It was a case decided on the circuit.

The safeguard when Congress thus acts outside of those limitations to which I am going now to refer, and which I regard as applicable, is what Chief Justice Marshall refers to in *Gibbons* v. *Ogden*, 9 Wheaton, 1, where, meeting the objection that, according to the position taken by counsel for the Government, despotic power was given by the clause authorizing Congress to regulate commerce among the several States, he said (p. 197): "The wisdom and the discretion of Congress, their identity with the people, and the influence which their constituents possess at elections are, in this, as in many other instances—as that, for example, of declaring war—the sole restraints on which they have relied to secure them from its abuse. They are the restraints on which the people must often rely solely in all representative governments.

But there are limitations which apply to Congress in exercising the territorial grant. Obviously those limitations which are laid upon the exercise by Congress of a special power, irrespective of the place where exercised, do apply, such as those forbidding Congress to pass any bill of attainder, or any *ex post facto* law, or confer any title of nobility. These, as Madison said in No. 43 of the Federalist, are contrary to the first principles of the social compact. The prohibition of slavery operates by express provision everywhere. But these are not the only limitations. It is always to be borne in mind that this is a Government framed by the people, among other things, to establish justice and to secure the blessings of liberty. A Government thus dedicated to liberty and justice is based on fundamental principles, and at all times must show respect for fundamental rights. This, I take it, is what Mr. Justice Bradley meant when he said in the *Mormon Church Case*, 136 U. S. 44—
"Doubtless Congress, in legislating for the Territories, would be subject to those fundamental limitations in favor of personal rights which are formulated in the Constitution and its amendments; but these limitations would exist rather by inference and the general spirit of the Constitution from which Congress derives all its powers, than by any express and direct application of its provisions."

And obviously it was to this that Mr. Justice Harlan, speaking for the court, referred in *McAllister* v. *United States*. 141 U. S. 188, when he said:

"How far the exercise of that power [the power to govern the Territories] is restrained by the essential principles upon which our system of government rests, and which are embodied in the Constitution, we need not stop to inquire."

MR. JUSTICE BROWN. Can Congress take private property for public use without compensation in the Territories?

MR. SOLICITOR GENERAL. Well, I suppose the court will have to

define the fundamental limitations. I do not think I can. The
court has not categorically stated them as yet. The court has
contented itself with saying there are fundamental principles
embodied in the Constitution.

MR. JUSTICE BROWN. You prefer the court should define the
limitations and do not care to state them yourself ? [Laughter.]

MR. SOLICITOR GENERAL. I prefer to have the court define the
limitations rather than try to do so myself. I think it would
be presumptuous in me to act as pioneer in this matter. I am
content to follow the court.

The Government has never asserted, and does not believe,
that Congress has the power of a despot in Porto Rico. The
fundamental limitations in favor of personal rights which are
formulated in the Constitution and its amendments, referred to
by Mr. Justice Bradley, stand in the way of everything suggested which shocks the moral sense. Congress could not pass
any *ex post facto* law, or declare an attainder, or grant any title
of nobility, or provide for the trial or punishment of treason in
any other way than that marked out in the Constitution, all
these things being prohibited by direct and applicable provisions.
If the first ten Amendments do not limit by direct application
Congress in legislating for our new possessions—I put this as a
possible case—neither do they operate within the States which
compose the Union. As this court, speaking by Mr. Justice
Waite, said in *United States* v. *Cruikshank*, 92 U. S. 552: "The
first Amendment to the Constitution prohibits Congress from
abridging "the right of the people to assemble and to petition
the Government for a redress of grievances." This, like the
other Amendments proposed and adopted at the same time, was
not intended to limit the powers of the state governments in
respect to their own citizens, but to operate upon the National
Government alone."

" Protection to life, liberty, and property rests primarily with
the States," as Chief Justice Fuller said in *In re Kemmler*, 136
U. S. 448. " The Constitution makes no provision for protecting the citizens of the different States in their religious liberties;
this is left to the state constitutions and laws," said Mr. Justice
Catron, speaking for the court in *Permoli* v. *First Municipality*, 3 How. 609.

The Constitution forbids the States to pass any bill of attainder, *ex post facto* law, or law impairing the obligation of contracts, or to grant any title of nobility, and the Fourteenth Amendment provides that "no State shall deprive any person of life, liberty, or property without due process of law, nor deny to any person within its jurisdiction the equal protection of the laws;" but outside the range of these limitations the people of the State, through its constitution and laws, are supreme. They can define treason against the State as they see fit; they can limit the freedom of speech and of the press; they can restrict the bearing of arms; they can provide for the quartering of troops.

MR. JUSTICE HARLAN. Could a State have an established religion?

MR. SOLICITOR GENERAL. I have already read what the court said in regard to that in connection with the First Amendment. That question came before this court in the *Permoli* case, and the court said that the Constitution makes no provision for protecting the citizens of the respective States in their religious liberties.

MR. JUSTICE HARLAN. What does the word "liberty" in the Fourteenth Amendment mean?

MR. SOLICITOR GENERAL. That is a broad question which the court has not yet fully answered. I stand by the decision of the court upon a specific point, and if that is overruled by a general expression, I must yield.

MR. JUSTICE HARLAN. What would you say as to an act of Congress which absolutely forbade all trade between Porto Rico and the States? If Congress could not do that, what is the provision of the Federal Constitution that would stand in the way?

MR. SOLICITOR GENERAL. I think Congress could, if it saw fit, prohibit all trade.

MR. JUSTICE HARLAN. And could prohibit the people in that country from coming here at all, to the States?

MR. SOLICITOR GENERAL. I am disposed to think that goes along with the other. I will, however, discuss that phase of the question later. But let me say here, with respect to these ex-

treme illustrations of what might be done under a claimed
power, that I understand this court has repeatedly taken the position that although a certain thing is not expressly prohibited,
still if it is arbitrary and tyrannical, destructive of fundamental
rights, and, therefore, opposed to fundamental principles, the
court will find a way to protect the people against it.   In the
opinions of this court, where power in Congress has been upheld, carefully guarded language has been used, so as to leave
the court free to protect the people, in case Congress should
exercise such power in a way destructive of fundamental rights.
Thus, in the case of *Knowlton* v. *Moore*, in which the court upheld the graded feature of the legacy tax, the following language is used, (178 U. S. 109): "The grave consequences which it
is asserted must arise in the future if the right to levy a progressive tax be recognized, involves in its ultimate aspect the
mere assertion that free and representative government is a
failure, and that the grossest abuses of power are foreshadowed
unless the courts usurp a purely legislative function.   If a case
should ever arise where an arbitrary and confiscatory exaction
is imposed, bearing the guise of a progressive or any other form
of tax, it will be time enough to consider whether the judicial
power can afford a remedy by applying inherent and fundamental principles for the protection of the individual, even
though there be no express authority in the Constitution to do
so."

The people of the State, through its constitution and laws,
can provide for the trial of capital or otherwise infamous crimes,
upon information and without indictment, and without a jury,
and they have done so; and they can do away with the trial
by jury in civil cases, and they have done so; and they can do
many other things which I need not enumerate.

In other words, the right of the people of the States to change
their laws and system of procedure so as to conform them to
changed views of administration, or the developing exigencies of
their social life, has been sustained.   And now, I ask the question, if the Constitutional guarantees relating to indictment by
a grand jury and trial by a petit jury do not tie the hands of
the inhabitants of a Territory when organizing a State, why

should they be held to tie the hands of the President and Congress in preserving order and protecting life and property in our new possessions?

It is a strange contention that as soon as the treaty went into effect the power of the President and Congress to preserve order in the new possessions ceased. There were no grand juries, no petit juries, no machinery for punishing crime by the processes of the Anglo-Saxon law; and yet, according to the contention of the other side, if all the limitations of the Constitution apply everywhere throughout the scope of its authority, crime could be punished in no other way. The Constitution which gave the United States power to acquire territory by treaty and imposed upon Congress the duty of disposing of and governing it, did not leave the National Government helpless by demanding impossibilities. Until the progress of the people of the newly acquired territory will permit of the organization of courts and juries after our system, these guarantees must be held inoperative, or the preservation of peace and order, and the protection of life and property under the civil government be abandoned. The situation resembles that discussed in the case of *In re Ross*, 140 U. S. 453, which I commend to opposing counsel, who contend that everywhere throughout the scope of authority of the United States under the Constitution, all limitations apply. In that case, a conviction of murder by a consular court in Japan, acting under an act of Congress, and therefore under authority of the Constitution, without a jury, and upon information, was sustained. Mr. Justice Field said, respecting these guarantees of an indictment and trial by jury in criminal cases (p. 464): " And, besides, their enforcement abroad in numerous places, where it would be highly important to have consuls invested with judicial authority, would be impracticable from the impossibility of obtaining a competent grand or petit jury. The requirement of such a body to accuse and to try an offender would, in a majority of cases, cause an abandonment of all prosecution."

Having discussed the general question, I pass to the consideration of the Porto Rican act. This act provides that on and after a certain date the duties imposed by the Dingley law on goods brought into the United States shall be levied and collected on all articles imported into Porto Rico from ports other than those of the United States, with three exceptions :

A duty of 5 cents a pound is levied on coffee. This is in order to protect the coffee industries there against the cheap coffee of South America.

Spanish scientific, literary, and artistic works are to be admitted free of duty for ten years. This is to carry out the provision of the treaty.

American publications are placed upon the same footing with Spanish.

Now, of course, these duties are not involved in this case, but as a temporary measure to provide revenue for Porto Rico until a system of local taxation could be framed by a provisional government—a local government created by the act—it was provided that, upon all goods coming into Porto Rico from the United States and coming into the United States from Porto Rico, a duty equivalent to 15 per cent of the duties levied by the Dingley law should be imposed. In addition, on goods brought into the United States from Porto Rico which had been manufactured in Porto Rico, the internal revenue tax imposed by the laws of the United States on similar articles manufactured here should be imposed; and on articles manufactured in the United States and taken into Porto Rico, the internal revenue tax which might be imposed there upon similar goods should be collected. This internal revenue tax is to be levied and collected by the imposition of stamps under regulations to be promulgated by the Commissioner of Internal Revenue. The revenues collected from this tax are to be applied for the use and benefit of Porto Rico. It was also provided, as I have indicated, that just as soon as the legislative assembly of Porto Rico, created by this act, should put in operation a system of taxation sufficient to meet the local needs, and the President should make proclamation of that fact, all tariff duties on goods coming into Porto Rico from the United States and coming into the United States from Porto Rico should cease. And it further provided that in no event shall any duties be collected after the 1st day of March, 1902, on merchandise and articles going into Porto Rico from the United States or coming into the United States from Porto Rico.

I have in my brief, on page 74 and the succeeding pages, quoted from a speech of Senator Foraker, who had charge of the bill in the Senate, in which he stated with clearness the situation in Porto Rico which led to the enactment of the measure, and epitomizes its provisions. In this he says: " The committee found upon investigation that a civil government should be at once established in Porto Rico, and found that this government would require for its support not less than about $3,000,000 annually. They also found that an additional million dollars would be required to support the municipal governments of the island, making an aggregate of not less than $4,000,000."

They found that the total valuation of property of all kinds situated in the island would not exceed for taxation purposes $100,000,000. They found that this property was already burdened with a private debt, evidenced by mortgages on record, to the amount of about $26,000,000 of principal, with an accumulation of several years' interest, at extravagant rates, which swelled the sum to probably $30,000,000.

The committee further found that no system of property taxation was in force in the island, or ever had been, and that it would require at least a year, and probably two years, to inaugurate one and secure returns from it, and that, inasmuch as the people had no familiarity with such a system, it would be difficult, probably, to enforce it, at least for a time.

The committee also found that the public revenues of the island, except only such as were raised by a burdensome excise tax on incomes and business vocations, had always been chiefly received from duties on imports and exports—a system with which the people were therefore familiar.

The committee further found that this system was already in operation, and that revenues were then constantly being collected, upon which, so far as they went, the Government could at once depend.

The committee further found that our internal revenue law,

if applied in that island, would prove oppressive and ruinous to many people and interests.

To collect our heavy internal revenue taxes—far heavier than Spain ever imposed—on these products and vocations would be to invite violations of law so innumerable as to make prosecutions impossible, and to almost certainly alienate and destroy the friendship and good will of that people for the United States.

Now, it was in view of those considerations, and in order to find some way to exempt the people of Porto Rico both from the direct taxation of their property—such taxation as is imposed in every State and organized Territory of the United States—and also from the onerous burdens of an immediate application of our internal revenue laws, that this temporary system of taxing the exports from the island and the imports into the island was framed and put in operation. Manifestly, by the passage of the Porto Rican act, not only because of these temporary fiscal provisions, but also because of other provisions to which I call attention in my brief, Congress did not intend to recognize or treat the island as a part of the United States, but as a possession thereof, with a political existence under the sovereignty, but outside of the limits, of the United States, legislatively treated. The inhabitants are made citizens of Porto Rico, and as such entitled to the protection of the United States. A temporary civil government is provided, with a revenue system quite separate and distinct from that of the United States. The duties provided by the act, both on goods coming into the United States from Porto Rico and coming into Porto Rico from the United States, "shall be used for the government and benefit of Porto Rico." The taxation, therefore, is of a purely local nature. It cannot be said that the revenues derived from these duties were to be used "to pay the debts and provide for the common defence and welfare of the United States."

These duties are not laid by Congress under the general grant of the taxing power contained in the first clause of section 8 of article I, but under the power to dispose of and make all needful rules and regulations respecting the territory or other property belonging to the United States. The fact that the limitation in the first clause of section 8 of article I, and indeed the provisions of that clause generally, only apply to taxes which are levied to pay the debts and provide for the common defence and general welfare of the United States, is supported by what Mr. Justice Miller says in his work on the Constitution, page 230, and what Chief Justice Marshall says in *Gibbons* v. *Ogden*, 9 Wheaton, 199. In that case, with reference to the taxing power, Chief Justice Marshall says: " Congress is authorized to lay and collect taxes, etc., to pay the debts, and provide for the common defence and general welfare of the United States. This does not interfere with the power of the States to tax for the support of their own governments; nor is the exercise of that power by the States an exercise of any portion of the power granted to the United States."

But if the contention of the other side is correct, and because the duties on exports from Porto Rico into the United States are collected in this country, although the proceeds are applied for the benefit of the Porto Rican governments, if because of the collection here this clause applies, and these duties must be uniform throughout the United States, then my answer is that they are uniform throughout the United States, being uniformly collected in the ports of every State into which goods may be brought from Porto Rico.

Now, Congress has determined that this temporary local revenue measure is for the welfare of Porto Rico, and I submit that that determination is conclusive, unless there is some other limitation or prohibition which prevents. The only other provision suggested as applicable is that which provides " that no tax or duty shall be laid on articles exported from any State." The only goods which could possibly be regarded as articles exported from any State are the goods which are imported into Porto Rico from the United States. But these goods are not exports from any State. They are imports into Porto Rico. A duty laid on exports is a duty laid upon the goods at the time they are shipped abroad, and because of that fact. When goods are received at the port of destination, they cease to be exports and become imports, and a tax then laid upon them because of their importation is not a tax upon exports, but a duty upon imports.

Whether the tax shall be considered as a tax upon exports or as a duty upon imports may depend upon the application of the revenue collected.   In this case the revenue is all to be applied for the benefit of Porto Rico.   The revenue collected in Porto Rico on what the other side claim are exports from the United States, is applied to the use of Porto Rico, and I say that fact is sufficient, in testing these two views, to determine that the goods are to be regarded as imports into Porto Rico.

MR. JUSTICE HARLAN.   As far as the question of power is concerned, it would be the same, would it or not, if the duties collected upon Porto Rican products were paid into the Treasury of the United States and remained here?

MR. SOLICITOR GENERAL.   I think it makes a material difference as to whether the revenue is to be paid to the United States or Porto Rico.

MR. JUSTICE HARLAN.   As to the question of power?

MR. SOLICITOR GENERAL.   As to the authority to levy this particular duty.

MR. JUSTICE HARLAN.   I do not say it does not.   I want to get your views.

MR. SOLICITOR GENERAL.   I contend that this is, in a sense, a local revenue measure.   It is not a case where Congress exercises the Federal power of taxation to raise revenue to pay the debts and to provide for the general welfare and the common defence, under that section of the Constitution, but it is a measure providing local revenue for Porto Rico, under the provision which authorizes Congress to pass all needful rules and regulations for Porto Rico.   And what I am inquiring now is whether there is any other provision of the Constitution, any other limitation, which prevents.

MR. JUSTICE BREWER.   Under that power, would it be competent for Congress to pass an act requiring a duty to be paid on all goods shipped from the other States into New Mexico, for the support of New Mexico?

MR. SOLICITOR GENERAL.   New Mexico might be placed, as I take it, by Congress, if Congress saw fit, in the exact position of Porto Rico.   I think logically I would have to so contend. Alaska might, if circumstances demanded, be placed in the exact position of Porto Rico. I believe Congress has full power over them, subject, however, I should say, to certain provisions which protect citizens of the United States in the enjoyment of certain rights. Now, whether the vested rights and privileges which follow citizenship would prevent what you suggest, I confess I am not able at once to state. I believe that Congress could sell Alaska if it saw fit. I think that so long as territory remains under the plenary power marked out in the Constitution, it is for Congress to say whether that territory shall be taken into the Union as a State, and so indissolubly become a part of the United States, or whether the general welfare would be better subserved by parting with the territory, making, at the same time, due provision for safeguarding all rights of citizenship, and all rights of property belonging to citizens of the United States residing there.

Mʀ. Jᴜsᴛɪᴄᴇ Bʀᴇᴡᴇʀ. Does not the effect of that argument come to this, that the uniformity clause of the Constitution in respect of duties, etc., applies solely to the States?

Mʀ. Soʟɪᴄɪᴛᴏʀ Gᴇɴᴇʀᴀʟ. The uniformity clause does, I insist, apply solely to the States, unless Congress has seen fit to provide otherwise.

Mʀ.. Jᴜsᴛɪᴄᴇ Bʀᴇᴡᴇʀ. Unless Congress has extended the power?

Mʀ. Soʟɪᴄɪᴛᴏʀ Gᴇɴᴇʀᴀʟ. Yes, unless Congress has enlarged the boundaries of the United States—I mean within the meaning of the taxing laws.

Mʀ. Jᴜsᴛɪᴄᴇ Bʀᴇᴡᴇʀ. If it enlarges, it can restrict?

Mʀ. Soʟɪᴄɪᴛᴏʀ Gᴇɴᴇʀᴀʟ. Certainly, unless vested rights intervene to prevent.

Mʀ. Jᴜsᴛɪᴄᴇ Wʜɪᴛᴇ. You say Congress would have the right in your judgment to dispose of Arizona and New Mexico, provided it made provision in the treaty to protect the rights of citizenship, and so on?

Mʀ. Soʟɪᴄɪᴛᴏʀ Gᴇɴᴇʀᴀʟ. Yes.

Mʀ. Jᴜsᴛɪᴄᴇ Wʜɪᴛᴇ. But how would those rights of citizenship come into being and require protection, unless Arizona, for instance, has become a part of the United States and citizenship has resulted?

Mr. Solicitor General. Congress has entire authority over the matter of naturalization, and it may naturalize not only by a law applying uniformly, but collectively, by special acts, and it has done so. It has naturalized Indians who lived in the Indian Territory, although the Indian Territory has not been regarded as a part of the United States in the imposition of our excise taxes. Many instances of collective naturalization might be given. And so I say, that if we have conferred citizenship, why, then, in disposing of territory that belongs to the United States, but has not become an inseparable part of the Union, doubtless the treaty-making power or Congress would provide for the safeguarding and protection of all personal and property rights flowing from citizenship in such territory.

I believe that the Government can dispose of the Philippines if it deems best to do so. The power that can acquire, can sell or exchange. I do not occupy the position from which the other side cannot escape, that the cession made the Philippines an integral part of the United States, inseparably incorporated under the Constitution, and with rights unalterably fixed by the Constitution. I believe they are but a possession—territory belonging to the United States—which we can part with whenever it becomes apparent that their interests or our welfare demands a separation.

It may be further suggested that within the decision of *Woodruff* v. *Parham*, 8 Wallace, 123, the goods shipped into Porto Rico from the United States are not exports from the States, because not shipped to a foreign country. The commerce, I take it, between Porto Rico and the United States since the passage of the Porto Rican act is not foreign commerce, but domestic commerce. It is commerce passing between countries under the sovereignty of the United States, commerce which is regulated by Congress, possibly under the power to regulate commerce either among the several States or with foreign nations—I say possibly, having in mind the opinion in the case of *Stoutenburgh* v. *Hennick*, 129 U. S. 141, in which the court held that the action of the local authorities of the District of Columbia in taxing a commercial traveler was in violation of the commerce clause—or under the power, as I have said, to make all needful rules and regulations respecting the territory or other property belonging to the United States.

I submit that the authority to regulate these insular possessions includes authority to regulate their commerce, both with foreign countries and with the United States. Commerce is always a rightful subject of regulation by a governing body. It is true that the Constitution places certain limitations upon the power of Congress to regulate the commerce of the States. While Congress is given express power to regulate commerce with the foreign nations, and among the several States and with the Indian tribes, it is provided that no preference shall be given by any regulation of commerce or revenue to the ports of one State over those of another. But obviously this Porto Rican act gives no preference to the ports of one State over those of another. All States are treated alike. Goods going into Porto Rico pay a certain duty there, no matter from what State or port shipped; and goods coming into the United States from Porto Rico pay a certain duty here, no matter to what port or State shipped. It is true that the Constitution declares that the citizens of each State are entitled to all the privileges and immunities of citizens of the several States. That is what I referred to a moment ago in answering the question of Mr. Justice Brewer with reference to Arizona; but I fail to see in what way the rights of a citizen of any State can be infringed by the Porto Rican act. All citizens are treated alike.

MR. JUSTICE HARLAN. Suppose they are not treated alike. Suppose this act had given a preference to the commerce coming to this country to the ports of one State over the ports of another. Under your view, what clause of the Constitution would forbid Congress from doing that?

MR. SOLICITOR GENERAL. The very clause I have read.

MR. JUSTICE HARLAN. You call that a regulation of commerce, do you?

MR. SOLICITOR GENERAL. I do, most emphatically. But the clause applies also to any " regulation of revenue." Moreover, no privilege or immunity granted to the people of Porto Rico by the treaty of Paris is infringed by this legislation, for the treaty itself expressly provided that their civil rights and polit-

ical *status* should be determined by Congress; and Congress
has declined to make them citizens of the United States, restricting their *status* to citizens of Porto Rico, entitled to the
protection of the United States.   As such, Congress has framed
a measure peculiarly adapted to raise the insular revenues in
the easiest way, thus avoiding the imposition upon them of burdens which would become intolerable if our internal revenue
taxes were extended to them.

Before the adoption of the Constitution—and I will now direct myself, possibly, to something that is in the mind of Mr.
Justice Harlan—the States had severally the power to lay duties and imposts on imports and exports, and they exercised it.
The Constitution forbade the further exercise of this power
without the consent of Congress and unless the net proceeds
of all duties and imposts so laid should be applied for the use of
the Treasury of the United States, the clause reading as follows:

"No State shall, without the consent of the Congress, lay any
imposts or duties on imports or exports, except what may be
absolutely necessary for executing its inspection laws; and the
net produce of all duties and imposts laid by any State on imports or exports shall be for the use of the Treasury of the
United States; and all such laws shall be subject to the revision
and control of the Congress."

"Now, this seeming prohibition—I should not say seeming—
this prohibition, is equivalent to an implied grant of authority
to a State, or a recognition of authority existing in a State, to
lay imposts or duties on imports or exports, providing Congress
shall consent, and upon the condition that the net produce of
such duties shall be for the use of the Treasury of the United
States.   And it is a recognition of the fact that the needs of
both a State and of the United States might be promoted by
special duties on the imports or exports of a State.   The condition thus recognized and provided for in the case of a State
has, in this particular instance, been legislated for by Congress,
which possesses both state and Federal jurisdiction in the case
of Porto Rico.   I might say in this connection, respecting the
levying of duties by a State on imports with the consent of

Congress, that the same limitation and grant applies in the case
of tonnage duties, and that the legislative history of the country shows that Congress has given its consent to a great many
measures where a State levied duties, either on tonnage or on
imports. With reference to tonnage duties, Chief Justice Marshall said (9 Wheaton, 202): "A duty on tonnage is as much
a tax, as a duty on imports or exports; and the reason which
induced the prohibition of those taxes extends to this also.
This tax may be imposed by a State, with the consent of Congress."

I have here a list of thirty acts, passed from 1790 to 1847, in
which the assent of Congress was given to the acts of States
levying duties on imports or tonnage for harbor improvements
or other local purposes.

It may be insisted that the constitutional provision which requires all duties, imposts, and excises to be uniform throughout
the United States lays down a fundamental rule of taxation applicable everywhere; that no special mode of taxation, to meet
the needs of a particular territory, can be framed by Congress,
but that all duties and excises must be laid uniformly throughout all the territory over which the sovereignty of the United
States extends. With respect to this, I beg to say that there
was a good reason for requiring duties and excises to be uniform
throughout the States of the Union, and that reason is stated
clearly in the opinion of the court in *Knowlton* v. *Moore.* But
there is neither reason nor justice in requiring the same taxes to
be imposed wherever the flag flies. The collection of our internal
revenue taxes is impossible and impracticable in Porto Rico
and the Philippines. They were framed to meet conditions
here; they would be ruinous there. We are not engaged at
present in collecting taxes in Porto Rico for the benefit of the
United States. The only taxes collected are used for the benefit of Porto Rico. Of course Porto Rico receives the benefit
of the general revenues to a certain degree, for the General
Government is there with its agencies supported at the general
expense, and it would be only fair, if Porto Rico could stand
it, to make her bear her fair share of the national burdens in
return for the benefits she receives. But, after all, the entire

matter is left with Congress, and the uniform imposition in Porto Rico of the national system of taxation would not relieve the island from the necessity of responding to further exactions, should Congress deem them necessary in order to meet the local expenses of the government of the island. Congress possesses over Porto Rico, to use the language of Mr. Justice Gray in *Shively* v. *Bowlby*, "the entire dominion and sovereignty, national and municipal, Federal and state." What good purpose could be served, then, by attempting to apply in Porto Rico the provision that Federal taxes shall be uniform throughout the States. It is all right to require Federal taxes to be uniform throughout the States. This secures a uniform contribution from the States for a uniform benefit. Only the national taxes are raised in the States by the Federal authority. The States raise their own state, county, and municipal taxes. They regulate these to suit themselves. Congress has no say about them. But in Porto Rico Congress has power to raise not merely national but all insular revenues, everything needed to carry on the local government. It is not necessary, as I understand it, that in raising taxes for a Territory Congress should distinguish between the purposes to which the taxes are to be applied and levy specific taxes for national purposes and other taxes for other purposes. Especially is this true before a territorial government has been organized and has established and put in operation a system of local taxation. Congress may and must necessarily combine the sources of revenue and apply the proceeds as the circumstances require. The power and the necessity of doing this prevents any just comparison between the revenue system established by Congress in a Territory and that in force for purely Federal purposes in the States.

Respecting the territorial governments, with their courts and laws, Mr. Justice Nelson, speaking for the court, said in *Benner* v. *Porter*, 9 How. 242: "They are legislative governments, and their courts legislative courts, Congress, in the exercise of its powers in the organization and government of the Territories, combining the powers of both the Federal and state authorities. There is but one system of government, or of laws operating within their limits, as neither is subject to the constitutional provisions in respect to state and Federal jurisdiction."

With regard to the matter of taxation in Porto Rico, it is quite pertinent to put the question which Mr. Justice Harlan, speaking for the court, put in the case of *McAllister* v. *United States*, 141 U. S. 190, respecting the power of Congress over the courts of a Territory :

" Has Congress, under 'the general right of sovereignty' existing in the Government of the United States as to all matters committed to its exclusive control, including the making of needful rules and regulations respecting the Territories of the United States, any less power over the judges of the Territories than a State, if unrestricted by its own organic law, might exercise over judges of its own creation ? "

In other words, to paraphrase this, has Congress, under "the general right of sovereignty" existing in the Government of the United States as to all matters committed to its exclusive control, including the making of needful rules and regulations respecting the Territories of the United States, any less power in raising territorial revenue than a State, if unrestrained by its own organic law, might exercise in raising revenue within its borders ?

In the argument of counsel on the other side, reference was made to the ordinance of 1787, as showing that the term "the United States" includes the territory belonging to the United States. Counsel called attention to the fact that in the treaty between this country and Great Britain the description of the United States included the vast expanse outside of the limits of the thirteen Colonies, but claimed by them as the successors of the royal power, stretching into the great West, and insisted that that constituted the United States. I think a careful reading of the ordinance of 1787 and the history of the release by the Colonies, which composed the United States under the Confederation, of their claims to the territory covered by the ordinance of 1787 shows conclusively that a distinction was drawn between the United States under the Confederation and the territory belonging to them which lay northwest of the Ohio. The ordinance itself says that it is an ordinance " for the government of the territory of the United States northwest of the Ohio River." This territory had been ceded by certain of the Colonies—Virginia, New York and others—who claimed it, to the United States, because the Colonies properly claimed that unless they succeeded in the war with Great Britain the title would amount to nothing. It was being won by the blood and treasure of all, and therefore should belong to all, and the Colonies conceded this to be a fact, and therefore turned over their title and claim to the United States. And then this ordinance for the government of the territory was passed, and it says it is an ordinance for the government of the territory of the United States northwest of the Ohio River.

With respect to members of the general assembly it provides that no person shall be eligible unless he shall have been "a citizen of one of the United States three years." Did that mean a citizen of the Northwest Territory? Evidently not, because it goes on to provide, "and be a resident in the district, or unless he shall have resided in the district three years." In other words, a citizen of one of the United States was eligible if he resided in the district, while a person not a citizen of one of the United States must have resided in the district three years to be eligible.

"For extending the fundamental principles of civil and religious liberty, which form the basis whereon these republics, their laws and constitutions, are erected," it was provided and declared that certain articles should be considered "as articles of compact" between the original States (that is, the United States under the Confederation) "and the people and States in the said territory, and forever remain unalterable, unless by common consent." Here is a distinct recognition that the Northwest Territory was not a part of the United States. The ordinance forms a compact between the United States under the Confederation and the people and States to be formed in the Northwest Territory.

In the fourth article it is provided that the navigable waters leading into the Mississippi and St. Lawrence, etc., shall be common highways, and forever free, "as well to the inhabitants of the said Territory as to the citizens of the United States, and those of any other States that may be admitted into the Confederacy."

As I have sat and listened to these elaborate arguments, whereby counsel, ignoring the plain and simple provisions of the Constitution, seek, by a refinement of reasoning, to induce this court to take away from the President and Congress the power to govern newly acquired territory according to its nature and needs—a power which has been exercised, from the days of the founders of the Republic, by the nation which then, to use the words of the Declaration, assumed, " among the powers of the earth, the separate and *equal* station to which the laws of nature and of nature's God entitled it," I cannot but recall the impressive language of the great Chief Justice Marshall, at the close of the remarkable opinion which he delivered in the case of *Gibbons* v. *Ogden:*

" Powerful and ingenious minds, taking, as postulates, that the powers expressly granted to the Government of the Union, are to be contracted by construction, into the narrowest possible compass, . . . may, by a course of well-digested but refined and metaphysical reasoning, founded on these premises, explain away the Constitution of our country, and leave it, a magnificent structure, indeed, to look at, but totally unfit for use. They may so entangle and perplex the understanding, as to obscure principles, which were before thought quite plain, and induce doubts where, if the mind were to pursue its own course, none would be perceived."

We have the new territories. We are responsible for them, responsible to their people, to ourselves, to the world. We must provide them a government. May we not give them a government adopted to their needs? May we not in governing them carry out the solemn stipulations of the treaty through which we acquired sovereignty over them? The path of duty is plain. May we not walk in it? Does the Constitution prevent? Is the Constitution a stumbling block, or a trap, caught in which we shall excite the pity of our friends and the derision of our foes? I refuse to believe so. The Constitution is no mere declaration of denials. It created a nation to which was intrusted the full power asserted in the Declaration of Independence—"to levy war, conclude peace, contract alliances, establish commerce, and to do all other acts and things which independent States may of right do." When it conferred power, it took care not to cripple action. It still remains the most perfect instrument ever struck off at a given time by the brain and purpose of man, under which we are armed for every emergency, and able to cope with every condition.

Mr. JUSTICE BROWN delivered the opinion of the court.

This case raises the single question whether territory acquired by the United States by cession from a foreign power remains a "foreign country" within the meaning of the tariff laws.

1. Did the question of jurisdiction raised by the demurrer involve only the jurisdiction of the Circuit Court as a Federal court, we should be obliged to say that the defendant was not in a position to make this claim, since the case was removed to the Federal court upon his own petition. It is no infringment upon the ancient maxim of the law that consent cannot confer jurisdiction, to hold that, where a party has procured the removal of a cause from a state court upon the ground that he is lawfully entitled to a trial in a Federal court, he is estopped to deny that such removal was lawful, if the Federal court could take jurisdiction of the case or that the Federal court did not have the same right to pass upon the questions at issue that the state court would have had, if the cause had remained there. Defendant neither gains nor loses by the removal, and the case proceeds as if no such removal had taken place. *Cowley* v. *Northern Pacific Railroad Co.*, 159 U. S. 569, 583 ; *Mansfield Railway Co.* v. *Swan*, 111 U. S. 379; *Mexican Nat. Railroad* v. *Davidson*, 157 U. S. 201.

This, however, is more a matter of words than of substance, as the defendant unquestionably has the right to show that the state court had no jurisdiction, or that the complaint did not set forth facts sufficient to constitute a cause of action. This we understand to be the substance of the defence in this connection.

By Rev. Stat. sec. 2931, it was enacted that the decision of the collector "as to the rate and amount of duties" to be paid upon imported merchandise should be final and conclusive, unless the owner or agent entered a protest, and within thirty days appealed therefrom to the Secretary of the Treasury; and, further, that the decision of the Secretary should be final and conclusive, unless suit were brought within ninety days after the decision of the Secretary. By Rev. Stat. sec. 3011, any person having made payment under such protest was given the right to bring an action at law and recover back any excess of duties so paid.

The law stood in this condition until June 10, 1890, when an act known as the Customs Administrative Act was passed, 26 Stat. 131, c. 407, by which the above sections Rev. Stat. secs. 2931, 3011, were repealed and new regulations established, by which an appeal was given from the decision of the collector "as to the rate and amount of the duties chargeable upon imported merchandise," if such duties were paid under protest, to a Board of General Appraisers, whose decision should be final and conclusive (sec. 14) "as to the construction of the law and the facts respecting the classification of such merchandise and the rate of duties imposed thereon under such classification," unless within thirty days one of the parties applied to the Circuit Court of the United States for a review of the questions of law and fact involved in such decision. Sec. 15. It was further provided that the decision of such court should be final, unless the court were of opinion that the question involved was of such importance as to require a review by this court, which was given power to affirm, modify or reverse the decision of the Circuit Court.

The effect of the Customs Administrative Act was considered by this court in *In re Fassett, Petitioner,* 142 U. S. 479, in which we held that the decision of the collector that a yacht was an imported article might be reviewed upon a libel for possession filed by the owner, notwithstanding the Customs Administrative Act. It was held that the review of the decision of the Board of General Appraisers, provided for by section fifteen of that act, was limited to decisions of the board "as to the construction of the law and the facts respecting the classification " of imported merchandise "and the rate of duties imposed thereon under such classification," and that it did not bring up for review the question whether an article be imported merchandise or not, nor, under section fifteen, is the ascertainment of that fact such a decision as is provided for. Said Mr. Justice Blatchford : " Nor can the court of review pass upon any question which the collector had not original authority to determine. The collector has no authority to make any determination regarding any article which is not imported merchandise ; and if the vessel in question here is not imported merchandise, the court of review would have no jurisdiction to determine any matter regarding that question, and could not determine the very fact which is in issue under the libel in the District Court, on which the rights of the libellant depend."

" Under the Customs Administrative Act, the libellant, in order to have the benefit of the proceedings thereunder, must concede that the vessel is imported merchandise, which is the very question put in contention under the libel, and must make entry of her as imported merchandise, with an invoice and consular certificate to that effect." It was held that the libel was properly filed.

The question involved in this case is not whether the sugars were importable articles under the tariff laws, but whether, coming as they did from a port alleged to be domestic, they were imported from a foreign country—in other words, whether they were *imported* at all as that word is defined in *Woodruff* v. *Parham*, 8 Wall. 123, 132. We think the decision in the *Fassett* case is conclusive to the effect that, if the question be whether the sugars were imported or not, such question could not be raised before the Board of General Appraisers ; and that whether they were imported merchandise for the reasons given in the *Fassett* case that a vessel is not an importable article, or because the merchandise was not brought from a foreign country, is immaterial. In either case the article is not *imported*.

Conceding then that section 3011 has been repealed, and that no remedy exists under the Customs Administrative Act, does it follow that no action whatever will lie ? If there be an admitted wrong, the courts will look far to supply an adequate remedy. If an action lay at common law the repeal of sections 2931 and 3011, regulating proceedings in customs cases, (that is, turning upon the classification of merchandise,) to make way for another proceeding before the Board of General Appraisers in the same class of cases, did not destroy any right of action that might have existed as to other than customs cases; and the fact that by section 25 no collector shall be liable "for or on account of any rulings or decisions as to the classification of such merchandise or the duties charged thereon, or the collection of any dues, charges or duties on or on account of any such merchandise," or any other matter which the importer might have brought before the Board of General Appraisers, does not restrict the right which the owner of the merchandise might have against the collector in cases not falling within the Customs Administrative Act. If the position of the Government be correct, the plaintiff would be remediless; and if a collector should seize and hold for duties goods brought from New Orleans, or any other concededly domestic port, to New York, there would be no method of testing his right to make such seizure. It is hardly possible that the owner could be placed in this position. But we are not without authority upon this point.

The case of *Elliott* v. *Swartwout*, 10 Pet. 137, 154, was an action of assumpsit against the collector of the port of New York to recover certain duties upon goods alleged to have been improperly classified. It was held that as the payment was purely voluntary, by a mutual mistake of law, no action would lie to recover them back, although it would have been different if they had been paid under protest. Said Mr. Justice Thompson: "Here, then, is the true distinction: when the money is paid voluntarily and by mistake to the agent, and he has paid it over to his principal, he cannot be made personally responsible; but if, before paying it over, he is apprised of the mistake, and required not to pay it over, he is personally liable." If the payment of the money be accompanied by a notice to the collector that the duties charged are too high, and that the person paying intends to sue to recover back the amount erroneously paid, it was held that such action must lie "unless the broad proposition can be maintained, that no action will lie against a collector to recover back an excess of duties paid him, but that recourse must be had to the Government for redress." The case recognized the fact that, with respect to money paid under a mistake of law, the collector stood in the position of an ordinary agent and could be made personally liable in case the money were paid under protest.

This decision was made in 1836. Apparently in consequence of it an act was passed in 1839 requiring moneys collected for duties to be deposited to the credit of the Treasurer of the United States; and it was made the duty of the Secretary of the Treasury to draw his warrant upon the Treasurer in case he found more money had been paid to the collector than the law required. It was held by a majority of this court in *Cary* v. *Curtis*, 3 How. 236, that this act precluded an action of assumpsit for money had and received against the collector for duties received by him, and that the act of 1839 furnished the sole remedy. It was said of that case in *Arnson* v. *Murphy*, 109 U. S. 238, 240 : "Congress, being in session at the time that the decision was announced, passed the explanatory act of February 26, 1845, which, by legislative construction of the act of 1839, restored to the claimant his right of action against the collector, but required the protest to be made in writing at the time of payment of the duties alleged to have been illegally exacted, and took from the Secretary of the Treasury the authority to refund conferred by the act of 1839. 5 Stat. 349, 727. This act of 1845 was in force, as was decided in *Barney* v. *Watson*, 92 U. S. 449, until repealed by implication by the act of June 30, 1864," c. 171, 13 Stat. 202, 214, carried into the Revised Statutes as sections 2931 and 3011. In the same case of *Arnson* v. *Murphy*, 109 U. S. 238, it was decided that the common-law right of action against the collector to recover back duties illegally collected was taken away by statute, and a remedy given, based upon these sections, which was exclusive. The decision in *Elliott* v. *Swartwout* was recognized, but so far as respected *customs cases* (*i. e.*, classification cases) was held to be superseded by the statutes. So in *Schoenfeld* v. *Hendricks*, 152

U. S. 691, it was held that an action could not be maintained against the collector, either at common law or under the statutes, to recover duties alleged to have been exacted, in 1892, upon an importation of merchandise, the remedy given through the Board of General Appraisers being exclusive.

The criticism to be made upon the applicability of these cases is, that they dealt only with *imported* merchandise and with the duties collected thereon, and have no reference whatever to exactions made by a collector, under color of the revenue laws, upon goods which have never been imported at all. With respect to these the collector stands as if, under color of his office, he had seized a ship or its equipment, or any other article not comprehended within the scope of the tariff laws. Had the sugars involved in this case been admittedly imported, that is, brought into New York from a confessedly foreign country, and the question had arisen whether they were dutiable, or belonged to the free list, the case would have fallen within the Customs Administrative Act, since it would have turned upon a question of classification.

The fact that the collector may have deposited the money in the Treasury is no bar to a judgment against him, since Rev. Stat. sec. 989 provides that, in case of a recovery of any money exacted by him and paid into the Treasury, if the court certifies that there was probable cause for the act done, no execution shall issue against him, but the amount of the judgment shall be paid out of the proper appropriation from the Treasury.

We are not impressed by the argument that, if the plaintiffs insisted that these sugars were not imported merchandise, they should have stood upon their rights, refused to enter the goods, and brought an action of replevin to recover their possession. It is true that, to prevent the seizure of the sugars, plaintiffs did enter them as imported merchandise; but any admission derivable from that fact is explained by their protest against the exaction of duties upon them as such. They waived nothing by taking this course. The collector lost nothing, since he was apprised of the course they would probably take. It is true that in the *Fassett Case*, 142 U. S. 479, the proceeding was

by libel for possession of the vessel, which is analogous to an
action of replevin at common law; but it would appear that Rev.
Stat. sec. 934 would stand in the way of such a remedy here,
since by that section "all property taken or detained by any
officer or other person under authority of any revenue law of
the United States shall be irrepleviable, and shall be deemed
to be in the custody of the law and subject only to the orders
and decrees of the courts of the United States having jurisdiction thereof." If the words "under authority of any revenue
law" are to be construed as if they read "under color of any
revenue law," it would seem that these sugars could not be
made the subject of a replevin; but even conceding that replevin would lie, we consider it merely a choice of remedies,
and that the plaintiffs were at liberty to waive the tort and
proceed in assumpsit.

We are all of opinion that this action was properly brought.

2. Whether these cargoes of sugar were subject to duty depends solely upon the question whether Porto Rico was a "foreign country" at the time the sugars were shipped, since the
tariff act of July 24, 1897, c. 11, 30 Stat. 151, commonly known
as the Dingley act, declares that "there shall be levied, collected
and paid upon all articles imported from foreign countries"
certain duties therein specified. A foreign country was defined
by Mr. Chief Justice Marshall and Mr. Justice Story to be one
exclusively within the sovereignty of a foreign nation, and
without the sovereignty of the United States. *The Boat Eliza*,
2 Gall. 4; *Taber* v. *United States*, 1 Story, 1; *The Ship Adven-
ture*, 1 Brock. 235, 241.

The *status* of Porto Rico was this: The island had been for
some months under military occupation by the United States
as a conquered country, when, by the second article of the treaty
of peace between the United States and Spain, signed December 10, 1898, and ratified April 11, 1899, Spain ceded to the
United States the island of Porto Rico, which has ever since
remained in our possession, and has been governed and administered by us. If the case depended solely upon these facts, and
the question were broadly presented whether a country which
had been ceded to us, the cession accepted, possession delivered,

and the island occupied and administered without interference by Spain or any other power, was a foreign country or domestic territory, it would seem that there could be as little hesitation in answering this question as there would be in determining the ownership of a house deeded in fee simple to a purchaser, who had accepted the deed, gone into possession, paid taxes and made improvements without let or hindrance from his vendor. But it is earnestly insisted by the Government that it never could have been the intention of Congress to admit Porto Rico into a customs union with the United States, and that, while the island may be to a certain extent domestic territory, it still remains a "foreign country" under the tariff laws, until Congress has embraced it within the general revenue system.

We shall consider this subject more at length hereafter, but for the present call attention to certain cases in this court and certain regulations of the executive departments which are supposed to favor this contention.

In *United States* v. *Rice*, 4 Wheat. 246, which was an action of debt brought by the United States upon a bond for duties upon goods imported into Castine, in the district (now State) of Maine, during its temporary occupation by the British troops in the war of 1812, it was held the action would not lie, though Castine was subsequently evacuated by the enemy and restored to the United States. The court said that, by the military occupation of Castine, the enemy acquired a possession which enabled him to exercise the fullest rights of sovereignty; that the sovereignty of the United States was suspended, and our laws could be no longer rightfully enforced there, or be obligatory upon the inhabitants; that by the surrender the inhabitants passed under a temporary allegiance to the British government, and were only bound by the laws of that government, and that Castine was during this period to be deemed a foreign port; that goods brought there were subject to duties which the British government chose to impose, and were in no correct sense imported into the United States; and that the subsequent evacuation by the enemy did not change the character of the transaction, since the goods were not liable to American duties when imported. In that case the character of the port, as foreign or domestic, was held to depend upon the question of actual occupation, and the right of the defendant determinable by the facts then existing, and further, that the subsequent reoccupation of the port by the United States was ineffectual to change the right of the defendant or to vest a new right in the United States.

A case, somewhat to the converse of this, was that of *Fleming* v. *Page*, 9 How. 603, which was an action against the collector at Philadelphia, to recover back duties upon merchandise imported from Tampico, in Mexico, during a temporary military occupation of that place by the United States. It was held that, although Tampico was within the military occupation of the United States, it had not ceased to be a foreign country, in the sense in which these words are used in the acts of Congress. In delivering the opinion of the court, Mr. Chief Justice Taney observed: "The United States, it is true, may extend its boundaries by conquest or treaty, and may demand the cession of territory as the condition of peace, in order to indemnify its citizens for the injuries they have suffered, or to reimburse the government for the expenses of the war. But this can be done only by the treaty-making power or the legislative authority, and is not a part of the power conferred upon the President by the declaration of war. . . . While it was occupied by our troops, they were in an enemy's country, and not in their own; the inhabitants were still foreigners and enemies, and owed to the United States nothing more than a submission and obedience, sometimes called temporary allegiance, which is due from a conquered enemy, when he surrenders to a force which he is unable to resist."

This was clearly a sufficient reason for disposing of the case adversely to the importer, but the learned Chief Justice proceeded to put the case upon another ground, that "there was no act of Congress establishing a custom house at Tampico, nor authorizing the appointment of a collector; and consequently there was no officer of the United States authorized by law to grant the clearance and authenticate the coasting manifest of the cargo in the manner directed by law, where the voyage is from one port of the United States to another;" that the only collector was one appointed by the military commander, and that a coasting manifest granted by him could not be recognized in the United States as the document required by law, when the vessel is engaged in the coasting trade, nor exempt the cargo from the payment of duties. He states that this construction of the tariff laws had been uniformly given by the administrative department of the Government, and cited the case of Florida, after it had been ceded to the United States and the military forces had taken possession of Pensacola: "That is, that, although Florida had, by cession, actually become a part of the United States, and was in our possession, yet, under our revenue laws, its ports must be regarded as foreign until they were established as domestic, by acts of Congress. And it appears that this decision was sanctioned at the time by the Attorney General of the United States, the law officer of the Government. And, although not so directly applicable to the case before us, yet the decisions of the Treasury Department in relation to Amelia Island, and certain ports in Louisiana, after that province had been ceded to the United States, were both made upon the same grounds. And in the later case, after a custom house had been established by law, (2 Stat. 418,) at New Orleans, the collector at that place was instructed to regard as foreign ports Baton Rouge and other settlements still in the possession of Spain, whether on the Mississippi, Iberville, or the seacoast. The department, in no instance that we are aware of, since the establishment of the Government, has ever recognized a place in a newly acquired country as a domestic port, from which the coasting trade might be carried on, unless it had been previously made so by act of Congress."

While we see no reason to doubt the conclusion of the court that the port of Tampico was still a foreign port, it is not perceived why the fact that there was no act of Congress establishing a custom house there or authorizing the appointment of a collector, should have prevented the collector appointed by the military commander from granting the usual documents required to be issued to a vessel engaged in the coasting trade. A collector, though appointed by a military commander, may be presumed to have the ordinary power of a collector under an act of Congress, with authority to grant clearances to ports within the United States, though, of course, he would have no power to make a domestic port of what was in reality a foreign port.

It is not intended to intimate that the cases of *United States* v. *Rice* and *Fleming* v. *Page* are not harmonious. In fact, they are perfectly consistent with each other. In the first case it was merely held that duties could not be collected upon goods brought into a domestic port during a temporary occupation by the enemy, though the enemy subsequently evacuated it; in the latter case, that the temporary military occupation by the United States of a foreign port did not make it a domestic port, and that goods imported into the United States from that port were still subject to duty. It would have been obviously unjust in the *Rice* case to impose a duty upon goods which might already have paid a duty to the British commander. It would have been equally unjust in the *Fleming* case to exempt the goods from duty by reason of our temporary occupation of the port without a formal cession of such port to the United States.

The next case is that of *Cross* v. *Harrison*, 16 How. 164. This was an action of assumpsit to recover back moneys paid to Harrison while acting as collector at the port of San Francisco for tonnage and duties upon merchandise imported from foreign countries into California between February 2, 1848,—the date of the treaty of peace between the United States and Mexico— and November 13, 1849, when the collector appointed by the President (according to an act of Congress passed March 3, 1849,) entered upon his duties. Plaintiffs insisted that, until such collector had been appointed, California was and continued to be after the date of the treaty a foreign territory, and hence that no duties were payable as upon an importation into the United States. The plaintiffs proceeded upon the theory, stated in the *dictum* in *Fleming* v. *Page*, that duties had never been held to accrue to the United States in her newly acquired territories until provision was made by act of Congress for their collection, and that the revenue laws had always been held to speak only as to the United States and its territories existing at the time when the several acts were passed. The collector had been appointed by the military governor of California, and duties were assessed, after the treaty, according to the United States tariff act of 1846. In holding that these duties were properly assessed, Mr. Justice Wayne cited with apparent approval a despatch written by Mr. Buchanan, then Secretary of State, and a circular letter issued by the Secretary of the Treasury, Mr. Robert J. Walker, holding that from the necessities of the case the military government established in California did not cease to exist with the treaty of peace, but continued as a government *de facto* until Congress should provide a territorial government. "The great law of necessity," says Mr. Buchanan, "justifies this conclusion. The consent of the people is irresistibly inferred from the fact that no civilized community could possibly desire to abrogate an existing government, when the alternative presented would be to place themselves in a state of anarchy, beyond the protection of all laws, and reduce them to the unhappy necessity of submitting to the dominion of the strongest." These letters will be alluded to hereafter in treating of the action of the executive departments.

The court further held in this case that "after the ratification of the treaty, California became a part of the United States, or a ceded, conquered, territory;" that, "as there is nothing differently stipulated in the treaty with respect to commerce, it became instantly bound and privileged by the laws which Congress had passed to raise a revenue from duties on imports and tonnage;" that (p. 193) "the territory had been ceded as a conquest, and was to be preserved and governed as such until the sovereignty to which it had passed had legislated for it. That sovereignty was the United States, under the Constitution, by which power had been given to Congress to dispose of and make all needful rules and regulations respecting the territory or other property belonging to the United States. . . . That the civil government of California, organized as it was from a right of conquest, did not cease or become defunct in consequence of the signature of the treaty, or from its ratification, . . . and that until Congress legislated for it, the duty upon foreign goods imported into San Francisco were legally demanded and lawfully received by Mr. Harrison."

To the objection that no collection districts had been established in California, and in apparent dissent from the views of the Chief Justice in *Fleming* v. *Page*, he added (p. 196): "It was urged that our revenue laws covered only so much of the territory of the United States as had been divided into collection districts, and that out of them no authority had been given to prevent the landing of foreign goods or to charge duties upon them, though such landing had been made within the territorial limits of the United States. To this it may be successfully replied, that collection districts and ports of entry are no more than designated localities within and at which Congress had extended a liberty of commerce in the United States, and that so much of its territory as was not within any collection district must be considered as having been withheld from that liberty. It is very well understood to be a part of the law of nations that each nation may designate, upon its own terms, the ports and places within its territory for foreign commerce, and that any attempt to introduce foreign goods elsewhere, within its jurisdiction, is a violation of its sovereignty. It is not necessary that such should be declared in terms, or by any decree or enactment, the expressed allowance being the limit of the liberty given to foreigners to trade with such nation."

The court also cited the cases of Louisiana and Florida, and seemed to take an entirely different view of the facts connected with the admission of those territories from what had been taken in *Fleming* v. *Page*. The opinion, which is quite a long one, establishes the three following propositions : (1) That under the war power the military governor of California was authorized to prescribe a scale of duties upon importations from foreign countries to San Francisco, and to collect the same through a collector appointed by himself, until the ratification of the treaty of peace. (2) That after such ratification duties were legally exacted under the tariff laws of the United States, which took effect immediately. (3) That the civil government established in California continued from the necessities of the case until Congress provided a territorial government.

It will be seen that the three propositions involve a recognition of the fact that California became domestic territory immediately upon the ratification of the treaty, or, to speak more accurately, as soon as this was officially known in California. The doctrine that a port ceded to and occupied by us does not lose its foreign character until Congress has acted, and a collector is appointed, was distinctly repudiated with the apparent acquiescence of Chief Justice Taney, who wrote the opinion in *Fleming* v. *Page*, and still remained the Chief Justice of the court. The opinion does not involve directly the question at issue in this case: whether goods carried from a port in a ceded territory directly to New York are subject to duties, since the duties in *Cross* v. *Harrison* were exacted upon foreign goods imported into San Francisco as an American port; but it is impossible to escape the logical inference from that case that goods carried from San Francisco to New York after the ratification of the treaty would not be considered as imported from a foreign country.

The practice and rulings of the executive departments with respect to the *status* of newly acquired territories, prior to such *status* being settled by acts of Congress, is, with a single exception, strictly in line with the decision of this court in *Cross* v. *Harrison, supra*. The only possessions in connection with which the question has arisen are Louisiana, Florida, Texas, California and Alaska. We take these up in their order.

LOUISIANA: By treaty between France and Spain, October 1, 1800, 8 Stat. 202, His Catholic Majesty promised to cede to the French Republic the colony or province of Louisiana; and by treaty between the United States and the French Republic of April 30, 1803, France ceded to the United States, "forever and in full sovereignty, the said territory with all its rights and appurtenances," with a provision, (Art. 3,) "that the inhabitants of the ceded territory shall be incorporated in the Union of the United States, and admitted as soon as possible, according to the principles of the Federal Constitution." This treaty was ratified October 21, 1803. Possession of the territory was not delivered by Spain to France until November 30, 1803, and by France to the United States, December 20, 1803. In the meantime, and on October 31, 1803, Congress authorized the President to take possession of the territory, and to administer it until Congress had further acted upon the subject. 2 Stat. 245. On February 24, 1804, Congress passed another act, 2 Stat. 251, taking Louisiana within the Customs Union, and repealing certain special laws laying duties upon goods imported from that territory into the United States. This act was to take effect March 25, 1804. We are then concerned only with the interval between December 20, 1803, when possession was delivered to the United States, and March 25, 1804, when the act of February 24 took effect.

In a letter to President Jefferson of July 9, 1803, Mr. Gallatin, then Secretary of the Treasury, expressed the opinion that all the duties on exports, now payable at New Orleans by Spanish laws, should cease, and all articles the growth of Louisiana, which, when imported into the United States, now pay duty, should continue to pay the same, or at least such rates as would on the whole not affect the revenue. Writings of Gallatin, vol. 1, p. 127.

The instructions of the Treasury Department with respect to this interval are contained in a letter by Mr. Gallatin to Governor Claiborne, who was about to start for his post as governor of the new province, under date of October 3, 1803, in which he says: "It is understood that the existing duties on imports and exports, which by the Spanish law are now levied within the province, will continue until Congress shall have otherwise provided." On November 14, 1803, Mr. Gallatin issued an order directed to Mr. Trist, who had been designated as collector of the port of New Orleans, as follows: "You will also be pleased to observe, first, that the taxes and the duties to be collected under your direction are precisely the same which by the existing laws and regulations of Louisiana were demandable under the Spanish government at the time of taking possession. . . . 10. That until otherwise provided for, the same duties are to be collected on the importation of goods in the Mississippi district, from New Orleans and *vice versa*, as heretofore."

On February 28, 1804, Mr. Gallatin issued a circular letter notifying the collectors of the passage of the act of February 24, and that the same would go into effect March 25, and "that by the third section of said act so much of any law or laws imposing duties on the importations into the United States of goods, wares and merchandise from New Orleans, which is the only port of entry in said territories, has been repealed."

These instructions undoubtedly show that Mr. Gallatin treated New Orleans as a foreign port until Congress, by the act of February 24, 1804, admitted it within the Customs Union, and, so far, is an authority in favor of the position taken by the collector in this case. But it should be borne in mind in this connection, that his instructions to collect duties levied by the *Spanish law* upon foreign importations into New Orleans, is manifestly inconsistent with the position subsequently taken by this court in *Cross* v. *Harrison, supra,* wherein it is said (p. 189) of the action of Mr. Harrison in California: "That war tariff, however, was abandoned as soon as the military governor had received from Washington information of the exchange and ratification of the treaty with Mexico, and duties were afterwards levied in conformity with such as Congress had imposed upon foreign merchandise imported into other ports of the United States, Upper California having been ceded by the treaty to the United States." After saying that this action had been recognized by the President, Mr. Justice Wayne adds: "We think it was a rightful and correct recognition under all the circumstances, and when we say rightful we mean that it was constitutional, although Congress had not passed an act to extend the collection of tonnage and import duties to the ports of California." Indeed, it is quite evident from this case that the court took an entirely different view of the relations of California to the Union from that which had been taken by Mr. Gallatin as to Louisiana in his instructions to the collector of New Orleans.

FLORIDA: Florida was ceded by Spain to the United States by treaty signed February 22, 1819, but not ratified until October 29, 1820. 8 Stat. 252. By act of March 3, 1821, 3 Stat. 637, Congress authorized the President to take possession of the Floridas and extend thereto the revenue laws of the United States. Possession of East Florida was not delivered until July 10, 1821; nor of West Florida until July 17. It is true that certain ports of Florida were in the military occupation of the United States prior to the actual delivery of possession by

Spain, but the cession did not take effect until there had been a voluntary and complete delivery under the treaty. As the act extending the revenue laws to the Floridas was passed before the surrender of the province to the United States, there was no interval of time upon which the Treasury Department could act, the provinces, immediately upon the surrender, becoming subject to the act of March 3, 1821.

An opinion of Mr. Wirt, then Attorney General, of August 20, 1821, in the case of *The Olive Branch*, 1 Ops. Atty. Gen. 314, 483, is instructive in this connection as illustrating the views of the administration. After stating that possession of East Florida was not delivered until July 17, (a mistake for July 10,) he held that the cargo of the Olive Branch, which had cleared from the port of St. Augustine, July 14, was imported into Philadelphia from a foreign port or place, and consequently subject to duty, because possession had not been delivered, citing the case of *The Fama*, 5 Ch. Rob. 97, and adding: " On the other hand, I apprehend that goods imported into a port of Florida before the delivery, remaining in port on shipboard until after the delivery, and then brought into the United States in the same vessel, or by transhipment into others, having never been entered in the Spanish customs houses, nor landed, nor the duties thereon paid or secured, but having continued all the while water-borne, would be subject to our revenue laws. . . . Our laws impose duties only on goods imported into the United States from some foreign port or place. If, therefore, in the case put, the importation be, in contemplation of law, an importation from the Floridas, the case is not within our laws; because at the time of the importation the Floridas were not foreign ports or places." The learned Attorney General evidently took the view that the Floridas ceased to be a foreign country upon a delivery of possession under the treaty. In a subsequent letter of January 24, 1823, 5 Ops. Atty. Gen. 748, Mr. Wirt admits that he had been misled by the newspapers in the belief that East Florida had been surrendered prior to July 14, on which day the Olive Branch left St. Augustine, and recommended that the case be sent to the President, as it seemed to involve a dispute with Great Britain.

Tᴇxᴀs: On March 1, 1845, Congress adopted a joint resolution consenting to the annexation of Texas upon certain conditions, 5 Stat. 797, but it was not until December 25, 1845, that it was formally admitted as a State. 9 Stat. 108. In this interval, and on July 29, 1845, the Secretary of the Treasury issued a circular letter directing the collectors to collect duties upon all imports from Texas into the United States until Congress had further acted. Of course, there could be no question that Texas remained a foreign state until December 25, when she was formally admitted. The circular, therefore, is of no pertinence to the question here involved.

Cᴀʟɪꜰᴏʀɴɪᴀ : California was ceded by Mexico to the United States by treaty signed February 2, 1848, ratifications of which were exchanged May 30, 1848, and proclamation made July 4. 9 Stat. 922. On March 3, 1849, an act was passed, 9 Stat. 400, including San Francisco within one of the collection districts, and on November 13 the collector appointed by the President entered upon his duties. California had been in our military possession since August, 1847. There was therefore an interval of one year and nine months between the date of the treaty, February 3, 1848, and November 13, 1849, when the collector entered upon his duties.

On October 7, 1848, Mr. Buchanan, then Secretary of State, addressed a letter to Mr. Vorhies, already referred to, in which he states that, although the military government ceased to exist with the conclusion of the treaty of peace, it would continue with the presumed consent of the people until Congress should provide for them a territorial government, and then adds: "This government *de facto* will, of course, exercise no power inconsistent with the provisions of the Constitution of the United States, which is the supreme law of the land. For this reason no import duties can be levied in California on articles of growth, produce or manufacture of the United States, as no such duties can be imposed in any other port of our Union on the productions of California. Nor can new duties be charged in California upon such foreign productions as have already paid duties in any of our ports of entry, for the obvious reason that California is within the territory of the United

States. I shall not enlarge upon this subject, however, as the Secretary of the Treasury will perform that duty." Ex. Docs. 2d Sess. 30th Cong. vol. 1, p. 47.

Mr. Walker, then Secretary of the Treasury, did perform that duty in a circular letter of the same date to the collectors, in which he instructed the collectors as follows: "First, All articles of the growth, produce or manufacture of California, shipped therefrom at any time since the 30th day of May last," (the date when the ratifications were exchanged), "are entitled to admission free of duty into all the ports of the United States; and, second, all articles of the growth, produce or manufacture of the United States are entitled to admission free of duty into California, as are also all foreign goods which are exempt from duty by the laws of Congress, or on which goods the duties prescribed by those laws have been paid to any collector of the United States previous to their introduction into California." Ibid. p. 45. He adds that foreign goods imported into California, not paying duties there, will be subject to duty if shipped thence to any port or place in the United States. In a letter from Mr. Marcy, Secretary of War, to Colonel Mason, the military commander, of October 9, 1848, he uses the same language.

These letters are cited with approval by this court in *Cross* v. *Harrison*, 16 How. 184, and although the question there related only to duties on goods imported from foreign countries, the tenor of the opinion, as already stated, is a virtual indorsement of the position taken by the executive departments. It is evident that the administration took an entirely different view of the law from what had been taken by Mr. Gallatin in his instructions regarding Louisiana, and established a practice which has never since been departed from, of treating territory ceded to the United States and occupied by its troops as being domestic and not foreign territory.

This correspondence with reference to California took place in 1848. The decision in *Fleming* v. *Page*, 9 How. 603, was pronounced in 1850, yet as appears from the list of documents submitted by Mr. Johnson upon the argument of that case, (p. 611,) the attention of the court was not called to these instructions, though other letters and circulars were introduced bearing date of 1846 and 1847, as well as the treaty of peace of February 2, 1848. Had the correspondence above cited been laid before the court it is incredible that the Chief Justice should have said "that the department in no instance that we are aware of, since the establishment of the government, has ever recognized a place in a newly acquired country as a domestic port, from which the coasting trade might be carried on, unless it had been previously made so by act of Congress."

ALASKA : This territory was ceded to us by Russia by treaty ratified June 20, 1867, 15 Stat. 539, and possession was delivered to us at the same time. No act of Congress extending the revenue laws to Alaska and erecting a collection district was passed until July 27, 1868. 15 Stat. 240, c. 273. A period of thirteen months then elapsed before Alaska was formally recognized by Congress as within the Customs Union, yet during that period goods from Alaska were, under a decision of the Secretary of the Treasury, admitted free of duty. By letter of Mr. McCullough, then Secretary of the Treasury, to the collector of the port of New York, dated April 6, 1868, he acknowledges receipt of a request from the Russian Minister for the free entry of certain oil shipped from Sitka to San Francisco and reshipped to New York. He states: "The request for the free entry of said oil was made on the ground that the oil was shipped from Sitka after the ratification of the treaty, by which the territory of Alaska became the property of the United States. The treaty in question was ratified on the 20th of June, 1867, and the collector at San Francisco has reported that the manifest of the vessel shows the oil to have been shipped from Alaska on the 6th day of July, 1867, and that the shipment consisted of fifty-two packages. Under these circumstances you are hereby authorized to admit the said fifty-two packages of oil free of duty."

This position was indorsed by the Secretary of State, Mr. Seward, in a letter dated January 30, 1869, in which he said: "I understand the decision of the Supreme Court in the case of *Harrison* v. *Cross*, 16 How. 164, to declare its opinion that, upon the addition to the United States of new territory by conquest and cession, the acts regulating foreign commerce attach to and take effect within such territory *ipso facto*, and without any fresh act of legislation expressly giving such extension to the preëxisting laws.   I can see no reason for a discrimination in this effect between acts regulating foreign commerce and the laws regulating intercourse with the Indian tribes."

As showing the construction put upon this question by the legislative department, we need only to add that sec. 2 of the Foraker act makes a distinction between foreign countries and Porto Rico, by enacting that the same duties shall be paid upon "all articles imported into Porto Rico from ports other than those of the United States, which are required by law to be collected upon articles imported into the United States from foreign countries."

From this *résumé* of the decisions of this court, the instructions of the executive departments, and the above act of Congress, it is evident that, from 1803, the date of Mr. Gallatin's letter, to the present time, there is not a shred of authority, except the *dictum* in *Fleming* v. *Page*, (practically overruled in *Cross* v. *Harrison*,) for holding that a district ceded to and in the possession of the United States remains for any purpose a foreign country.   Both these conditions must exist to produce a change of nationality for revenue purposes.   Possession is not alone sufficient, as was held in *Fleming* v. *Page;* nor is a treaty ceding such territory sufficient without a surrender of possession. *Keene* v. *McDonough*, 8 Pet. 308 ; *Pollard's Heirs* v. *Kibbe*, 14 Pet. 353, 406 ; *Hallett* v. *Hunt*, 7 Ala. 882, 899 ; *The Fama*, 5 Ch. Rob. 97.   The practice of the executive departments, thus continued for more than half a century, is entitled to great weight, and should not be disregarded nor overturned except for cogent reasons, and unless it be clear that such construction be erroneous.   *United States* v. *Johnston*, 124 U. S. 236, and other cases cited.

But were this presented as an original question we should be impelled irresistibly to the same conclusion.

By Article II, section 2, of the Constitution, the President is given power, "by and with the advice and consent of the Senate, to make treaties, provided that two-thirds of the senators present concur;" and by Art. VI, "this Constitution and the laws of the United States, which shall be made in pursuance thereof; and all treaties made or which shall be made, under the authority of the United States, shall be the supreme law of the land." It will be observed that no distinction is made as to the question of supremacy between laws and treaties, except that both are controlled by the Constitution. A law requires the assent of both houses of Congress, and, except in certain specified cases, the signature of the President. A treaty is negotiated and made by the President, with the concurrence of two thirds of the Senators present, but each of them is the supreme law of the land.

As was said by Chief Justice Marshall in *The Peggy*, 1 Cranch, 103, 110: "Where a treaty is the law of the land, and as such affects the rights of parties litigating in court, that treaty as much binds those rights, and is as much to be regarded by the court as an act of Congress." And in *Foster* v. *Neilson*, 2 Pet. 253, 314, he repeated this in substance: "Our Constitution declares a treaty to be the law of the land. It is, consequently, to be regarded in courts of justice as equivalent to an act of the legislature, whenever it operates of itself without the aid of any legislative provision." So in *Whitney* v. *Robertson*, 124 U. S. 190: "By the Constitution a treaty is placed on the same footing, and made of like obligation, with an act of legislation. Both are declared by that instrument to be the supreme law of the land, and no superior efficacy is given to either over the other. When the two relate to the same subject, the courts will always endeavor to construe them so as to give effect to both, if that can be done without violating the language of either; but if the two are inconsistent, the one last in date will control the other, provided always that the stipulation of the treaty on the subject is self-executing." To the same effect are the *Cherokee Tobacco*, 11 Wall. 616, and the *Head Money Cases*, 112 U. S. 580.

One of the ordinary incidents of a treaty is the cession of territory. It is not too much to say it is the rule, rather than the exception, that a treaty of peace, following upon a war, provides for a cession of territory to the victorious party. It was said by Chief Justice Marshall in *American Ins. Co.* v. *Canter*, 1 Pet. 511, 542: "The Constitution confers absolutely upon the Government of the Union the powers of making war and of making treaties; consequently that Government possesses the power of acquiring territory, either by conquest or by treaty." The territory thus acquired is acquired as absolutely as if the annexation were made, as in the case of Texas and Hawaii, by an act of Congress.

It follows from this that by the ratification of the treaty of Paris the island became territory of the United States—although not an organized territory in the technical sense of the word.

It is true Mr. Chief Justice Taney held in *Scott* v. *Sandford*, 19 How. 393, that the territorial clause of the Constitution was confined, and intended to be confined, to the territory which at that time belonged to or was claimed by the United States, and was within their boundaries, as settled by the treaty with Great Britain; and was not intended to apply to territory subsequently acquired. He seemed to differ in this construction from Chief Justice Marshall in the *American &c. Ins. Co.* v. *Canter*, 1 Pet. 511, 542, who, in speaking of Florida before it became a State, remarked that it continued to be a Territory of the United States, governed by the territorial clause of the Constitution.

But whatever be the source of this power, its uninterrupted exercise by Congress for a century, and the repeated declarations of this court, have settled the law that the right to acquire territory involves the right to govern and dispose of it. That was stated by Chief Justice Taney in the *Dred Scott* case. In the more recent case of *National Bank* v. *County of Yankton*, 101 U. S. 129, it was said by Mr. Chief Justice Waite that Congress "has full and complete legislative authority over the people of the Territories and all the departments of the territorial governments. It may do for the Territories what the people, under the Constitution of the United States, may do for the States." Indeed, it is scarcely too much to say that there has not been a session of Congress since the Territory of Louisiana was purchased, that that body has not enacted legislation based upon the assumed authority to govern and control the Territories. It is an authority which arises, not necessarily from the territorial clause of the Constitution, but from the necessities of the case, and from the inability of the States to act upon the subject. Under this power Congress may deal with territory acquired by treaty; may administer its government as it does that of the District of Columbia; it may organize a local territorial government; it may admit it as a State upon an equality with other States; it may sell its public lands to individual citizens or may donate them as homesteads to actual settlers. In short, when once acquired by treaty, it belongs to the United States, and is subject to the disposition of Congress.

Territory thus acquired can remain a foreign country under the tariff laws only upon one of two theories: either that the word "foreign" applies to such countries as were foreign at the time the statute was enacted, notwithstanding any subsequent change in their condition, or that they remain foreign under the tariff laws until Congress has formally embraced them within the customs union of the States. The first theory is obviously untenable. While a statute is presumed to speak from the time of its enactment, it embraces all such persons or things as subsequently fall within its scope, and ceases to apply to such as thereafter fall without its scope. Thus, a statute forbidding the sale of liquors to minors applies not only to minors in existence at the time the statute was enacted, but to all who are subsequently born; and ceases to apply to such as thereafter reach their majority. So, when the Constitution of the United States declares in Art. I, sec. 10, that the States shall not do certain things, this declaration operates not only upon the thirteen original States, but upon all who subsequently become such; and when Congress places certain restrictions upon the powers of a territorial legislature, such restrictions cease to operate the moment such Territory is admitted as a State. By parity of reasoning a country ceases to be foreign the instant it becomes domestic. So, too, if Congress saw fit to cede one of its newly acquired territories (even assuming that it had the right to do so) to a foreign power, there could be no doubt that from the day of such cession and the delivery of possession, such territory would become a foreign country, and be reinstated as such under the tariff laws. Certainly no act of Congress would be necessary in such case to declare that the laws of the United States had ceased to apply to it.

The theory that a country remains foreign with respect to the tariff laws until Congress has acted by embracing it within the Customs Union, presupposes that a country may be domestic for one purpose and foreign for another. It may undoubtedly become necessary for the adequate administration of a domestic territory to pass a special act providing the proper machinery and officers, as the President would have no authority, except under the war power, to administer it himself; but no act is necessary to make it domestic territory if once it has been ceded to the United States. We express no opinion as to whether Congress is bound to appropriate the money to pay for it. This has been much discussed by writers upon constitutional law, but it is not necessary to consider it in this case, as Congress made prompt appropriation of the money stipulated in the treaty. This theory also presupposes that territory may be held indefinitely by the United States; that it may be treated in every particular, except for tariff purposes, as domestic territory; that laws may be enacted and enforced by officers of the United States sent there for that purpose; that insurrections may be suppressed, wars carried on, revenues collected, taxes imposed; in short, that everything may be done which a government can do within its own boundaries, and yet that the territory may still remain a foreign country. That this state of things may continue for years, for a century even, but that until Congress enacts otherwise, it still remains a foreign country. To hold that this can be done as matter of law we deem to be pure judicial legislation. We find no warrant for it in the Constitution or in the powers conferred upon this court. It is true the nonaction of Congress may occasion a temporary inconvenience; but it does not follow that courts of justice are authorized to remedy it by inverting the ordinary meaning of words.

If an act of Congress be necessary to convert a foreign country into domestic territory, the question at once suggests itself, what is the character of the legislation demanded for this purpose? Will an act appropriating money for its purchase be sufficient? Apparently not. Will an act appropriating the duties collected upon imports to and from such country for the benefit of its government be sufficient? Apparently not. Will acts making appropriations for its postal service, for the establishment of lighthouses, for the maintenance of quarantine stations, for erecting public buildings, have that effect? Will an act establishing a complete local government, but with the reservation of a right to collect duties upon commerce, be adequate for that purpose? None of these, nor all together, will be sufficient, if the contention of the Government be sound, since acts embracing all these provisions have been passed in connection with Porto Rico, and it is insisted that it is still a foreign country within the meaning of the tariff laws. We are unable to acquiesce in this assumption that a territory may be at the same time both foreign and domestic.

A single further point remains to be considered : It is insisted that an act of Congress, passed March 24, 1900, c. 339, 31 Stat. 151, applying for the benefit of Porto Rico the amount of the customs revenue received on importations by the United States from Porto Rico since the evacuation of Porto Rico by the Spanish forces, October 18, 1898, to January 1, 1900, together with any further customs revenues collected on importations from Porto Rico since January 1, 1900, or that shall hereafter be collected under existing law, is a recognition by Congress of the right to collect such duties as.upon importations from a foreign country, and a recognition of the fact that Porto Rico continued to be a foreign country until Congress embraced it within the Customs Union. It may be seriously questioned whether this is anything more than a recognition of the fact that there were moneys in the Treasury not subject to existing appropriation laws. Perhaps we may go farther and say that, so far as these duties were paid voluntarily and without protest, the legality of the payment was intended to be recognized ; but it can clearly have no retroactive effect as to moneys theretofore paid under protest, for which an action to recover back had already been brought. As the action in this case was brought March 13, 1900, eleven days before the act was passed, the right to recover the money sued for could not be taken away by a subsequent act of Congress. Plaintiffs sue in assumpsit for money which the collector has in his hands, justly and equitably belonging to them. To say that Congress could by a subsequent act deprive them of the right to prosecute this action, would be beyond its power. In any event, it should not be interpreted so as to make it retroactive. *Kennett's Petition*, 24 N. H. 139; *Alter's Appeal*, 67 Penn. St. 341; *Norman* v. *Heist*, 5 W. & S. 171; *Donavan* v. *Pitcher*, 53 Ala. 411; *Palairet's Appeal*, 67 Penn. St. 479; *State* v. *Warren*, 28 Maryland, 338.

We are therefore of opinion that at the time these duties were levied Porto Rico was not a foreign country within the meaning of the tariff laws but a territory of the United States, that the duties were illegally exacted and that the plaintiffs are entitled to recover them back.

The judgment of the Circuit Court for the Southern District of New York is therefore reversed and the case remanded to that court for further proceedings in consonance with this opinion.

MR. JUSTICE McKENNA, (with whom concurred MR. JUSTICE SHIRAS and MR. JUSTICE WHITE,) dissenting.

MR. JUSTICE SHIRAS, MR. JUSTICE WHITE and myself are unable to concur in the conclusion of the court, and the importance of the case justifies an expression of the grounds of our dissent.

Settle whether Porto Rico is "foreign country" or "domestic territory," to use the antithesis of the opinion of the court, and, it is said, you settle the controversy in this litigation. But in what sense, foreign or domestic? Abstractly and unqualifiedly—to the full extent that those words imply—or limitedly, in the sense that the word foreign is used in the customs laws of the United States? If abstractly, the case turns upon a definition, and the issue becomes single and simple, presenting no difficulty, and yet the arguments at bar have ranged over all the powers of government, and this court divides in opinion. If at the time the duties, which are complained of, were levied, Porto Rico was as much a foreign country as it was before the war with Spain; if it was as much domestic territory as New York now is, there would be no serious controversy in the case. If the former, the terms and the intention of the Dingley act would apply. If the latter, whatever its words or intention, it could not be applied. Between these extremes there are other relations, and that Porto Rico occupied one of them and its products hence were subject to duties under the Dingley Tariff act can be demonstrated. Indeed, we have the authority of a member of the majority of the court, and the organ of the court's opinion in this case, that even if Porto Rico were domestic territory, its products could be legally subjected to tariff duties. This principle is expressed by him in *Downes v. Bidwell.* The other members of the court, though agreeing with him in the case at bar, do not agree with him in *Downes v. Bidwell.* They assert that Porto Rico, being a territory of the United States, tariff duties on its products are inhibited by the Constitution of the United States. Their judgment and his only unite in the case at bar, and, we may assume, that the reasoning of the opinion just announced is the road which has brought them together, and, assuming further, that such reasoning is the best judicial support of the conclusion it is presented to establish, we address ourselves to the consideration of that reasoning.

(1) The statement of the opinion is that whether the cargoes of sugar were subject to duty depends solely upon the question whether Porto Rico was a foreign country at the time they were shipped, and a foreign country is defined to be, following Chief Justice Marshall, "'one exclusively within the sovereignty of a foreign nation' and without the sovereignty of the United States." This makes sovereignty the test and gives a rule as sure and exact in its application as it is clear and simple in its expression. There is no difficulty in applying it. Difficulty comes with attempts to limit it. The difference between our country and one not ours would seem to be of substance, not needing words to explain the difference, but defying words to confound it, and having the consequence of carrying, not only one law, but all laws. The court does not go so far, and why? Is there weakness in the logic or do its consequences repel? The argument of the court certainly proceeds as if the test is universal—illustrations are used to make it unmistakable.

Under the effect of the treaty of cession and our government of Porto Rico, it is said, if the question was broadly presented

whether it was "a foreign country or domestic territory," there
would be as little hesitation in answering the question "as
there would be in determining the ownership of a house deeded
in fee simple to a purchaser, after he had gone into possession,
paid taxes and made improvements, without let or hindrance,
from his vendor." And we would have as little hesitation in
applying all of the consequences and concomitants of ownership. But we do not care to join issue on an illustration, although it may suggest wrong principles. We submit that the
administration of a government has more complexity—must
consider more things—than the management of a piece of real
estate. But even the conveyance of real estate may be conditional, all of the incidents of ownership not immediately applying. However, we need not dwell on insufficient analogies.
There are better ones. The history of our country has examples of the acquisition of foreign territory—examples of what
relation such territory bears to the United States—authorities,
executive, legislative and judicial, as to what was wise in states-
manship, as well as what was legal and constitutional, in withholding or extending, our laws to such territory ; and finding
these examples and authorities in the way the opinion of the
court attempts to answer or distinguish or overrule them.

United States v. *Rice*, 4 Wheat. 246, is reviewed. In that
case, Castine, a port of the United States, was in temporary
occupation by the British during the war of 1812, and it was
declared to be a foreign country within the meaning of our
customs laws ; as much, the court said by Mr. Justice Story, as
if "Castine had been a foreign territory ceded by treaty to the
United States, and the goods had been previously imported
there." In other words, not a cession to another country, but
the accidental occupation by the armed forces of another country made a port in the State of Maine foreign territory. The
conclusion had the sanction of great names and the authority
of this court. Temporary sovereignty, not permanent dominion, was seemingly made the test.

Fleming v. *Page*, 9 How. 603, is also reviewed. The case involved the legality of duties levied in Philadelphia upon goods
imported from Tampico. Tampico was a port of Mexico, tem-

porarily occupied by the United States forces—the exact condition which, in the *Rice* case, made a port in one of the States of our Union English territory. Tampico was nevertheless held to be a foreign country within the meaning of our revenue laws. In other words, the military occupation and the sovereignty which attended it, which determined in the *Rice* case, was rejected in the *Fleming* case. There is apparent antagonism between the cases, and the court in the case at bar observe it. And strangely enough, that which is "somewhat of the converse" (to quote the court in the case at bar) of the *Rice* case is held sufficient for the judgment in the *Fleming* case, and other grounds of decision are declared to be *dicta*.

An attempt is made, however, to reconcile the cases, and we think they can be reconciled, but not upon the grounds stated by the court in the opinion in the case at bar. Harmony cannot be established between them by that which in the *Fleming* case is the converse of the *Rice* case, and by rejecting as *dicta* all other grounds as unnecessary to the judgment in the *Fleming* case. However, we will proceed to the consideration of the latter case.

Delivering the opinion of the court, Chief Justice Taney substantially said that the boundaries of our country could not be enlarged or diminished by the advance or retreat of armies, and based his opinion besides and the judgment of the case on the absence of an act of Congress establishing a custom house at Tampico, and authorizing the appointment of a collector, " and, consequently, there was no officer of the United States authorized by law to grant the clearance and authenticate the coasting manifest of the cargo, in the manner directed by law, where the voyage is from one port of the United States to another," and the necessity of a legal permit and coasting manifest was expressly asserted. He further said :

" This construction of the revenue laws has been uniformly given by the administrative department of the government in every case that has come before it. And it has, indeed, been given in cases where there appears to have been stronger ground for regarding the place of shipment as a domestic port. For after Florida had been ceded to the United States, and the forces of the United States had taken possession of Pensacola, it was decided by the Treasury Department that goods imported from Pensacola before an act of Congress was passed erecting it into a collection district, and authorizing the appointment of a collector, were liable to duty. That is that although Florida had, by cession, actually become a part of the United States, and was in our possession, yet, under our revenue laws, its ports must be regarded as foreign until they were established as domestic, by act of Congress; and it appears that this decision was sanctioned at the time by the Attorney General of the United States, the law officer of the government. And although not so directly applicable to the case before us, yet the decisions of the Treasury Department in relation to Amelia Island, and certain ports in Louisiana, after that province had been ceded to the United States, were both made upon the same grounds. And in the latter case, after a custom house had been established by law at New Orleans, the collector at that place was instructed to regard as foreign ports Baton Rouge and other settlements still in the possession of Spain, whether on the Mississippi, Iberville, or the seacoast. The department in no instance that we are aware of since the establishment of the government, has ever recognized a place in a newly acquired country as a domestic port, from which the coasting trade might be carried on, unless it had been previously made so by act of Congress."

The opinion in the case at bar disregards this reasoning and the conclusion from it, and says: "While we see no reason to doubt the conclusion of the court (in *Fleming* v. *Page*) that the port of Tampico was still a foreign port, it is not perceived why the fact that there was no act of Congress establishing a custom house there and authorizing the appointment of a collector should have prevented the collector appointed by the military commander from granting the usual documents required to be issued to the vessel engaged in the coasting trade." Such power, it was said, "a military commander may be presumed to have," but, "of course, he would have no power to make a domestic port of what was in reality a foreign port." But why did it remain a foreign port? Castine did not remain a domestic port. We, however, need not dwell any longer on this point for, under the latest utterances of this court, the test of dominion breaks down. Cuba is under the dominion of the United States. We held in the *Neely Case*, 180 U. S. 109, that it is a foreign country.

We think that *Fleming* v. *Page* is disposed of too summarily by the majority in the case at bar, and we have shown that it is not antagonistic to the *Castine* case. Both cases recognized inevitable conditions. At Castine the instrumentalities of the custom laws had been divested; at Tampico they had not been invested, and hence the language of the court: " The department, in no instance that we are aware of, since the establishment of the government, has ever recognized a place in a newly acquired country as a domestic port, from which the coasting trade might be carried on, unless it had been previously made so by act of Congress."

We submit that the principle upon which *Fleming* v. *Page* was based is still a proper principle for judicial application. Does it not make government provident, not haphazard, ignoring circumstances and producing good or ill accidentally? Does it not leave to the executive and the legislative departments that which pertains to them? Did it not stand as a guide to the executive—a warrant of action, so far as action might affect private rights? Indeed, what is of greater concern—so far as action might affect great public interests? It should, we submit, be accepted as a precedent. It is wise in practice; considerate of what government must regard, and of the different functions of the executive, legislative and judicial departments and of their independence. Why should it then be discarded as *dictum?* If constancy of judicial decision is necessary to regulate the relations and property rights of individuals, is not constancy of decision the more necessary when it may influence or has influenced the action of a nation? If the other departments of the government must look to the judicial for light, that light should burn steadily. It should not, like the exhalations of a marsh, shine to mislead.

The case of *Cross* v. *Harrison*, 16 How. 164, is relied on especially. The curiosity of that case is that all parties cite it, and this court even finds it as convenient and as variously adapt-

ive. It therefore challenges the application of the wise maxim
expressed by Chief Justice Marshall, " that general expressions
in every opinion are to be taken in connection with the case in
which those expressions are used."   And certainly to ascertain
the meaning of the court we must see what was before the court,
and interpret its opinion by that, and, if there is confusion in
its language, it may resolve itself into satisfactory meaning.

It is cited to sustain the proposition that immediately upon
the cession of territory it becomes a part of the United States,
"instantly bound and privileged by the laws which Congress
has passed to raise a revenue from duties on imports and tonnage."   This is the strongest expression of the case. It is attempted to be made its controlling one—the point decided.   It
was neither the point decided nor was it the controlling expression.   It was immediately accompanied by the qualification
"as there is nothing differently stipulated in the treaty in respect to commerce."   The effect of the qualification the opinion
in the present case does not explicitly notice, and we shall attempt to show with what meaning the expression was used, and
what was decided.

The case involved the legality of duties on imports into California between the 3d of February, 1848, and the 13th of November, 1849.   The time was divided by the plaintiffs in the
case "into two portions," the court said, "to each of which
they supposed that different rules of law attached;" and further,
that "the claim covered various amounts of money which were
paid at intervals between the 3d of February, 1848, and the
13th of November, 1849."   The first of those dates was that of
the treaty of peace between the United States and Mexico, and
the latter when Mr. Collier, a person who had been regularly
appointed collector at that port, entered upon the performance
of the duties of his office.   " During the whole of this period it
was alleged by the plaintiffs that there existed no legal authority
to receive or collect any duty whatever accruing upon goods imported from foreign countries."

Meeting the contention and replying to it fully, the court held
that the duties were legally levied and collected during the whole
of the period—from the 3d of February, 1848, until some time

in the following fall under the war tariff instituted by Governor
Mason; after that under the Walker tariff.. In other words,
before and after cession, under the war tariff. Speaking of
that tariff, the court said: "They (duties) were paid until some
time in the fall of 1848, at the rate of the war tariff, which had
been established early in the year before, by the direction of the
President of the United States." And speaking of the action
of Governor Mason, and the law which sanctioned it, it was
further said:

"He may not have comprehended fully the principle applicable to what he might rightly do in such a case, but he felt
rightly, and acted accordingly. He determined, in the absence of all instruction, to maintain the existing government.
The territory had been ceded as a conquest, and was to be preserved and governed as such until the sovereignty to which it
had passed had legislated for it. That sovereignty was the
United States, under the Constitution, by which power had
been given to Congress to dispose of and make all needful
rules and regulations respecting the territory or other property belonging to the United States, with the power also to
admit new States into this Union, with only such limitations
as are expressed in the section in which this power is given.
The government, of which Colonel Mason was the executive,
had its origin in the lawful exercise of a belligerent right over
a conquered territory. It had been instituted during the war
by the command of the President of the United States. It
was the government when the territory was ceded as a conquest, and it did not cease, as a matter of course, or as a necessary consequence of the restoration of peace. The President
might have dissolved it by withdrawing the army and navy officers who administered it, but he did not do so. Congress could
have put an end to it, but that was not done. The right inference from the inaction of both is that it was meant to be continued until it had been legislatively changed. No presumption
of a contrary intention can be made. Whatever may have been
the causes of delay, it must be presumed that the delay was consistent with the true policy of the government. And the more
so, as it was continued until the people of the territory met in

convention to form a state government, which was subsequently
recognized by Congress under its power to admit new States into
the Union."

And further replying to the contention that there was neither
treaty nor law permitting the collection of duties, "it having
been shown that the ratification of the treaty made California
a part of the United States, and that as soon as it became so
the territory became subject to the acts which were in force
to regulate foreign commerce with the United States, after
those had ceased which had been instituted for its regulation
as a belligerent right."

An important inquiry is, when did the laws cease "which
had been instituted for the regulation of the territory as a belligerent right," and how did they cease? The answer is instant—they ceased when the President withdrew them and
because he withdrew them. The laws of Congress did not instantly apply upon the cession. There was an interval of time,
during which they did not apply, and if there can be such interval, who is to judge of what duration it shall be? Who can
but the political department of the government, and how impracticable any other ruling would be. It is not for the judiciary
to question it. It involves circumstances which the judiciary
can take no account of or estimate. It is essentially a political
function.

We have quoted largely from *Cross* v. *Harrison* because it
is made the pivot of the opinion of the court in the present case,
and we will recur to it again. But it should be said now that
some of the expressions may be accounted for and understood
by the state of precedent opinion.

It is a matter of some surprise that the only explicit provision of the Constitution of the United States in regard to the
territory not embraced within the jurisdiction of a State is expressed in the following provision : "The Congress shall have
power to dispose of and make all needful rules and regulations
respecting the territory or other property of the United States."
What was meant by it, what its relation was to other provisions of the Constitution, was the subject of discussion. Gou-
veneur Morris, who wrote the provision, subsequently declared

that it was intended to confer power to govern acquisitions of territory as "provinces and allow them no voice in our councils." He admitted, however, that it was not expressed more pointedly in order to avert opposition. In his mind it certainly contemplated the government of after-acquired territory. In *Scott* v. *Sandford*, 19 How. 393, however, the provision was declared to be confined, and was intended to be confined, to the territory which at that time belonged to the United States. "It was a special provision for a known and particular territory, and to meet a present emergency, and nothing more." This conclusion was claimed to be established by the history of the times, "as well as the careful terms in which the article is framed." We will not stop to reconcile this conflict between him who wrote the provision and the court who interpreted it. The conflict was but an incident in the evolution of opinion. And there were other conflicts, or rather diversities of view, caused or encouraged by the silence of the Constitution. That instrument contained no provision for acquiring new territory. The power was derived from the powers of making war and of making peace, and might be accomplished by conquest or by treaty. There was a question, however, of the effect of an acquisition. It is certain that Mr. Jefferson doubted the power of incorporating new territory into the Union without an amendment to the Constitution, and the debates in Congress exhibit the diverse views held by public men on the relation which such territory would bear to the United States, the application of the laws to and the power of Congress over the acquired territory under the Constitution. We shall not stop to quote the debates. That will be done in a subsequent case, and the conclusion which they demonstrate expressed. It is only necessary for us to observe that distinctions always existed between territory which might be acquired (whether by purchase or by conquest) and that which was within the acknowledged limits of the United States, and also that which might be acquired by the establishment of a disputed line. These distinctions were conspicuous in the opinion of Mr. Justice Johnson, at circuit, in the case of *American Insurance Company* v. *Canter*, 1 Pet. 511. In that case the relation of Florida to the United States was necessary to be considered, and of that relation the learned Justice said:

"It is obvious that there is a material distinction between the territory now under consideration and that which is acquired from the aborigines, (whether by purchase or conquest,) within the acknowledged limits of the United States, as also that which is acquired by the establishment of a disputed line. As to both these, there can be no question that the sovereignty of the State or territory within which it lies, and of the United States, immediately attach, producing a complete subjection to all the laws and institutions of the two governments, local and general, unless modified by treaty. The question now to be considered relates to territories previously subject to the acknowledged jurisdiction of another sovereign; such as was Florida to the crown of Spain. *And on this subject we have the most explicit proof that the understanding of our public functionaries is, that the government and laws of the United States do not extend to such territory by the mere act of cession.*" The italics are ours.

All the history and utterances of the past declare the same way.

And how important those utterances and decisive of the present controversy! They were not the utterances of inattention and ignorance, and therefore to be discarded. They were the utterances of men whose actions illustrated them. They were the utterances of men (to borrow the thought of Benton) whose sacrifices made the Constitution possible, whose genius conceived and wrote it. Shall it be said that the farther time separates us from them the better we understand them—better than they understood themselves?

American Insurance Co. v. *Canter* came to this court and was argued by Mr. Webster. We may quote what he said. His views were more than those of an advocate. He expressed them elsewhere when a different, if not higher, duty demanded reflection, consideration and sincerity. "What is Florida?" he asked. "It is no part of the United States. How can it be? How is it represented? Do the laws of the United States reach Florida? Not unless by particular provision." And, responding to the argument, the court decided through Chief Justice

JUSTICES McKENNA, SHIRAS and WHITE, dissenting.

Marshall that the judicial power of the United States, as declared by the Constitution, did not extend to Florida, and the title to one hundred and fifty-six bales of cotton was held to pass by a sale under the order of a court, which consisted of a notary and five jurors, established by an act of the governor and council of Florida.

From the light of previous opinions the language of Mr. Justice Wayne, in *Cross* v. *Harrison*, receives explanation. The treaty with Mexico, following the war, defined the "boundaries of the United States," and made the reclaimed territory, which included California, a part of the United States. In other words, the acquisition (if it can be called such) of California was in recognition of boundaries, and hence the learned justice called it a part of the United States. But not uniformly. Mark this sentence: "But after the ratification of the treaty, California became a part of the United States or a ceded conquered territory." That his language marked a distinction there can be no doubt, but it was of no consequence to observe. The principle enforced did not need it. In either case the action of the president was the potent thing.

2. The line of judicial precedents relied upon in the opinion of the court in the case at bar ends with *Cross* v. *Harrison*, and the practice and rulings of the executive departments of the government are considered. They are said to be in accordance with the ruling ascribed to *Cross* v. *Harrison*, with but a single exception. If there is one legal exception the rule is gone. It is not a case where an exception can prove the rule; it is one where the exception destroys the rule. The exception was Louisiana. Between December 20, 1803, when possession was delivered to the United States, and March 25, 1804, when the act of February 24 became effective, Louisiana was treated as a foreign country under the customs laws; but this the court in the opinion just announced says "it is manifestly inconsistent with the position subsequently taken by this court in *Cross* v. *Harrison*, wherein it is said of the action of Mr. Harrison in California: 'That war tariff, however, was abandoned as soon as the military governor had received from Washington information of the exchange and ratification of the treaty with Mexico,

and duties were afterwards levied in conformity with such as Congress had imposed upon foreign merchandise imported into the other ports of the United States, Upper California having been ceded by the treaty to the United States. This last was done with the assent of the executive of the United States or without any interference to prevent it. Indeed, from the letter from the then Secretary of the Treasury, we cannot doubt that the action of the military governor of California was recognized as allowable and lawful by Mr. Polk and his cabinet.' After saying that, and this action having been recognized by the President, Mr. Justice Wayne adds : ' We think it was rightful and correct recognition under all circumstances, and when we say rightful we mean that it was constitutional, although Congress had not passed an act to extend the collection of tonnage and import duties to the ports of California. ' "

If the laws of Congress instantly applied, why was the recognition of the President necessary ? They could gain no legal efficacy from such recognition which they did not have without it, under the supposition that they applied on cession by their own force. Surely so obvious a consequence would have occurred to the court in *Cross* v. *Harrison,* and we cannot believe that the court used its language carelessly or uselessly. If the assent and recognition of the President were not necessary, why dwell upon them ? Why so confuse the statement of a simple principle—simple in application and expression—and cast doubt upon it by unnecessary qualifications ? The case, therefore, is not inconsistent with the ruling in regard to Louisiana. For a period of time, after the cession of Louisiana, President Jefferson treated it as foreign territory under the custom laws, and duties were levied upon its products, and no one disputed the legality of it. If the instance was not the same as in *Cross* v. *Harrison,* the principle was the same. There was not an immediate change upon the cession of either California or Louisiana. In California, duties were levied for a time under the war tariff, and afterwards under the act of Congress; and of the latter it was said : " This last was done either with the assent of the executive of the United States, or without any interference to prevent it." And this, it was further said, was "recognized as allowable and lawful by Mr. Polk and his cabinet." We are disposed to ask again, was the language inadvertent? Did not the court use it with full consciousness of its meaning and its necessity? Was the court in confusion as to the principles which applied and jumbled them together without seeing or making a distinction between the force of the act of Congress of itself and the action of the President in giving it efficacy, the necessity of its being recognized as "allowable and lawful by Mr. Polk and his cabinet?" Surely not. Rights were involved which depended upon the legality of the war tariff both before and after cession, and that legality was intended to be and was passed upon and sustained. An automatic effect was not given to the act of Congress as it is given in the case at bar. The act was applied by the President—not in simple execution of it, but as giving it legal effect. And it was this that the court said "was a rightful and correct recognition under all the circumstances." "Rightful," because "it was constitutional, although Congress had not passed an act to extend the collection of tonnage and import duties to the ports of California." In other words, an act of Congress was not necessary to extend the collection of duties; the power of the President was sufficient, and of that power the court left no doubt. Speaking of the duties which were collected under the war tariff after the cession, it was observed, " but after the ratification of the treaty, California became a part of the United States, or a ceded, conquered territory. Our inquiry here is to be, whether or not the cession gave any right to the plaintiffs to have the duties restored to them, which they may have paid between the ratification, and exchange of the treaty and the notification of that fact by our government to the military governor of California. It was not received by him until two months after the ratification, and not then with any instructions or even remote intimation from the President that the civil and military government which had been instituted during the war was discontinued. Up to that time, whether such an intimation had or had not been given, duties had been collected under the war tariff, strictly in conformity with the instructions which had been received from Washington."

Comment would seem to be unnecessary to make this passage clear. If the act of Congress applied by cession, it applied immediately. It could not be delayed by taking time for notice. Besides, it would by its own force displace all other provisions, and would not need for operation upon rights or the creation of rights, that the President give instructions or intimations, near or remote, " that the civil and military government, which had been instituted during the war, was discontinued." But we need not comment further. We may use the language of the court in summarizing its conclusion :

" Our conclusion from what has been said is that the civil government of California, organized as it was from a right of conquest, did not cease or become defunct in consequence of the signature of the treaty or from its ratification. We think it was continued over a ceded conquest, without any violation of the Constitution or laws of the United States, and that until Congress legislated for it the duties upon foreign goods imported into San Francisco were legally demanded and lawfully received by Mr. Harrison, the collector of the port, who received his appointment, according to instructions from Washington, from Governor Mason."

This explicit statement, as well as the analysis and review which have first been made, leaves no ground to sustain the conclusion that *Cross* v. *Harrison* held that the tariff laws of the United States were immediately operative in California without regard to the exercise of the President's discretion putting them in force. But purely for argument sake we may concede the contrary. The decision must have been, in any conception, based on the provisions of the treaty with Mexico. The court said so. But the treaty with Spain, instead of providing for incorporating the ceded territory into the United States, as did the treaty with Mexico, expressly declares that the *status* of the ceded territory is to be determined by Congress. This difference in the treaties removes *Cross* v. *Harrison* as a factor in the judgment of the case at bar, supposing its interpretation, in the opinion we are reviewing, be correct.

3. The opinion of the court says: " On March 1, 1845, Congress adopted a joint resolution consenting to the annexation

of Texas upon certain conditions, 5 Stat. 797, but it was not until December 25, 1845, that it was formally admitted as a State. 9 Stat. 108. In this interval, and on July 29, 1845, the Secretary of the Treasury issued a circular letter directing the collectors to collect duties upon all imports from Texas into the United States until Congress had further acted. Of course, there could be no question that Texas remained a foreign state until December 25, when she was formally admitted. The circular, therefore, is of no pertinence to the question here involved." We think otherwise. Even after her admission as a State it was deemed necessary to extend the laws of the United States to her. 9 Stat. 1. She was an example, as Florida was, as to what Congress believed to be necessary, and Oregon and Alaska are like examples. The simple rule of the automatic action of the custom and revenue laws seemingly did not occur to anybody; not even as to incorporated territory nor to a new State formed from foreign territory. Nor, as we have seen, did such theory seem to be sustainable when Chief Justice Taney announced in *Fleming* v. *Page* a contrary conclusion.

4. But independent of precedent the court says it is "irresistibly impelled to the same conclusion." The argument is mainly based upon the treaty-making power invested in the President and Senate. A treaty made by that power is said to be the supreme law of the land—as efficacious as an act of Congress; and if subsequent to and inconsistent with an act of Congress, repeals it. This must be granted, and also that "one of the ordinary incidents of a treaty is the cession of territory," and that "the territory thus acquired is acquired as absolutely as if the annexation were made, as in the case of Texas and Hawaii, by an act of Congress." But to tell us of the sources of the treaty-making power and to define the extent of that power helps us very little to the solution of the present problem.

The question occurs, What has the treaty-making power done? Is the treaty with Spain inconsistent with the Dingley act, and was it intended to work the repeal of that act? That act when passed was undoubtedly intended to apply to products from Porto Rico, and, we suppose, it will not be contended in determining whether the treaty has rendered the act inoperative, the terms of the treaty are not to be looked at? Assuredly the treaty cannot have an automatic force contrary to its terms. That is, it cannot be contended, that the automatic force of the treaty is greater than the force of the treaty itself.

This court said, speaking by Mr. Justice Brown, in *Holden* v. *Hardy*, 169 U. S. 366:

"In the future growth of the nation, as heretofore, it is not impossible that Congress may see fit to annex territories whose jurisprudence is that of the civil law. One of the considerations moving to such annexation might be the very fact that the territory so annexed should enter the Union with its traditions, laws and systems of administration unchanged. It would be a narrow construction of the Constitution to require them to abandon these, or to substitute for a system, which represents the growth of generations of inhabitants, a jurisprudence with which they had had no previous acquaintance or sympathy."

The statement being accepted, may not a fiscal system be as important as other matters of administration? May not a change of taxation, new burdens of taxation suddenly imposed, be worthy of consideration?

The opinion of the case at bar has not discussed the treaty. It takes it for granted that the cession of Porto Rico was absolute, and the conclusion that it is not a foreign country, within the meaning of the revenue laws, is deduced from that. But necessarily that depends upon the treaty, and interpretation is called for. The power of Congress over ceded territory is asserted in the opinion in somewhat absolute terms—it "involves the right to govern and dispose of it." This being so, it would seem to be certain that the treaty-making power would not forestall Congress or accept with the cession of territory the destruction of the fiscal and industrial policies of the country. We should hesitate to so pronounce for reasons which must occur to every one, except upon the compulsion of the clearest expression.

The opinion of the court further says " territory thus acquired (by treaty) can remain a foreign country under the tariff laws only on one of two theories: either that the word 'foreign' applies to such countries as were foreign at the time the statute was enacted, notwithstanding any change in their condition, or that they remain foreign under the tariff laws until Congress has formally embraced them within the customs union of the States." Both theories are rejected as untenable. The first because, "while a statute is presumed to speak from the time of its enactment, it embraces all such persons or things as subsequently fall within its scope." But what constitutes the scope of a statute—its letter inevitably, or may its spirit be regarded as interpreting and applying its letter? In other words, shall the purpose of its enactment be executed or defeated? There can be but one answer to these questions, nor can confidence in the answer be lessened by the analogies used by the court.

The law against selling liquors to minors, it is said, contemplates all minors—those existing and those which may come into being afterwards. Very true, but the purpose of the law is that. The same with territories (to use another illustration of the opinion) being bound as States when they come into the Union. But these illustrations assume that the territory referred to was incorporated by the treaty into the United States, an ever-recurring and misleading fallacy, in our judgment.

Let us, however, look at the argument under the wrong assumption of incorporation. The provisions of the Constitution for the admission of new States contemplate the consequences of statehood—contemplate territories ceasing to be bound as such and becoming bound as States. In other words, those provisions regard the future, and have their purpose fulfilled, not defeated, by territories becoming States. But a tariff law does not contemplate additions to or subtractions from itself. It may be said to be occasional. It regards certain conditions, and may be dependent upon them, whether it be enacted for revenue only or for protection and revenue. Its entire plan may be impaired or be destroyed by change in any part. The revenues of the government may be lessened, even taken away by change; the industrial policy of the country may be destroyed by change. We are repelled by the argument which leads to such consequences, whether regarding our own country or the foreign country made "domestic." If "domestic" as to what comes from it, it is "domestic" as to what goes to it, and its custom laws as well

as our custom laws may be cast into confusion, and its business
and affairs deranged before there is possibility of action.

As we have already said, to set the word foreign in antithesis
to the word domestic proves nothing. Their opposition does
not express the controversy. The controversy is narrower. It
is whether a particular tariff law applies. That, indeed, may
be the consequence of the principle that all laws apply. Or
that customs laws apply by reason of the provision of the Constitution which requires duties, imposts and excises to be uniform throughout the United States, and the treaty-making
power cannot prevent the application of that provision. That
principle is asserted by counsel and is very simple, but applied,
as counsel apply it, is fraught with grave consequences. It
takes this great country out of the world and shuts it up within
itself. It binds and cripples the power to make war and peace.
It may take away the fruits of victory, and, if we may contemplate the possibility of disaster, it may take away the means of
mitigating that. All those great and necessary powers, are, as
a consequence of the argument, limited by the necessity to make
some impost or excise " uniform throughout the United States."

The treaty-making power is as much a constitutional power
as the legislative or judicial powers. It is a supreme attribute
of sovereignty, but often less determined in its exercise than
others—more dependent on contingency, and may be less optional. It may precede war or follow war—command or be
commanded by war. The kind or direction of its exercise cannot always be predicted or marked. There can be no verbal
limitations upon it, and, wisely, none were attempted. Whatever restraints should be put upon it might have to yield to the
greater restraints of life or death—not only material prosperity,
but national existence. These, of course, are extreme contingencies, but they are not impossible, and are necessary to be regarded when limitations are urged which take no account of
them. We do not mean to say that there are no limitations.
They are certainly not those which counsel urge. Besides, the
contention of counsel is answered by the *Canter* case. The difference between military occupation of a territory and its cession at the treaty of peace was noted. " If ceded by the treaty,"

the court said, "the acquisition is confirmed, and the ceded territory becomes a part of the nation to which it is annexed, either on the terms stipulated in the treaty of cession or such as its new master may impose." What is the significance of this? It would seem like useless language; its purpose often defeated if the Constitution and laws of the conqueror, and, to drop from the abstract and supposing this country the conqueror, if our Constitution and laws immediately apply on cession of territory. The terms which may be granted or received would be, to a certain and important extent, predetermined. Neither we nor the conquered nation would have any choice in the new situation—could make no accommodation to exigency, would stand bound in a helpless fatality. Whatever might be the interests, temporary or permanent, whatever might be the condition or fitness of the ceded territory, the effect on it or on us, the territory would become a part of the United States with all that implies. It is only true to say that counsel shrink somewhat from the consequences of their contention, or if "shrink" be too strong an expression, deny that it can be carried to the nationalization of uncivilized tribes. Whether that limitation can be logically justified we are not called upon to say. There may be no ready test of the civilized and uncivilized, between those who are capable of self-government and those who are not, available to the judiciary, or could be applied or enforced by the judiciary. Upon what degree of civilization could civil and political rights under the Constitution be awarded by courts? The question suggests the difficulties, and how essentially the whole matter is legislative, not judicial. Nor can those difficulties be put out of contemplation, under the assumption that the principles which we may declare will have no other consequence than to affect duties upon a cargo of sugar. We need not, however, dwell on this part of the discussion. From our construction of the powers of the government and of the treaty with Spain the danger of the nationalization of savage tribes cannot arise.

These views answer, in our judgment, the chief arguments of the opinion, but to make a complete reply and to justify a different conclusion we should consider and interpret the treaty with Spain. We will, however, not do so now. It has been done in the concurring opinion in *Downes* v. *Bidwell*, and it is not necessary to anticipate the statements and reasoning of that opinion.

We said at the outset that it could be demonstrated that Porto Rico occupied a relation to the United States between that of being a foreign country absolutely and of being domestic territory absolutely, and because of that relation its products were subject to the duties imposed by the Dingley act. And, concluding, we say, we believe that, in this opinion and the one referred to, we have made that demonstration; made it from the Constitution itself, the immediate and continued practice under the Constitution, judicial authority and the treaty with Spain. And that demonstration does more than declare the legality of the duties which were levied upon the sugars of the plaintiff in error. It vindicates the government from national and international weakness. It exhibits the Constitution as a charter of great and vital authorities, with limitations indeed, but with such limitations as serve and assist government, not destroy it; which, though fully enforced, yet enable the United States to have—what it was intended to have—" an equal station among the Powers of the earth," and to do all "Acts and Things which Independent States may of right do." And confidently do, able to secure the fullest fruits of their performance. All powers of government, placed in harmony under the Constitution; the rights and liberties of every citizen secured—put to no hazard of loss or impairment; the power of the nation also secured in its great station, enabled to move with strength and dignity and effect among the other nations of the earth to such purpose as it may undertake or to such destiny as it may be called.

The judgment of the Circuit Court should be affirmed.

Mr. Justice Gray, dissenting.

I am compelled to dissent from the judgment in this case. It appears to me irreconcilable with the unanimous opinion of this court in *Fleming* v. *Page*, 9 How. 603, and with the opinions of the majority of the Justices in the case, this day decided, of *Downes* v. *Bidwell*.

GOETZE *v.* UNITED STATES.
CROSSMAN *v.* UNITED STATES.

APPEALS FROM THE CIRCUIT COURT OF THE UNITED STATES FOR THE SOUTHERN DISTRICT OF NEW YORK.

No. 340 was argued December 17, 18, 19, 20, 1900; No. 515 was argued January 14, 15, 1901. The two were decided together May 27, 1901.

De Lima v. *Bidwell, ante,* 1, followed by reversing the action of the general appraisers.

THESE were petitions for a review of two decisions of the board of general appraisers, holding subject to duty certain merchandise, imported, in one case from Porto Rico, and in the other, from Honolulu, in the Hawaiian Islands. The action of the board of general appraisers in each case was affirmed.

Mr. Edward C. Perkins and *Mr. Everit Brown* for appellant in No. 340. *Mr. J. B. Henderson* also filed a brief for same. *Mr. E. Ham, Mr. Alexander Porter Morse* and *Mr. Charles F. Manderson* filed a brief in this case on behalf of industrial interests in the States.

Mr. Attorney General for the United States.

Mr. W. Wickham Smith for appellants in No. 515. *Mr. Charles Curie* was on his brief.

Mr. Solicitor General for the United States.

MR. JUSTICE BROWN, after making the above statement, delivered the opinion of the court.

As the sole question presented by the record in these cases was whether Porto Rico and the Hawaiian Islands were foreign countries within the meaning of the tariff laws, we must hold, for the reasons stated in *De Lima* v, *Bidwell*, just decided, that the board of general appraisers had no jurisdiction of the cases. *The judgments of the Circuit Court are therefore reversed, and the cases remanded to that court with instructions to reverse the action of the board of general appraisers.*

DOOLEY *v.* UNITED STATES.

ERROR TO THE CIRCUIT COURT OF THE UNITED STATES FOR THE SOUTHERN DISTRICT OF NEW YORK.

No. 501. Argued January 8, 9, 10, 11, 1901.—Decided May 27, 1901.

The Court of Claims, and the Circuit Courts, acting as such, have jurisdiction of actions for the recovery of duties illegally exacted upon merchandise, alleged not to have been imported from a foreign country.

Duties upon imports from the United States to Porto Rico, collected by the military commander and by the President as Commander-in-Chief, from the time possession was taken of the island until the ratification of the treaty of peace, were legally exacted under the war power.

As the right to exact duties upon importations from Porto Rico to New York ceased with the ratification of the treaty of peace, the correlative right to exact duties upon imports from New York to Porto Rico also ceased at the same time.

THIS was an action begun in the Circuit Court, as a Court of Claims, by the firm of Dooley, Smith & Co., engaged in trade and commerce between Porto Rico and New York, to recover back certain duties to the amount of $5374.68, exacted and paid under protest at the port of San Juan, Porto Rico, upon several consignments of merchandise imported into Porto Rico from New York between July 26, 1898, and May 1, 1900, viz.:

1. From July 26, 1898, until August 19, 1898, under the terms of the proclamation of General Miles, directing the exaction of the former Spanish and Porto Rican duties.

2. From August 19, 1898, until February 1, 1899, under the customs tariff for Porto Rico, proclaimed by order of the President.

8. From February 1, 1899, to May 1, 1900, under the amended tariff customs promulgated January 20, 1899, by order of the President.

It thus appears that the duties were collected partly before and partly after the ratification of the treaty, but in every instance prior to the taking effect of the Foraker act. The revenues thus collected were used by the military authorities for the benefit of the provisional government.

A demurrer was interposed upon the ground of the want of jurisdiction, and the insufficiency of the complaint. The Circuit Court sustained the demurrer upon the second ground, and dismissed the petition. Hence this writ of error.

Mr. Henry M. Ward and *Mr. John G. Carlisle* for plaintiffs in error. *Mr. William Edmond Curtis* was on *Mr. Ward's* brief. *Mr. William G. Choate* and *Mr. Joseph Larocque* filed a separate brief for plaintiffs in error.

Mr. Solicitor General and *Mr. Attorney General* for defendants in error.

MR. JUSTICE BROWN, after making the above statement, delivered the opinion of the court.

1. The jurisdiction of the court in this case is attacked by the government upon the ground that the Circuit Court, as a Court of Claims, cannot take cognizance of actions for the recovery of duties illegally exacted.

By an act passed March 3, 1887, to provide for the bringing of suits against the government, known as the Tucker act, 24 Stat. 505, c. 359, the Court of Claims was vested with jurisdiction over "first, all claims founded upon the Constitution of the United States or any law of Congress, except for pensions, or upon any regulation of an Executive Department, or upon any contract, express or implied, with the government of the United States, or for damages, liquidated or unliquidated, in cases not sounding in tort, in respect of which claims the party would be entitled to redress against the United States either in a court of law, equity, or admiralty, if the United States were suable;"

and by section 2 the District and Circuit Courts were given concurrent jurisdiction to a certain amount.

The first section evidently contemplates four distinct classes of cases: (1) those founded upon the Constitution or any law of Congress, with an exception of pension cases; (2) cases founded upon a regulation of an Executive Department; (3) cases of contract, express or implied, with the government; (4) actions for damages, liquidated or unliquidated, in cases *not sounding in tort.* The words "not sounding in tort" are in terms referable only to the fourth class of cases.

The exception to the jurisdiction is based upon two grounds: First, that the court has no jurisdiction of cases arising under the revenue laws; and, second, that it has no jurisdiction in actions for tort.

In support of the first proposition we are cited to the case of *Nichols* v. *United States,* 7 Wall. 122, in which it was broadly stated that "cases arising under the revenue laws are not within the jurisdiction of the Court of Claims." The action in that case was brought to recover an excess of duties paid upon certain liquors which had leaked out during the voyage, and, being thus lost, were never imported in fact into the United States. Plaintiffs paid the duties, as exacted, but made no protest, and subsequently brought suit in the Court of Claims for the overpayment. The act in force at that time gave the Court of Claims power to hear and determine "all claims founded upon any law of Congress, or upon any regulation of an Executive Department, or upon any contract, express or implied, with the government of the United States." The court held, first, that the duties could not be recovered because they were not paid under protest, and, second, that Congress did not intend to confer upon the Court of Claims jurisdiction of cases arising under the revenue laws, inasmuch as, by the act of February 26, 1845, 5 Stat. 727, c. 22, Congress had given a right of action against the collector in favor of persons "who have paid, or shall hereafter pay, money, as and for duties, under protest . . . in order to obtain goods, wares, or merchandise imported by him or them, or on his or their account, which duties are not authorized or payable in part or in whole by law," provided that protests were duly made in writing. It was held that this remedy was exclusive, and that Congress, after having carefully constructed a revenue system, with ample provisions to redress wrong, did not intend to give to the taxpayer and importer a different and further remedy.

Subsequent statutes, however, have so far modified that special remedy that it can no longer be made available, and the broad statement in the *Nichols* case, that revenue cases are not within the cognizance of the Court of Claims, if still true, must be accepted with material qualifications. By the Customs Administrative act of 1890, as we have just held in *De Lima* v. *Bidwell,* an appeal is given from the decision of the collector " as to the rate and amount of the duties chargeable upon imported merchandise," to a board of general appraisers, whose decision shall be final and conclusive " as to the construction of the law and the facts respecting the classification of such merchandise and the rate of duties imposed thereon under such classification," unless application be made for a review to the Circuit Court of the United States. This remedy is doubtless exclusive as applied to customs cases ; but, as we then held, it has no application to actions against the collector for duties exacted upon goods which were not imported at all. Such cases, although arising under the revenue laws, are not within the purview of the Customs Administrative act ; as for such cases there is still a common-law right of action against the collector, and we think also by application to the Court of Claims. There would seem to be no doubt about plaintiffs' remedy against the collector at San Juan.

In the *Nichols* case, it was held that, as there was a remedy by action against the collector, expressly provided by statute, that remedy was exclusive. In *De Lima* v. *Bidwell* we held that although no other remedy was given expressly by statute than that provided by the Customs Administrative act, there was still a common law remedy against the collector for duties exacted upon goods not imported at all ; but it does not therefore follow that this remedy is exclusive, and that the importer may not avail himself of his right of action in the Court of Claims.

But conceding that the *Nichols* case does not stand in the way of a suit in the Court of Claims, the government takes the position that a suit in the United States to recover back duties illegally exacted by a collector of customs is really an action "sounding in tort," though not an action "for damages, liquidated or unliquidated," within the fourth class of cases enumerated in the Tucker act.

There are a number of authorities in this court upon that subject which require examination. The question is, whether any claim sounding in tort can be prosecuted in the Court of Claims, notwithstanding the words "not sounding in tort," in the Tucker act, are apparently limited to claims for damages, liquidated or unliquidated. The question was first considered in *Langford* v. *United States*, 101 U. S. 341, under the statute above cited, giving the Court of Claims power to hear and determine "all claims found upon any law of Congress, or upon any regulation of an Executive Department, or upon any contract, express or implied, with the government of the United States." The suit was brought to recover for the use and occupation of certain lands and buildings of which possession had been forcibly taken by agents of the government, against the will of Langford, who claimed title to the lands. It was held that the act of the United States in taking and holding possession was an unequivocal tort, and a distinction was drawn between such a case and one where the government takes for public use lands to which it asserts no claim of title, but admits the ownership to be private or individual, in which class there arises an implied obligation to pay the owner its just value. "It is a very different matter where the government claims it is dealing with its own, and recognizes no title superior to its own. In such case the government, or the officers who seize such property, are guilty of a tort, if it be in fact private property." It was held that the limitation of the act to cases of contract, express or implied, "was established in reference to the distinction between actions arising out of contracts, as distinguished between those founded on torts, which is inherent in the essential nature of judicial remedies under all systems, and especially under the system of the common law."

The case was rested largely upon that of *Gibbons* v. *United States*, 8 Wall. 269, in which an army contractor who had agreed to furnish certain oats at a fixed price had, after the delivery of part of the amount, been released from the obligation to deliver the balance. He was, however, carried before the military authority, and, influenced by threats, agreed to deliver, and did deliver, the full quantity of oats specified in the contract. He brought suit for the difference between the contract price and the market price of the oats at the time of delivery. It was said that "if such pressure was brought to bear upon him as would make the renewal of the contract void, as being obtained by duress, then there was no contract, and the proceeding was a tort for which the officer may have been personally liable," but that it was not within the Court of Claims act.

The act of March 3, 1887, the Tucker act, was first considered by this court in *United States* v. *Jones*, 131 U. S. 1, in which it was held not to confer upon the Court of Claims jurisdiction in equity to compel the issue and entry of a patent for public land, following *United States* v. *Alire*, 6 Wall. 573, and *Bonner* v. *United States*, 9 Wall. 156. In delivering the opinion, Mr. Justice Bradley compared the original act with the Tucker act, and held that there was no such difference in language as to justify an equitable jurisdiction to compel the issue of a patent.

In *Hill* v. *United States*, 149 U. S. 593, it was held that a claim for damages for the use and occupation of land under tidewater, for the erection and maintenance of a lighthouse, without the consent of the owner, but not showing that the United States had acknowledged any right of property in him as against them, was a case sounding in tort, of which the Circuit Court had no jurisdiction under the Tucker act. It was said that "the United States cannot be sued in their own courts without their consent, and have never permitted themselves to be sued in any court for torts committed in their name by their officers. Nor can the settled distinction, in this respect, between contract and tort be evaded by framing the claim as upon an implied contract." "An action in the nature of assumpsit for the use and occupation of real estate will never lie where there has been no relation of contract between the parties, and where the possession has been acquired and maintained under a different or adverse title, or where it is tortious and makes the defendant a trespasser." No distinction was noticed between the phraseology of the original act and the Tucker act, though it seems to have been assumed that the case was one for the recovery of "damages" sounding in tort.

In *Schillinger* v. *United States*, 155 U. S. 163, it was held that the Court of Claims had no jurisdiction of an action upon a claim against the government for the wrongful appropriation of a patent by the United States, against the protest of the patentee. It was said to be an action for damages sounding in tort, and therefore not maintainable. "Not only does the petition count upon a tort, but also the findings show a tort. That is the essential fact underlying the transaction and upon which rests every pretense of a right to recover. There was no suggestion of a waiver of the tort or a pretense of any implied contract until after the decision of the Court of Claims that it had no jurisdiction over an action to recover for the tort."

In the cases under consideration the argument is made that the money was tortiously exacted; that the alternative of payment to the collector was a seizure and sale of the merchandise for the non-payment of duties; and that it mattered not that at common law an action for money had and received would have lain against the collector to recover them back. But whether the exactions of these duties were tortious or not; whether it was within the power of the importer to waive the tort and bring suit in the Court of Claims for money had and received, as upon an implied contract of the United States to refund the money in case it was illegally exacted, we think the case is one within the first class of cases specified in the Tucker act of claims founded upon a law of Congress, namely, a revenue law, in respect to which class of cases the jurisdiction of the Court of Claims, under the Tucker act, has been repeatedly sustained.

Thus, in *United States* v. *Kaufman*, 96 U. S. 567, a brewer who had been illegally assessed for a special tax upon his business, was held entitled to bring suit in the Court of Claims to recover back the amount, upon the ground that no special remedy had been provided for the enforcement of the payment, and consequently the general laws which govern the Court of Claims may be resorted to for relief, if any can be found applicable to such a case. This is upon the principle that a liability created by statute without a remedy may be enforced by a common-law action. The *Nichols* case was distinguished upon the ground that the statute there *had* provided a special remedy.

So, too, in *United States* v. *Savings Bank*, 104 U. S. 728, the Court of Claims was held to have jurisdiction of a suit to recover back certain taxes and penalties assessed upon a savings bank.

In *Campbell* v. *United States*, 107 U. S. 407, it was held that a party claiming to be entitled to a drawback of duties upon manufactured articles exported might, when payment thereof has been refused, maintain a suit in the Court of Claims, because the facts found raised an implied contract that the United States would refund to the importer the amount he had paid to the government. There was here no question of tort.

In *United States* v. *Great Falls Manufacturing Co.*, 112 U. S. 645, it was held, following the observation of Mr. Justice Miller in *Langford* v. *United States*, that where property to which the United States *asserts no title* is taken by their officers or agents, pursuant to an act of Congress, as private property for public use, there was an implied obligation to compensate the owner, which might be enforced by suit in the Court of Claims.

So, too, in *Hollister* v. *Benedict & Burnham Mfg. Co.*, 113 U. S. 59, it was held that a suit might be maintained in the Court of Claims to recover for the use of a patented invention, if the right of the patentee were acknowledged. To the same effect are *United States* v. *Palmer*, 128 U. S. 262, and *United States* v. *Berdan Fire-Arms Co.*, 156 U. S. 552.

In *Medbury* v. *United States*, 173 U. S. 492, it was held the Court of Claims had jurisdiction of an action to recover an excess of payment for lands within the limits of a railroad grant, which grant was, subsequent to the payment, forfeited by act of Congress for non-construction of the road.

In *Swift* v. *United States*, 111 U. S. 22, the same right was treated as existing in favor of a party who sued for a commission upon the amount of certain adhesive stamps, which he had at one time purchased for his own use from the Bureau of Internal Revenue. See also *United States* v. *Lawson*, 101 U. S. 164; *Mosby* v. *United States*, 133 U. S. 273.

2. In their legal aspect, the duties exacted in this case were of three classes: (1) the duties prescribed by General Miles under order of July 26, 1898, which merely extended the existing regulations; (2) the tariffs of August 19, 1898, and February 1, 1899, prescribed by the President as Commander-in-Chief, which continued in effect until April 11, 1899, the date of the ratification of the treaty and the cession of the island to the United States; (3) from the ratification of the treaty to May 1, 1900, when the Foraker act took effect.

There can be no doubt with respect to the first two of these classes, namely, the exaction of duties under the war power, prior to the ratification of the treaty of peace. While it is true the treaty of peace was signed December 10, 1898, it did not take effect upon individual rights, until there was an exchange of ratifications. *Haver* v. *Yaker*, 9 Wall. 32. Upon the occupation of the country by the military forces of the United States, the authority of the Spanish Government was superseded, but the necessity for a revenue did not cease. The government must be carried on, and there was no one left to administer its functions but the military forces of the United States. Money is requisite for that purpose, and money could only be raised by order of the military commander. The most natural method was by the continuation of existing duties. In adopting this method, General Miles was fully justified by the laws of war. The doctrine upon this subject is thus summed up by Halleck in his work on International Law, (vol. 2, page 444): "The right of one belligerent to occupy and govern the territory of the enemy while in its military possession, is one of the incidents of war, and flows directly from the right to conquer. We, therefore, do not look to the Constitution or political institutions of the conquerer, for authority to establish a government for the territory of the enemy in his possession, during its

military occupation, nor for the rules by which the powers of
such government are regulated and limited. Such authority
and such rules are derived directly from the laws of war, as
established by the usage of the world, and confirmed by the
writings of publicists and decisions of courts—in fine, from the
law of the nations. . . . The municipal laws of a conquered
territory, or the laws which regulate private rights, continue in
force during military occupation, except so far as they are suspended or changed by the acts of the conqueror. . . . He,
nevertheless, has all the powers of a *de facto* government, and
can at his pleasure either change the existing laws or make new
ones."

In *New Orleans* v. *Steamship Co.*, 20 Wall. 387, 393, it was
said, with respect to the powers of the military government
over the city of New Orleans after its conquest, that it had
"the same power and rights in territory held by conquest as if
the territory had belonged to a foreign country and had been
subjugated in a foreign war. In such cases the conquering
power has the right to displace the preëxisting authority, and
to assume to such extent as it may deem proper the exercise by
itself of all the powers and functions of government. It may
appoint all the necessary officers and clothe them with designated powers, larger or smaller, according to its pleasure. It
may prescribe the revenues to be paid, and apply them to its
own use or otherwise. It may do anything necessary to
strengthen itself and weaken the enemy. There is no limit to
the powers that may be exerted in such cases, save those which
are found in the laws and usages of war. These principles have
the sanction of all publicists who have considered the subject."
See also *Thirty Hogsheads of Sugar* v. *Boyle*, 9 Cr. 191; *Fleming* v. *Page*, 9 How. 603; *American Ins. Co.* v. *Canter*, 1 Pet.
511.

But it is useless to multiply citations upon this point, since
the authority to exact similar duties was fully considered and
affirmed by this court in *Cross* v. *Harrison*, 16 How. 164. This
case involved the validity of duties exacted by the military
commander of California upon imports from foreign countries,
from the date of the treaty of peace, February 3, 1848, to No-

vember 13, 1849, when the collector of customs appointed by the President entered upon the duties of his office. Prior to the treaty of peace, and from August, 1847, duties had been exacted by the military authorities, the validity of which does not seem to have been questioned. Page 189: " That war tariff, however, was abandoned as soon as the military governor had received from Washington information of the exchange and ratification of the treaty with Mexico, and duties were afterwards levied in conformity with such as Congress had imposed upon foreign merchandise imported into other ports of the United States, Upper California having been ceded by the treaty to the United States." The duties were held to have been legally exacted. Speaking of the duties exacted before the treaty of peace, Mr. Justice Wayne observed (p. 190): " No one can doubt that these orders of the President, and the action of our Army and Navy commanders in California, in conformity with them, was according to the law of arms and the right of conquest, or that they were operative until the ratification and exchange of a treaty of peace. Such would be the case upon general principles in respect to war and peace between nations." It was further held that the right to collect these duties continued from the date of the treaty up to the time when official notice of its ratification and exchange were received in California. Owing to the fact that no telegraphic communication existed at that time, the news of the ratification of this treaty did not reach California until August 7, 1848, during which time the war tariff was continued. The question does not arise in this case, as the ratifications of the treaty appear to have been known as soon as they were exchanged.

The court further held in *Cross* v. *Harrison* that the right of the military commander to exact the duties prescribed by the tariff laws of the United States continued until a collector of customs had been appointed. Said the court: " The government, of which Colonel Mason was the executive, had its origin in the lawful exercise of a belligerent right over a conquered territory. It had been instituted during the war by the command of the President of the United States. It was the government when the territory was ceded as a conquest, and it did not cease, as a matter of course, or as a necessary consequence, of the restoration of peace. The President might have dissolved it by withdrawing the army and navy officers who administered it, but he did not do so. Congress could have put an end to it, but that was not done. The right inference from the inaction of both is, that it was meant to be continued until it had been legislatively changed. . . . We think it was continued over a ceded conquest, without any violation of the Constitution or laws of the United States, and that, until Congress legislated for it, the duties upon foreign goods, imported into San Francisco, were legally demanded and lawfully received by Mr. Harrison, the collector of the port, who received his appointment, according to instructions from Washington, from Governor Mason."

Upon this point that case differs from the one under consideration only in the particular that the duties were levied in *Cross* v. *Harrison* upon goods imported from foreign countries into California, while in the present case they were imported from New York, a port of the conquering country. This, however, is quite immaterial. The United States and Porto Rico were still foreign countries with respect to each other, and the same right which authorized us to exact duties upon merchandise imported from Porto Rico to the United States authorized the military commander in Porto Rico to exact duties upon goods imported into that island from the United States. The fact that, notwithstanding the military occupation of the United States, Porto Rico remained a foreign country within the revenue laws is established by the case of *Fleming* v. *Page*, 9 How. 603, in which we held that the capture and occupation of a Mexican port during our war with that country did not make it a part of the United States, and that it still remained a foreign country within the meaning of the revenue laws. The right to exact duties upon goods imported into Porto Rico from New York arises from the fact that New York was still a foreign country with respect to Porto Rico, and from the correlative right to exact at New York duties upon merchandise imported from that island.

3. Different considerations apply with respect to duties levied

after the ratification of the treaty and the cession of the island to the United States. Porto Rico then ceased to be a foreign country, and, as we have just held in *De Lima* v. *Bidwell*, the right of the collector of New York to exact duties upon imports from that island ceased with the exchange of ratifications. We have no doubt, however, that, from the necessities of the case, the right to administer the government of Porto Rico continued in the military commander after the ratification of the treaty, and until further action by Congress. *Cross* v. *Harrison*, above cited. At the same time, while the right to administer the government continued, the conclusion of the treaty of peace and the cession of the island to the United States were not without their significance. By that act Porto Rico ceased to be a foreign country, and the right to collect duties upon imports from that island ceased. We think the correlative right to exact duties upon importations from New York to Porto Rico also ceased. The spirit as well as the letter of the tariff laws admit of duties being levied by a military commander only upon importations from foreign countries; and while his power is necessarily despotic, this must be understood rather in an administrative than in a legislative sense. While in legislating for a conquered country he may disregard the laws of that country, he is not wholly above the laws of his own. For instance, it is clear that while a military commander during the civil war was in the occupation of a Southern port, he could impose duties upon merchandise arriving from abroad, it would hardly be contended that he could also impose duties upon merchandise arriving from ports of his own country. His power to administer would be absolute, but his power to legislate would not be without certain restrictions— in other words, they would not extend beyond the necessities of the case. Thus in the case of *The Admittance; Jecker* v. *Montgomery*, 13 How. 498, it was held that neither the President, nor the military commander, could establish a court of prize, competent to take jurisdiction of a case of capture, whose judgments would be conclusive in other admiralty courts. It was said that the courts established in Mexico during the war "were nothing more than agents of the military power, to assist it in preserving order in the conquered territory, and to protect the inhabitants in their persons and property, while it was occupied by the American arms. They were subject to the military power, and their decisions under its control, whenever the commanding officer thought proper to interfere. They were not courts of the United States, and had no right to adjudicate upon a question of prize or no prize," although Congress, in the exercise of its general authority in relation to the national courts, would have power to validate their action. *The Grapeshot*, 9 Wall. 129, 133.

So, too, in *Mitchell* v. *Harmony*, 13 How. 115, it was held that, where the plaintiff entered Mexico during the war with that country, under a permission of the commander to trade with the enemy and under the sanction of the executive power of the United States, his property was not liable to seizure by law for such trading, and that the officer directing the seizure was liable to an action for the value of the property taken. To the same effect is *Mostyn* v. *Fabrigas*, 1 Cowp. 161.

In *Raymond* v. *Thomas*, 91 U. S. 712, a special order, by the officer in command of the forces in the State of South Carolina, annulling a decree rendered by a court of chancery in that State, was held to be void. In delivering the opinion, Mr. Justice Swayne observed: "Whether Congress could have conferred the power to do such an act is not the question we are called upon to consider. It is an unbending rule of law, that the exercise of military power, where the rights of the citizens are concerned, shall never be pushed beyond what the exigency requires."

Without questioning at all the original validity of the order imposing duties upon goods imported into Porto Rico from foreign countries, we think the proper construction of that order is, that it ceased to apply to goods imported from the United States from the moment the United States ceased to be a foreign country with respect to Porto Rico, and that until Congress otherwise constitutionally directed, such merchandise was entitled to free entry.

An unlimited power on the part of the Commander-in-Chief to exact duties upon imports from the States might have placed

Porto Rico in a most embarrassing situation. The ratification of the treaty and the cession of the island to us severed her connection with Spain, of which the island was no longer a colony, and with respect to which she had become a foreign country. The wall of the Spanish tariff was raised against her exports, the wall of the military tariff against her imports, from the mother country. She received no compensation from her new relations with the United States. If her exports, upon arriving there, were still subject to the same duties as merchandise arriving from other foreign countries, while her imports from the United States were subjected to duties prescribed by the Commander-in-Chief, she would be placed in a position of practical isolation, which could not fail to be disastrous to the business and finances of an island. It had no manufactures or markets of its own, and was dependent upon the markets of other countries for the sale of her productions of coffee, sugar and tobacco. - In our opinion the authority of the President as Commander-in-Chief to exact duties upon imports from the United States ceased with the ratification of the treaty of peace, and her right to the free entry of goods from the ports of the United States continued until Congress should constitutionally legislate upon the subject.

The judgment of the Circuit Court is therefore reversed and the case remanded to that court for further proceedings in consonance with this opinion.

MR. JUSTICE WHITE, (with whom concurred MR. JUSTICE GRAY, MR. JUSTICE SHIRAS and MR. JUSTICE McKENNA,) dissenting.

The question involved in this case is the validity of certain impost duties laid on goods coming from the United States into Porto Rico under the tariff imposed by the military commander and under tariffs proclaimed by the President as Commander-in-Chief. The duties collected prior to the ratification of the treaty of peace are now decided to have been valid ; those collected after the ratification of the treaty are decided to have been unlawfully imposed, upon the doctrine announced in the case of *De Lima* v. *Bidwell*, just previously decided. I concur in so far as it is held that the duties collected prior to the ratification were validly collected, but dissent in so far as it is decided that the duties collected after the ratification were illegal. I might content myself with referring to the dissent in the *De Lima* case as expressing the grounds which prevent me from concurring in this case; but the importance of the subject and the grave consequences which I think are to be entailed by the decision now announced leads me to refer to some additional considerations.

As a prelude to doing so, however, let me briefly resume the propositions which seem to me to have been hitherto established.

1. There is a *non sequitur* involved in stating that the question is whether Porto Rico was a foreign country *within the meaning of the tariff laws*, and then discussing, not the question thus stated, but a different subject, that is, whether the territory ceded by the treaty with Spain came under the sovereignty of the United States by the effect of the cession.

2. And the confusion which arises from stating one question and then analyzing and expressing opinions on another and different one, is additionally demonstrated when it is considered that most of the authorities now relied upon in relation to the extension of the sovereignty of the United States over territory were cited to the court in *Fleming* v. *Page*, to establish that the dominancy of the sovereignty of the United States over a territory was the proper test by which to determine whether, under all circumstances, the revenue laws of the United States were applicable, and the court decided adversely to such contention. *Fleming* v. *Page*, 9 How. 603.

3. As the treaty with Spain provided "that the civil rights and political *status* of the native inhabitants should be determined by Congress," in reason this provision should not be controlled by conclusions deduced from treaties made by the United States in the past with other countries which did not contain such a provision, but expressly stipulated to the contrary.

4. In view of the terms of the treaty with Spain, to hold that the *status* of the ceded territory as previously existing was *ipso facto* changed, within the meaning of the tariff laws of the

United States, without action by Congress, is to deprive that body of the rights which the stipulations of the treaty sedulously sought to preserve.

5. Even ignoring the terms of the treaty, the conclusion that the *status* of the ceded territory, within the meaning of the tariff laws, was changed by the treaty before Congress could act on the subject, can only be upheld by disregarding the opinion of the court expressed by Mr. Chief Justice Taney in *Fleming* v. *Page*, and treating the important declarations on this subject by him in that case as mere *dicta*.

6. The result also cannot be supported without a misconception of the case of *Cross* v. *Harrison*, since that decision enforced the payment of a tariff duty levied after the ratification of the treaty with Mexico at a different rate from that imposed by the existing tariff laws of the United States, and since, moreover, that case can only be harmoniously interpreted by recalling the fact that several months after the notification of the ratification of the treaty with Mexico was received in California the President ordered the tariff laws of the United States to be enforced in California, and this authority may well have been treated as not only a direction for the future, but as a ratification of the act of the military officials in enforcing the tariff laws of the United States after they had learned of the ratification of the treaty.

7. In no single case from the foundation of the government except, if it can be called an exception, in the brief period prior to the President's order enforcing the tariff laws in California, as above stated, have the revenue laws of the United States been enforced in acquired territory without the action of the President or the consent of Congress, express or implied.

8. The rule of the immediate bringing, by the self-operating force of a treaty, ceded territory inside of the line of the tariff laws of the United States denies the existence of powers which the Constitution expressly bestows, overthrows the authority conferred on Congress by the Constitution, and is impossible of execution.

Having thus imperfectly summarized the propositions which are more lucidly stated in the dissent in the *De Lima* case, I

come to express the additional thoughts which have been previously adverted to.

Before the outbreak of the war with Spain it cannot be disputed that Porto Rico was embraced within the words "foreign country," as used in the tariff laws. Why was that island
so embraced without specific reference to it in such laws? is
the question which naturally arises. To answer this question
it is essential to determine what is the import of the words
"foreign country," not internationally, but within the meaning
of the tariff laws. It is settled that the power of Congress to
lay an impost duty does not give the right to levy such a duty
on merchandise coming from one part of the United States to
the other. *Woodruff* v. *Parham,* 8 Wall. 123. It follows,
therefore, that when, in the exercise of its power to lay impost
duties, Congress specifies such duties are to be collected on
merchandise from foreign countries, those words but gener-
ically embody the declaration of Congress that it is exerting
its taxing power conformably to the Constitution; that is,
it is causing the taxes which are levied to be applicable to the
entire area to which they may be extended under the Constitution. The command, then, in tariff laws, that impost duties
when laid shall be collected on all merchandise coming from
"foreign countries," is but a provision that they are to be levied
on merchandise arriving from countries which are not a part
of the United States, *within the meaning of the tariff laws,*
and which are hence subject to such duties. It must follow
that, as long as a locality is in a position where it is subject to
the power of Congress to levy an impost tariff duty on merchandise coming from that country into the United States,
such country must be a foreign country *within the meaning of
the tariff laws.* Now, this court has just decided in *Downes* v.
Bidwell that, despite the treaty of cession, Porto Rico remained
in a position where Congress could impose a tariff duty on
goods coming from that island into the United States. If,
however, it remained in that position, how then can it be now
declared that it ceased to be in that relation because it was no
longer foreign country within the meaning of the tariff laws?
But, it is said, although when the treaty was ratified, the coun-

try at once ceased to be foreign within the meaning of the tariff laws it yet subsequently became foreign for the purpose of the tariff laws when the act of Congress imposing a duty on goods from Porto Rico took effect. To what, in reason, does this proposition come? In my opinion only to this: Congress, under the Constitution, may not impose a tariff duty on goods brought from a country which has ceased to be foreign, but, although a country has so ceased to be foreign within the meaning of the tariff laws, nevertheless Congress may thereafter cause it to become foreign within such intendment by levying an impost upon its products coming into the United States. This is but to say an act of Congress can have the effect of changing the *status* of a territory from not foreign within the meaning of the tariff laws to foreign within such meaning, although a law attempting to so do would be plainly in violation of the Constitution, if the principle announced in this case be true, that the treaty from the moment of its ratification by its own force caused the ceded territory to be no longer foreign within the meaning of the tariff laws.

The only escape my mind can point out from this deduction is to say that territory which has become domestic, and therefore ceases to be foreign within the meaning of the tariff law, can yet be constitutionally treated by Congress as if it had not ceased to be foreign and had not become domestic. But this would expressly overrule *Woodruff* v. *Parham,* 8 Wall. 123, and cannot therefore be the rule of decision now announced, since that case is referred to and cited approvingly in the opinion of my brethren who dissent in the *Downes* case, and who do not dissent from the opinion of the court now announced.

Passing these considerations, it is impossible for me to conceive that Porto Rico ceased to be subject to the tariff laws, for the reasons fully stated by me in my concurring opinion in *Downes* v. *Bidwell,* which need not be reiterated. But, for the purposes of this case and *arguendo* only, let me now admit that the treaty incorporated Porto Rico into the United States despite the provisions which were contained in that instrument. Does it follow that such territory at once ceased to be subject to the tariff laws before Congress had the time to act? I am constrained to think not.

The power to originate revenue laws is lodged by the Constitution in the House of Representatives. When a tariff bill is drawn the revenue to arise from it must depend upon the sum of the articles which are to be imported and which are to pay the duty provided in the law. Let me illustrate it: Suppose a tariff law is so adjusted that the greater portion of the revenue which it seeks to provide is drawn from a few articles of general consumption. The duties to be paid on these articles, when imported, will, therefore, largely furnish the revenues essential to carry on the government. Suppose a treaty of cession which embraces territory producing in large quantities the articles upon which the existing tariff laws mainly rely for revenue to sustain the government. If, instantly, on the ratification of the treaty, before Congress can remodel or change the laws so as to provide for the support of the government, the articles stated coming into the United States from the country in question would be within the tariff line, and thereby entitled to free entry into the United States, what would become of the power of the House of Representatives and of the Congress on the subject of revenue as provided in the Constitution? It may be said in answer to this suggestion that Congress could make the change, and whilst of course a brief interval of disaster would ensue, during which there would be no revenue, the country must suffer the consequences during such interval. But does this follow? Suppose the political state of the country should be such that there was a difference of opinion as to the policy to be embodied in a tariff law, analogous to that which existed when California was acquired from Mexico, where, in consequence of division on the subject of the slavery question between the different branches of Congress, it was impossible to enact legislation conferring a territorial government upon California, what would be the situation then? Look at it practically from another point of view. Certainly before revenue laws can be made operative in a district or country it is essential that the situation be taken into account, for the purpose of establishing ports of entry, collection districts and the necessary machinery to enforce them. Of course, it is patent that such investigations cannot be made prior to acquisition. But, as the laws immediately extend, without action of Congress, as the result of acquisition, it must follow that they extend, although none of the means and instrumentalities for their successful enforcement can possibly be devised until the acquisition is completed. This must be, unless it be held that there is power in the government of the United States to enter a foreign country, examine its situation and enact legislation for it before it has passed under the sovereignty of the United States. From the point of view of the United States, then, it seems to me that the doctrine of the immediate placing of the tariff laws outside the line of newly acquired territory, however extreme may be the opinion entertained of the doctrine of immediate incorporation, is inadmissible and in conflict with the Constitution.

Let me look at and illustrate it from the point of view of the ceded territory. In doing so let me take for granted the accuracy of suggestions which have been advanced in argument. It is said that the public revenues of the Island of Porto Rico, except only such as were raised by a burdensome and complicated excise tax on incomes and business vocations, had always been chiefly obtained by duties on imports and exports; that our internal revenue laws, if applied in the island, would prove oppressive and ruinous to many people and interests; that one of the staple productions of the island—coffee—had always been protected by a tariff duty, whereas under our tariff laws coffee was admitted into the United States free of duty; that there was no system of direct taxation of property in operation when the island was ceded, there was no time to establish one, and such a system, moreover, would have entailed upon the people burdens incapable of being borne. I cannot conceive that under the provisions of the Constitution conferring upon Congress the power to raise revenue that consequences such as would flow from immediately putting in force in Porto Rico the revenue laws of the United States could constitutionally be brought about without affording to the Congress the opportunity to adjust the revenue laws of the United States to meet the new situation.

All these suggestions, however, it is argued, but refer to expediency, and are entitled to no weight as against the theory that, under the Constitution, the tariff laws of the United States took effect of their own force immediately upon the cession. But this is fallacious. For, if it be demonstrated that a particular result cannot be accomplished without destroying the revenue power conferred upon Congress by the Constitution, and without annihilating the conceded authority of the government in other respects, such demonstration shows the unsoundness of the argument which magnifies the results flowing from the exercise by the treaty-making power of its authority to acquire, to the detriment and destruction of that balanced and limited government which the Constitution called into being.

ARMSTRONG *v.* UNITED STATES.

APPEAL FROM THE COURT OF CLAIMS.

No. 509. Argued January 8, 9, 10 and 11, 1901.—Decided May 27, 1901.

Dooley v. *United States, ante,* 222, followed.

THIS was a petition to the Court of Claims by a British subject, to recover duties exacted by the collector of the port of San Juan, and paid under protest, upon goods, wares, and merchandise of the growth, produce, or manufacture of the United States, between August 12, 1898, and December 5, 1899.

The same demurrer was filed and the same judgment was entered as in the preceding case.

Mr. Alphonso Hart and *Mr. John G. Carlisle* for appellant. *Mr. John C. Chaney* and *Mr. Charles C. Leeds* were on *Mr. Hart's* brief.

Mr. Solicitor General and *Mr. Attorney General* for appellee.

MR. JUSTICE BROWN delivered the opinion of the court.

This case is controlled by the case of *Dooley* v. *United States*, No. 501, just decided. So far as the duties were exacted upon goods imported prior to the ratification of the treaty of April 11, 1899, they were properly exacted. So far as they were imposed upon importations after that date and prior to December 5, 1899, plaintiff is entitled to recover them back.

The judgment of the Court of Claims is therefore reversed and the case remanded to that court for further proceedings not inconsistent with this opinion.

DOWNES v. BIDWELL.

ERROR TO THE CIRCUIT COURT OF THE UNITED STATES FOR THE SOUTHERN DISTRICT OF NEW YORK.

No. 507. Argued January 8, 9, 10, 11, 1901.—Decided May 27, 1901.[1]

By MR. JUSTICE BROWN, in announcing the conclusion and judgment of the court.

The Circuit Courts have jurisdiction, regardless of amount, of actions against a collector of customs for duties exacted and paid under protest upon merchandise alleged not to have been imported.

The island of Porto Rico is not a part of the United States within that provision of the Constitution which declares that "all duties, imposts, and excises shall be uniform throughout the United States."

---

[1] In announcing the conclusion and judgment of the court in this case, MR. JUSTICE BROWN delivered an opinion. MR. JUSTICE WHITE delivered a concurring opinion which was also concurred in by MR. JUSTICE SHIRAS and MR. JUSTICE MCKENNA. MR. JUSTICE GRAY also delivered a concurring opinion. The Chief Justice, MR. JUSTICE HARLAN, MR. JUSTICE BREWER, and MR. JUSTICE PECKHAM dissented. Thus it is seen that there is no opinion in which a majority of the court concurred. Under these circumstances I have, after consultation with MR. JUSTICE BROWN, who announced the judgment, made headnotes of each of the sustaining opinions, and placed before each the names of the justices or justice who concurred in it.

There is a clear distinction between such prohibitions of the Constitution as go to the very root of the power of Congress to act at all, irrespective of time or place, and such as are operative only throughout the United States, or among the several States.

A long continued and uniform interpretation, put by the executive and legislative departments of the government, upon a clause in the Constitution should be followed by the judicial department, unless such interpretation be manifestly contrary to its letter or spirit.

By MR. JUSTICE WHITE, with whom MR. JUSTICE SHIRAS and MR. JUSTICE McKENNA concurred.

The government of the United States was born of the Constitution, and all powers which it enjoys or may exercise must be either derived expressly or by implication from that instrument. Ever then, when an act of any department is challenged, because not warranted by the Constitution, the existence of the authority is to be ascertained by determining whether the power has been conferred by the Constitution, either in express terms or by lawful implication, to be drawn from the express authority conferred or deduced as an attribute which legitimately inheres in the nature of the powers given, and which flows from the character of the government established by the Constitution. In other words, whilst confined to its constitutional orbit, the government of the United States is supreme within its lawful sphere.

Every function of the government being thus derived from the Constitution, it follows that that instrument is everywhere and at all times potential in so far as its provisions are applicable.

Hence it is that wherever a power is given by the Constitution and there is a limitation imposed on the authority, such restriction operates upon and confines every action on the subject within its constitutional limits.

Consequently it is impossible to conceive that where conditions are brought about to which any particular provision of the Constitution applies its controlling influence may be frustrated by the action of any or all of the departments of the government. Those departments, when discharging, within the limits of their constitutional power, the duties which rest on them, may of course deal with the subjects committed to them in such a way as to cause the matter dealt with to come under the control of provisions of the Constitutions which may not have been previously applicable. But this does not conflict with the doctrine just stated, or presuppose that the Constitution may or may not be applicable at the election of any agency of the government.

The Constitution has undoubtedly conferred on Congress the right to create such municipal organizations as it may deem best for all the territories of the United States whether they have been incorporated or not, to give to the inhabitants as respects the local governments such degree of representation as may be conducive to the public well-being, to deprive such territory of representative government if it is considered just to do so, and to change such local governments at discretion.

As Congress in governing the territories is subject to the Constitution, it results that all the limitations of the Constitution which are applicable to Congress in exercising this authority necessarily limit its power on this subject. It follows also that every provision of the Constitution which is applicable to the territories is also controlling therein. To justify a departure from this elementary principle by a criticism of the opinion of Mr. Chief Justice Taney in *Scott* v. *Sandford*, 19 How. 393, is unwarranted. Whatever may be the view entertained of the correctness of the opinion of the court in that case, in so far as it interpreted a particular provision of the Constitution concerning slavery and decided that as so construed it was in force in the territories, this in no way affects the principle which that decision announced, that the applicable provisions of the Constitution were operative.

In the case of the territories, as in every other instance, when a provision of the Constitution is invoked, the question which arises is, not whether the Constitution is operative, for that is self-evident, but whether the provision relied on is applicable.

As Congress derives its authority to levy local taxes for local purposes within the territories, not from the general grant of power to tax as expressed in the Constitution, it follows that its right to locally tax is not to be measured by the provision empowering Congress "To lay and collect Taxes, Duties, Imposts, and Excises," and is not restrained by the requirement of uniformity throughout the United States. But the power just referred to, as well as the qualification of uniformity, restrains Congress from imposing an impost duty on goods coming into the United States from a territory which has been incorporated into and forms a part of the United States. This results because the clause of the Constitution in question does not confer upon Congress power to impose such an impost duty on goods coming from one part of the United States to another part thereof, and such duty besides would be repugnant to the requirement of uniformity throughout the United States.

BY MR. JUSTICE GRAY.

The civil government of the United States cannot extend immediately, and of its own force, over territory acquired by war. Such territory must necessarily, in the first instance, be governed by the military power under the control of the President as commander in chief. Civil government cannot take effect at once, as soon as possession is acquired under military authority, or even as soon as that possession is confirmed by treaty. It can only be put in operation by the action of the appropriate political department of the government, at such time and in such degree as that department may determine.

In a conquered territory, civil government must take effect, either by the action of the treaty-making power, or by that of the Congress of the United States. The office of a treaty of cession ordinarily is to put an end to all authority of the foreign government over the territory; and to subject the territory to the disposition of the Government of the United States.

The government and disposition of territory so acquired belong to the Government of the United States, consisting of the President, the Senate, elected by the States, and the House of Representatives, chosen by and immediately representing the people of the United States.

So long as Congress has not incorporated the territory into the United States, neither military occupation nor cession by treaty makes the conquered territory domestic territory, in the sense of the revenue laws. But those laws concerning "foreign countries" remain applicable to the conquered territory, until changed by Congress.

If Congress is not ready to construct a complete government for the conquered territory, it may establish a temporary government, which is not subject to all the restrictions of the Constitution.

THIS was an action begun in the Circuit Court by Downes, doing business under the firm name of S. B. Downes & Co., against the collector of the port of New York, to recover back duties to the amount of $659.35 exacted and paid under protest upon certain oranges consigned to the plaintiff at New York, and brought thither from the port of San Juan in the Island of Porto Rico during the month of November, 1900, after the passage of the act temporarily providing a civil government and revenues for the Island of Porto Rico, known as the Foraker act.

The District Attorney demurred to the complaint for the want of jurisdiction in the court, and for insufficiency of its averments. The demurrer was sustained, and the complaint dismissed. Whereupon plaintiff sued out this writ or error.

Mr. Frederic R. Coudert, Jr., and *Mr. John G. Carlisle* for plaintiff in error. *Mr. Paul Fuller* was on Mr. Coudert's brief.

Mr. Solicitor General and *Mr. Attorney General* for defendants in error.

MR. JUSTICE BROWN, after making the above statement, announced the conclusion and judgment of the court.

This case involves the question whether merchandise brought into the port of New York from Porto Rico since the passage of the Foraker act, is exempt from duty, notwithstanding the third section of that act, which requires the payment of "fifteen per centum of the duties which are required to be levied, collected and paid upon like articles of merchandise imported from foreign countries."

1. The exception to the jurisdiction of the court is not well taken. By Rev. Stat. sec. 629, subdivision 4, the Circuit Courts are vested with jurisdiction "of all suits at law or equity arising under any act providing for a revenue from imports or tonnage," irrespective of the amount involved. This section should be construed in connection with sec. 643, which provides for the removal from state courts to Circuit Courts of the United States of suits against revenue officers "on account of any act done under color of his office, or of any such [revenue] law, or on account of any right, title or authority claimed by such officer or other person under any such law." Both these sections are taken from the act of March 2, 1833, c. 57, 4 Stat. 632, commonly known as the Force Bill, and are evidently intended to include all actions against customs officers acting under color of their office. While, as we have held in *De Lima* v. *Bidwell,* actions against the collector to recover back duties assessed upon non-importable property are not "customs cases" in the sense of the Administrative Act, they are, nevertheless, actions arising under an act to provide for a revenue from imports, in the sense of section 629, since they are for acts done by a collector under color of his office. This subdivision of sec. 629 was not repealed by the Jurisdictional Act of 1875, or the subsequent act of August 13, 1888, since these acts were "not intended to interfere with the prior statutes conferring jurisdiction upon the Circuit or District Courts in special cases, and over particular subjects." *United States* v. *Mooney,* 116 U. S. 104, 107. See also *Ins. Co.* v. *Ritchie,* 5 Wall. 541; *Philadelphia* v. *The Collector,* 5 Wall. 720; *Hornthall* v. *The Collector,* 9 Wall. 560. As the case "involves the construction or application of the Constitution" as well as the constitutionality of a law of the United States, the writ of error was properly sued out from this court.

2. In the case of *De Lima* v. *Bidwell,* just decided, we held that upon the ratification of the treaty of peace with Spain, Porto Rico ceased to be a foreign country, and became a terri-

tory of the United States, and that duties were no longer collectible upon merchandise brought from that island. We are now asked to hold that it became a part of the *United States* within that provision of the Constitution which declares that "all duties, imposts and excises shall be uniform throughout the United States." Art. I, sec. 8. If Porto Rico be a part of the United States, the Foraker act imposing duties upon its products is unconstitutional, not only by reason of a violation of the uniformity clause, but because by section 9 " vessels bound to or from one State" cannot "be obliged to enter, clear or pay duties in another."

The case also involves the broader question whether the revenue clauses of the Constitution extend of their own force to our newly acquired territories. The Constitution itself does not answer the question. Its solution must be found in the nature of the government created by that instrument, in the opinion of its contemporaries, in the practical construction put upon it by Congress and in the decisions of this court.

The Federal government was created in 1777 by the union of thirteen colonies of Great Britain in " certain articles of confederation and perpetual union," the first one of which declared that " the stile of this confederacy shall be the United States of America." Each member of the confederacy was denominated a *State*. Provision was made for the representation of each State by not less than two nor more than seven delegates; but no mention was made of territories or other lands, except in Art. XI, which authorized the admission of Canada, upon its "acceding to this confederation," and of other colonies if such admission were agreed to by nine States. At this time several States made claims to large tracts of land in the unsettled West, which they were at first indisposed to relinquish. Disputes over these lands became so acrid as nearly to defeat the confederacy, before it was fairly put in operation. Several of the States refused to ratify the articles, because the convention had taken no steps to settle the titles to these lands upon principles of equity and sound policy; but all of them, through fear of being accused of disloyalty, finally yielded their claims, though Maryland held out until 1781. Most of these States in the mean time having ceded their interests in these lands, the confederate Congress, in 1787, created the first territorial government northwest of the Ohio River, provided for local self-government, a bill of rights, a representation in Congress by a delegate, who should have a seat "with a right of debating, but not of voting," and for the ultimate formation of States therefrom, and their admission into the Union on an equal footing with the original States.

The confederacy, owing to well-known historical reasons, having proven a failure, a new Constitution was formed in 1787 by "the people of the United States" "for the United States of America," as its preamble declares. All legislative powers were vested in a Congress consisting of representatives from the several States, but no provision was made for the admission of delegates from the territories, and no mention was made of territories as separate portions of the Union, except that Congress was empowered "to dispose of and make all needful rules and regulations respecting the territory or other property belonging to the United States." At this time all of the States had ceded their unappropriated lands except North Carolina and Georgia. It was thought by Chief Justice Taney in the *Dred Scott* case, 19 How. 393, 436, that the sole object of the territorial clause was "to transfer to the new government the property then held in common by the States, and to give to that government power to apply it to the objects for which it had been destined by mutual agreement among the States before their league was dissolved;" that the power "to make needful rules and regulations" was not intended to give the powers of sovereignty, or to authorize the establishment of territorial governments—in short, that these words were used in a proprietary and not in a political sense. But, as we observed in *De Lima* v. *Bidwell*, the power to establish territorial governments has been too long exercised by Congress and acquiesced in by this court to be deemed an unsettled question. Indeed, in the *Dred Scott* case it was admitted to be the inevitable consequence of the right to acquire territory.

It is sufficient to observe in relation to these three fundamental instruments that it can nowhere be inferred that the territories were considered a part of the United States. The Constitution was created by the people of the *United States*, as a union of *States*, to be governed solely by representatives of the *States ;* and even the provision relied upon here, that all duties, imposts, and excises shall be uniform " throughout the United States," is explained by subsequent provisions of the Constitution, that "no tax or duty shall be laid on articles exported from any *State*," and "no preference shall be given by any regulation of commerce or revenue to the ports of one *State* over those of another ; nor shall vessels bound to or from one *State* be obliged to enter, clear or pay duties in another." In short, the Constitution deals with *States*, their people, and their representatives.

The Thirteenth Amendment to the Constitution, prohibiting slavery and involuntary servitude "within the United States, or in any place subject to their jurisdiction," is also significant as showing that there may be places within the jurisdiction of the United States that are no part of the Union. To say that the phraseology of this amendment was due to the fact that it was intended to prohibit slavery in the seceded States, under a possible interpretation that those States were no longer a part of the Union, is to confess the very point in issue, since it involves an admission that, if these States were not a part of the Union they were still subject to the jurisdiction of the United States.

Upon the other hand, the Fourteenth Amendment, upon the subject of citizenship, declares only that "all persons born or naturalized *in the United States*, and subject to the jurisdiction thereof, are citizens of the United States, and of the *State* wherein they reside." Here there is a limitation to persons born or naturalized in the United States which is not extended to persons born in any place "subject to their jurisdiction."

The question of the legal relations between the States and the newly acquired territories first became the subject of public discussion in connection with the purchase of Louisiana in 1803. This purchase arose primarily from the fixed policy of Spain to exclude all foreign commerce from the Mississippi. This restriction became intolerable to the large number of immigrants who were leaving the Eastern States to settle in the fertile val-

ley of that river and its tributaries. After several futile attempts to secure the free navigation of that river by treaty,
advantage was taken of the exhaustion of Spain in her war with
France, and a provision inserted in the treaty of October 27,
1795, by which the Mississippi River was opened to the commerce of the United States. 8 Stat. 138, 140, Art. IV. In
October, 1800, by the secret treaty of San Ildefonso, Spain
retroceded to France the territory of Louisiana. This treaty
created such a ferment in this country that James Monroe
was sent as minister extraordinary with discretionary powers
to coöperate with Livingston, then minister to France, in the
purchase of New Orleans, for which Congress appropriated
$2,000,000. To the surprise of the negotiators, Bonaparte in-
vited them to make an offer for the whole of Louisiana at a
price finally fixed at $15,000,000. It is well known that Mr.
Jefferson entertained grave doubts as to his power to make
the purchase, or, rather, as to his right to annex the territory
and make it part of the United States; and had instructed Mr.
Livingston to make no argeement to that effect in the treaty,
as he believed it could not be legally done. Owing to a new
war between England and France being upon the point of break-
ing out, there was need for haste in the negotiations, and Mr.
Livingston took the responsibility of disobeying his instructions,
and, probably owing to the insistence of Bonaparte, consented
to the third article of the treaty, which provided that " the
inhabitants of the ceded territory shall be incorporated in the
Union of the United States, and admitted as soon as possible,
according to the principles of the Federal Constitution, to the
enjoyment of all the rights, advantages and immunities of citizens of the United States; and in the meantime they shall be
maintained and protected in the free enjoyment of their liberty,
property and the religion which they profess." This evidently
committed the government to the ultimate, but not to the immediate, admission of Louisiana as a State, and postponed its
incorporation into the Union to the pleasure of Congress. In
regard to this, Mr. Jefferson, in a letter to Senator Breckinridge
of Kentucky, of August 12, 1803, used the following language:
" This treaty must, of course, be laid before both houses, because

both have important functions to exercise respecting it. They, I presume, will see their duty to their country in ratifying and paying for it, so as to secure a good which would otherwise probably be never again in their power. But I suppose they must then appeal to the nation for an additional article to the Constitution approving and confirming an act which the nation had not previously authorized. The Constitution has made no provision for holding foreign territory, still less for incorporating foreign nations into our Union. The Executive, in seizing the fugitive occurrence which so much advances the good of their country, has done an act beyond the Constitution."

To cover the questions raised by this purchase Mr. Jefferson prepared two amendments to the Constitution, the first of which declared that "the province of Louisiana is incorporated with the United States and made part thereof;" and the second of which was couched in a little different language, viz.: "Louisiana, as ceded by France to the United States, is made a part of the United States. Its white inhabitants shall be citizens, and stand, as to their rights and obligations, on the same footing as other citizens in analogous situations." But by the time Congress assembled, October 17, 1803, either the argument of his friends or the pressing necessity of the situation seems to have dispelled his doubts regarding his power under the Constitution, since in his message to Congress he referred the whole matter to that body, saying that "with the wisdom of Congress it will rest to take those ulterior measures which may be necessary for the immediate occupation and temporary government of the country; for its incorporation into the Union." Jefferson's Writings, vol. 8, p. 269.

The raising of money to provide for the purchase of this territory and the act providing a civil government gave rise to an animated debate in Congress, in which two questions were prominently presented: First, whether the provision for the ultimate incorporation of Louisiana into the Union was constitutional; and, second, whether the seventh article of the treaty admitting the ships of Spain and France for the next twelve years "into the ports of New Orleans, and in all other legal ports of entry within the ceded territory, in the same manner as the ships of

the United States coming directly from France or Spain, or any
of their colonies, without being subject to any other or greater
duty on merchandise or other or greater tonnage than that paid
by the citizens of the United States," was an unlawful discrimination in favor of those ports and an infringement upon Art. I,
sec. 9, of the Constitution, that "no preference shall be given by
any regulation of commerce or revenue to the ports of one State
over those of another." This article of the treaty contained
the further stipulation that "during the space of time above
mentioned no other nation shall have a right to the same privileges in the ports of the ceded territory; . . . and it is well
understood that the object of the above article is to favor the
manufactures, commerce, freight and navigation of France and
Spain."

It is unnecessary to enter into the details of this debate. The
arguments of individual legislators are no proper subject for judicial comment. They are so often influenced by personal or
political considerations, or by the assumed necessities of the situation, that they can hardly be considered even as the deliberate
views of the persons who make them, much less as dictating the
construction to be put upon the Constitution by the courts.
United States v. *Union Pac. Railroad,* 91 U. S. 72, 79. Suffice
it to say that the administration party took the ground that,
under the constitutional power to make treaties, there was ample
power to acquire territory, and to hold and govern it under laws
to be passed by Congress; and that as Louisiana was incorporated into the Union as a territory, and not as a State, a stipulation for citizenship became necessary; that as a State they
would not have needed a stipulation for the safety of their liberty, property and religion, but as territory this stipulation would
govern and restrain the undefined powers of Congress to "make
rules and regulations" for territories. The Federalists admitted
the power of Congress to acquire and hold territory, but denied
its power to incorporate it into the Union under the Constitution as it then stood.

They also attacked the seventh article of the treaty, discriminating in favor of French and Spanish ships, as a distinct violation of the Constitution against preference being given to the

ports of one State over those of another. The administration party, through Mr. Elliott of Vermont, replied to this that " the States, as such, were equal and intended to preserve that equality; and the provision of the Constitution alluded to was calculated to prevent Congress from making any odious discrimination or distinctions between particular States. It was not contemplated that this provision would have application to colonial or territorial acquisitions." Said Mr. Nicholson of Maryland, speaking for the administration : " It [Louisiana] is in the nature of a colony whose commerce may be regulated without any reference to the Constitution. Had it been the Island of Cuba which was ceded to us, under a similar condition of admitting French and Spanish vessels for a limited time into Havana, could it possibly have been contended that this would be giving a preference to the ports of one State over those of another, or that the uniformity of duties, imposts and excises throughout the United States would have been destroyed? And because Louisiana lies adjacent to our own territory is it to be viewed in a different light?"

As a sequence to this debate two bills were passed, one October 31, 1803, 2 Stat. 245, authorizing the President to take possession of the territory, and to continue the existing government, and the other November 10, 1803, 2 Stat. 245, making provision for the payment of the purchase price. These acts continued in force until March 26, 1804, when a new act was passed providing for a temporary government, 2 Stat. 283, c. 38, and vesting all legislative powers in a governor and legislative council, to be appointed by the President. These statutes may be taken as expressing the views of Congress, first, that territory may be lawfully acquired by treaty, with a provision for its ultimate incorporation into the Union ; and, second, that a discrimination in favor of certain foreign vessels trading with the ports of a newly acquired territory is no violation of that clause of the Constitution, Art. 1, sec. 9, that declares that no preference shall be given to the ports of one State over those of another. It is evident that the constitutionality of this discrimination can only be supported upon the theory that ports of territories are not ports of States within the meaning of the Constitution.

The same construction was adhered to in the treaty with Spain for the purchase of Florida, 8 Stat. 252, the sixth article of which provided that the inhabitants should "be incorporated into the Union of the United States, as soon as may be consistent with the principles of the Federal Constitution;" and the fifteenth article of which agreed that Spanish vessels coming directly from Spanish ports and laden with productions of Spanish growth or manufacture, should be admitted, for the term of twelve years, to the ports of Pensacola and St. Augustine, "without paying other or higher duties on their cargoes, or of tonnage, than will be paid by the vessels of the United States," and that "during the said term no other nation shall enjoy the same privileges within the ceded territories."

So, too, in the act annexing the Republic of Hawaii, there was a provision continuing in effect the customs relations of the Hawaiian Islands with the United States and other countries, the effect of which was to compel the collection in those islands of a duty upon certain articles, whether coming from the United States or other countries, much greater than the duty provided by the general tariff law then in force. This was a discrimination against the Hawaiian ports wholly inconsistent with the revenue clauses of the Constitution, if such clauses were there operative.

The very treaty with Spain under discussion in this case contains similar discriminative provisions, which are apparently irreconcilable with the Constitution, if that instrument be held to extend to these islands immediately upon their cession to the United States. By Art. IV the United States agree "for the term of ten years from the date of the exchange of the ratifications of the present treaty, to admit Spanish ships and merchandise to the ports of the Philippine Islands on the same terms as ships and merchandise of the United States"—a privilege not extending to any other ports. It was a clear breach of the uniformity clause in question, and a manifest excess of authority on the part of the commissioners, if ports of the Philippine Islands be ports of the United States.

So, too, by Art. XIII, "Spanish scientific, literary and artistic works . . . shall be continued to be admitted free of duty in such territories, for the period of ten years, to be reck-oned from the date of the exchange of the ratifications of this treaty." This is also a clear discrimination in favor of Spanish literary productions into particular ports.

Notwithstanding these provisions for the incorporation of territories into the Union, Congress, not only in organizing the territory of Louisiana by act of March 26, 1804, but all other territories carved out of this vast inheritance, has assumed that the Constitution did not extend to them of its own force, and has in each case made special provision, either that their legis-latures shall pass no law inconsistent with the Constitution of the United States, or that the Constitution or laws of the United States shall be the supreme law of such territories. Finally, in Rev. Stat. sec. 1891, a general provision was enacted that "the Constitution and all laws of the United States which are not locally inapplicable shall have the same force and effect within all the organized territories, and in every territory here-after organized, as elsewhere within the United States."

So, too, on March 6, 1820, 3 Stat. 545, c. 22, in an act au-thorizing the people of Missouri to form a state government, after a heated debate, Congress declared that in the territory of Louisiana north of 36° 30′ slavery should be forever prohib-ited. It is true that, for reasons which have become historical, this act was declared to be unconstitutional in *Scott* v. *Sand-ford*, 19 How. 393, but it is none the less a distinct annuncia-tion by Congress of power over property in the territories which it obviously did not possess in the several States.

The researches of counsel have collated a large number of other instances, in which Congress has in its enactments recog-nized the fact that provisions intended for the States did not embrace the territories, unless specially mentioned. These are found in the laws prohibiting the slave trade with "the United States or territories thereof;" or equipping ships "in any port or place within the *jurisdiction* of the United States;" in the internal revenue laws, in the early ones of which no pro-vision was made for the collection of taxes in the territory not included within the boundaries of the existing States, and others of which extended them expressly to the territories, or "within the exterior boundaries of the United States;" and in the acts extending the internal revenue laws to the Territories of Alaska and Oklahoma. It would prolong this opinion unnecessarily to set forth the provisions of these acts in detail. It is sufficient to say that Congress has or has not applied the revenue laws to the territories, as the circumstances of each case seemed to require, and has specifically legislated for the territories whenever it was its intention to execute laws beyond the limits of the States. Indeed, whatever may have been the fluctuations of opinion in other bodies, (and even this court has not been exempt from them,) Congress has been consistent in recognizing the difference between the States and territories under the Constitution.

The decisions of this court upon this subject have not been altogether harmonious. Some of them are based upon the theory that the Constitution does not apply to the territories without legislation. Other cases, arising from territories where such legislation has been had, contain language which would justify the inference that such legislation was unnecessary, and that the Constitution took effect immediately upon the cession of the territory to the United States. It may be remarked, upon the threshold of an analysis of these cases, that too much weight must not be given to general expressions found in several opinions that the power of Congress over territories is complete and supreme, because these words may be interpreted as meaning only supreme under the Constitution; nor upon the other hand, to general statements that the Constitution covers the territories as well as the States, since in such cases it will be found that acts of Congress had already extended the Constitution to such territories, and that thereby it subordinated not only its own acts, but those of the territorial legislatures, to what had become the supreme law of the land. "It is a maxim not to be disregarded, that general expressions, in every opinion, are to be taken in connection with the case in which those expressions are used. If they go beyond the case, they may be respected, but ought not to control the judgment in a subsequent suit when the very point is presented for decision. The reason of this maxim is obvious. The question actually before the court is investigated with care, and considered in its full extent. Other principles which may serve to illustrate it, are considered in their relation to the case decided, but their possible bearing on all other cases is seldom completely investigated." *Cohens* v. *Virginia,* 6 Wheat. 264, 399.

. The earliest case is that of *Hepburn* v. *Ellzey,* 2 Cranch, 445, in which this court held that, under that clause of the Constitution limiting the jurisdiction of the courts of the United States to controversies between citizens of different *States,* a citizen of the District of Columbia could not maintain an action in the Circuit Court of the United States. It was argued that the word "State," in that connection, was used simply to denote a distinct political society. "But," said the Chief Justice, "as the act of Congress obviously used the word 'State' in reference to that term as used in the Constitution, it becomes necessary to inquire whether Columbia is a State in the sense of that instrument. The result of that examination is a conviction that the members of the American confederacy only are the States contemplated in the Constitution, . . . and excludes from the term the signification attached to it by writers on the law of nations." This case was followed in *Barney* v. *Baltimore City,* 6 Wall. 280, and quite recently in *Hooe* v. *Jamieson,* 166 U. S. 395. The same rule was applied to citizens of territories in *New Orleans* v. *Winter,* 1 Wheat. 91, in which an attempt was made to distinguish a territory from the District of Columbia. But it was said that "neither of them is a *State* in the sense in which that term is used in the Constitution." In *Scott* v. *Jones,* 5 How. 343, and in *Miners' Bank* v. *Iowa,* 12 How. 1, it was held that under the Judiciary Act, permitting writs of error to the Supreme Court of a State, in cases where the validity of a *state statute* is drawn in question, an act of a territorial legislature was not within the contemplation of Congress.

Loughborough v. *Blake,* 5 Wheat. 317, was an action of trespass (or, as appears by the original record, *replevin*) brought in the Circuit Court for the District of Columbia to try the right of Congress to impose a direct tax for general purposes on that District. 3 Stat. 216, c. 60, Feb. 17, 1815. It was insisted that Congress could act in a double capacity: in one as legislating for the States; in the other as a local legislature for the District of Columbia. In the latter character, it was admitted that the power of levying direct taxes might be exercised, but for District purposes only, as a state legislature might tax for state purposes; but that it could not legislate for the District under Art. I, sec. 8, giving to Congress the power " to lay and collect taxes, imposts and excises," which " shall be uniform throughout the United States," inasmuch as the District was no part of the United States. It was held that the grant of this power was a general one without limitation as to place, and consequently extended to all places over which the government extends; and that it extended to the District of Columbia as a constituent part of the United States. The fact that Art. I, sec. 20, declares that " representatives and direct taxes shall be apportioned among the several States . . . according to their respective numbers," furnished a standard by which taxes were apportioned; but not to exempt any part of the country from their operation. " The words used do not mean, that direct taxes shall be imposed on States only which are represented, or shall be apportioned to representatives; but that direct taxation, in its application to States, shall be apportioned to numbers." That Art. I, sec. 9, ¶ 4, declaring that direct taxes shall be laid in proportion to the census, was applicable to the District of Columbia, " and will enable Congress to apportion on it its just and equal share of the burden, with the same accuracy as on the respective States. If the tax be laid in this proportion, it is within the very words of the restriction. It is a tax in proportion to the census or enumeration referred to." It was further held that the words of the ninth section did not " in terms require that the system of direct taxation, when resorted to, shall be extended to the territories, as the words of the second section require that it shall be extended to all the States. They therefore may, without violence, be understood to give a rule when the territories shall be taxed without imposing the necessity of taxing them."

There could be no doubt as to the correctness of this conclusion, so far, at least, as it applied to the District of Columbia. This District had been a part of the States of Maryland and

Virginia. It had been subject to the Constitution, and was a
part of the United States. The Constitution had attached to
it irrevocably. There are steps which can never be taken back-
ward. The tie that bound the States of Maryland and Virginia
to the Constitution could not be dissolved, without at least the
consent of the Federal and state governments to a formal separation. The mere cession of the District of Columbia to the
Federal government relinquished the authority of the States, but
it did not take it out of the United States or from under the ægis
of the Constitution. Neither party had ever consented to that
construction of the cession. If, before the District was set off,
Congress had passed an unconstitutional act, affecting its inhabitants, it would have been void. If done after the District was
created, it would have been equally void; in other words, Congress could not do indirectly by carving out the District what
it could not do directly. The District still remained a part of
the United States, protected by the Constitution. Indeed, it
would have been a fanciful construction to hold that territory
which had been once a part of the United States ceased to be
such by being ceded directly to the Federal government.

In delivering the opinion, however, the Chief Justice made
certain observations which have occasioned some embarrassment in other cases. " The power," said he, " to lay and collect
duties, imposts, and excises may be exercised, and must be exercised, throughout the United States. Does this term designate
the whole, or any particular portion of the American empire ?
Certainly this question can admit but of one answer. It is the
name given to our great republic, which is composed of States
and territories. The District of Columbia, or the territory
west of the Missouri, is not less within the United States than
Maryland and Pennsylvania; and it is not less necessary, on
the principles of our Constitution, that uniformity in the imposition of imposts, duties and excises, should be observed in the
one, than in the other. Since, then, the power to lay and collect taxes, which includes direct taxes, is obviously coextensive
with the power to lay and collect duties, imposts and excises,
and since the latter extends throughout the United States, it follows, that the power to impose direct taxes also extends through-

out the United States." So far as applicable to the District of
Columbia, these observations are entirely sound. So far as they
apply to the territories, they were not called for by the exigencies of the case.

In line with *Loughborough* v. *Blake* is the case of *Callan* v.
Wilson, 127 U. S. 540, in which the provisions of the Constitution relating to trial by jury were held to be in force in the
District of Columbia. Upon the other hand, in *Geofroy* v.
Riggs, 133 U. S. 258, the District of Columbia, as a political
community, was held to be one of "the States of the Union"
within the meaning of that term as used in a consular convention of February 23, 1853, with France. The seventh article
of that convention provided that in all the States of the Union,
whose existing laws permitted it, Frenchmen should enjoy the
right of holding, disposing of and inheriting property in the
same manner as citizens of the United States; and as to the
States of the Union, by whose existing laws aliens were not
permitted to hold real estate, the President engaged to recommend to them the passage of such laws as might be necessary
for the purpose of conferring this right. The court was of opinion that if these terms, "States of the Union," were held to
exclude the District of Columbia and the territories, our government would be placed in the inconsistent position of stipulating that French citizens should enjoy the right of holding,
disposing of and inheriting property in like manner as citizens
of the United States, in States whose laws permitted it, and
engaging that the President should recommend the passage of
laws conferring that right in States whose laws did not permit
aliens to hold real estate, while at the same time refusing to
citizens of France, holding property in the District of Columbia
and in some of the territories, where the power of the United
States is in that respect unlimited, a like release from the disa-
bilities of alienage, "thus discriminating against them in favor
of citizens of France holding property in States having similar
legislation. No plausible motive can be assigned for such discrimination. A right which the government of the United
States apparently desires that citizens of France should enjoy
in all the States it would hardly refuse to them in the district

embracing its capital, or in any of its own territorial dependencies."

This case may be considered as establishing the principle that, in dealing with foreign sovereignties, the term "United States" has a broader meaning than when used in the Constitution, and includes all territories subject to the jurisdiction of the Federal government, wherever located. In its treaties and conventions with foreign nations this government is a unit. This is so not because the territories comprised a part of the government established by the people of the States in their Constitution, but because the Federal government is the only authorized organ of the territories, as well as of the States, in their foreign relations. By Art. I, sec. 10, of the Constitution, "no State shall enter into any treaty, alliance or confederation, . . . or enter into any agreement or compact with another State, or with a foreign power." It would be absurd to hold that the territories, which are much less independent than the States, and are under the direct control and tutelage of the general government, possess a power in this particular which is thus expressly forbidden to the States.

It may be added in this connection that, to put at rest all doubts regarding the applicability of the Constitution to the District of Columbia, Congress by the act of February 21, 1871, c. 62, 16 Stat. 419, 426, sec. 34, specifically extended the Constitution and laws of the United States to this District.

The case of *American Ins. Co.* v. *Canter*, 1 Pet. 511, originated in a libel filed in the District Court of South Carolina, for the possession of 356 bales of cotton, which had been wrecked on the coast of Florida, abandoned to the insurance companies, and subsequently brought to Charleston. Canter claimed the cotton as *bona fide* purchaser at a marshal's sale at Key West, by virtue of a decree of a territorial court consisting of a notary and five jurors, proceeding under an act of the governor and legislative council of Florida. The case turned upon the question whether the sale by that court was effectual to divest the interest of the underwriters. The District Judge pronounced the proceedings a nullity, and rendered a decree from which both parties appealed to the Circuit Court. The Circuit Court

reversed the decree of the District Court upon the ground that
the proceedings of the court at Key West were legal, and transferred the property to Canter, the alleged purchaser.

The opinion of the Circuit Court was delivered by Mr. Justice
Johnson of the Supreme Court, and is published in full in a note
in Peters' Reports. It was argued that the Constitution vested
the admiralty jurisdiction exclusively in the general government; that the legislature of Florida had exercised an illegal
power in organizing this court, and that its decrees were void.
On the other hand, it was insisted that this was a court of separate and distinct jurisdiction from the courts of the United
States, and as such its acts were not to be reviewed in a foreign
tribunal, such as was the court of South Carolina; "that the
District of Florida was not part of the United States, but only
an acquisition or dependency, and as such the Constitution per
se had no binding effect in or over it." "It becomes," said the
court "indispensable to the solution of these difficulties, that
we should conceive a just idea of the relation in which Florida
stands to the United States. . . . And, first, it is obvious
that there is a material distinction between the territory now
under consideration, and that which is acquired from the aborigines (whether by purchase or conquest) within the acknowledged
limits of the United States, as also that which is acquired by the
establishment of a disputed line. As to both these there can be
no question, that the sovereignty of the State or territory within
which it lies, and of the United States, immediately attach, producing a complete subjection to all the laws and institutions of
the two governments, local and general, unless modified by
treaty. The question now to be considered, relates to territories
previously subject to the acknowledged jurisdiction of another
sovereign, such as was Florida to the crown of Spain. And on
this subject, we have the most explicit proof, that the understanding of our public functionaries, is, that the government
and laws of the United States do not extend to such territory
by the mere act of cession. For, in the act of Congress of
March 30, 1822, section nine, we have an enumeration of the
acts of Congress, which are to be held in force in the territory;
and in the tenth section an enumeration, in the nature of a bill

of rights, of privileges and immunities, which could not be denied to the inhabitants of the territory, if they came under the Constitution by the mere act of cession. . . . These States, this territory, and future *States* to be admitted into the Union are the sole objects of the Constitution; there is no express provision whatever made in the Constitution for the acquisition or government of territories beyond those limits." He further held that the right of acquiring territory was altogether incidental to the treaty-making power; that their government was left to Congress; that the territory of Florida did "not stand in the relation of a State to the United States;" that the acts establishing a territorial government were the constitution of Florida; that while, under these acts, the territorial legislature could enact nothing inconsistent with what Congress had made inherent and permanent in the territorial government, it had not done so in organizing the court at Key West.

From the decree of the Circuit Court the underwriters appealed to this court, and the question was argued whether the Circuit Court was correct in drawing a distinction between territories existing at the date of the Constitution and territories subsequently acquired. The main contention of the appellants was that the Superior Courts of Florida had been vested by Congress with exclusive jurisdiction in all admiralty and maritime cases; that salvage was such a case, and therefore any law of Florida giving jurisdiction in salvage cases to any other court was unconstitutional. On behalf of the purchaser it was argued that the Constitution and laws of the United States were not *per se* in force in Florida, nor the inhabitants citizens of the United States; that the Constitution was established by the people of the United States *for* the United States; that if the Constitution were in force in Florida it was unnecessary to pass an act extending the laws of the United States to Florida. "What is Florida?" said Mr. Webster. "It is no part of the United States. How can it be? How is it represented? Do the laws of the United States reach Florida? Not unless by particular provisions."

The opinion of Mr. Chief Justice Marshall in this case should be read in connection with Art. III, secs. 1 and 2, of the Constitution, vesting " the judicial power of the United States " in " one Supreme Court and in such inferior courts as Congress may from time to time ordain and establish. The judges both of the Supreme Court and the inferior courts shall hold their offices during good behavior," etc. He held that the court "should take into view the relation in which Florida stands to the United States;" that territory ceded by treaty "becomes a part of the nation to which it is annexed; either on the terms stipulated in the treaty of cession, or upon such as its new master shall impose." That Florida, upon the conclusion of the treaty, became a territory of the United States and subject to the power of Congress under the territorial clause of the Constitution. The acts providing a territorial government for Florida were examined in detail. He held that the judicial clause of the Constitution, above quoted, did not apply to Florida; that the judges of the Superior Courts of Florida held their office for four years; that " these courts are not constitutional courts in which the judicial power conferred by the Constitution on the general government, can be deposited;" that " they are legislative courts, created in virtue of the general right of sovereignty which exists in the government," or in virtue of the territorial clause of the Constitution; that the jurisdiction with which they are invested is not a part of judicial power of the Constitution, but is conferred by Congress, in the exercise of those general powers which that body possesses over the territories of the United States; and that in legislating for them Congress exercises the combined powers of the general and of a state government. The act of the territorial legislature, creating the court in question, was held not to be "inconsistent with the laws and Constitution of the United States," and the decree of the Circuit Court was affirmed.

As the only judicial power vested in Congress is to create courts whose judges shall hold their offices during good behavior, it necessarily follows that, if Congress authorizes the creation of courts and the appointment of judges for a limited time, it must act independently of the Constitution, and upon territory which is not part of the United States within the meaning of the Constitution. In delivering his opinion in this case Mr. Chief Justice Marshall made no reference whatever to the prior case of *Loughborough* v. *Blake*, 5 Wheat. 317, in which he had intimated that the territories were part of the United States. But if they be a part of the United States, it is difficult to see how Congress could create courts in such territories, except under the judicial clause of the Constitution. The power to make needful rules and regulations would certainly not authorize anything inconsistent with the Constitution if it applied to the territories. Certainly no such court could be created within a State, except under the restrictions of the judicial clause. It is sufficient to say that this case has ever since been accepted as authority for the proposition that the judicial clause of the Constitution has no application to courts created in the territories, and that with respect to them Congress has a power wholly unrestricted by it. We must assume as a logical inference from this case that the other powers vested in Congress by the Constitution have no application to these territories, or that the judicial clause is exceptional in that particular.

This case was followed in *Benner* v. *Porter*, 9 How. 235, in which it was held that the jurisdiction of these territorial courts ceased upon the admission of Florida into the Union, Mr. Justice Nelson remarking of them (p. 242) that "they are not organized under the Constitution, nor subject to its complex distribution of the powers of government, as the organic law; but are the creations, exclusively, of the legislative department, and subject to its supervision and control. Whether, or not, there are provisions in that instrument which extend to and act upon these territorial governments, it is not now material to examine. We are speaking here of those provisions that refer particularly to the distinction between Federal and State jurisdiction. . . . (p. 244.) Neither were they organized by Congress under the Constitution, as they were invested with powers and jurisdiction which that body were incapable of conferring upon a court within the limits of a State." To the same effect are *Clinton* v. *Englebrecht*, 13 Wall. 434; *Good* v. *Martin*, 95 U. S. 90, 98, and *McAllister* v. *United States*, 141 U. S. 174.

That the power over the territories is vested in Congress

without limitation, and that this power has been considered the foundation upon which the territorial governments rest, was also asserted by Chief Justice Marshall in *McCulloch* v. *Maryland*, 4 Wheat. 316, 422, and in *United States* v. *Gratiot*, 14 Pet. 526. So, too, in *Mormon Church* v. *United States*, 136 U. S. 1, in holding that Congress had power to repeal the charter of the church, Mr. Justice Bradley used the following forceful language: "The power of Congress over the territories of the United States is general and plenary, arising from and incidental to the right to acquire the territory itself, and from the power given by the Constitution to make all needful rules and regulations respecting the territory or other property belonging to the United States. It would be absurd to hold that the United States has power to acquire territory, and no power to govern it when acquired. The power to acquire territory, other than the Territory northwest of the Ohio River, (which belonged to the United States at the adoption of the Constitution,) is derived from the treaty-making power and the power to declare and carry on war. The incidents of these powers are those of national sovereignty, and belong to all independent governments. The power to make acquisitions of territory by conquest, by treaty and by cession is an incident of national sovereignty. The territory of Louisiana, when acquired from France, and the territories west of the Rocky Mountains, when acquired from Mexico, became the absolute property and domain of the United States, subject to such conditions as the government, in its diplomatic negotiations, had seen fit to accept relating to the rights of the people then inhabiting those territories. Having rightfully acquired said territories, the United States government was the only one which could impose laws upon them, and its sovereignty over them was complete. . . . Doubtless Congress, in legislating for the territories would be subject to those fundamental limitations in favor of personal rights which are formulated in the Constitution and its amendments; but these limitations would exist rather by inference and the general spirit of the Constitution from which Congress derives all its powers, than by any express and direct application of its provisions." See also, to the same effect, *National Bank* v. *County of Yankton,* 101 U. S. 129; *Murphy* v. *Ramsey,* 114 U. S. 15.

In *Webster* v. *Reid,* 11 How. 437, it was held that a law of the Territory of Iowa, which prohibited the trial by jury of certain actions at law, founded on contract to recover payment for services, was void; but the case is of little value as bearing upon the question of the extension of the Constitution to that Territory, inasmuch as the organic law of the Territory of Iowa, by express provision and by reference, extended the laws of the United States, including the ordinance of 1787, (which provided expressly for jury trials,) so far as they were applicable; and the case was put upon this ground. 5 Stat. 235, 239, sec. 12.

In *Reynolds* v. *United States,* 98 U. S. 145, a law of the Territory of Utah, providing for grand juries of fifteen persons, was held to be constitutional, though Rev. Stat. sec. 808 required that a grand jury empanelled before any Circuit or District Court of the United States shall consist of not less than sixteen nor more than twenty-three persons. Section 808 was held to apply only to the Circuit and District Courts. The territorial courts were free to act in obedience to their own laws.

In *Ross's Case,* 140 U. S. 453, petitioner had been convicted by the American Consular Tribunal in Japan, of a murder committed upon an American vessel in the harbor of Yokohama, and sentenced to death. There was no indictment by a grand jury, and no trial by a petit jury. This court affirmed the conviction, holding that the Constitution had no application, since it was ordained and established "for the United States of America," and not for countries outside of their limits. "The guarantees it affords against accusation of capital or infamous crimes, except by indictment or presentment by a grand jury, and for an impartial trial by a jury when thus accused, apply only to citizens and others within the United States, or who are brought there for trial for alleged offences committed elsewhere, and not to residents and temporary sojourners abroad."

In *Springville* v. *Thomas,* 166 U. S. 707, it was held that a verdict returned by less than the whole number of jurors was invalid, because in contravention of the Seventh Amendment to the Constitution and the act of Congress of April 7, 1874, c.

80, 18 Stat. 27, which provide " that no party has been or shall be deprived of the right of trial by jury in cases cognizable at common law." It was also intimated that Congress "could not impart the power to change the constitutional rule," which was obviously true with respect to Utah, since the organic act of that Territory had expressly extended to it the Constitution and laws of the United States. As we have already held, that provision once made could not be withdrawn. If the Constitution could be withdrawn directly, it could be nullified indirectly by acts passed inconsistent with it. The Constitution would thus cease to exist as such, and become of no greater authority than an ordinary act of Congress. In *American Pub. Co.* v. *Fisher*, 166 U. S. 464, a similar law providing for majority verdicts was put upon the express ground above stated, that the organic act of Utah extended the Constitution over that Territory. These rulings were repeated in *Thompson* v. *Utah*, 170 U. S. 343, and applied to felonies committed before the Territory became a State, although the state constitution continued the same provision.

Eliminating, then, from the opinions of this court all expressions unnecessary to the disposition of the particular case, and gleaning therefrom the exact point decided in each, the following propositions may be considered as established:

1. That the District of Columbia and the territories are not States, within the judicial clause of the Constitution giving jurisdiction in cases between citizens of different States;

2. That territories are not States, within the meaning of Revised Statutes, sec. 709, permitting writs of error from this court in cases where the validity of a *state* statute is drawn in question;

3. That the District of Columbia and the territories are States, as that word is used in treaties with foreign powers, with respect to the ownership, disposition and inheritance of property;

4. That the territories are not within the clause of the Constitution providing for the creation of a Supreme Court and such inferior courts as Congress may see fit to establish;

5. That the Constitution does not apply to foreign countries or to trials therein conducted, and that Congress may lawfully

provide for such trials before consular tribunals, without the intervention of a grand or petit jury;

6. That where the Constitution has been once formally extended by Congress to territories, neither Congress nor the territorial legislature can enact laws inconsistent therewith.

The case of *Dred Scott* v. *Sandford*, 19 How. 393, remains to be considered. This was an action of trespass *vi et armis* brought in the Circuit Court for the District of Missouri by Scott, alleging himself to be a citizen of Missouri, against Sandford, a citizen of New York. Defendant pleaded to the jurisdiction that Scott was not a citizen of the State of Missouri, because a negro of African descent, whose ancestors were imported as negro slaves. Plaintiff demurred to this plea and the demurrer was sustained; whereupon, by stipulation of counsel and with leave of the court, defendant pleaded in bar the general issue, and specially that the plaintiff was a slave and the lawful property of defendant, and, as such, he had a right to restrain him. The wife and children of the plaintiff were also involved in the suit.

The facts in brief were, that plaintiff had been a slave belonging to Dr. Emerson, a surgeon in the army; that, in 1834, Emerson took the plaintiff from the State of Missouri to Rock Island, Illinois, and subsequently to Fort Snelling, Minnesota, (then known as Upper Louisiana,) and held him there until 1838. Scott married his wife there, of whom the children were subsequently born. In 1838 they returned to Missouri.

Two questions were presented by the record: First, whether the Circuit Court had jurisdiction; and, second, if it had jurisdiction, was the judgment erroneous or not? With regard to the first question, the court stated that it was its duty "to decide whether the facts stated in the plea are or are not sufficient to show that the plaintiff is not entitled to sue as a citizen in a court of the United States," and that the question was whether "a negro, whose ancestors were imported into this country, and sold as slaves, became a member of the political community formed and brought into existence by the Constitution of the United States, and as such entitled to all the rights and privileges and immunities guaranteed by that instrument to the citizen, one of which rights is the privilege of suing in a court

of the United States." It was held that he was not, and was
not included under the words "citizens" in the Constitution,
and therefore could claim "none of the rights and privileges
which that instrument provides for and secures to citizens of
the United States;" that it did not follow because he had all
the rights and privileges of a citizen of a State, he must be a citizen of the United States; that no State could by any law of its
own "introduce a new member into the political community
created by the Constitution;" that the African race was not intended to be included, and formed no part of the people who
framed and adopted the Declaration of Independence. The
question of the *status* of negroes in England and the several
States was considered at great length by the Chief Justice, and
the conclusion reached that Scott was not a citizen of Missouri,
and that the Circuit Court had no jurisdiction of the case.

This was sufficient to dispose of the case without reference to
the question of slavery; but, as the plaintiff insisted upon his
title to freedom and citizenship by the fact that he and his wife,
though born slaves, were taken by their owner and kept four
years in Illinois and Minnesota, they thereby became free, and
upon their return to Missouri became citizens of that State, the
Chief Justice proceeded to discuss the question whether Scott
was still a slave. As the court had decided against his citizenship upon the plea in abatement, it was insisted that further
decision upon the question of his freedom or slavery was extra-
judicial and mere *obiter dicta*. But the Chief Justice held that
the correction of one error in the court below did not deprive
the appellate court of the power of examining further into the
record and correcting any other material error which may have
been committed; that the error of an inferior court in actually
pronouncing judgment for one of the parties, in a case in which
it had no jurisdiction, can be looked into or corrected by this
court, even though it had decided a similar question presented
in the pleadings.

Proceeding to decide the case upon the merits, he held that
the territorial clause of the Constitution was confined to the territory which belonged to the United States at the time the Con-

stitution was adopted, and did not apply to territory subsequently acquired from a foreign government.

In further examining the question as to what provision of the Constitution authorizes the Federal government to acquire territory outside of the original limits of the United States and what powers it may exercise therein over the person or property of a citizen of the United States, he made use of the following expressions, upon which great reliance is placed by the plaintiff in this case (p. 446): "There is certainly no power given by the Constitution to the Federal government to establish or maintain colonies bordering on the United States or at a distance, to be ruled and governed at its own pleasure; . . . and if a new State is admitted, it needs no further legislation by Congress, because the Constitution itself defines the relative rights and powers and duties of the State, and the citizens of the State, and the Federal government. But no power is given to acquire a territory to be held and governed permanently in that character."

He further held that citizens who migrate to a territory cannot be ruled as mere colonists, and that while Congress had the power of legislating over territories until States were formed from them, it could not deprive a citizen of his property merely because he brought it into a particular territory of the United States, and that this doctrine applied to slaves as well as to other property. Hence, it followed that the act of Congress which prohibited a citizen from holding and owning slaves in territories north of 36° 30′ (known as the Missouri Compromise) was unconstitutional and void, and the fact that Scott was carried into such territory, referring to what is now known as Minnesota, did not entitle him to his freedom.

He further held that, whether he was made free by being taken into the free State of Illinois and being kept there two years, depended upon the laws of Missouri and not those of Illinois, and that by the decisions of the highest court of that State his *status* as a slave continued, notwithstanding his residence of two years in Illinois.

It must be admitted that this case is a strong authority in favor of the plaintiff, and if the opinion of the Chief Justice be taken at its full value it is decisive in his favor. We are not, however, bound to overlook the fact that, before the Chief Justice gave utterance to his opinion upon the merits, he had already disposed of the case adversely to the plaintiff upon the question of jurisdiction, and that, in view of the excited political condition of the country at the time, it is unfortunate that he felt compelled to discuss the question upon the merits, particularly so in view of the fact that it involved a ruling that an act of Congress, which had been acquiesced in for thirty years, was declared unconstitutional. It would appear from the opinion of Mr. Justice Wayne that the real reason for discussing these constitutional questions was that "there had become such a difference of opinion" about them "that the peace and harmony of the country required the settlement of them by judicial decision." (p. 455.) The attempt was not successful. It is sufficient to say that the country did not acquiesce in the opinion, and that the civil war, which shortly thereafter followed, produced such changes in judicial, as well as public sentiment, as to seriously impair the authority of this case.

While there is much in the opinion of the Chief Justice which tends to prove that he thought all the provisions of the Constitution extended of their own force to the territories west of the Mississippi, the question actually decided is readily distinguishable from the one involved in the cause under consideration. The power to prohibit slavery in the territories is so different from the power to impose duties upon territorial products, and depends upon such different provisions of the Constitution, that they can scarcely be considered as analogous, unless we assume broadly that every clause of the Constitution attaches to the territories as well as to the States—a claim quite inconsistent with the position of the court in the *Canter* case. If the assumption be true, that slaves are indistinguishable from other property, the inference from the *Dred Scott* case is irresistible that Congress had no power to prohibit their introduction into a territory. It would scarcely be insisted that Congress could with one hand invite settlers to locate in the territories of the United States, and with the other deny them the right to take their property and belongings with them. The two

are so inseparable from each other that one could scarcely be
granted and the other withheld without an exercise of arbitrary power inconsistent with the underlying principles of a
free government. It might indeed be claimed with great plausibility that such a law would amount to a deprivation of property within the Fourteenth Amendment. The difficulty with
the *Dred Scott* case was that the court refused to make a distinction between property in general, and a wholly exceptional
class of property. Mr. Benton tersely stated the distinction
by saying that the Virginian might carry his slave into the
territories, but he could not carry with him the Virginian law
which made him a slave.

In his history of the *Dred Scott* case, Mr. Benton states
that the doctrine of the Constitution extended to territories as
well as to States, first made its appearance in the Senate in the
session of 1848–1849, by an attempt to amend a bill giving territorial government to California, New Mexico and Utah, (itself
"hitched on" to a general appropriation bill,) by adding the
words "that the Constitution of the United States and all and
singular the several acts of Congress (describing them,) be and
the same hereby are extended and given full force and efficacy
in said territories." Says Mr. Benton: "The novelty and
strangeness of this proposition called up Mr. Webster, who re-
pulsed as an absurdity and as an impossibility the scheme of
extending the Constitution to the territories, declaring that instrument to have been made for States, not territories; that
Congress governed the territories independently of the Constitution and incompatibly with it; that no part of it went to a
territory but what Congress chose to send; that it could not
act of itself anywhere, not even in the States for which it was
made, and that it required an act of Congress to put it in operation before it had effect anywhere. Mr. Clay was of the same
opinion and added: 'Now, really, I must say the idea that *eo
instanti*, upon the consummation of the treaty, the Constitution
of the United States spread itself over the acquired territory
and carried along with it the institution of slavery, is so irreconcilable with my comprehension, or any reason I possess, that
I hardly know how to meet it.' Upon the other hand, Mr. Cal-

houn boldly avowed his intent to carry slavery into them under the wing of the Constitution, and denounced as enemies of the South all who opposed it."

The amendment was rejected by the House, and a contest brought on which threatened the loss of the general appropriation bill in which this amendment was incorporated, and the Senate finally receded from its amendment. "Such," said Mr. Benton, "were the portentous circumstances under which this new doctrine first revealed itself in the American Senate, and then as needing legislative sanction requiring an act of Congress to carry the Constitution into the territories and to give it force and efficacy there." Of the *Dred Scott* case he says: "I conclude this introductory note with recurring to the great fundamental error of the court, (father of all the political errors,) that of assuming the extension of the Constitution to the territories. I call it assuming, for it seems to be a naked assumption without a reason to support it, or a leg to stand upon, condemned by the Constitution itself, and the whole history of its formation and administration. Who were the parties to it? The States alone. Their delegates framed it in the Federal convention; their citizens adopted it in the state conventions. The Northwest Territory was then in existence and it had been for three years; yet it had no voice either in the framing or adopting of the instrument, no delegate at Philadelphia, no submission of it to their will for adoption. The preamble shows it made by States. Territories are not alluded to in it."

Finally, in summing up the results of the decisions holding the invalidity of the Missouri Compromise and the self-extension of the Constitution to the territories, he declares "that the desions conflict with the uniform action of all the departments of the Federal government from its foundation to the present time, and cannot be received as rules governing Congress and the people without reversing that action, and admitting the political supremacy of the court, and accepting an altered Constitution from its hands and taking a new and portentous point of departure in the working of the government."

To sustain the judgment in the case under consideration it by no means becomes necessary to show that none of the articles of the Constitution apply to the Island of Porto Rico. There is a clear distinction between such prohibitions as go to the very root of the power of Congress to act at all, irrespective of time or place, and such as are operative only "throughout the United States" or among the several States.

Thus, when the Constitution declares that "no bill of attainder or *ex post facto* law shall be passed," and that "no title of nobility shall be granted by the United States," it goes to the competency of Congress to pass a bill *of that description.* Perhaps, the same remark may apply to the First Amendment, that "Congress shall make no law respecting an establishment of religion, or prohibiting the free exercise thereof; or abridging the freedom of speech, or of the press; or the right of the people to peacefully assemble, and to petition the government for a redress of grievances." We do not wish, however, to be understood as expressing an opinion how far the bill of rights contained in the first eight amendments is of general and how far of local application.

Upon the other hand, when the Constitution declares that all duties shall be uniform "throughout the United States," it becomes necessary to inquire whether there be any territory over which Congress has jurisdiction which is not a part of the "United States," by which term we understand the *States* whose people *united* to form the Constitution, and such as have since been admitted to the Union upon an equality with them. Not only did the people in adopting the Thirteenth Amendment thus recognize a distinction between the United States and "any place subject to their jurisdiction," but Congress itself, in the act of March 27, 1804, c. 56, 2 Stat. 298, providing for the proof of public records, applied the provisions of the act not only to "every court and office within the United States," but to the "courts and offices of the respective territories of the United States, and countries subject to the jurisdiction of the United States," as to the courts and offices of the several States. This classification, adopted by the Eighth Congress, is carried into the Revised Statutes as follows:

"SEC. 905. The acts of the legislature of any State or Territory, or of any country subject to the jurisdiction of the United States, shall be authenticated," etc.

"SEC. 906. All records and exemplifications of books, which may be kept in any public office of any State or Territory, or any country subject to the jurisdiction of the United States," etc.

. Unless these words are to be rejected as meaningless, we must treat them as a recognition by Congress of the fact that there may be territories subject to the jurisdiction of the United States, which are not *of* the United States.

In determining the meaning of the words of Article I, section 6, "uniform throughout the United States," we are bound to consider not only the provisions forbidding preference being given to the ports of one State over those of another, (to which attention has already been called,) but the other clauses declaring that no tax or duty shall be laid on articles exported from any State, and that no State shall, without the consent of Congress, lay any imposts or duties upon imports or exports, nor any duty on tonnage. The object of all of these was to protect the States which united in forming the Constitution from discriminations by Congress, which would operate unfairly or injuriously upon some States and not equally upon others. The opinion of Mr. Justice White in *Knowlton* v. *Moore*, 178 U. S. 41, contains an elaborate historical review of the proceedings in the convention, which resulted in the adoption of these different clauses and their arrangement, and he there comes to the conclusion (p. 105) that " although the provision as to preference between ports and that regarding uniformity of duties, imposts and excises were one in purpose, one in their adoption," they were originally placed together, and " became separate only in arranging the Constitution for the purpose of style." Thus construed together, the purpose is irresistible that the words " throughout the United States " are indistinguishable from the words " among or between the several States," and that these prohibitions were intended to apply only to commerce between ports of the several States as they then existed or should thereafter be admitted to the Union.

Indeed, the practical interpretation put by Congress upon the Constitution has been long continued and uniform to the effect that the Constitution is applicable to territories acquired by purchase or conquest only when and so far as Congress shall so direct. Notwithstanding its duty to "guarantee to every State in this Union a republican form of government," Art. IV, sec. 4, by which we understand, according to the definition of Webster, "a government in which the supreme power resides in the whole body of the people, and is exercised by representatives elected by them," Congress did not hesitate, in the original organization of the territories of Louisiana, Florida, the Northwest Territory, and its subdivisions of Ohio, Indiana, Michigan, Illinois and Wisconsin, and still more recently in the case of Alaska, to establish a form of government bearing a much greater analogy to a British crown colony than a republican State of America, and to vest the legislative power either in a governor and council, or a governor and judges, to be appointed by the President. It was not until they had attained a certain population that power was given them to organize a legislature by vote of the people. In all these cases, as well as in Territories subsequently organized west of the Mississippi, Congress thought it necessary either to extend the Constitution and laws of the United States over them, or to declare that the inhabitants should be entitled to enjoy the right of trial by jury, of bail, and of the privilege of the writ of *habeas corpus*, as well as other privileges of the bill of rights.

We are also of opinion that the power to acquire territory by treaty implies not only the power to govern such territory, but to prescribe upon what terms the United States will receive its inhabitants, and what their *status* shall be in what Chief Justice Marshall termed the "American Empire." There seems to be no middle ground between this position and the doctrine that if their inhabitants do not become, immediately upon annexation, citizens of the United States, their children thereafter born, whether savages or civilized, are such, and entitled to all the rights, privileges and immunities of citizens. If such be their *status*, the consequences will be extremely serious. Indeed, it is doubtful if Congress would ever assent to the annexation of territory upon the condition that its inhabitants, however foreign they may be to our habits, traditions and modes of life, shall become at once citizens of the United States. In all its treaties hitherto the treaty-making power has made special provision for this subject; in the cases of Louisiana and Florida, by stipulating that "the inhabitants shall be incorporated into the Union of the United States and admitted as soon as possible . . . to the enjoyment of all the rights, advantages and immunities of citizens of the United States;" in the case of Mexico, that they should "be incorporated into the Union, and be admitted at the proper time, (to be judged of by the Congress of the United States,) to the enjoyment of all the rights of citizens of the United States;" in the case of Alaska, that the inhabitants who remained three years, "with the exception of uncivilized native tribes, shall be admitted to the enjoyment of all the rights," etc.; and in the case of Porto Rico and the Philippines, "that the civil rights and political *status* of the native inhabitants . . . shall be determined by Congress." In all these cases there is an implied denial of the right of the inhabitants to American citizenship until Congress by further action shall signify its assent thereto.

Grave apprehensions of danger are felt by many eminent men—a fear lest an unrestrained possession of power on the part of Congress may lead to unjust and oppressive legislation, in which the natural rights of territories, or their inhabitants, may be engulfed in a centralized despotism. These fears, however, find no justification in the action of Congress in the past century, nor in the conduct of the British Parliament towards its outlying possessions since the American Revolution. Indeed, in the only instance in which this court has declared an act of Congress unconstitutional as trespassing upon the rights of territories, (the Missouri Compromise,) such action was dictated by motives of humanity and justice, and so far commanded popular approval as to be embodied in the Thirteenth Amendment to the Constitution. There are certain principles of natural justice inherent in the Anglo-Saxon character which need no expression in constitutions or statutes to give them effect or to secure dependencies against legislation manifestly hostile to their real interests. Even in the Foraker act itself, the constitutionality of which is so vigorously assailed, power was given to the legislative assembly of Porto Rico to repeal the very tariff in question in this case, a power it has not seen fit to exercise. The words of Chief Justice Marshall in *Gibbons* v. *Ogden*, 9 Wheat. 1, with respect to the power of Congress to regulate commerce, are pertinent in this connection: " This power," said he, " like all others vested in Congress, is complete in itself, may be exercised to its utmost extent, and acknowledges no limitations other than are prescribed in the Constitution. . . . The wisdom and discretion of Congress, their identity with the people, and the influence which their constituents possess at elections, are, in this, as in many other instances, as that, for example, of declaring war, the sole restraints on which they have relied to secure them from its abuse. They are the restraints on which the people must often rely on solely, in all representative governments."

So, too, in *Johnson* v. *McIntosh*, 8 Wheat. 543, 589, it was said by him :

" The title by conquest is acquired and maintained by force. The conqueror prescribes its limits. Humanity, however, acting on public opinion, has established, as a general rule, that the conquered shall not be wantonly oppressed, and that their condition shall remain as eligible as is compatible with the objects of the conquest. Most usually, they are incorporated with the victorious nation, and become subjects or citizens of the government with which they are connected. The new and old members of the society mingle with each other; the distinction between them is gradually lost, and they make one people. Where this incorporation is practicable, humanity demands, and a wise policy requires, that the rights of the conquered to property should remain unimpaired; that the new subjects should be governed as equitably as the old, and that confidence in their security should gradually banish the painful sense of being separated from their ancient connections, and united by force to strangers.

" When the conquest is complete, and the conquered inhabitants can be blended with the conquerors, *or safely governed as a distinct people*, public opinion, which not even the conqueror can disregard, imposes these restraints upon him; and he cannot neglect them without injury to his fame, and hazard to his power."

The following remarks of Mr. Justice White in the case of *Knowlton* v. *Moore*, 178 U. S. 41, 109, in which the court upheld the progressive features of the legacy tax, are also pertinent:

"The grave consequences which it is asserted must arise in the future if the right to levy a progressive tax be recognized involves in its ultimate aspect the mere assertion that free and representative government is a failure, and that the grossest abuses of power are foreshadowed unless the courts usurp a purely legislative function. If a case should ever arise, where an arbitrary and confiscatory exaction is imposed bearing the guise of a progressive or any other form of tax, it will be time enough to consider whether the judicial power can afford a remedy by applying inherent and fundamental principles for the protection of the individual, even though there be no express authority in the Constitution to do so."

It is obvious that in the annexation of outlying and distant possessions grave questions will arise from differences of race, habits, laws and customs of the people, and from differences of soil, climate and production, which may require action on the part of Congress that would be quite unnecessary in the annexation of contiguous territory inhabited only by people of the same race, or by scattered bodies of native Indians.

We suggest, without intending to decide, that there may be a distinction between certain natural rights, enforced in the Constitution by prohibitions against interference with them, and what may be termed artificial or remedial rights, which are peculiar to our own system of jurisprudence. Of the former class are the rights to one's own religious opinion and to a public expression of them, or, as sometimes said, to worship God according to the dictates of one's own conscience; the right to personal liberty and individual property; to freedom of speech and of the press; to free access to courts of justice, to due process of law and to an equal protection of the laws; to immunities from unreasonable searches and seizures, as well as cruel and unusual punishments; and to such other immunities as are indispensable to a free government. Of the latter class are the rights to citizenship, to suffrage, *Minor* v. *Happersett*, 21 Wall. 162, and to the particular methods of procedure pointed out in the Constitution, which are peculiar to Anglo-Saxon jurisprudence, and some of which have already been held by the States to be unnecessary to the proper protection of individuals.

Whatever may be finally decided by the American people as to the *status* of these islands and their inhabitants—whether they shall be introduced into the sisterhood of States or be permitted to form independent governments—it does not follow that, in the meantime, awaiting that decision, the people are in the matter of personal rights unprotected by the provisions of our Constitution, and subject to the merely arbitrary control of Congress. Even if regarded as aliens, they are entitled under the principles of the Constitution to be protected in life, liberty and property. This has been frequently held by this court in respect to the Chinese, even when aliens, not possessed of the political rights of citizens of the United States. *Yick Wo* v. *Hopkins*, 118 U. S. 356; *Fong Yue Ting* v. *United States*, 149 U. S. 698; *Lem Moon Sing* v. *United States*, 158 U. S. 538, 547; *Wong Wing* v. *United States*, 163 U. S. 228. We do not desire, however, to anticipate the difficulties which would naturally arise in this connection, but merely to disclaim any intention to hold that the inhabitants of these territories are subject to an unrestrained power on the part of Congress to deal with them upon the theory that they have no rights which it is bound to respect.

Large powers must necessarily be entrusted to Congress in dealing with these problems, and we are bound to assume that they will be judiciously exercised. That these powers may be abused is possible. But the same may be said of its powers under the Constitution as well as outside of it. Human wisdom has never devised a form of government so perfect that it may not be perverted to bad purposes. It is never conclusive to argue against the possession of certain powers from possible abuses of them. It is safe to say that if Congress should venture upon legislation manifestly dictated by selfish interests, it would receive quick rebuke at the hands of the people. Indeed, it is scarcely possible that Congress could do a greater injustice to these islands than would be involved in holding that it could not impose upon the States taxes and excises without extending the same taxes to them. Such requirement would bring them at once within our internal revenue system, including stamps, licenses, excises and all the paraphernalia of that system, and applying it to territories which have had no experience of this kind, and where it would prove an intolerable burden.

This subject was carefully considered by the Senate committee in charge of the Foraker bill, which found, after an examination of the facts, that property in Porto Rico was already burdened with a private debt amounting probably to $30,000,000; that no system of property taxation was or ever had been in force in the island, and that it probably would require two years to inaugurate one and secure returns from it; that the revenues had always been chiefly raised by duties on imports and exports, and that our internal revenue laws, if applied in that island, would prove oppressive and ruinous to many people and interests; that to undertake to collect our heavy internal revenue tax, far heavier than Spain ever imposed upon their products and vocations, would be to invite violations of the law so innumerable as to make prosecutions impossible, and to almost certainly alienate and destroy the friendship and good will of that people for the United States.

In passing upon the questions involved in this case and kindred cases, we ought not to overlook the fact that, while the Constitution was intended to establish a permanent form of government for the States which should elect to take advantage of its conditions, and continue for an indefinite future, the vast possibilities of that future could never have entered the minds of its framers. The States had but recently emerged from a war with one of the most powerful nations of Europe; were disheartened by the failure of the confederacy, and were doubtful as to the feasibility of a stronger union. Their territory was confined to a narrow strip of land on the Atlantic coast from Canada to Florida, with a somewhat indefinite claim to territory beyond the Alleghenies, where their sovereignty was disputed by tribes of hostile Indians supported, as was popularly believed, by the British, who had never formally delivered possession under the treaty of peace. The vast territory beyond the Mississippi, which formerly had been claimed by France, since 1762 had belonged to Spain, still a powerful nation, and the owner of a great part of the Western Hemisphere. Under these circumstances it is little wonder that the question of annexing these territories was not made a subject of debate. The difficulties of bringing about a union of the States were so great, the objections to it seemed so formidable, that the whole thought of the convention centered upon surmounting these obstacles. The question of territories was dismissed with a single clause, apparently applicable only to the territories then existing, giving Congress the power to govern and dispose of them.

Had the acquisition of other territories been contemplated as a possibility, could it have been foreseen that, within little more than one hundred years, we were destined to acquire not only the whole vast region between the Atlantic and Pacific Oceans, but the Russian possessions in America and distant islands in the Pacific, it is incredible that no provision should have been made for them, and the question whether the Constitution should or should not extend to them have been definitely settled. If it be once conceded that we are at liberty to acquire foreign territory, a presumption arises that our power with respect to such territories is the same power which other nations have been accustomed to exercise with respect to territories acquired by them. If, in limiting the power which Congress was to exercise within the United States, it was also intended to limit it with regard to such territories as the people of the United States should thereafter acquire, such limitations should have been expressed. Instead of that, we find the Constitution speaking only to States, except in the territorial clause, which is absolute in its terms, and suggestive of no limitations upon the power of Congress in dealing with them. The States could only delegate to Congress such powers as they themselves possessed, and as they had no power to acquire new territory they had none to delegate in that connection. The logical inference from this is, that if Congress had power to acquire new territory, which is conceded, that power was not hampered by the constitutional provisions. If, upon the other hand, we assume

that the territorial clause of the Constitution was not intended
to be restricted to such territory as the United States then possessed, there is nothing in the Constitution to indicate that the
power of Congress in dealing with them was intended to be
restricted by any of the other provisions.

There is a provision that "new States may be admitted by
the Congress into this Union." These words, of course, carry
the Constitution with them, but nothing is said regarding the
acquisition of new territories or the extension of the Constitution over them. The liberality of Congress in legislating the
Constitution into all our contiguous territories has undoubtedly
fostered the impression that it went there by its own force, but
there is nothing in the Constitution itself, and little in the interpretation put upon it, to confirm that impression. There is
not even an analogy to the provisions of an ordinary mortgage
for its attachment to after-acquired property, without which it
covers only property existing at the date of the mortgage. In
short, there is absolute silence upon the subject. The executive
and legislative departments of the government have for more
than a century interpreted this silence as precluding the idea
that the Constitution attached to these territories as soon as acquired, and unless such interpretation be manifestly contrary
to the letter or spirit of the Constitution, it should be followed
by the judicial department. Cooley's Consti. Lim. secs. 81 to
85. *Burrow-Giles Lithographic Co.* v. *Sarony,* 111 U. S. 53,
57; *Field* v. *Clark,* 143 U. S. 649, 691.

Patriotic and intelligent men may differ widely as to the
desireableness of this or that acquisition, but this is solely a
political question. We can only consider this aspect of the case
so far as to say that no construction of the Constitution should
be adopted which would prevent Congress from considering
each case upon its merits, unless the language of the instrument imperatively demand it. A false step at this time might
be fatal to the development of what Chief Justice Marshall
called the American Empire. Choice in some cases, the natural gravitation of small bodies towards large ones in others, the
result of a successful war in still others, may bring about conditions which would render the annexation of distant posses-

sions desirable. If those possessions are inhabited by alien races, differing from us in religion, customs, laws, methods of taxation and modes of thought, the administration of government and justice, according to Anglo-Saxon principles, may for a time be impossible; and the question at once arises whether large concessions ought not to be made for a time, that, ultimately, our own theories may be carried out, and the blessings of a free government under the Constitution extended to them. We decline to hold that there is anything in the Constitution to forbid such action.

We are therefore of opinion that the Island of Porto Rico is a territory appurtenant and belonging to the United States, but not a part of the United States within the revenue clauses of the Constitution; that the Foraker act is constitutional, so far as it imposes duties upon imports from such island, and that the plaintiff cannot recover back the duties exacted in this case.

The judgment of the Circuit Court is therefore

Affirmed.

MR. JUSTICE WHITE, with whom concurred MR. JUSTICE SHIRAS and MR. JUSTICE MCKENNA, uniting in the judgment of affirmance.

MR. JUSTICE BROWN, in announcing the judgment of affirmance, has in his opinion stated his reasons for his concurrence in such judgment. In the result I likewise concur. As, however, the reasons which cause me to do so are different from, if not in conflict with, those expressed in that opinion, if its meaning is by me not misconceived, it becomes my duty to state the convictions which control me.

The recovery sought is the amount of duty paid on merchandise which came into the United States from Porto Rico after July 1, 1900. The exaction was made in virtue of the act of Congress approved April 12, 1900, entitled "An act temporarily to provide revenue and a civil government for Porto Rico, and for other purposes." 31 Stat. 77, c. 191. The right to recover is predicated on the assumption that Porto Rico, by the ratification of the treaty with Spain, became incorporated into the

United States, and therefore the act of Congress which imposed
the duty in question is repugnant to Article I, sec. 8, clause 1,
of the Constitution providing that "The Congress shall have
Power To lay and collect Taxes, Duties, Imposts and Excises, to
pay the Debts and provide for the common Defence and general Welfare of the United States; but all Duties, Imposts and
Excises shall be Uniform throughout the United States." Subsidiarily, it is contended that the duty collected was also repugnant to the export and preference clauses of the Constitution.
But as the case concerns no duty on goods going from the United
States to Porto Rico, this proposition must depend also on the
hypothesis that the provisions of the Constitution referred to
apply to Porto Rico because that island has been incorporated
into the United States. It is hence manifest that this latter contention is involved in the previous one, and need not be separately considered.

The arguments at bar embrace many propositions which seem
to me to be irrelevant, or, if relevant, to be so contrary to reason and so in conflict with previous decisions of this court as to
cause them to require but a passing notice. To eliminate all controversies of this character, and thus to come to the pivotal contentions which the case involves, let me state and concede the
soundness of some principles, referring, in doing so, in the
margin to the authorities by which they are sustained, and
making such comment on some of them as may to me appear
necessary.

First. The government of the United States was born of the
Constitution, and all powers which it enjoys or may exercise
must be either derived expressly or by implication from that
instrument. Ever then, when an act of any department is challenged, because not warranted by the Constitution, the existence of the authority is to be ascertained by determining whether
the power has been conferred by the Constitution, either in express terms or by lawful implication, to be drawn from the express authority conferred or deduced as an attribute which legitimately inheres in the nature of the powers given, and which
flows from the character of the government established by the
Constitution. In other words, whilst confined to its constitu-

tional orbit, the government of the United States is supreme within its lawful sphere.[1]

Second. Every function of the government being thus derived from the Constitution, it follows that that instrument is everywhere and at all times potential in so far as its provisions are applicable.[2]

Third. Hence it is that wherever a power is given by the Constitution and there is a limitation imposed on the authority, such restriction operates upon and confines every action on the subject within its constitutional limits.[3]

Fourth. Consequently it is impossible to conceive that where conditions are brought about to which any particular provision of the Constitution applies, its controlling influence may be frustrated by the action of any or all of the departments of the government. Those departments, when discharging, within the limits of their constitutional power, the duties which rest on them, may of course deal with the subjects committed to them in such a way as to cause the matter dealt with to come under the control of provisions of the Constitution which may not have been previously applicable. But this does not conflict with the doctrine just stated, or presuppose that the Constitution may or may not be applicable at the election of any agency of the government.

Fifth. The Constitution has undoubtedly conferred on Congress the right to create such municipal organizations as it may deem best for all the territories of the United States whether they have been incorporated or not, to give to the inhabitants as respects the local governments such degree of representation as may be conducive to the public well-being, to deprive such

---

[1] *Marbury* v. *Madison,* 1 Cranch, 137, 176 *et. seq.* ; *Martin* v. *Hunter,* 1 Wheat. 304, 326; *New Orleans* v. *United States,* 10 Pet. 662, 736; *Geofroy* v. *Riggs,* 133 U. S. 258, 266; *United States* v. *Gettysburg Electric Railway,* 160 U. S. 668, 679, and cases cited.

[2] *The City of Panama,* 101 U. S. 453, 460; *Fong Yue Ting* v. *United States,* 149 U. S. 698, 716, 738.

[3] *Monongahela Navigation Company* v. *United States,* 148 U. S. 312, 336; *Interstate Commerce Commission* v. *Brimson,* 154 U. S. 447, 479; *United States* v. *Joint Traffic Association,* 171 U. S. 505, 571.

JUSTICES WHITE, SHIRAS and McKENNA, concurring.

territory of representative government if it is considered just
to do so, and to change such local governments at discretion.[1]

The plentitude of the power of Congress as just stated is conceded by both sides to this controversy.   It has been manifest
from the earliest days and so many examples are afforded of it
that to refer to them seems superfluous.   However, there is an
instance which exemplifies the exercise of the power substantially
in all its forms, in such an apt way that reference is made to it.
The instance referred to is the District of Columbia, which has
had from the beginning different forms of government conferred
upon it by Congress, some largely representative, others only
partially so, until, at the present time, the people of the District
live under a local government totally devoid of local representation, in the elective sense, administered solely by officers appointed by the President, Congress, in which the District has
no representative in effect, acting as the local legislature.

In some adjudged cases the power to locally govern at discretion has been declared to arise as an incident to the right to
acquire territory.   In others it has been rested upon the clause
of section 3, Article IV, of the Constitution, which vests Congress with the power to dispose of and make all needful rules
and regulations respecting the territory or other property of
the United States.[2]   But this divergence, if not conflict of
opinion, does not imply that the authority of Congress to govern
the territories is outside of the Constitution, since in either case
the right is founded on the Constitution, although referred to
different provisions of that instrument.

Whilst, therefore, there is no express or implied limitation on
Congress in exercising its power to create local governments for

---

[1] *United States* v. *Kagama*, 118 U. S. 375, 378; *Shively* v. *Bowlby*, 152 U. S.
1, 48.

[2] *Sere* v. *Pitot*, 6 Cranch, 332, 336; *McCulloch* v. *Maryland*, 4 Wheat. 316,
421; *American Ins. Co.* v. *Canter*, 1 Pet. 511, 542; *United States* v. *Gratiot*,
14 Pet. 526, 537; *Dred Scott* v. *Sandford*, 19 How. 393, 448; *Clinton* v. *Englebrecht*, 13 Wall. 434, 447; *Hamilton* v. *Dillin*, 21 Wall. 73, 93; *National Bank*
v. *County of Yankton*, 101 U. S. 129, 132; *The City of Panama*, 101 U. S.
453, 457; *Murphy* v. *Ramsey*, 114 U. S. 15, 44; *United States* v. *Kagama*,
118 U. S. 375, 380; *Mormon Church* v. *United States*, 136 U. S. 1, 42; *Boyd*
v. *Thayer*, 143 U. S. 135, 169.

any and all of the territories, by which that body is restrained from the widest latitude of discretion, it does not follow that there may not be inherent, although unexpressed, principles which are the basis of all free government which cannot be with impunity transcended. But this does not suggest that every express limitation of the Constitution which is applicable has not force, but only signifies that even in cases where there is no direct command of the Constitution which applies, there may nevertheless be restrictions of so fundamental a nature that they cannot be transgressed, although not expressed in so many words in the Constitution.

Sixth. As Congress in governing the territories is subject to the Constitution, it results that all the limitations of the Constitution which are applicable to Congress in exercising this authority necessarily limit its power on this subject. It follows also that every provision of the Constitution which is applicable to the territories is also controlling therein. To justify a departure from this elementary principle by a criticism of the opinion of Mr. Chief Justice Taney in *Scott* v. *Sandford*, 19 How. 393, appears to me to be unwarranted. Whatever may be the view entertained of the correctness of the opinion of the court in that case, in so far as it interpreted a particular provision of the Constitution concerning slavery and decided that as so construed it was in force in the territories, this in no way affects the principle which that decision announced, that the applicable provisions of the Constitution were operative. That doctrine was concurred in by the dissenting judges, as the following excerpts demonstrate. Thus Mr. Justice McLean, in the course of his dissenting opinion, said, (19 How. 542):

"In organizing the government of a territory, Congress is limited to means appropriate to the attainment of the constitutional object. No powers can be exercised which are prohibited by the Constitution, or which are contrary to its spirit."

---

[1] *Mormon Church* v. *United States*, 136 U. S. 1, 44.

Mr. Justice Curtis, also in the dissent expressed by him, said (p. 614):

"If, then, this clause does contain a power to legislate respecting the territory, what are the limits of that power?

"To this I answer that, in common with all the other legislative powers of Congress, it finds limits in the express prohibitions on Congress not to do certain things; that, in the exercise of the legislative power, Congress cannot pass an *ex post facto* law or bill of attainder; and so in respect to each of the other prohibitions contained in the Constitution."

Seventh. In the case of the territories, as in every other instance, when a provision of the Constitution is invoked, the question which arises is, not whether the Constitution is operative, for that is self-evident, but whether the provision relied on is applicable.

Eighth. As Congress derives its authority to levy local taxes for local purposes within the territories, not from the general grant of power to tax as expressed in the Constitution, it follows that its right to locally tax is not to be measured by the provision empowering Congress "To lay and collect Taxes, Duties, Imposts and Excises," and is not restrained by the requirement of uniformity throughout the United States. But the power just referred to, as well as the qualification of uniformity, restrains Congress from imposing an impost duty on goods coming into the United States from a territory which has been incorporated into and forms a part of the United States. This results because the clause of the Constitution in question does not confer upon Congress power to impose such an impost duty on goods coming from one part of the United States to another part thereof, and such duty besides would be repugnant to the requirement of uniformity throughout the United States.[1]

To question the principle above stated on the assumption that the rulings on this subject of Mr. Chief Justice Marshall in *Loughborough* v. *Blake* were mere *dicta*, seems to me to be entirely inadmissible. And, besides, if such view was justified,

---

[1] *Loughborough* v. *Blake*, 5 Wheat. 317, 322; *Woodruff* v. *Parham*, 8 Wall. 123, 133; *Brown* v. *Houston*, 114 U. S. 622, 628; *Fairbank* v. *United States*, 181 U. S. 283.

the principle would still find support in the decision in *Woodruff* v. *Parham*, and that decision, in this regard, was affirmed by this court in *Brown* v. *Houston* and *Fairbank* v. *United States*, *supra*.

From these conceded propositions it follows that Congress in legislating for Porto Rico was only empowered to act within the Constitution and subject to its applicable limitations, and that every provision of the Constitution which applied to a country situated as was that island, was potential in Porto Rico.

And the determination of what particular provision of the Constitution is applicable, generally speaking, in all cases, involves an inquiry into the situation of the territory and its relations to the United States. This is well illustrated by some of the decisions of this court which are cited in the margin.[1] Some of these decisions hold on the one hand that, growing out of the presumably ephemeral nature of a territorial government, the provisions of the Constitution relating to the life tenure of judges is inapplicable to courts created by Congress, even in territories which are incorporated into the United States, and some on the other hand decide that the provisions as to common-law juries found in the Constitution are applicable under like conditions; that is to say, although the judge presiding over a jury need not have the constitutional tenure, yet the jury must be in accordance with the Constitution. And the application of the provision of the Constitution relating to juries has been also considered in a different aspect, the case being noted in the margin.[2]

The question involved was the constitutionality of the statutes of the United States conferring power on ministers and consuls

---

[1] *American Insurance Co.* v. *Canter*, 1 Pet. 511; *Benner* v. *Porter*, 9 How. 235; *Webster* v. *Reid*, 11 How. 437, 460; *Clinton* v. *Englebrecht*, 13 Wall. 434; *Reynolds* v. *United States*, 98 U. S. 145; *Cullen* v. *Wilson*, 127 U. S. 540; *McAllister* v. *United States*, 141 U. S. 174; *Springville* v. *Thomas*, 166 U. S. 707; *Bauman* v. *Ross*, 167 U. S. 548; *Thompson* v. *Utah*, 170 U. S. 343; *Capital Traction Co.* v. *Hof*, 174 U. S. 1; *Black* v. *Jackson*, 177 U. S. 349, 363.

[2] *In re Ross*, 140 U. S. 453, 461, 462, 463.

to try American citizens for crimes committed in certain foreign countries. Rev. Stat. secs. 4083–4086. The court held the provisions in question not to be repugnant to the Constitution, and that a conviction for a felony without a previous indictment by a grand jury, or the summoning of a petty jury, was valid.

It was decided that the provisions of the Constitution relating to grand and petty juries were inapplicable to consular courts exercising their jurisdiction in certain countries foreign to the United States. But this did not import that the government of the United States in creating and conferring jurisdiction on consuls and ministers acted outside of the Constitution, since it was expressly held that the power to call such courts into being and to confer upon them the right to try, in the foreign countries in question, American citizens was deducible from the treaty-making power as conferred by the Constitution. The court said (p. 463):

"The treaty-making power vested in our government extends to all proper subjects of negotiation with foreign governments. It can, equally with any of the former or present governments of Europe, make treaties providing for the exercise of judicial authority in other countries by its officers appointed to reside therein."

In other words, the case concerned not the question of a power outside the Constitution, but simply whether certain provisions of the Constitution were applicable to the authority exercised under the circumstances which the case presented.

Albeit, as a general rule, the *status* of a particular territory has to be taken in view when the applicability of any provision of the Constitution is questioned, it does not follow when the Constitution has absolutely withheld from the government all power on a given subject, that such an inquiry is necessary. Undoubtedly, there are general prohibitions in the Constitution in favor of the liberty and property of the citizen which are not mere regulations as to the form and manner in which a conceded power may be exercised, but which are an absolute denial of all authority under any circumstances or conditions to do particular acts. In the nature of things, limitations of this character cannot be under any circumstances transcended, because of the complete absence of power

The distinction which exists between the two characters of restrictions, those which regulate a granted power and those which withdraw all authority on a particular subject, has in effect been always conceded, even by those who most strenuously insisted on the erroneous principle that the Constitution did not apply to Congress in legislating for the territories, and was not operative in such districts of country. No one had more broadly asserted this principle than Mr. Webster. Indeed, the support which that proposition receives from expressions of that illustrious man have been mainly relied upon to sustain it, and yet there can be no doubt that, even whilst insisting upon such principle, it was conceded by Mr. Webster that those positive prohibitions of the Constitution which withhold all power on a particular subject were always applicable. His views of the principal proposition and his concession as to the existence of the qualification are clearly shown by a debate which took place in the Senate on February 24, 1849, on an amendment offered by Mr. Walker extending the Constitution and certain laws of the United States over California and New Mexico. Mr. Webster, in support of his conception that the Constitution did not, generally speaking, control Congress in legislating for the territories or operate in such districts, said as follows (20 Cong. Globe, App. p. 272):

"Mr. President, it is of importance that we should seek to have clear ideas and correct notions of the question which this amendment of the member from Wisconsin has presented to us; and especially that we should seek to get some conception of what is meant by the proposition, in a law, to 'extend the Constitution of the United States to the territories.' Why, sir, the thing is utterly impossible. All the legislation in the world, in this general form, could not accomplish it. There is no cause for the operation of the legislative power in such a matter as that. The Constitution, what is it—we extend the Constitution of the United States by law to a territory? What is the Constitution of the United States? Is not its very first principle that all within its influence and comprehension shall be represented in the legislature which it establishes, with not only the right of debate and the right to vote in both houses of Congress, but a right to partake in the choice of the President and Vice President? And can we by law extend these rights, or any of them, to a territory of the United States? Everybody will see that it is altogether impracticable."

Thereupon, the following colloquy ensued between Mr. Underwood and Mr. Webster:

"Mr. Underwood: 'The learned Senator from Massachusetts says, and says most appropriately and forcibly, that the principles of the Constitution are obligatory upon us even while legislating for the territories. That is true, I admit, in its fullest force, but if it is obligatory upon us while legislating for the territories, is it possible that it will not be equally obligatory upon the officers who are appointed to administer the laws in these territories?'

"Mr. Webster: 'I never said it was not obligatory upon them. What I said was, that in making. laws for these territories it was the high duty of Congress to regard those great principles in the Constitution intended for the security of personal liberty and for the security of property.'

"Mr. Underwood: ' . . . Suppose we provide by our legislation that nobody shall be appointed to an office there who professes the Catholic religion. What do we do by an act of this sort?'

"Mr. Webster: 'We violate the Constitution, which says that no religious test shall be required as qualification for office.'"

And this was the state of opinion generally prevailing in the Free Soil and Republican parties, since the resistance of those parties to the extension of slavery into the territories, whilst in a broad sense predicated on the proposition that the Constitution was not generally controlling in the territories, was sustained by express reliance upon the Fifth Amendment to the Constitution forbidding Congress from depriving any person of life, liberty or property without due process of law. Every platform adopted by those parties down to and including 1860, whilst propounding the general doctrine, also in effect declared the rule just stated. I append in the margin an excerpt from the platform of the Free Soil Party adopted in 1842.[1]

The conceptions embodied in these resolutions were in almost identical language reiterated in the platform of the Liberty Party in 1843, in that of the Free Soil Party in 1852 and in the platform of the Republican Party in 1856. Stanwood, Hist. of Presidency, pp. 218, 253, 254 and 271. In effect, the same thought was repeated in the declaration of principles made by the Republican Party convention in 1860, when Mr. Lincoln was nominated, as will be seen from an excerpt therefrom set out in the margin.[2]

The doctrine that those absolute withdrawals of power which

---

[1] Extract from the Free Soil Party platform of 1842 (Stanwood, Hist. of Presidency, p. 240):

"*Resolved*, That our fathers ordained the Constitution of the United States in order, among other great national objects, to establish justice, promote the general welfare, and secure the blessings of liberty, but expressly denied to the Federal government, which they created, all constitutional power to deprive any person of life, liberty or property without due legal process.

"*Resolved*, That, in the judgment of this convention, Congress has no more power to make a slave than to make a king; no more power to institute or establish slavery than to institute or establish a monarchy. No such power can be found among those specifically conferred by the Constitution, or derived by any just implication from them.

"*Resolved*, That it is the duty of the Federal government to relieve itself from all responsibility for the existence or continuance of slavery wherever the government possesses constitutional authority to legislate on that subject, and is thus responsible for its existence.

"*Resolved*, That the true and in the judgment of this convention the only safe means of preventing the extension of slavery into territory now free is to prohibit its existence in all such territory by an act of Congress."

[2] Excerpt from declarations made in the platform of the Republican Party in 1860 (Stanwood, Hist. of Presidency, p. 293):

"8. That the normal condition of all the territory of the United States is that of freedom; that as our republican fathers, when they had abolished slavery in all our national territory, ordained that no person should be deprived of life, liberty or property without due process of law, it becomes our duty, by legislation, whenever such legislation is necessary, to maintain this provision of the Constitution against all attempts to violate it; and we deny the authority of Congress, of a territorial legislature or of any individual to give legal existence to slavery in any territory of the United States."

the Constitution has made in favor of human liberty are applicable to every condition or *status* has been clearly pointed out
by this court in *Chicago, Rock Island &c. Railway* v. *McGlinn*,
(1885) 114 U. S. 542, where, speaking through Mr. Justice Field,
the court said (p. 546):

" It is a general rule of public law, recognized and acted upon
by the United States, that whenever political jurisdiction and
legislative power over any territory are transferred from one
nation or sovereign to another, the municipal laws of the country
—that is, laws which are intended for the protection of private
rights—continue in force until abrogated or changed by the
new government or sovereign. By the cession public property
passes from one government to the other, but private property
remains as before, and with it those municipal laws which are
designed to secure its peaceful use and enjoyment. As a matter
of course, all laws, ordinances, and regulations in conflict with
the political character, institutions, and constitution of the new
government are at once displaced. Thus, upon a cession of political jurisdiction and legislative power—and the latter is involved in the former—to the United States, the laws of the
country in support of an established religion, or abridging the
freedom of the press, or authorizing cruel and unusual punishments, and the like, would at once cease to be of obligatory
force without any declaration to that effect ; and the laws of
the country on other subjects would necessarily be superseded
by existing laws of the new government upon the same matters.
But with respect to other laws affecting the possession, use and
transfer of property, and designed to secure good order and
peace in the community, and promote its health and prosperity,
which are strictly of a municipal character, the rule is general
that a change of government leaves them in force until, by direct action of the new government, they are altered or repealed.
Amer. Ins. Co. v. *Canter*, 1 Pet. 542 ; Halleck, Int. Law, chap. 34,
§ 14."

There is in reason then no room in this case to contend that
Congress can destroy the liberties of the people of Porto Rico
by exercising in their regard powers against freedom and justice which the Constitution has absolutely denied. There can

also be no controversy as to the right of Congress to locally govern the island of Porto Rico as its wisdom may decide and in so doing to accord only such degree of representative government as may be determined on by that body. There can also be no contention as to the authority of Congress to levy such local taxes in Porto Rico as it may choose, even although the amount of the local burden so levied be manifold more onerous than is the duty with which this case is concerned. But as the duty in question was not a local tax, since it was levied in the United States on goods coming from Porto Rico, it follows that if that island was a part of the United States, the duty was repugnant to the Constitution, since the authority to levy an impost duty conferred by the Constitution on Congress, does not, as I have conceded, include the right to lay such a burden on goods coming from one to another part of the United States. And, besides, if Porto Rico was a part of the United States the exaction was repugnant to the uniformity clause.

The sole and only issue, then, is not whether Congress has taxed Porto Rico without representation—for, whether the tax was local or national, it could have been imposed, although Porto Rico had no representative local government and was not represented in Congress—but is, whether the particular tax in question was levied in such form as to cause it to be repugnant to the Constitution. This is to be resolved by answering the inquiry, Had Porto Rico, at the time of the passage of the act in question, been incorporated into and become an integral part of the United States?

On the one hand, it is affirmed that, although Porto Rico had been ceded by the treaty with Spain to the United States, the cession was accompanied by such conditions as prevented that island from becoming an integral part of the United States, at least, temporarily, and until Congress had so determined. On the other hand, it is insisted that by the fact of cession to the United States alone, irrespective of any conditions found in the treaty, Porto Rico became a part of the United States, and was incorporated into it. It is incompatible with the Constitution, it is argued, for the government of the United States to accept a cession of territory from a foreign country without

complete incorporation following as an immediate result, and
therefore it is contended that it is immaterial to inquire what
were the conditions of the cession, since if there were any
which were intended to prevent incorporation they were repugnant to the Constitution and void.   The result of the argument is that the Government of the United States is absolutely
without power to acquire and hold territory as property or as
appurtenant to the United States.   These conflicting contentions are asserted to be sanctioned by many adjudications of
this court and by various acts of the executive and legislative
branches of the government; both sides, in many instances, referring to the same decisions and to the like acts, but deducing
contrary conclusions from them.   From this it comes to pass
that it will be impossible to weigh the authorities relied upon
without ascertaining the subject-matter to which they refer, in
order to determine their proper influence.   For this reason, in
the orderly discussion of the controversy, I propose to consider
the subject from the Constitution itself, as a matter of first impression, from that instrument as illustrated by the history of
the government, and as construed by the previous decisions of
this court.   By this process, if accurately carried out, it will
follow that the true solution of the question will be ascertained,
both deductively and inductively, and the result, besides, will
be adequately proven.

It may not be doubted that by the general principles of the
law of nations every government which is sovereign within its
sphere of action possesses as an inherent attribute the power to
acquire territory by discovery, by agreement or treaty, and by
conquest.   It cannot also be gainsaid that as a general rule
wherever a government acquires territory as a result of any of
the modes above stated, the relation of the territory to the new
government is to be determined by the acquiring power in the
absence of stipulations upon the subject.   These general principles of the law of nations are thus stated by Halleck in his
treatise on International Law, page 126:

"A state may acquire property or domain in various ways;
its title may be acquired originally by mere occupancy, and
confirmed by the presumption arising from the lapse of time;

or by discovery and lawful possession; or by conquest, confirmed by treaty or tacit consent; or by grant, cession, purchase or exchange; in fine, by any of the recognized modes by which private property is acquired by individuals.   It is not our object to enter into any general discussion of these several modes of acquisition, any further than may be necessary to distinguish the character of certain rights of property which are the peculiar objects of international jurisprudence.   Wheaton, Elm. Int. Law, pt. 2, ch. 4, secs. 1, 4, 5; Phillimore on Int. Law, vol. 1, secs. 221–217; Grotius, de Jur. Bel. ac. Pac. lib. 2, cap. 4; Vattel, Droit des Gens, liv. 2, chs. 7 and 11; Rutherford, Institutes, b. 1, ch. 3; b. 2, ch. 9; Puffendorf, de Jur. Nat. et Gent. lib. 4, chs. 4, 5, 6; Moser, Versuch, etc., b. 5, cap. 9; Martens, Précis du Droit des Gens, sec 35, *et seq.;* Schmaltz, Droit des Gens, liv. 4, ch. 1; Kluber, Droit des Gens, secs. 125, 126; Heffter, Droit International, sec. 76; Ortolan, Domaine International, sec. 53, *et seq.;* Bowyer, Universal Public Law, ch. 28, Bello, Derecho Internacional, pt. 1, cap. 4; Riquelme, Derecho Pub. Int. lib. 1, tit. 1, cap. 2; Burlamaqui, Droit de la Nat. et des Gens, tome 4, pt. 3, ch. 5.''

Speaking of a change of sovereignty, Halleck says (pp. 76, 814):

"Ch. III, Sec. 23. The sovereignty of a state may be lost in various ways.   It may be vanquished by a foreign power, and become incorporated into the conquering state *as a province* or as one of its component parts; or it may voluntarily unite itself with another in such a way that its independent existence as a state will entirely cease."

*   *   *   *   *   *   *   *

"Ch. XXXIII, Sec. 3. If the hostile nation be subdued and the entire state conquered, a question arises as to the manner in which the conqueror may treat it without transgressing the just bounds established by the rights of conquest.   If he simply replaces the former sovereign, and, on the submission of the people, governs them according to the laws of the State, they can have no cause of complaint.   Again, if he incorporates them with his former states, giving to them the rights, privileges and immunities of his own subjects, he does for them all that is due from a humane and equitable conqueror to his vanquished foes. But if the conquered are a fierce, savage and restless people, he may, according to the degree of their indocility, govern them with a tighter rein, so as to curb their 'impetuosity, and to keep them under subjection.' Moreover, the rights of conquest may, in certain cases, justify him in imposing a tribute or other burthen, either a compensation for the expenses of the war or as a punishment for the injustice he has suffered from them. . . . Vattel, Droit des Gens, liv. 3, ch. 13, § 201; Curtis, History, etc, liv. 7, cap. 8; Grotius, de Jur. Bel. ac. Pac. lib. 3, caps. 8, 15; Puffendorf, de Jur. Nat. et Gent., lib. 8, cap. 6, § 24; Real, Science du Gouvernement, tome 5, ch. 2, § 5; Heffter, Droit International, § 124; Abegg, Untersuchungen, etc., p. 86."

In *American Ins. Co.* v. *Canter*, 1 Pet. 511, the general doctrine was thus summarized, in the opinion delivered by Mr. Chief Justice Marshall (p. 542):

"If it (conquered territory) be ceded by the treaty, the acquisition is confirmed, and the ceded territory becomes a part of the nation to which it is annexed, either on the terms stipulated in the treaty of cession, or *on such as its new master shall impose.*"

When our forefathers threw off their allegiance to Great Britain and established a republican government, assuredly they deemed that the nation which they called into being was endowed with those general powers to acquire territory which all independent governments in virtue of their sovereignty enjoyed. This is demonstrated by the concluding paragraph of the Declaration of Independence, which reads as follows:

"As free and independent States, they [the United States of America] have full power to levy war, conclude peace, contract alliances, establish commerce, and to do all other acts and things which independent States may of right do."

That under the confederation it was considered that the government of the United States had authority to acquire territory like any other sovereignty, is clearly established by the eleventh of the articles of confederation.

The decisions of this court leave no room for question that, under the Constitution, the government of the United States, in virtue of its sovereignty, supreme within the sphere of its delegated power, has the full right to acquire territory enjoyed by every other sovereign nation.

In *American Insurance Co.* v. *Canter,* 1 Pet. 511, the court, by Mr. Chief Justice Marshall, said (p. 542): " The Constitution confers absolutely on the government of the Union, the powers of making war, and of making treaties; consequently, that government possesses the power of acquiring territory, either by conquest or by treaty."

In *United States* v. *Huckabee,* (1872) 16 Wall. 414, the court, speaking through Mr. Justice Clifford, said (p. 434): " Power to acquire territory either by conquest or treaty is vested by the Constitution in the United States. Conquered territory, however, is usually held as a mere military occupation until the fate of the nation from which it is conquered is determined, but if the nation is entirely subdued, or in case it be destroyed and ceases to exist, the right of occupation becomes permanent, and the title vests absolutely in the conqueror. *American Ins. Co.* v. *Canter,* 1 Pet. 511; *Hogsheads of Sugar* v. *Boyle,* 9 Cranch, 195; *Shanks* v. *Dupont,* 3 Pet. 246; *United States* v. *Rice,* 4 Wheat. 254; *The Amy Warwick,* 2 Sprague, 143; *Johnson* v. *McIntosh,* 8 Wheat. 588. Complete conquest, by whatever mode it may be perfected, carries with it all the rights of the former government, or, in other words, the conqueror, by the completion of his conquest, becomes the absolute owner of the property conquered from the enemy, nation or state. His rights are no longer limited to mere occupation of what he has taken into his actual possession, but they extend to all the property and rights of the conquered state, including even debts as well as personal and real property. Halleck, International Law, 839; *Elphinstone* v. *Bedreechund,* 1 Knapp's Privy Council Cases, 329; Vattel, 365; 3 Phillimore's International Law, 505."

In *Mormon Church* v. *United States,* (1889) 136 U. S. 1, Mr. Justice Bradley, announcing the opinion of the court, declared (p. 42): " The power to acquire territory, other than the territory northwest of the Ohio River, (which belonged to the United States at the adoption of the Constitution,) is derived from the treaty-making power and the power to declare and carry

on war. The incidents of these powers are those of national
sovereignty, and belong to all independent governments. The
power to make acquisitions of territory by conquest, by treaty
and by cession is an incident of national sovereignty. The
Territory of Louisiana, when acquired from France, and the
territories west of the Rocky Mountains, when acquired from
Mexico, became the absolute property and domain of the United
States, subject to such conditions as the government, in its
diplomatic negotiations, had seen fit to accept relating to the
rights of the people then inhabiting those territories."

Indeed, it is superfluous to cite authorities establishing the
right of the government of the United States to acquire territory, in view of the possession of the Northwest Territory when
the Constitution was framed and the cessions to the general
government by various States subsequent to the adoption of
the Constitution, and in view also of the vast extension of the
territory of the United States brought about since the existence
of the Constitution by substantially every form of acquisition
known to the law of nations. Thus, in part at least, "the title
of the United States to Oregon was founded upon original discovery and actual settlement of citizens of the United States,
authorized or approved by the government of the United
States." *Shively* v. *Bowlby*, 152 U. S. 50. The Province of
Louisiana was ceded by France in 1803; the Floridas were
transferred by Spain in 1819; Texas was admitted into the
Union by compact with Congress in 1845; California and New
Mexico were acquired by the treaty with Mexico of 1848, and
other western territory from Mexico by the treaty of 1853;
numerous islands have been brought within the dominion of
the United States under the authority of the act of August 18,
1856, c. 164, usually designated as the Guano Islands act, re-
enacted in Revised Statutes, sections 5570–5578; Alaska was
ceded by Russia in 1867; Medway Island, the western end of
the Hawaiian group, 1200 miles from Honolulu, was acquired
in 1867, and $50,000 was expended in efforts to make it a naval
station; on the renewal of a treaty with Hawaii, November 9,
1887, Pearl Harbor was leased for a permanent naval station;
by joint resolution of Congress the Hawaiian Islands came un-

der the sovereignty of the United States in 1898; and on April 30, 1900, an act for the government of Hawaii was approved, by which the Hawaiian Islands were given the *status* of an incorporated territory; on May 21, 1890, there was proclaimed by the President an agreement, concluded and signed with Germany and Great Britain, for the joint administration of the Samoan Islands, 26 Stat. 1497; and, on February 16, 1900, 31 Stat. 67, there was proclaimed a convention between the United States, Germany and Great Britain, by which Germany and Great Britain renounced in favor of the United States all their rights and claims over and in respect to the Island of Tutuilla and all other islands of the Samoan group east of longitude 171° west of Greenwich. And finally the treaty with Spain which terminated the recent war was ratified.

It is worthy of remark that, beginning in the administration of President Jefferson, the acquisitions of foreign territory above referred to were largely made whilst that political party was in power, which announced, as its fundamental tenet, the duty of strictly construing the Constitution, and it is true to say that all shades of political opinion have admitted the power to acquire and lent their aid to its accomplishment. And the power has been asserted in instances where it has not been exercised. Thus, during the administration of President Pierce, in 1854, a draft of a treaty for the annexation of Hawaii was agreed upon, but, owing to the death of the King of the Hawaiian Islands, was not executed. The second article of the proposed treaty provided as follows (Ex. Doc. Senate, 55th Congress, 2d sess., Report No. 681, Calendar No. 747, p. 91):

"Article II.

"The Kingdom of the Hawaiian Islands shall be incorporated into the American Union as a State, enjoying the same degree of sovereignty as other States, and admitted as such as soon as it can be done in consistency with the principles and requirements of the Federal Constitution, to all the rights, privileges and immunities of a State as aforesaid, on a perfect equality with the other States of the Union."

It is insisted, however, that, conceding the right of the government of the United States to acquire territory, as all such territory when acquired becomes absolutely incorporated into the United States, every provision of the Constitution which would apply under that situation is controlling in such acquired territory. This, however, is but to admit the power to acquire and immediately to deny its beneficial existence.

The general principle of the law of nations, already stated, is that acquired territory, in the absence of agreement to the contrary, will bear such relation to the acquiring government as may be by it determined. To concede to the government of the United States the right to acquire and to strip it of all power to protect the birthright of its own citizens and to provide for the well-being of the acquired territory by such enactments as may in view of its condition be essential, is, in effect, to say that the United States is helpless in the family of nations, and does not possess that authority which has at all times been treated as an incident of the right to acquire. Let me illustrate the accuracy of this statement. Take a case of discovery. Citizens of the United States discover an unknown island, peopled with an uncivilized race, yet rich in soil, and valuable to the United States for commercial and strategic reasons. Clearly, by the law of nations, the right to ratify such acquisition and thus to acquire the territory would pertain to the government of the United States. *Johnson* v. *McIntosh*, 8 Wheat. 543, 595; *Martin* v. *Waddell*, 16 Pet. 367, 409; *Jones* v. *United States*, 137 U. S. 202, 212; *Shively* v. *Bowlby*, 152 U. S. 1, 50. Can it be denied that such right could not be practically exercised if the result would be to endow the inhabitants with citizenship of the United States and to subject them not only to local but also to an equal proportion of national taxes, even although the consequence would be to entail ruin on the discovered territory and to inflict grave detriment on the United States to arise both from the dislocation of its fiscal system and the immediate bestowal of citizenship on those absolutely unfit to receive it?

The practice of the government has been otherwise. As early as 1856 Congress enacted the Guano Islands act, heretofore referred to, which, by section 1, provided that, when any citizen of the United States shall "discover a deposit of guano on any island, rock or key, not within the lawful jurisdiction of any other government, and not occupied by the citizens of any other government, and shall take peaceable possession thereof, and occupy the same, said island, rock or key may, at the discretion of the President of the United States, be considered *as appertaining* to the United States." 11 Stat. 119, c. 164; Rev. Stat. § 5570. Under the act referred to, it was stated in argument, that the government now holds and protects American citizens in the occupation of some seventy islands. The statute came under consideration in *Jones* v. *United States*, 137 U. S. 202, where the question was whether or not the act was valid and it was decided that the act was a lawful exercise of power, and that islands thus acquired were "appurtenant" to the United States. The court, in the course of the opinion, speaking through Mr. Justice Gray, said (p. 212):

"By the law of nations, recognized by all civilized states, dominion of new territory may be acquired by discovery and occupation, as well as by cession or conquest; and when citizens or subjects of one nation, in its name, and by its authority or with its assent, take and hold actual, continuous and useful possession, (although only for the purpose of carrying on a particular business, such as catching and curing fish or working mines,) of territory unoccupied by any other government or its citizens, the nation to which they belong may exercise such jurisdiction and for such period as it sees fit over territory so acquired. This principle affords ample warrant for the legislation of Congress concerning guano islands. Vattel, lib. 1, chap. 18; Wheaton on International Law (8th ed.), §§ 161, 165, 176, note 104; Halleck on International Law, chap. 6, §§ 7, 15; 1 Phillimore on International Law (3d ed.), §§ 227, 229, 230, 242; 1 Calvo, Droit International (4th ed.), §§ 266, 277, 300; *Whiton* v. *Albany County Ins. Co.*, 109 Mass. 24, 31."

And these considerations concerning discovery are equally applicable to ownership resulting from conquest. A just war is declared and in its prosecution the territory of the enemy is invaded and occupied. Would not the war, even if waged successfully, be fraught with danger if the effect of occupation was to necessarily incorporate an alien and hostile people into the United States? Take another illustration. Suppose at the termination of a war the hostile government had been overthrown and the entire territory or a portion thereof was occupied by the United States, and there was no government to treat with or none willing to cede by treaty, and thus it became necessary for the United States to hold the conquered country for an indefinite period, or at least until such time as Congress deemed that it should be either released or retained because it was apt for incorporation into the United States. If holding was to have the effect which is now claimed for it, would not the exercise of judgment respecting the retention be so fraught with danger to the American people that it could not be safely exercised?

Yet, again. Suppose the United States, in consequence of outrages perpetrated upon its citizens, was obliged to move its armies or send its fleets to obtain redress, and it came to pass that an expensive war resulted and culminated in the occupation of a portion of the territory of the enemy, and that the retention of such territory—an event illustrated by examples in history—could alone enable the United States to recover the pecuniary loss it had suffered. And suppose further that to do so would require occupation for an indefinite period, dependent upon whether or not payment was made of the required indemnity. It being true that incorporation must necessarily follow the retention of the territory, it would result that the United States must abandon all hope of recouping itself for the loss suffered by the unjust war, and, hence, the whole burden would be entailed upon the people of the United States. This would be a necessary consequence, because if the United States did not hold the territory as security for the needed indemnity it could not collect such indemnity, and on the other hand if incorporation must follow from holding the territory the uniformity provision of the Constitution would prevent the assessment of the cost of the war solely upon the newly acquired country. In this, as in the case of discovery, the traditions and practices of the government demonstrate the unsoundness of the contention. Congress, on May 13, 1846, declared that war existed with Mexico. In the summer of that year New Mexico and California were subdued by the American arms and the military occupation which follöwed continued until after the treaty of peace was ratified, in May, 1848. Tampico, a Mexican port, was occupied by our forces on November 15, 1846, and possession was not surrendered until after the ratification. In the spring of 1847 President Polk, through the Secretary of the Treasury, prepared a tariff of duties on imports and tonnage which was put in force in the conquered country. 1 Senate Documents, First Session, 30th Congress, pp. 562, 569. By this tariff, *duties were laid as well on merchandise exported from the United States* as from other countries, except as to supplies for our army, and on May 10, 1847, an exemption from tonnage duties was accorded to "all vessels chartered by the United States to convey supplies of any and all descriptions to our army and navy, and actually laden with supplies." Ib. 583. An interesting debate respecting the constitutionality of this action of the President is contained in 18 Cong. Globe, First Session, 30th Congress, at pp. 478, 479, 484–489, 495, 498, etc.

In *Fleming* v. *Page*, 9 How. 603, it was held that the revenue officials properly treated Tampico as a port of a foreign country during the occupation by the military forces of the United States, and that duties on imports into the United States from Tampico were lawfully levied under the general tariff act of 1846. Thus, although Tampico was in the possession of the United States, and the court expressly held that in an international sense the port was a part of the territory of the United States, yet it was decided that, in the sense of the revenue laws, Tampico was a foreign country. The special tariff act promulgated by President Polk was in force in New Mexico and California until after notice was received of the ratification of the treaty of peace. In *Cross* v. *Harrison*, 16 How. 164, certain collections of impost duties on goods brought from foreign countries into California prior to the time when official notification had been received in California that the treaty of cession had been ratified, as well as impost duties levied after the receipt of such notice, were called in question. The duties collected prior to the receipt of notice were laid at the rate fixed by the tariff promulgated by the Presi-

dent; those laid after the notification conformed to the general
tariff laws of the United States. The court decided that all
the duties collected were valid. The court undoubtedly in the
course of its opinion said that immediately upon the ratification
of the treaty California became a part of the United States and
subject to its revenue laws. However, the opinion pointedly
referred to a letter of the Secretary of the Treasury directing
the enforcement of the tariff laws of the United States, upon the
express ground that Congress had enacted laws which recognized
the treaty of cession. Besides, the decision was expressly placed
upon the conditions of the treaty, and it was stated, in so many
words, that a different rule would have been applied had the
stipulations in the treaty been of a different character.

But, it is argued, all the instances previously referred to may
be conceded, for they but illustrate the rule *inter arma silent
leges*. Hence, they do not apply to acts done after the cessation
of hostilities when a treaty of peace has been concluded. This
not only begs the question, but also embodies a fallacy. A case
has been supposed in which it was impossible to make a treaty
because of the unwillingness or disappearance of the hostile government, and, therefore, the occupation necessarily continued,
although actual war had ceased. The fallacy lies in admitting
the right to exercise the power, if only it is exerted by the military arm of the government, but denying it wherever the civil
power comes in to regulate and make the conditions more in
accord with the spirit of our free institutions. Why it can be
thought, although under the Constitution the military arm of
the government is in effect the creature of Congress, that such
arm may exercise a power without violating the Constitution,
and yet Congress—the creator—may not regulate, I fail to comprehend.

This further argument, however, is advanced. Granting that
Congress may regulate without incorporating, where the military arm has taken possession of foreign territory, and where
there has been or can be no treaty, this does not concern the
decision of this case, since there is here involved no regulation
but an actual cession to the United States of territory by treaty.
The general rule of the law of nations, by which the acquiring

government fixes the *status* of acquired territory, it is urged, does not apply to the government of the United States, because it is incompatible with the Constitution that that government should hold territory under a cession and administer it as a dependency without its becoming incorporated. This claim, I have previously said, rests on the erroneous assumption that the United States under the Constitution is stripped of those powers which are absolutely inherent in and essential to national existence. The certainty of this is illustrated by the examples already made use of in the supposed cases of discovery and conquest.

If the authority by treaty is limited as suggested, then it will be impossible to terminate a successful war by acquiring territory through a treaty, without immediately incorporating such territory into the United States. Let me, however, eliminate the case of war and consider the treaty-making power as subserving the purposes of the peaceful evolution of national life. Suppose the necessity of acquiring a naval station or a coaling station on an island inhabited with people utterly unfit for American citizenship and totally incapable of bearing their proportionate burden of the national expense. Could such island, under the rule which is now insisted upon, be taken? Suppose again the acquisition of territory for an interoceanic canal, where an inhabited strip of land on either side is essential to the United States for the preservation of the work. Can it be denied that, if the requirements of the Constitution as to taxation are to immediately control, it might be impossible by treaty to accomplish the desired result?

Whilst no particular provision of the Constitution is referred to to sustain the argument that it is impossible to acquire territory by treaty without immediate and absolute incorporation, it is said that the spirit of the Constitution excludes the conception of property or dependencies possessed by the United States, and which are not so completely incorporated as to be in all respects a part of the United States; that the theory upon which the Constitution proceeds is that of confederated and independent States, and that no territory therefore can be acquired which does not contemplate statehood, and excludes the acquisition of any territory which is not in a position to be treated as an integral part of the United States. But this reasoning is based on political and not judicial considerations. Conceding that the conception upon which the Constitution proceeds is that no territory as a general rule should be acquired unless the territory may reasonably be expected to be worthy of statehood, the determination of when such blessing is to be bestowed is wholly a political question, and the aid of the judiciary cannot be invoked to usurp political discretion in order to save the Constitution from imaginary or even real dangers. The Constitution may not be saved by destroying its fundamental limitations.

Let me come, however, to a consideration of the express powers which are conferred by the Constitution to show how unwarranted is the principle of immediate incorporation, which is here so strenuously insisted on. In doing so it is conceded at once that the true rule of construction is not to consider one provision of the Constitution alone, but to contemplate all, and therefore to limit one conceded attribute by those qualifications which naturally result from the other powers granted by that instrument, so that the whole may be interpreted by the spirit which vivifies, and not by the letter which killeth. Undoubtedly, the power to carry on war and to make treaties implies also the exercise of those incidents which ordinarily inhere in them. Indeed, in view of the rule of construction which I have just conceded—that all powers conferred by the Constitution must be interpreted with reference to the nature of the government and be construed in harmony with related provisions of the Constitution—it seems to me impossible to conceive that the treaty-making power by a mere cession can incorporate an alien people into the United States without the express or implied approval of Congress. And from this it must follow that there can be no foundation for the assertion that where the treaty-making power has inserted conditions which preclude incorporation until Congress has acted in respect thereto, such conditions are void and incorporation results in spite thereof. If the treaty-making power can absolutely, without the consent of Congress, incorporate territory, and if that power may not insert conditions against incorporation, it must follow that the treaty-making power is endowed by the Constitution with the most unlimited right, susceptible of destroying every other provision of the Constitution ; that is, it may wreck our institutions. If the proposition be true, then millions of inhabitants of alien territory, if acquired by treaty, can, without the desire or consent of the people of the United States speaking through Congress, be immediately and irrevocably incorporated into the United States, and the whole structure of the government be overthrown. While thus aggrandizing the treaty-making power on the one hand, the construction at the same time minimizes it on the other, in that it strips that authority of any right to acquire territory upon any condition which would guard the people of the United States from the evil of immediate incorporation. The treaty-making power then, under this contention, instead of having the symmetrical functions which belong to it from its very nature, becomes distorted —vested with the right to destroy upon the one hand and deprived of all power to protect the government on the other.

And, looked at from another point of view, the effect of the principle asserted is equally antagonistic, not only to the express provisions but to the spirit of the Constitution in other respects. Thus, if it be true that the treaty-making power has the authority which is asserted, what becomes of that branch of Congress which is peculiarly the representative of the people of the United States, and what is left of the functions of that body under the Constitution ?   For, although the House of Representatives might be unwilling to agree to the incorporation of alien races, it would be impotent to prevent its accomplishment, and the express provisions conferring upon Congress the power to regulate commerce, the right to raise revenue—bills for which, by the Constitution, must originate in the House of Representatives—and the authority to prescribe uniform naturalization laws would be in effect set at naught by the treaty-making power. And the consequent result—incorporation—would be beyond all future control of or remedy by the American people, since, at once and without hope of redress or power of change, incorporation by the treaty would have been brought about.

The inconsistency of the position is at once manifest.   The basis
of the argument is that the treaty must be considered to have
been incorporated, because acquisition presupposes the exercise
of judgment as to fitness for immediate incorporation.   But the
deduction drawn is, although the judgment exercised is against
immediate incorporation and this result is plainly expressed, the
conditions are void because no judgment against incorporation
can be called into play.

All the confusion and dangers above indicated, however, it is
argued, are more imaginary than real, since, although it be conceded that the treaty-making power has the right by cession to
incorporate without the consent of Congress, that body may
correct the evil by availing itself of the provision of the Constitution giving to Congress the right to dispose of the territory
and other property of the United States.   This assumes that
there has been absolute incorporation by the treaty-making power
on the one hand, and yet asserts that Congress may deal with
the territory as if it had not been incorporated into the United
States.   In other words, the argument adopts conflicting theories of the Constitution and applies them both at the same time.
I am not unmindful that there has been some contrariety of decision on the subject of the meaning of the clause empowering
Congress to dispose of the territories and other property of the
United States, some adjudged cases treating that article as referring to property as such and others deriving from it the general grant of power to govern territories.   In view, however,
of the relations of the territories to the government of the United States at the time of the adoption of the Constitution, and
the solemn pledge then existing that they should forever " remain
a part of the confederacy of the United States of America," I
cannot resist the belief that the theory that the disposing clause
relates as well to a relinquishment or cession of sovereignty
as to a mere transfer of rights of property, is altogether erroneous.

Observe again the inconsistency of this argument.   It considers, on the one hand, that so vital is the question of incorporation that no alien territory may be acquired by a cession without absolutely endowing the territory with incorporation and

the inhabitants with resulting citizenship, because, under our system of government, the assumption that a territory and its inhabitants may be held by any other title than one incorporating is impossible to be thought of. And yet to avoid the evil consequences which must follow from accepting this proposition, the argument is that all citizenship of the United States is precarious and fleeting, subject to be sold at any moment like any other property. That is to say, to protect a newly acquired people in their presumed rights, it is essential to degrade the whole body of American citizenship.

The reasoning which has sometimes been indulged in by those who asserted that the Constitution was not at all operative in the territories is that, as they were acquired by purchase, the right to buy included the right to sell. This has been met by the proposition that if the country purchased and its inhabitants became incorporated into the United States, it came under the shelter of the Constitution, and no power existed to sell American citizens. In conformity to the principles which I have admitted it is impossible for me to say at one and the same time that territory is an integral part of the United States protected by the Constitution, and yet the safeguards, privileges, rights and immunities which arise from this situation are so ephemeral in their character that by a mere act of sale they may be destroyed. And, applying this reasoning to the provisions of the treaty under consideration, to me it seems indubitable that if the treaty with Spain incorporated all the territory ceded into the United States, it resulted that the millions of people to whom that treaty related were, without the consent of the American people as expressed by Congress, and without any hope of relief, indissolubly made a part of our common country.

Undoubtedly, the thought that under the Constitution power existed to dispose of people and territory and thus to annihilate the rights of American citizens was contrary to the conceptions of the Constitution entertained by Washington and Jefferson. In the written suggestions of Mr. Jefferson, when Secretary of State, reported to President Washington in March, 1792, on the subject of proposed negotiations between the United States and Spain, which were intended to be communicated by way of in-

struction to the commissioners of the United States appointed
to manage such negotiations, it was observed, in discussing the
possibility as to compensation being demanded by Spain "for
the ascertainment of our right" to navigate the lower part of
the Mississippi, as follows:

"We have nothing else" (than a relinquishment of certain
claims on Spain) "to give in exchange. For as to territory, we
have neither the right nor the disposition to alienate an inch of
what belongs to any member of our Union. Such a proposition
therefore is totally inadmissible, and not to be treated for a moment." Ford's Writings of Jefferson, vol. v, p. 476.

The rough draft of these observations was submitted to Mr.
Hamilton, then Secretary of the Treasury, for suggestions, previously to sending it to the President, some time before March 5,
and Hamilton made the following (among other) notes upon it:

"Page 25. Is it true that the United States have no right to
alienate an inch of the territory in question, except in the case
of necessity intimated in another place? Or will it be useful
to avow the denial of such a right? It is apprehended that the
doctrine which restricts the alienation of territory to cases of
extreme necessity is applicable rather to *peopled* territory than
to waste and uninhabited districts. Positions restraining the
right of the United States to accommodate to exigencies which
may arise ought ever to be advanced with great caution." Ford's
Writings of Jefferson, vol. v, p. 443.

Respecting this note, Mr. Jefferson commented as follows:

"The power to alienate the *unpeopled* territories of any State
is not among the enumerated powers, given by the Constitution
to the general government, and if we may go out of that instrument and *accommodate to exigencies which may arise* by alienating the *unpeopled* territory of a State, we may accommodate
ourselves a little more by alienating that which is *peopled*, and
still a little more by selling the *people* themselves. A shade or
two more in the degree of exigency is all that will be requisite,
and of that degree we shall ourselves be the judges. However,
may it not be hoped that these questions are forever laid to rest
by the Twelfth Amendment once made a part of the Constitution, declaring expressly that ' the powers not delegated to the

United States by the Constitution are reserved to the States respectively?' And if the general government has no power to alienate the territory of a State, it is too irresistible an argument to deny ourselves the use of it on the present occasion." Ib.

The opinions of Mr. Jefferson, however, met the approval of President Washington. On March 18, 1792, in inclosing to the commissioners to Spain their commission, he said, among other things:

"You will herewith receive your commission; as also observations on these several subjects reported to the President and approved by him, which will therefore serve as instructions for you. These expressing minutely the sense of our government, and what they wish to have done, it is unnecessary for me to do more here than desire you to pursue these objects unremittingly," etc. Ford's Writings of Jefferson, vol. v, p. 456.

When the subject-matter to which the negotiation related is considered it becomes evident that the word "State" as above used related merely to territory which was either claimed by some of the States, as Mississippi Territory was by Georgia, or to the Northwest Territory embraced within the ordinance of 1787, or the territory south of the Ohio (Tennessee), which had also been endowed with all the rights and privileges conferred by that ordinance, and all which territory had originally been ceded by States to the United States under express stipulations that such ceded territory should be ultimately formed into States of the Union. And this meaning of the word "State" is absolutely in accord with what I shall hereafter have occasion to demonstrate was the conception entertained by Mr. Jefferson of what constituted the United States.

True, from the exigency of a calamitous war or the necessity of a settlement of boundaries, it may be that citizens of the United States may be expatriated by the action of the treaty-making power, impliedly or expressly ratified by Congress.

But the arising of these particular conditions cannot justify the general proposition that territory which is an integral part of the United States may, as a mere act of sale, be disposed of. If however the right to dispose of an incorporated American territory and citizens by the mere exertion of the power to sell be conceded, *arguendo*, it would not relieve the dilemma. It is ever true that where a malign principle is adopted, as long as the error is adhered to it must continue to produce its baleful results. Certainly, if there be no power to acquire subject to a condition, it must follow that there is no authority to dispose of subject to conditions, since it cannot be that the mere change of form of the transaction could bestow a power which the Constitution has not conferred. It would follow then that any conditions annexed to a disposition which looked to the protection of the people of the United States or to enable them to safeguard the disposal of territory would be void; and thus it would be that either the United States must hold on absolutely or must dispose of unconditionally.

A practical illustration will at once make the consequences clear. Suppose Congress should determine that the millions of inhabitants of the Philippine Islands should not continue appurtenant to the United States, but that they should be allowed to establish an autonomous government, outside of the Constitution of the United States, coupled, however, with such conditions providing for control as far only as essential to the guarantee of life and property and to protect against foreign encroachment. If the proposition of incorporation be well founded, at once the question would arise whether the ability to impose these conditions existed, since no power was conferred by the Constitution to annex conditions which would limit the disposition. And if it be that the question of whether territory is immediately fit for incorporation when it is acquired is a judicial and not a legislative one, it would follow that the validity of the conditions would also come within the scope of judicial authority, and thus the entire political policy of the government be alone controlled by the judiciary.

The theory as to the treaty-making power upon which the argument which has just been commented upon rests, it is now proposed to be shown, is refuted by the history of the government from the beginning. There has not been a single cession made from the time of the Confederation up to the present day, excluding the recent treaty with Spain, which has not contained stipulations to the effect that the United States through Con-

gress would either not disincorporate or would incorporate the
ceded territory into the United States. There were such conditions in the deed of cession by Virginia when it conveyed the
Northwest Territory to the United States. Like conditions
were attached by North Carolina to the cession whereby the
territory south of the Ohio, now Tennessee, was transferred.
Similar provisions were contained in the cession by Georgia
of the Mississippi territory, now the States of Alabama and
Mississippi. Such agreements were also expressed in the treaty
of 1803, ceding Louisiana; that of 1819, ceding the Floridas,
and in the treaties of 1848 and 1853, by which a large extent
of territory was ceded to this country, as also in the Alaska
treaty of 1867. To adopt the limitations on the treaty-making
power now insisted upon would presuppose that every one of
these conditions thus sedulously provided for was superfluous,
since the guaranties which they afforded would have obtained,
although they were not expressly provided for.

When the various treaties by which foreign territory has been
acquired are considered in the light of the circumstances which
surrounded them, it becomes to my mind clearly established
that the treaty-making power was always deemed to be devoid
of authority to incorporate territory into the United States
without the assent, express or implied, of Congress, and that no
question to the contrary has ever been even mooted. To appreciate this it is essential to bear in mind what the words " United
States" signified at the time of the adoption of the Constitution.
When by the treaty of peace with Great Britain the independence of the United States was acknowledged, it is unquestioned
that all the territory within the boundaries defined in that treaty,
whatever may have been the disputes as to title, substantially
belonged to particular States. The entire territory was part of
the United States, and all the native white inhabitants were
citizens of the United States and endowed with the rights and
privileges arising from that relation. When, as has already
been said, the Northwest Territory was ceded by Virginia, it
was expressly stipulated that the rights of the inhabitants in
this regard should be respected. The ordinance of 1787, providing for the government of the Northwest Territory, fulfilled

this promise on behalf of the Confederation. Without undertaking to reproduce the text of the ordinance, it suffices to say
that it contained a bill of rights, a promise of ultimate statehood,
and it provided (italics mine) that "The said territory and the
States which may be formed therein *shall ever remain a part of
this confederacy of the United States of America*, subject to the articles of confederation, and to such alterations therein as shall
be constitutionally made, and to all the acts and ordinances of
the United States in Congress assembled, conformably thereto."
It submitted the inhabitants to a liability for a tax to pay their
proportional part of the public debt and the expenses of the government to be assessed by the rule of apportionment which governed the States of the confederation. It forbade slavery
within the Territory, and contained a stipulation that the provisions of the ordinance should ever remain unalterable unless
by common consent.

Thus it was that, at the adoption of the Constitution, the United States, as a geographical unit and as a governmental conception both in the international and domestic sense, consisted not
only of States, but also of territories, all the native white inhab-
tiants being endowed with citizenship, protected by pledges of a
common union, and, except as to political advantages, all enjoying equal rights and freedom, and safeguarded by substantially
similar guaranties, all being under the obligation to contribute
their proportionate share for the liquidation of the debt and
future expenses of the general government.

The opinion has been expressed that the ordinance of 1787
became inoperative and a nullity on the adoption of the Constitution (Taney, C. J., in *Scott* v. *Sandford*, 19 How. 393, 438,)
while, on the other hand, it has been said that the ordinance of
1787 was "the most solemn of all engagements," and became
a part of the Constitution of the United States by reason of
the sixth article, which provided that "all debts contracted and
engagements entered into, before the adoption of this Constitution, shall be as valid against the United States under this
Constitution as under the confederation." Per Baldwin, J.,
concurring opinion in *Lessee of Pollard's Heirs* v. *Kibbe*, 14
Pet. 353, 417, and per Catron, J., in dissenting opinion in *Stra-*

der v. *Graham*,10 How. 82, 98. Whatever view may be taken of this difference of legal opinion, my mind refuses to assent to the conclusion that under the Constitution the provision of the Northwest Territory ordinance making such territory forever a part of the confederation was not binding on the government of the United States when the Constitution was formed. When it is borne in mind that large tracts of this territory were reserved for distribution among the continental soldiers, it is impossible for me to believe that it was ever considered that the result of the cession was to take the Northwest Territory out of the Union, the necessary effect of which would have been to expatriate the very men who by their suffering and valor had secured the liberty of their united country. Can it be conceived that North Carolina, after the adoption of the Constitution, would cede to the general government the territory south of the Ohio River, intending thereby to expatriate those dauntless mountaineers of North Carolina who had shed lustre upon the Revolutionary arms by the victory of King's Mountain? And the rights bestowed by Congress after the adoption of the Constitution, as I shall proceed to demonstrate, were utterly incompatible with such a theory.

Beyond question, in one of the early laws enacted at the first session of the First Congress, the binding force of the ordinance was recognized, and certain of its provisions concerning the appointment of officers in the territory were amended to conform the ordinance to the new Constitution. c. 8, August 7, 1789, 1 Stat. 50.

In view of this it cannot, it seems to me, be doubted that the United States continued to be composed of States and territories, all forming an integral part thereof and incorporated therein, as was the case prior to the adoption of the Constitution. Subsequently, the territory now embraced in the State of Tennessee was ceded to the United States by the State of North Carolina. In order to insure the rights of the native inhabitants, it was expressly stipulated that the inhabitants of the ceded territory should enjoy all the rights, privileges, benefits and advantages set forth in the ordinance " of the late Congress for the government of the western territory of the United

States." A condition was, however, inserted in the cession that no regulation should be made by Congress tending to emancipate slaves. By act of April 2, 1790, 1 Stat. 106, c. 6, this cession was accepted. . And, at the same session, on May 26, 1790, an act was passed for the government of this territory, under the designation of "the territory of the United States south of the Ohio River." 1 Stat. 123, c. 14. This act, except as to the prohibition which was found in the Northwest Territory ordinance as to slavery, in express terms declared that the inhabitants of the territory should enjoy all the rights conferred by that ordinance.

A government for the Mississippi Territory was organized on April 7, 1798. 1 Stat. 549, c. 28. The land embraced was claimed by the State of Georgia, and her rights were saved by the act. The sixth section thereof provided as follows:

"SEC. 6. *And be it further enacted*, That from and after the establishment of the said government, the people of the aforesaid territory, shall be entitled to and enjoy, all and singular the rights, privileges and advantages granted to the people of the territory of the United States northwest of the river Ohio, in and by the aforesaid ordinance of the thirteenth day of July, in the year one thousand seven hundred and eighty-seven, in as full and ample a manner as the same are possessed and enjoyed by the people of the said last-mentioned territory."

Thus clearly defined by boundaries, by common citizenship, by like guarantees, stood the United States when the plan of acquiring by purchase from France the Province of Louisiana was conceived by President Jefferson. Naturally, the suggestion which arose, was the power on the part of the government of the United States, under the Constitution, to incorporate into the United States—a Union then composed, as I have stated, of States and Territories—a foreign province inhabited by an alien people, and thus make them partakers in the American commonwealth. Mr. Jefferson, not doubting the power of the United States to acquire, consulted Attorney General Lincoln as to the right by treaty to stipulate for incorporation. By that officer Mr. Jefferson was, in effect, advised that the power to incorporate, that is, to share the privileges and im-

munities of the people of the United States with a foreign population, required the consent of the people of the United States,
and it was suggested, therefore, that if a treaty of cession were
made containing such agreements it should be put in the form
of a change of boundaries instead of a cession, so as thereby to
bring the territory within the United States.   The letter of
Mr. Lincoln was sent by President Jefferson to Mr. Gallatin,
the Secretary of the Treasury.   Mr. Gallatin did not agree as
to the propriety of the expedient suggested by Mr. Lincoln.
In a letter to President Jefferson, in effect so stating, he said:

"But, does any constitutional objection really exist?   To me
it would appear (1) that the United States as a nation have. an
inherent right to acquire territory; (2) that whenever that acquisition is by treaty, the same constituted authorities in which
the treaty-making power is vested have a constitutional right
to sanction the acquisition; (3) that whenever the territory has
been acquired Congress have the power either of admitting into
the Union as a new State or of annexing to a State, with the
consent of that State, or of making regulations for the government of such territory."   Gallatin's Writings, vol. 1, p. 11, etc.

To this letter President Jefferson replied in January, 1803,
clearly showing that he thought there was no question whatever of the right of the United States to acquire, but that he
did not believe incorporation could be stipulated for and carried
into effect without the consent of the people of the United
States.   He said (italics mine):

"You are right, in my opinion, as to Mr. L.'s proposition:
*There is no constitutional difficulty as to the acquisition of territory, and whether when acquired it may be taken into the Union
by the Constitution as it now stands* will become a question of
expediency.   I think it will be safer not to permit the enlargement of the Union but by amendment of the Constitution."
Gallatin's Writings, vol. 1, p. 115.

And the views of Mr. Madison, then Secretary of State, exactly
conformed to those of President Jefferson, for, on March 2,
1803, in a letter to the commissioners who were negotiating
the treaty, he said:

"To incorporate the inhabitants of the hereby ceded territory with the citizens of the United States, being a provision which cannot now be made, it is to be expected from the character and policy of the United States that such incorporation will take place without unnecessary delay." State Papers, II, 540.

Let us pause for a moment to accentuate the irreconcilable conflict which exists between the interpretation given to the Constitution at the time of the Louisiana treaty by Jefferson and Madison, and the import of that instrument as now insisted upon. You are to negotiate, said Madison to the commissioners, to obtain a cession of the territory, but you must not under any circumstances agree " *to incorporate the inhabitants of the hereby ceded territory with the citizens of the United States, being a provision which cannot now be made.*" Under the theory now urged, Mr. Madison should have said : You are to negotiate for the cession of the territory of Louisiana to the United States, and if deemed by you expedient in accomplishing this purpose, you may provide for the immediate incorporation of the inhabitants of the acquired territory into the United States. This you can freely do because the Constitution of the United States has conferred upon the treaty-making power the absolute right to bring all the alien people residing in acquired territory into the United States and thus divide with them the rights which peculiarly belong to the citizens of the United States. Indeed, it is immaterial whether you make such agreements, since by the effect of the Constitution without reference to any agreements which you may make for that purpose, all the alien territory and its inhabitants will instantly become incorporated into the United States if the territory is acquired.

Without going into details, it suffices to say that a compliance with the instructions given them would have prevented the negotiators on behalf of the United States from inserting in the treaty any provision looking even to the ultimate incorporation of the acquired territory into the United States. In view of the emergency and exigencies of the negotiations, however, the commissioners were constrained to make such a stipulation, and the treaty provided as follows :

" Art. III. The inhabitants of the ceded territory shall be incorporated into the Union of the United States, and admitted as soon as possible, according to the principles of the Federal Constitution, to the enjoyment of all the rights, advantages and immunities of citizens of the United States; and in the mean time they shall be maintained and protected in the free enjoyment of their liberty, property and the religion which they profess." 8 Stat. 202.

Weighing the provisions just quoted, it is evident they refute the theory of incorporation arising at once from the mere force of a treaty, even although such result be directly contrary to any provisions which a treaty may contain. Mark the language. It expresses a promise: "The inhabitants of the ceded territory *shall be incorporated into the Union of the United States. . . .*" Observe how guardedly the fulfillment of this pledge is postponed until its accomplishment is made possible by the will of the American people, since it is to be executed only "*as soon as possible according to the principles of the Federal Constitution.*" If the view now urged be true, this wise circumspection was unnecessary, and, indeed, as I have previously said, the entire proviso was superfluous, since everything which it assured for the future was immediately and unalterably to arise.

It is said, however, that the treaty for the purchase of Louisiana took for granted that the territory ceded would be immediately incorporated into the United States, and hence the guarantees contained in the treaty related, not to such incorporation, but was a pledge that the ceded territory was to be made a part of the Union as a State. The minutest analysis, however, of the clauses of the treaty fails to disclose any reference to a promise of statehood, and hence it can only be that the pledges made referred to incorporation into the United States. This will further appear when the opinions of Jefferson and Madison and their acts on the subject are reviewed. The argument proceeds upon the theory that the words of the treaty "shall be incorporated into the Union of the United States," could only have referred to a promise of statehood, since the then existing and incorporated Territories were not a part of the Union of the United States, as that Union consisted only of the States. But this has been shown to be unfounded,

since the "Union of the United States" was composed of States
and Territories, both having been embraced within the boundaries fixed by the treaty of peace between Great Britain and the
United States which terminated the Revolutionary war, the
latter, the Territories, embracing districts of country which
were ceded by the States to the United States under the express
pledge that they should forever remain a part thereof.   That
this conception of the Union composing the United States was
the understanding of Jefferson and Madison, and indeed of all
those who participated in the events which preceded and led
up to the Louisiana treaty, results from what I have already
said, and will be additionally demonstrated by statements to be
hereafter made.   Again, the inconsistency of the argument is
evident.   Thus, whilst the premise upon which it proceeds is
that foreign territory, when acquired, becomes at once a part
of the United States, despite conditions in the treaty expressly
excluding such consequence, it yet endeavors to escape the refu-
tation of such theory which arises from the history of the government by the contention that the territories which were a
part of the United States were not component constituents of
the Union which composed the United States.   I do not understand how foreign territory which has been acquired by treaty
can be asserted to have been absolutely incorporated into the
United States as a part thereof despite conditions to the contrary inserted in the treaty, and yet the assertion be made that
the territories which, as I have said, were in the United States
originally as a part of the States and which were ceded by them
upon express condition that they should forever so remain a
part of the United States, were not a part of the Union composing the United States.   The argument, indeed, reduces itself
to this, that for the purpose of incorporating foreign territory
into the United States domestic territory must be disincorpo-
rated.   In other words, that the Union must be, at least in theory, dismembered for the purpose of maintaining the doctrine
of the immediate incorporation of alien territory.

That Mr. Jefferson deemed the provision of the treaty relating to incorporation to be repugnant to the Constitution is
unquestioned.   Whilst he conceded, as has been seen, the right

to acquire, he doubted the power to incorporate the territory into the United States without the consent of the people by a constitutional amendment. In July, 1803, he proposed two drafts of a proposed amendment, which he thought ought to be submitted to the people of the United States to enable them to ratify the terms of the treaty. The first of these, which is dated July, 1803, is printed in the margin.[1]

The second and revised amendment was as follows:

" Louisiana, as ceded by France to the United States, is made a part of the United States. Its white inhabitants shall be citizens, and stand, as to their rights and obligations, on the same footing with other citizens of the United States, in analogous situations. Save only that, as to the portion thereof lying north of the latitude of the mouth of Arcana River, no new State shall be established nor any grants of land made therein other than to Indians in exchange for equivalent portions of lands occupied by them until an amendment of the Constitution shall be made for those purposes.

" Florida also, whensoever it may be rightfully obtained, shall become a part of the United States. Its white inhabitants shall thereupon become citizens, and shall stand, as to their rights and obligations, on the same footing with other citizens of the United States in analogous situations." Ford's Writings of Jefferson, vol. 8, p. 241.

It is strenuously insisted that Mr. Jefferson's conviction on the subject of the repugnancy of the treaty to the Constitution was

---

[1] First draft of Mr. Jefferson's proposed amendment to the Constitution. "The Province of Louisiana is incorporated with the United States and made part thereof. The rights of occupancy in the soil and of self-government are confirmed to Indian inhabitants as they now exist." It then proceeded with other provisions relative to Indian rights and possession and exchange of lands, and forbidding Congress to dispose of the lands otherwise than is therein provided without further amendment to the Constitution. This draft closes thus: "Except as to that portion thereof which lies south of the latitude of 31°, which, whenever they deem expedient, they may enact into a territorial government, either separate or as making part with one on the eastern side of the river, vesting the inhabitants thereof with all rights possessed by other territorial citizens of the United States." Writings of Jefferson, edited by Ford, vol. 8, p. 241.

based alone upon the fact that he thought the treaty exceeded the limits of the Constitution, because he deemed that it provided for the admission, according to the Constitution, of the acquired territory as a new State or States into the Union, and hence, for the purpose of conferring this power, he drafted the amendment. The contention is refuted by two considerations; the first, because the two forms of amendment which Mr. Jefferson prepared did not purport to confer any power upon Congress to admit new States ; and, second, they absolutely forbade Congress from admitting a new State out of a described part of the territory without a further amendment to the Constitution. It cannot be conceived that Mr. Jefferson would have drafted an amendment to cure a defect which he thought existed and yet say nothing in the amendment on the subject of such defect. And, moreover, it cannot be conceived that he drafted an amendment to confer a power he supposed to be wanting under the Constitution, and thus ratify the treaty, and yet in the very amendment withhold in express terms, as to a part of the ceded territory, the authority which it was the purpose of the amendment to confer.

I excerpt in the margin [1] two letters from Mr. Jefferson, one

---

[1] Letter to William Dunbar of July 7, 1803:

" Before you receive this you will have heard through the channel of the public papers of the cession of Louisiana by France to the United States. The terms as stated in the National Intelligencer are accurate. That the treaty may be ratified in time, I have found it necessary to convene Congress on the 17th of October, and it is very important for the happiness of the country that they should possess all information which can be obtained respecting it, that they make the best arrangements practicable for its good government. It is most necessary because they will be obliged to ask from the people an amendment of the Constitution authorizing their receiving the province into the Union and providing for its government, and limitations of power which shall be given by that amendment will be unalterable but by the same authority." Jefferson's Writings, vol. 8, p. 254.

Letter to Wilson Cary Nicholas of September 7, 1803:

" I am aware of the force of the observations you make on the power given by the Constitution to Congress to admit new States into the Union without restraining the subject to the territory then constituting the United States. But when I consider that the limits of the United States are precisely fixed by the treaty of 1783, that the Constitution expressly declares itself to be made for the United States, I cannot help believing written under date of July 7, 1803, to William Dunbar, and the other dated September 7, 1803, to Wilson Cary Nicholas, which show clearly the difficulties which were in the mind of Mr. Jefferson, and which remove all doubt concerning the meaning of the amendment which he wrote and the adoption of which he deemed necessary to cure any supposed want of power concerning the treaty would be provided for.

These letters show that Mr. Jefferson bore in mind the fact that the Constitution in express terms delegated to Congress the power to admit new States, and, therefore, no further authority on this subject was required. But he thought this power in Congress was confined to the area embraced within the limits of the United States, as existing at the adoption of the Constitution. To fulfill the stipulations of the treaty so as to cause the ceded territory to become a part of the United States, Mr. Jefferson deemed an amendment to the Constitution to be essential. For this reason the amendment which he formulated declared that the territory ceded was to be "*a part of the United States,* and its white inhabitants shall be citizens, and stand, as to their rights and obligations, on the same footing with other citizens of the United States, *in analogous situations.*" What these words meant is not open to doubt when it is observed that they were but the paraphrase of the following words, which were contained in the first proposed amendment which Mr. Jefferson wrote: "Vesting the inhabitants thereof with all rights possessed *by other territorial citizens of the United States,*" which clearly show that it was the want of power to incorporate the ceded country into the United States as a territory which was in Mr. Jefferson's mind, and to accomplish which re-

---

that the intention was to permit Congress to admit into the Union new States which should be formed out of the territory for which and under whose authority alone they were then acting. I do not believe it was meant that they might receive England, Ireland, Holland, etc., into it, which would be the case under your construction. When an instrument admits two constructions, the one safe, the other dangerous, the one precise, the other indefinite, I prefer that which is safe and precise. I had rather ask an enlargement of power from the nation where it is found necessary than to assume it by a construction which would make our powers boundless." Writings of Jefferson, vol. 8, p. 247.

sult he thought an amendment to the Constitution was required. This provision of the amendment applied to all of the territory ceded, and, therefore, brought it all into the United States, and hence placed it in a position where the power of Congress to admit new States would have attached to it. As Mr. Jefferson deemed that every requirement of the treaty would be fulfilled by incorporation, and that it would be unwise to form a new State out of the upper part of the new territory, after thus providing for the complete execution of the treaty by incorporation of all the territory into the United States, he inserted a provision *forbidding Congress from admitting a new State out of a part of the territory.*

With the debates which took place on the subject of the treaty I need not particularly concern myself. Some shared Mr. Jefferson's doubts as to the right of the treaty-making power to incorporate the territory into the United States without an amendment of the Constitution; others deemed that the provision of the treaty was but a promise that Congress would ultimately incorporate as a territory, and until by the action of Congress this latter result was brought about full power of legislation to govern as deemed best was vested in Congress. This latter view prevailed. Mr. Jefferson's proposed amendment to the Constitution, therefore, was never adopted by Congress, and hence was never submitted to the people.

An act was approved on October 31, 1803, 2 Stat. 245, "to enable the President of the United States to take possession of the territories ceded by France to the United States by the treaty concluded at Paris on the 30th of April last, and for the temporary government thereof." The provisions of this act were absolutely incompatible with the conception that the territory had been incorporated into the United States by virtue of the cession. On November 10, 1803, 2 Stat. 245, an act was passed providing for the issue of stock to raise the funds to pay for the territory. On February 24, 1804, 2 Stat. 251, an act was approved which expressly extended certain revenue and other laws over the ceded country. On March 26, 1804, 2 Stat. 283, an act was passed dividing the "Province of Louisiana" into Orleans Territory on the south and the District of Louisiana to the north. This act extended over the Territory of Orleans a large number of the general laws of the United States and provided a form of government. For the purposes of government the District of Louisiana was attached to the Territory of Indiana, which had been carved out of the Northwest Territory. Although the area described as Orleans Territory was thus under the authority of a territorial government and many laws of the United States had been extended by act of Congress to it, it was manifest that Mr. Jefferson thought that the requirement of the treaty that it should be incorporated into the United States had not been complied with.

In a letter written to Mr. Madison on July 14, 1804, Mr. Jefferson, speaking of the treaty of cession, said (Ford's Writings of Jefferson, vol. 8, p. 313):

"The enclosed reclamations of Girod & Chote against the claims of Bapstroop to a monopoly of the Indian commerce supposed to be under the protection of the third article of the Louisiana convention, as well as some other claims to abusive grants, will probably force us to meet that question. The article has been worded with remarkable caution on the part of our negotiators. It is that the inhabitants shall be admitted as soon as possible, according to the principles of our Constitution, to the enjoyment of all the rights of citizens, and, in the mean time, *en attendant*, shall be maintained in their liberty, property and religion. That is, that they shall continue under the protection of the treaty, until the principles of our Constitution can be extended to them, when the protection of the treaty is to cease, and that of our own principles to take its place. But as this could not be done at once, it has been provided to be as soon as our rules will admit. Accordingly, Congress has begun by extending about twenty particular laws by their titles, to Louisiana. Among these is the act concerning intercourse with the Indians, which establishes a system of commerce with them admitting no monopoly. That class of rights therefore are now taken from under the treaty and placed under the principles of our laws. I imagine it will be necessary to express an opinion to Governor Claiborne on this subject, after you shall have made up one."

In another letter to Mr. Madison, under date of August 15, 1804, Mr. Jefferson said (Ib. p. 315):

"I am so much impressed with the expediency of putting a termination to the right of France to patronize the rights of Louisiana, which will cease with their complete adoption as citizens of the United States, that I hope to see that take place on the meeting of Congress."

At the following session of Congress, on March 2, 1805, 2 Stat. 322, c. 23, an act was approved, which, among other purposes, doubtless was intended to fulfill the hope expressed by Mr. Jefferson in the letter just quoted. That act, in the first section, provided that the inhabitants of the Territory of Orleans "*shall be entitled to and enjoy all the rights, privileges and advantages secured by the said ordinance,*" (that is, the ordinance of 1787,) "*and now enjoyed by the people of the Mississippi Territory.*" As will be remembered, the ordinance of 1787 had been extended to that territory. 1 Stat. 550, c. 28. Thus, strictly in accord with the thought embodied in the amendments contemplated by Mr. Jefferson, citizenship was conferred, and the Territory of Orleans was incorporated into the United States to fulfill the requirements of the treaty, by placing it exactly in the position which it would have occupied had it been within the boundaries of the United States as a territory at the time the Constitution was framed. It is pertinent to recall that the treaty contained stipulations giving certain preferences and commercial privileges for a stated period to the vessels of French and Spanish subjects, and that even after the action of Congress above stated this condition of the treaty continued to be enforced, thus demonstrating that even after the incorporation of the territory the express provisions conferring a temporary right which the treaty had stipulated for and which Congress had recognized were not destroyed, the effect being that incorporation as to such matter was for the time being in abeyance.

The upper part of the Province of Louisiana, designated by the act of March 26, 1804, 2 Stat. 283, c. 38, as the District of Louisiana, and by the act of March 3, 1805, 2 Stat. 331, c. 27, as the Territory of Louisiana, was created the Territory of Missouri on June 4, 1812. 2 Stat. 743, c 95. By this latter act, though the ordinance of 1787 was not in express terms extended over the territory—probably owing to the slavery agitation—the inhabitants of the territory were accorded substantially all the rights of the inhabitants of the Northwest Territory. Citizenship was in effect recognized in the ninth section, whilst the fourteenth section contained an elaborate declaration of the rights secured to the people of the territory.

Pausing to analyze the practical construction which resulted from the acquisition of the vast domain covered by the Louisiana purchase, it indubitably results, first, that it was conceded by every shade of opinion that the government of the United States had the undoubted right to acquire, hold and govern the territory as a possession, and that incorporation into the United States could under no circumstances arise solely from a treaty of cession, even although it contained provisions for the accomplishment of such result; second, it was strenuously denied by many eminent men that in acquiring territory, citizenship could be conferred upon the inhabitants within the acquired territory; in other words, that the territory could be incorporated into the United States without an amendment to the Constitution; and, third, that the opinion which prevailed was that, although the treaty might stipulate for incorporation and citizenship under the Constitution, such agreements by the treaty-making power were but promises depending for their fulfillment on the future action of Congress. In accordance with this view the territory acquired by the Louisiana purchase was governed as a mere dependency, until, conformably to the suggestion of Mr. Jefferson, it was by the action of Congress incorporated as a Territory into the United States and the same rights were conferred in the same mode by which other Territories had previously been incorporated, that is, by bestowing the privileges of citizenship and the rights and immunities which pertained to the Northwest Territory.

Florida was ceded by treaty signed on February 2, 1819. 8 Stat. 252. Whilst drafted in accordance with the precedent afforded by the treaty ceding Louisiana, the Florida treaty was slightly modified in its phraseology, probably to meet the view that under the Constitution Congress had the right to determine the time when incorporation was to arise. Acting under the precedent afforded by the Louisiana case Congress adopted a plan of government which was wholly inconsistent with the theory that the territory had been incorporated. General Jackson was appointed governor under this act, and exercised a degree of authority entirely in conflict with the conception that the territory was a part of the United States, in the sense of incorporation, and that those provisions of the Constitution which would have been applicable under that hypothesis were then in force. It will serve no useful purpose to go through the gradations of legislation adopted as to Florida. Suffice it to say that in 1822, (3 Stat. 654, c. 13,) an act was passed, as in the case of Missouri, and presumably for the same reason, which, whilst not referring to the Northwest Territory ordinance, *in effect endowed the inhabitants of that territory with the rights granted by such ordinance.*

This treaty also, it is to be remarked, contained discriminatory commercial provisions incompatible with the conception of immediate incorporation arising from the treaty, and they were enforced by the executive officers of the government.

The intensity of the political differences which existed at the outbreak of hostilities with Mexico, and at the termination of the war with that country, and the subject around which such conflicts of opinion centered probably explains why the treaty of peace with Mexico departed from the form adopted in the previous treaties concerning Florida and Louisiana. That treaty, instead of expressing a cession in the form previously adopted, whether intentionally or not I am unable, of course, to say, resorted to the expedient suggested by Attorney General Lincoln to President Jefferson, and accomplished the cession *by changing the boundaries of the two countries;* in other words, *by bringing the acquired territory within the described boundaries of the United States.* The treaty, besides, contained a stipulation for rights of citizenship; in other words, a provision equivalent in terms to those used in the previous treaties to which I have referred. The controversy which was then flagrant on the subject of slavery prevented the passage of a bill giving California a territorial form of government, and California after considerable delay was therefore directly admitted into the Union as a State. After the ratification of the treaty various laws were enacted by Congress, which in effect treated the territory as acquired by the United States, and the executive officers of the government, conceiving that these acts were an implied or express ratification of the provisions of the treaty by Congress, acted upon the assumption that the provisions of the treaty were thus made operative, and hence incorporation had thus become efficacious.

Ascertaining the general rule from the provisions of this latter treaty and the practical execution which it received, it will be seen that the precedents established in the cases of Louisiana and Florida were departed from to a certain extent; that is, the rule was considered to be that where the treaty, in express terms, brought the territory within the boundaries of the United States and provided for incorporation, and the treaty was expressly or impliedly recognized by Congress, the provisions of the treaty ought to be given immediate effect. But this did not conflict with the general principles of the law of nations which I have at the outset stated, but enforced it, since the action taken assumed, not that incorporation was brought about by the treaty-making power wholly without the consent of Congress, but only that as the treaty provided for incorporation in express terms, and Congress had acted without repudiating it, its provisions should be at once enforced.

Without referring in detail to the acquisition from Russia of Alaska, it suffices to say that that treaty also contained provisions for incorporation and was acted upon exactly in accord with the practical construction applied in the case of the acquisitions from Mexico as just stated. However, the treaty ceding Alaska contained an express provision excluding from citizenship the uncivilized native tribes, and it has been nowhere contended that this condition of exclusion was inoperative because of the want of power under the Constitution in the treaty-making authority to so provide, which must be the case if the limitation on the treaty-making power, which is here asserted, be well founded. The treaty concerning Alaska, therefore, adds cogency to the conception established by every act of the government from the foundation—that the condition of a treaty, when expressly or impliedly ratified by Congress, becomes the measure by which the rights arising from the treaty are to be adjusted.

The demonstration which it seems to me is afforded by the review which has preceded is besides sustained by various other acts of the government which to me are wholly inexplicable except upon the theory that it was admitted that the government of the United States had the power to acquire and hold territory without immediately incorporating it. Take, for instance, the simultaneous acquisition and admission of Texas, which was admitted into the Union as a State by joint resolution of Congress instead of by treaty. To what grant of power under the Constitution can this action be referred, unless it be admitted that Congress is vested with the right to determine when incorporation arises? It cannot be traced to the authority conferred on Congress to admit new States, for to adopt that theory would be to presuppose that this power gave the prerogative of conferring statehood on wholly foreign territory. But this I have incidentally shown is a mistaken conception. Hence, it must be that the action of Congress at one and the same time fulfilled the function of incorporation; and this being so, the privilege of statehood was added. But I shall not prolong this opinion by occupying time in referring to the many other acts of the government which further refute the correctness of the propositions which are here insisted on and which I have previously shown to be without merit. In concluding my appreciation of the history of the government attention is called to the Thirteenth Amendment to the Constitution, which to my mind seems to be conclusive. The first section of the amendment, the italics being mine, reads as follows: "Sec. 1. Neither slavery nor involuntary servitude, except as a punishment for crime, whereof the party shall have been duly convicted, shall exist within the United States, *or any place subject to their jurisdiction.*" Obviously this provision recognized that there may be places subject to the jurisdiction of the United States but which are not incorporated into it, and hence are not within the United States in the completest sense of those words.

Let me proceed to show that the decisions of this court, without a single exception, are absolutely in accord with the true rule as evolved from a correct construction of the Constitution as a matter of first impression and as shown by the history of the government which has been previously epitomized. As it is appropriate here, I repeat the quotation which has heretofore been made from the opinion, delivered by Mr. Chief Justice Marshall, in *American Insurance Co.* v. *Canter*, 1 Pet. 511, where, considering the Florida treaty, the court said (p. 542):

" The usage of the world is, if a nation be not entirely subdued, to consider the holding of conquered territory as a mere military occupation, until its fate shall be determined at the treaty of peace. If it be ceded by the treaty, the acquisition is confirmed, and the ceded territory becomes a part of the nation to which it is annexed, either on the terms stipulated in the treaty of cession, or on such as its new master shall impose."

In *Fleming* v. *Page* the court, speaking through Mr. Chief Justice Taney, discussing the acts of the military forces of the United States while holding possession of Mexican territory, said (9 How. 603, 614):

" The United States, it is true, may extend its boundaries by conquest or treaty, and may demand the cession of territory as the condition of peace, in order to indemnify its citizens for the injuries they have suffered, or to reimburse the government for the expense of the war. But this can be done only by the treaty-making power or the legislative authority."

In *Cross* v. *Harrison*, 16 How. 164, the question for decision, as I have previously observed, was as to the legality of certain duties, collected both before and after the ratification of the treaty of peace, on foreign merchandise imported into California. Part of the duties collected were assessed upon importations made by local officials before notice had been received of the ratification of the treaty of peace, and when duties were laid under a tariff which had been promulgated by the President. Other duties were imposed subsequent to the receipt of notification of the ratification, and these latter duties were laid

according to the tariff as provided in the laws of the United
States. All the exactions were upheld. The court decided
that prior to and up to the receipt of notice of the ratification
of the treaty, the local government lawfully imposed the tariff
then in force in California, although it differed from that provided by Congress, and that subsequent to the receipt of notice
of the ratification of the treaty, the duty prescribed by the act
of Congress which the President had ordered the local officials
to enforce could be lawfully collected. The opinion undoubtedly expressed the thought that by the ratification of the treaty
in question, which, as I have shown, not only included the ceded
territory within the boundaries of the United States, but also expressly provided for incorporation, the territory had become a
part of the United States, and the body of the opinion quoted the
letter of the Secretary of the Treasury which referred to the enactment of laws of Congress by which the treaty had been impliedly ratified. The decision of the court as to duties imposed
subsequent to the receipt of notice of the ratification of the
treaty of peace undoubtedly took the fact I have just stated
into view and, in addition, unmistakably proceeded upon the
nature of the rights which the treaty conferred. No comment
can obscure or do away with the patent fact, namely, that
it was unequivocally decided that if different provisions had
been found in the treaty, a contrary result would have followed.
Thus, speaking through Mr. Justice Wayne, the court said (16
How. 197):

"By the ratification of the treaty, California became a part
of the United States. And, *as there is nothing differently stipulated in the treaty with respect to commerce*, it became instantly
bound and privileged by the laws which Congress had passed
to raise a revenue from duties on imports and tonnage."

It is then, as I think, indubitably settled by the principles of
the law of nations, by the nature of the government created
under the Constitution, by the express and implied powers conferred upon that government by the Constitution, by the mode
in which those powers have been executed, from the beginning, and by an unbroken line of decisions of this court, first
announced by Marshall and followed and lucidly expounded

by Taney, that the treaty-making power cannot incorporate
territory into the United States without the express or implied
assent of Congress, that it may insert in a treaty conditions
against immediate incorporation, and that on the other hand
when it has expressed in the treaty the conditions favorable to
incorporation, they will, if the treaty be not repudiated by
Congress, have the force of the law of the land, and therefore
by the fulfillment of such conditions cause incorporation to
result. It must follow, therefore, that where a treaty contains
no conditions for incorporation, and, above all, where it not
only has no such conditions but expressly provides to the contrary, incorporation does not arise until in the wisdom of Congress it is deemed that the acquired territory has reached
that state where it is proper that it should enter into and form
a part of the American family.

Does, then, the treaty in question contain a provision for
incorporation, or does it, on the contrary, stipulate that incorporation shall not take place from the mere effect of the treaty
and until Congress has so determined? is then the only question remaining for consideration.

The provisions of the treaty with respect to the *status* of
Porto Rico and its inhabitants are as follows:

"Article II.

"Spain cedes to the United States the Island of Porto Rico
and other islands now under Spanish sovereignty in the West
Indies, and the Island of Guam in the Marianas or Ladrones."

"Article IX.

"Spanish subjects, natives of the Peninsula, residing in the
territory over which Spain by the present treaty relinquishes
or cedes her sovereignty, may remain in such territory or may
remove therefrom, retaining in either event all their rights of
property, including the right to sell or dispose of such property or of its proceeds; and they shall also have the right to
carry on their industry, commerce and professions, being subject in respect thereof to such laws as are applicable to other
foreigners. In case they remain in the territory they may pre-

serve their allegiance to the crown of Spain by making, before a court of record, within a year from the date of the exchange of ratifications of this treaty, a declaration of their decision to preserve such allegiance; in default of which declaration they shall be held to have renounced it and to have adopted the nationality of the territory in which they may reside.

"The civil rights and political *status* of the native inhabitants of the territories hereby ceded to the United States shall be determined by the Congress.

"Article X.

"The inhabitants of the territories over which Spain relinquishes or cedes her sovereignty shall be secured in the free exercise of their religion."

It is to me obvious that the above quoted provisions of the treaty do not stipulate for incorporation, but on the contrary expressly provide that the "civil rights and political *status* of the native inhabitants of the territories hereby ceded," shall be determined by Congress. When the rights to which this careful provision refers are put in juxtaposition with those which have been deemed essential from the foundation of the government to bring about incorporation, all of which have been previously referred to, I cannot doubt that the express purpose of the treaty was not only to leave the *status* of the territory to be determined by Congress but to prevent the treaty from operating to the contrary. Of course, it is evident that the express or implied acquiescence by Congress in a treaty so framed cannot import that a result was brought about which the treaty itself—giving effect to its provisions —could not produce. And, in addition, the provisions of the act by which the duty here in question was imposed, taken as a whole, seem to me plainly to manifest the intention of Congress that for the present at least Porto Rico is not to be incorporated into the United States.

The fact that the act directs the officers to swear to support the Constitution does not militate against this view, for, as I have conceded, whether the island be incorporated or not, the applicable provisions of the Constitution are there in force. A further analysis of the provisions of the act seems to me not to be required in view of the fact that as the act was reported from the committee it contained a provision conferring citizenship upon the inhabitants of Porto Rico, and this was stricken out in the Senate. The argument, therefore, can only be that rights were conferred, which, after consideration, it was determined should not be granted. Moreover I fail to see how it is possible, on the one hand, to declare that Congress in passing the act had exceeded its powers by treating Porto Rico as not incorporated into the United States, and, at the same time, it be said that the provisions of the act itself amount to an incorporation of Porto Rico into the United States, although the treaty had not previously done so. It in reason cannot be that the act is void because it seeks to keep the island disincorpo-. rated, and, at the same time, that material provisions are not to be enforced because the act does incorporate. Two irreconcilable views of that act cannot be taken at the same time, the consequence being to cause it to be unconstitutional.

In what has preceded I have in effect considered every substantial proposition and have either conceded or reviewed every authority referred to as establishing that immediate incorporation resulted from the treaty of cession which is under consideration. Indeed, the whole argument in favor of the view that immediate incorporation followed upon the ratification of the treaty in its last analysis necessarily comes to this: Since it has been decided that incorporation flows from a treaty which provides for that result, when its provisions have been expressly or impliedly approved by Congress, it must follow that the same effect flows from a treaty which expressly stipulates to the contrary, even although the condition to that end has been approved by Congress. That is to say, the argument is this: Because a provision for incorporation when ratified incorporates, therefore a provision against incorporation must also produce the very consequence which it expressly provides against.

The result of what has been said is that whilst in an international sense Porto Rico was not a foreign country, since it was subject to the sovereignty of and was owned by the United States, it was foreign to the United States in a domestic sense, because the island had not been incorporated into the United States, but was merely appurtenant thereto as a possession. As a necessary consequence, the impost in question assessed on merchandise coming from Porto Rico into the United States after the cession was within the power of Congress, and that body was not, moreover, as to such imposts, controlled by the clause requiring that imposts should be uniform throughout the United States; in other words, the provision of the Constitution just referred to was not applicable to Congress in legislating for Porto Rico.

Incidentally I have heretofore pointed out that the arguments of expediency, pressed with so much earnestness and ability concern the legislative and not the judicial department of the government. But it may be observed that even if the disastrous consequences which are foreshadowed as arising from conceding that the government of the United States may hold property without incorporation were to tempt me to depart from what seems to me to be the plain line of judicial duty, reason admonishes me that so doing would not serve to prevent the grave evils which it is insisted must come, but, on the contrary, would only render them more dangerous. This must be the result, since, as already said, it seems to me it is not open to serious dispute, that the military arm of the government of the United States may hold and occupy conquered territory without incorporation for such length of time as may seem appropriate to Congress in the exercise of its discretion. The denial of the right of the civil power to do so would not therefore prevent the holding of territory by the United States if it was deemed best by the political department of the government, but would simply necessitate that it should be exercised by the military instead of by the civil power.

And to me it further seems apparent that another and more disastrous result than that just stated would follow as a consequence of an attempt to cause judicial judgment to invade the domain of legislative discretion. Quite recently one of the stipulations contained in the treaty with Spain which is now under consideration came under review by this court. By the provision in question Spain relinquished "all claim of sovereignty over and title to Cuba." It was further provided in the treaty as follows:

"And as the island is upon the evacuation by Spain to be occupied by the United States, the United States will so long as such occupation shall last assume and discharge the obligations that may under international law result from the fact of its occupation and for the protection of life and property."

It cannot, it is submitted, be questioned that, under this provision of the treaty, as long as the occupation of the United States lasts, the benign sovereignty of the United States extends over and dominates the Island of Cuba. Likewise, it is not, it seems to me, questionable that the period when that sovereignty is to cease is to be determined by the legislative department of the government of the United States in the exercise of the great duties imposed upon it and with the sense of the responsibility which it owes to the people of the United States and the high respect which it of course feels for all the moral obligations by which the government of the United States may, either expressly or impliedly, be bound. Considering the provisions of this treaty and reviewing the pledges of this government extraneous to that instrument, by which the sovereignty of Cuba is to be held by the United States for the benefit of the people of Cuba and for their account, to be relinquished to them when the conditions justify its accomplishment, this court unanimously held in *Neely* v. *Henkel*, 180 U. S. 109, that Cuba was not incorporated into the United States and was a foreign country. It follows from this decision that it is lawful for the United States to take possession of and hold in the exercise of its sovereign power a particular territory, without incorporating it into the United States, if there be obligations of honor and good faith which, although not expressed in the treaty, nevertheless sacredly bind the United States to terminate the dominion and control, when, in its political discretion, the situation is ripe to enable it to do so. Conceding, then, for the purpose of the argument, it to be true that it would be a violation of duty under the Constitution for the legislative department, in the exercise of its discretion, to accept a cession of and permanently hold territory which is not intended to be incorporated, the presumption necessarily must be that that department, which within its lawful sphere is but the expression of the political conscience of the people of the United States, will be faithful to its duty under the Constitution, and, therefore, when the unfitness of particular territory for incorporation is demonstrated the occupation will terminate. I cannot conceive how it can be held that pledges made to an alien people can be treated as more sacred than is that great pledge given by every member of every department of the government of the United States to support and defend the Constitution.

But if it can be supposed—which, of course, I do not think to be conceivable—that the judiciary would be authorized to draw to itself by an act of usurpation purely political functions, upon the theory that if such wrong is not committed a greater harm will arise, because the other departments of the government will forget their duty to the Constitution and wantonly transcend its limitations, I am further admonished that any judicial action in this case which would be predicated upon such an unwarranted conception would be absolutely unavailing. It cannot be denied that under the rule clearly settled in *Neely* v. *Henkel, supra*, the sovereignty of the United States may be extended over foreign territory to remain paramount until in the discretion of the political department of the government of the United States it be relinquished. This method, then, of dealing with foreign territory, would, in any event, be available. Thus, the enthralling of the treaty-making power, which would result from holding that no territory could be acquired by treaty of cession without immediate incorporation, would only result in compelling a resort to the subterfuge of relinquishment of sovereignty, and thus indirection would take the place of directness of action—a course which would be incompatible with the dignity and honor of the government.

I am authorized to say that Mr. Justice SHIRAS and Mr. Justice McKENNA concur in this opinion.

MR. JUSTICE GRAY, concurring.

Concurring in the judgment of affirmance in this case, and in substance agreeing with the opinion of Mr. Justice White, I will sum up the reasons for my concurrence in a few propositions, which may also indicate my position in other cases now standing for judgment.

The cases now before the court do not touch the authority of the United States over the Territories, in the strict and technical sense, being those which lie within the United States, as bounded by the Atlantic and Pacific Oceans, the Dominion of Canada and the Republic of Mexico, and the Territories of Alaska and Hawaii; but they relate to territory, in the broader sense, acquired by the United States by war with a foreign State.

As Chief Justice Marshall said: "The Constitution confers absolutely on the government of the Union the powers of making war, and of making treaties; consequently, that government possesses the power of acquiring territory, either by conquest or by treaty. The usage of the world is, if a nation be not entirely subdued, to consider the holding of conquered territory as a mere military occupation, until its fate shall be determined at the treaty of peace. If it be ceded by the treaty, the acquisition is confirmed, and the ceded territory becomes a part of the nation to which it is annexed; either on the terms stipulated in the treaty of cession, or on such as its new master shall impose." *American Insurance Co.* v. *Canter*, (1828) 1 Pet. 511, 542.

The civil government of the United States cannot extend immediately, and of its own force, over territory acquired by war. Such territory must necessarily, in the first instance, be governed by the military power under the control of the President as commander in chief. Civil government cannot take effect at once, as soon as possession is acquired under military authority, or even as soon as that possession is confirmed by treaty. It can only be put in operation by the action of the appropriate political department of the government, at such time and in such degree as that department may determine. There must, of necessity, be a transition period.

In a conquered territory, civil government must take effect, either by the action of the treaty-making power, or by that of the Congress of the United States. The office of a treaty of cession ordinarily is to put an end to all authority of the foreign government over the territory; and to subject the territory to the disposition of the Government of the United States.

The government and disposition of territory so acquired belong to the Government of the United States, consisting of the President, the Senate, elected by the States, and the House of Representatives, chosen by and immediately representing the people of the United States. Treaties by which territory is acquired from a foreign State usually recognize this.

It is clearly recognized in the recent treaty with Spain, especially in the ninth article, by which "The civil rights and political *status* of the native inhabitants of the territories hereby ceded to the United States shall be determined by the Congress."

By the fourth and thirteenth articles of the treaty, the United States agree that, for ten years, Spanish ships and merchandise shall be admitted to the ports of the Philippine Islands on the same terms as ships and merchandise of the United States, and Spanish scientific, literary and artistic works, not subversive of public order, shall continue to be admitted free of duty into all the ceded territories. Neither of the provisions could be carried out if the Constitution required the customs regulations of the United States to apply in those territories.

In the absence of Congressional legislation, the regulation of the revenue of the conquered territory, even after the treaty of cession, remains with the executive and military authority.

So long as Congress has not incorporated the territory into the United States, neither military occupation nor cession by treaty makes the conquered territory domestic territory, in the sense of the revenue laws. But those laws concerning "foreign countries" remain applicable to the conquered territory until changed by Congress. Such was the unanimous opinion of this court, as declared by Chief Justice Taney, in *Fleming* v. *Page*, 9 How. 603, 617.

If Congress is not ready to construct a complete government for the conquered territory, it may establish a temporary government, which is not subject to all the restrictions of the Constitution.

Such was the effect of the act of Congress of April 12, 1900, c. 191, entitled " An act temporarily to provide revenues and a civil government for Porto Rico, and for other purposes." By the third section of that act, it was expressly declared that the duties thereby established on merchandise and articles going into Porto Rico from the United States, or coming into the United States from Porto Rico, should cease in any event on March 1, 1902, and sooner if the legislative assembly of Porto Rico should enact and put into operation a system of local taxation to meet the necessities of the government established by that act.

The system of duties, temporarily established by that act during the transition period, was within the authority of Congress under the Constitution of the United States.

MR. CHIEF JUSTICE FULLER, (with whom concurred MR. JUSTICE HARLAN, MR. JUSTICE BREWER and MR. JUSTICE PECKHAM,) dissenting.

This is an action brought to recover moneys exacted by the collector of customs at the port of New York as import duties on two shipments of fruit from ports in the island of Porto Rico to the port of New York in November, 1900.

The treaty ceding Porto Rico to the United States was ratified by the Senate, February 6, 1899; Congress passed an act to carry out its obligations March 3, 1899; and the ratifications were exchanged, and the treaty proclaimed April 11, 1899. Then followed the act approved April 12, 1900. 31 Stat. 77, c. 191.

Mr. Justice Harlan, Mr. Justice Brewer, Mr. Justice Peckham and myself are unable to concur in the opinions and judgment of the court in this case. The majority widely differ in the reasoning by which the conclusion is reached, although there seems to be concurrence in the view that Porto Rico belongs to the United States, but nevertheless, and notwithstanding the act of Congress, is not a part of the United States, subject to the provisions of the Constitution in respect of the levy of taxes, duties, imposts and excises.

The inquiry is whether the act of April 12, 1900, so far as it requires the payment of import duties on merchandise brought from a port of Porto Rico as a condition of entry into other ports of the United States, is consistent with the Federal Constitution.

The act creates a civil government for Porto Rico, with a Governor, Secretary, Attorney General, and other officers, appointed by the President, by and with the advice and consent of the Senate, who, together with five other persons, likewise so appointed and confirmed, are constituted an executive council; local legislative powers are vested in a legislative assembly, consisting of the executive council and a house of delegates to be elected; courts are provided for, and, among other things, Porto Rico is constituted a judicial district, with a district judge, attorney and marshal to be appointed by the President for the term of four years. The district court is to be called the District Court of the United States for Porto Rico, and to possess, in addition to the ordinary jurisdiction of District Courts of the United States, jurisdiction of all cases cognizant in the Circuit Courts of the United States. The act also provides that "Writs of error and appeals from the final decisions of the Supreme Court of Porto Rico and the District Court of the United States shall be allowed and may be taken to the Supreme Court of the United States in the same manner and under the same regulations and in the same cases as from the Supreme Courts of the Territories of the United States; and such writs of error and appeal shall be allowed in all cases where the Constitution of the United States, or a treaty thereof, or an act of Congress is brought in question and the right claimed thereunder is denied."

It was also provided that the inhabitants continuing to reside in Porto Rico, who were Spanish subjects on April 11, 1899, and their children born subsequent thereto, (except such as should elect to preserve their allegiance to the Crown of Spain,) together with citizens of the United States, residing in Porto Rico, should "constitute a body politic under the name of The People of Porto Rico, with governmental powers as hereinafter conferred and with power to sue and be sued as such."

FULLER, C. J., HARLAN, BREWER and PECKHAM, JJ., dissenting.

All officials authorized by the act are required to "before entering upon the duties of their respective offices take an oath to support the Constitution of the United States and the laws of Porto Rico."

The second, third, fourth, fifth and thirty-eighth sections of the act are printed in the margin.[1]

---

[1] SEC. 2. That on and after the passage of this act the same tariffs, customs, and duties shall be levied, collected, and paid upon all articles imported into Porto Rico from ports other than those of the United States which are required by law to be collected upon articles imported into the United States from foreign countries: *Provided*, That on all coffee in the bean or ground imported into Porto Rico there shall be levied and collected a duty of five cents per pound, any law or part of law to the contrary notwithstanding: *And provided further*, That all Spanish scientific, literary, and artistic works, not subversive of public order in Porto Rico, shall be admitted free of duty into Porto Rico for a period of ten years, reckoning from the eleventh day of April, eighteen hundred and ninety-nine, as provided in said treaty of peace between the United States and Spain: *And provided further*, That all books and pamphlets printed in the English language shall be admitted into Porto Rico free of duty when imported from the United States.

SEC. 3. That on and after the passage of this act all merchandise coming into the United States from Porto Rico and coming into Porto Rico from the United States shall be entered at the several ports of entry upon payment of fifteen per centum of the duties which are required to be levied, collected, and paid upon like articles of merchandise imported from foreign countries; and in addition thereto upon articles of merchandise of Porto Rican manufacture coming into the United States and withdrawn for consumption or sale upon payment of a tax equal to the internal revenue tax imposed in the United States upon the like articles of merchandise of domestic manufacture; such tax to be paid by internal revenue stamp or stamps to be purchased and provided by the Commissioner of Internal Revenue and to be procured from the collector of internal revenue at or most convenient to the port of entry of said merchandise in the United States, and to be affixed under such regulations as the Commissioner of Internal Revenue, with the approval of the Secretary of the Treasury, shall prescribe; and on all articles of merchandise of United States manufacture coming into Porto Rico in addition to the duty above provided upon payment of a tax equal in rate and amount to the internal revenue tax imposed in Porto Rico upon the like articles of Porto Rican manufacture: *Provided*, That on and after the date when this act shall take effect, all merchandise and articles, except coffee, not dutiable under the tariff laws of the United States, and all merchandise and articles entered in Porto Rico free of duty under orders heretofore made by the Secretary of War, shall be admitted

It will be seen that duties are imposed upon "merchandise coming into Porto Rico from the United States;" "merchandise

---

into the several ports thereof, when imported from the United States, free of duty, all laws or parts of laws to the contrary notwithstanding; and whenever the legislative assembly of Porto Rico shall have enacted and put into operation a system of local taxation to meet the necessities of the government of Porto Rico, by this act established, and shall by resolution duly passed so notify the President, he shall make proclamation thereof, and thereupon all tariff duties on merchandise and articles going into Porto Rico from the United States or coming into the United States from Porto Rico shall cease, and from and after such date all such merchandise and articles shall be entered at the several ports of entry free of duty; and in no event shall any duties be collected after the first day of March, nineteen hundred and two, on merchandise and articles going into Porto Rico from the United States or coming into the United States from Porto Rico.

SEC. 4. That the duties and taxes collected in Porto Rico in pursuance of this act, less the cost of collecting the same, and the gross amount of all collections of duties and taxes in the United States upon articles of merchandise coming from Porto Rico, shall not be covered into the general fund of the Treasury, but shall be held as a separate fund, and shall be placed at the disposal of the President to be used for the government and benefit of Porto Rico until the government of Porto Rico herein provided for shall have been organized, when all moneys theretofore collected under the provisions hereof, then unexpended, shall be transferred to the local treasury of Porto Rico, and the Secretary of the Treasury shall designate the several ports and sub-ports of entry into Porto Rico and shall make such rules and regulations and appoint such agents as may be necessary to collect the duties and taxes authorized to be levied, collected, and paid in Porto Rico by the provisions of this act, and he shall fix the compensation and provide for the payment thereof of all such officers, agents, and assistants as he may find it necessary to employ to carry out the provisions hereof; *Provided, however,* That as soon as a civil government for Porto Rico shall have been organized in accordance with the provisions of this act and notice thereof shall have been given to the President he shall make proclamation thereof, and thereafter all collections of duties and taxes in Porto Rico under the provisions of this act shall be paid into the treasury of Porto Rico, to be expended as required by law for the government and benefit thereof instead of being paid into the Treasury of the United States.

SEC. 5. That on and after the day when this act shall go into effect all goods, wares, and merchandise previously imported from Porto Rico, for which no entry has been made, and all goods, wares, and merchandise previously entered without payment of duty and under bond for warehousing, transportation, or any other purpose, for which no permit of delivery to the importer or his agent has been issued, shall be subjected to the duties imposed by this act, and to no other duty, upon the entry or the withdrawal

FULLER, C. J., HARLAN, BREWER and PECKHAM, JJ., dissenting.

coming into the United States from Porto Rico;" taxes upon
"articles of merchandise of Porto Rican manufacture coming
into the United States and withdrawn from consumption or
sale" "equal to the internal-revenue tax imposed in the United
States upon like articles of domestic manufacture;" and "on
all articles of merchandise of United States manufacture coming
into Porto Rico," "a tax equal in rate and amount to the internal-revenue tax imposed in Porto Rico upon the like articles
of Porto Rican manufacture."

And it is also provided that all duties collected in Porto Rico
on imports from foreign countries and on "merchandise coming
into Porto Rico from the United States," and "the gross amount
of all collections of duties and taxes in the United States upon
articles of merchandise coming from Porto Rico," shall be
held as a separate fund and placed "at the disposal of the
President to be used for the government and benefit of Porto
Rico" until the local government is organized, when "all collections of taxes and duties under this act shall be paid into the
treasury of Porto Rico instead of being paid into the Treasury
of the United States."

The first clause of section 8 of Article I of the Constitution

---

thereof: *Provided*, That when duties are based upon the weight of merchandise deposited in any public or private bonded warehouse said duties shall
be levied and collected upon the weight of such merchandise at the time of
its entry.

 * * * * * * * *

SEC. 38. That no export duties shall be levied or collected on exports
from Porto Rico; but taxes and assessments on property, and license fees
for franchises, privileges, and concessions may be imposed for the purposes
of the insular and municipal governments, respectively as may be provided
and defined by act of the legislative assembly; and where necessary to anticipate taxes and revenues, bonds and other obligations may be issued by
Porto Rico or any municipal government therein as may be provided by law
to provide for expenditures authorized by law, and to protect the public
credit, and to reimburse the United States for any moneys which have been
or may be expended out of the emergency fund of the War Department for
the relief of the industrial conditions of Porto Rico caused by the hurricane
of August eighth, eighteen hundred and ninety-nine. *Provided, however*,
That no public indebtedness of Porto Rico or of any municipality thereof
shall be authorized or allowed in excess of seven per centum of the aggregate tax valuation of its property.

provides: "The Congress shall have power to levy and collect taxes, duties, imposts and excises, to pay the debts and provide for the common defence and general welfare of the United States; but all duties, imposts and excises shall be uniform throughout the United States."

Clauses four, five and six of section nine are:

"No capitation, or other direct, tax shall be laid, unless in proportion to the census or enumeration hereinbefore directed to be taken.

"No tax or duty shall be laid on articles exported from any State.

"No preference shall be given by any regulation of commerce or revenue to the ports of one State over those of another; nor shall vessels bound to, or from, one State, be obliged to enter, clear, or pay duties in another."

This act on its face does not comply with the rule of uniformity and that fact is admitted.

The uniformity required by the Constitution is a geographical uniformity, and is only attained when the tax operates with the same force and effect in every place where the subject of it is found. *Knowlton* v. *Moore*, 178 U. S. 41; *Head Money Cases*, 112 U. S. 580, 594. But it is said that Congress in attempting to levy these duties was not exercising power derived from the first clause of section 8, or restricted by it, because in dealing with the territories Congress exercises unlimited powers of government, and, moreover, that these duties are merely local taxes.

This court, in 1820, when Marshall was Chief Justice, and Washington, William Johnson, Livingston, Todd, Duvall and Story were his associates, took a different view of the power of Congress in the matter of laying and collecting taxes, duties, imposts and excises in the territories, and its ruling in *Loughborough* v. *Blake*, 5 Wheat. 317, has never been overruled.

It is said in one of the opinions of the majority that the Chief Justice "made certain observations which have occasioned some embarrassment in other cases." Manifestly this is so in this case, for it is necessary to overrule that decision in order to reach the result herein announced.

The question in *Loughborough* v. *Blake* was whether Congress had the right to impose a direct tax on the District of Columbia apart from the grant of exclusive legislation, which carried the power to levy local taxes. The court held that Congress had such power under the clause in question. The reasoning of Chief Justice Marshall was directed to show that the grant of the power " to lay and collect taxes, duties, imposts and excises," because it was general and without limitation as to place, consequently extended " to all places over which the government extends," and he declared that, if this could be doubted, the doubt was removed by the subsequent words, which modified the grant, " but all duties, imposts and excises shall be uniform throughout the United States." He then said : " It will not be contended that the modification of the power extends to places to which the power itself does not extend. The power then to lay and collect duties, imposts and excises may be exercised, and must be exercised throughout the United States. Does this term designate the whole, or any portion of the American empire? Certainly this question can admit of but one answer. It is the name given to our great republic, which is composed of States and territories. The District of Columbia, or the territory west of the Missouri, is not less within the United States, than Maryland or Pennsylvania; and it is not less necessary, on the principles of our Constitution, that uniformity in the imposition of imposts, duties and excises should be observed in the one, than in the other. Since, then, the power to lay and collect taxes, which includes direct taxes, is obviously coextensive with the power to lay and collect duties, imposts and excises, and since the latter extends throughout the United States, it follows that the power to impose direct taxes also extends throughout the United States."

It is wholly inadmissible to reject the process of reasoning by which the Chief Justice reached and tested the soundness of his conclusion as merely *obiter.*

Nor is there any intimation that the ruling turned on the theory that the Constitution irrevocably adhered to the soil of Maryland and Virginia, and, therefore, accompanied the parts which were ceded to form the District, or that "the tie" between those States and the Constitution " could not be dissolved, without at least the consent of the Federal and state governments to a formal separation," and that this was not given by the cession and its acceptance in accordance with the constitutional provision itself, and hence that Congress was restricted in the exercise of its powers in the District, while not so in the territories.

So far from that, the Chief Justice held the territories as well as the District to be part of the United States for the purposes of national taxation, and repeated in effect what he had already said in *McCulloch* v. *Maryland*, 4 Wheaton, 316, 408 : " Throughout this vast republic, from the St. Croix to the Gulf of Mexico, from the Atlantic to the Pacific, revenue is to be collected and expended, armies are to be marched and supported."

Conceding that the power to tax for the purposes of territorial government is implied from the power to govern territory, whether the latter power is attributed to the power to acquire or the power to make needful rules and regulations, these particular duties are nevertheless not local in their nature, but are imposed as in the exercise of national powers. The levy is clearly a regulation of commerce, and a regulation affecting the States and their people as well as this territory and its people. The power of Congress to act directly on the rights and interests of the people of the States can only exist if, and as, granted by the Constitution. And by the Constitution Congress is vested with power " to regulate commerce with foreign nations, and among the several States, and with the Indian tribes." The territories are indeed not mentioned by name, and yet commerce between the territories and foreign nations is covered by the clause, which would seem to have been intended to embrace the entire internal as well as foreign commerce of the country.

It is evident that Congress cannot regulate commerce between a territory and the States and other territories in the exercise of the bare power to govern the particular territory, and as this act was framed to operate and does operate on the people of the States, the power to so legislate is apparently rested on the assumption that the right to regulate commerce between the States and territories comes within the commerce clause by necessary implication. *Stoutenburgh* v. *Hennick*, 129 U. S. 141.

Accordingly the act of Congress of August 8, 1890, entitled "An act to limit the effect of the regulations of commerce between the several States and with foreign countries in certain cases," applied in terms to the territories as well as to the States.

In any point of view, the imposition of duties on commerce operates to regulate commerce, and is not a matter of local legislation; and it follows that the levy of these duties was in the exercise of the national power to do so, and subject to the requirement of geographical uniformity.

The fact that the proceeds are devoted by the act to the use of the territory does not make national taxes, local. Nobody disputes the power of Congress to lay and collect duties, geographically uniform, and apply the proceeds by a proper appropriation act to the relief of a particular territory, but the destination of the proceeds would not change the source of the power to lay and collect. And that suggestion certainly is not strengthened when based on the diversion of duties collected from all parts of the United States to a territorial treasury before reaching the Treasury of the United States. Clause 7 of section 9 of Article I provides that "no money shall be drawn from the Treasury, but in consequence of appropriations made by law," and the proposition that this may be rendered inapplicable if the money is not permitted to be paid in so as to be susceptible of being drawn out, is somewhat startling.

It is also urged that Chief Justice Marshall was entirely in fault because while the grant was general and without limitation as to place, the words, "throughout the United States," imposed a limitation as to place so far as the rule of uniformity was concerned, namely, a limitation to the States as such.

Undoubtedly the view of the Chief Justice was utterly inconsistent with that contention, and, in addition to what has been quoted, he further remarked: "If it be said that the principle of uniformity, established in the Constitution, secures the District from oppression in the imposition of indirect taxes, it is not less true that the principle of apportionment, also established in the Constitution, secures the District from any oppressive exercise of the power to lay and collect direct taxes." It must be borne in mind that the grant was of the absolute power of taxation for national purposes, wholly unlimited as to place, and subjected to only one exception and two qualifications. The exception was that exports could not be taxed at all. The qualifications were that direct taxes must be imposed by the rule of apportionment, and indirect taxes by the rule of uniformity. *License Tax Cases*, 5 Wall. 462. But as the power necessarily could be exercised throughout every part of the national domain, State, Territory, District, the exception and the qualifications attended its exercise. That is to say, the protection extended to the people of the States extended also to the people of the District and the Territories.

In *Knowlton* v. *Moore*, 178 U. S. 41, it is shown that the words "throughout the United States" are but a qualification introduced for the purpose .of rendering the uniformity prescribed, geographical, and not intrinsic, as would have resulted if they had not been used.

As the grant of the power to lay taxes and duties was unqualified as to place, and the words were added for the sole purpose of preventing the uniformity required from being intrinsic, the intention thereby to circumscribe the area within which the power could operate not only cannot be imputed, but the contrary presumption must prevail.

Taking the words in their natural meaning—in the sense in which they are frequently and commonly used—no reason is perceived for disagreeing with the Chief Justice in the view that they were used in this clause to designate the geographical unity known as "The United States," "our great republic, which is composed of States and territories."

Other parts of the Constitution furnish illustrations of the correctness of this view. Thus the Constitution vests Congress with the power "to establish an uniform rule of naturalization, and uniform laws on the subject of bankruptcies throughout the United States."

FULLER, C. J., HARLAN, BREWER and PECKHAM, JJ., dissenting.

This applies to the territories as well as the States, and has always been recognized in legislation as binding.

Aliens in the territories are made citizens of the United States, and bankrupts residing in the territories are discharged from debts owing citizens of the States pursuant to uniform rules and laws enacted by Congress in the exercise of this power.

The Fourteenth Amendment provides that "all persons born or naturalized in the United States, and subject to the jurisdiction thereof, are citizens of the United States and of the States wherein they reside;" and this court naturally held, in the *Slaughter House Cases*, 16 Wall. 36, that the United States included the District and the territories. Mr. Justice Miller observed : "It had been said by eminent judges that no man was a citizen of the United States, except as he was a citizen of one of the States composing the Union. Those, therefore, who had been born and resided always in the District of Columbia or in the territories, though within the United States, were not citizens. Whether this proposition was sound or not had never been judicially decided." And he said the question was put at rest by the Amendment, and the distinction between citizenship of the United States and citizenship of a State was clearly recognized and established. "Not only may a man be a citizen of the United States without being a citizen of a State, but an important element is necessary to convert the former into the latter. He must reside within the State to make him a citizen of it, but it is only necessary that he should be born or naturalized in the United States to be a citizen of the Union."

No person is eligible to the office of President unless he has "attained to the age of thirty-five years, and been fourteen years a resident within the United States." Clause 5, sec. 1, Art. II.

Would a native-born citizen of Massachusetts be ineligible if he had taken up his residence and resided in one of the territories for so many years that he had not resided altogether fourteen years in the States? When voted for he must be a citizen of one of the States (clause 3, sec. 1, Art. II; Art. XII), but as to length of time must residence in the territories be counted against him?

The Fifteenth Amendment declares that "the right of citizens of the United States to vote shall not be denied or abridged by the United States or by any State on account of race, color, or previous condition of servitude." Where does that prohibition on the United States especially apply if not in the territories?

The Thirteenth Amendment says that neither slavery nor involuntary servitude "shall exist within the United States or any place subject to their jurisdiction." Clearly this prohibition would have operated in the territories if the concluding words had not been added. The history of the times shows that the addition was made in view of the then condition of the country—the amendment passed the house January 31, 1865,—and it is moreover otherwise applicable than to the territories. Besides, generally speaking, when words are used simply out of abundant caution, the fact carries little weight.

Other illustrations might be adduced but it is unnecessary to prolong this opinion by giving them.

I repeat that no satisfactory ground has been suggested for restricting the words "throughout the United States," as qualifying the power to impose duties, to the States, and that conclusion is the more to be avoided when we reflect that it rests, in the last analysis, on the assertion of the possession by Congress of unlimited power over the territories.

The government of the United States is the government ordained by the Constitution, and possesses the powers conferred by the Constitution. "This original and supreme will organizes the government, and assigns to different departments their respective powers. It may either stop here, or establish certain limits not to be transcended by those departments. The government of the United States is of the latter description. The powers of the legislature are defined and limited; and that those limits may not be mistaken or forgotten, the Constitution is written. To what purpose are powers limited, and to what purpose is that limitation committed to writing, if these limits may, at any time, be passed by those intended to be restrained?" *Marbury* v. *Madison*, 1 Cranch, 137, 176. The opinion of the court, by Chief Justice Marshall, in that case, was delivered at the February term, 1803, and at the October term, 1885, the court, in *Yick Wo* v. *Hopkins*, 118 U. S. 356, speaking through Mr. Justice Matthews, said: "When we consider the nature and theory of our institutions of government, the principles upon which they are supposed to rest, and review the history of their development, we are constrained to conclude that they do not mean to leave room for the play and action of purely personal and arbitrary power. Sovereignty itself is, of course, not subject to law, for it is the author and source of law; but in our system, while sovereign powers are delegated to the agencies of government, sovereignty itself remains with the people, by whom and for whom all government exists and acts. And the law is the definition and limitation of power."

From *Marbury* v. *Madison* to the present day, no utterance of this court has intimated a doubt that in its operation on the people, by whom and for whom it was established, the national government is a government of enumerated powers, the exercise of which is restricted to the use of means appropriate and plainly adapted to constitutional ends, and which are "not prohibited, but consist with the letter and spirit of the Constitution."

The powers delegated by the people to their agents are not enlarged by the expansion of the domain within which they are exercised. When the restriction on the exercise of a particular power by a particular agent is ascertained, that is an end of the question.

To hold otherwise is to overthrow the basis of our constitutional law, and moreover, in effect, to reassert the proposition that the States and not the people created the government.

It is again to antagonize Chief Justice Marshall, when he said: "The government of the Union, then, (whatever may be the influence of this fact on the case,) is, emphatically, and truly, a government of the people. In form and in substance it emanates from them. Its powers are granted by them, and are to be exercised directly on them, and for their benefit. This government is acknowledged by all to be one of enumerated powers." 4 Wheat. 404.

The prohibitory clauses of the Constitution are many, and

they have been repeatedly given effect by this court in respect of the Territories and the District of Columbia.

The underlying principle is indicated by Chief Justice Taney, in *The Passenger Cases*, 7 How. 283, 492, where he maintained the right of the American citizen to free transit in these words: "Living as we do under a common government, charged with the great concerns of the whole Union, every citizen of the United States, from the most remote States or territories, is entitled to free access, not only to the principal departments established at Washington, but also to its judicial tribunals and public offices in every State and territory of the Union. . . . For all the great purposes for which the Federal government was formed, we are one people, with one common country. We are all citizens of the United States; and, as members of the same community, must have the right to pass and repass through every part of it without interruption, as freely as in our own States."

In *Cross* v. *Harrison*, 16 How. 164, 197, it was held that by the ratification of the treaty with Mexico "California became a part of the United States," and that: "The right claimed to land foreign goods within the United States at any place out of a collection district, if allowed, would be a violation of that provision in the Constitution which enjoins that all duties, imposts and excises shall be uniform throughout the United States."

In *Dred Scott* v. *Sandford*, 19 How. 393, the court was unanimous in holding that the power to legislate respecting a territory was limited by the restrictions of the Constitution, or, as Mr. Justice Curtis put it, by "the express prohibitions on Congress not to do certain things."

Mr. Justice McLean said: "No powers can be exercised which are prohibited by the Constitution, or which are contrary to its spirit."

Mr. Justice Campbell: "I look in vain, among the discussions of the time, for the assertion of a supreme sovereignty for Congress over the territory then belonging to the United States, or that they might thereafter acquire. I seek in vain for an annunciation that a consolidated power had been inaugurated, whose subject comprehended an empire, and which had no restriction but the discretion of Congress."

Chief Justice Taney : " The powers over persons and property of which we speak are not only not granted to Congress, but are in express terms denied, and they are forbidden to exercise them.　And this prohibition is not confined to the States, but the words are general, and extend to the whole territory over which the Constitution gives it power to legislate, including those portions of it remaining under territorial government, as well as that covered by States.　It is a total absence of power everywhere within the dominion of the United States, and places the citizens of a territory, so far as these rights are concerned, on the same footing with citizens of the States, and guards them as firmly and plainly against any inroads which the general government might attempt, under the plea of implied or incidental powers."

Many of the later cases were brought from territories over which Congress had professed to " extend the Constitution," or from the District after similar provision, but the decisions did not rest upon the view that the restrictions on Congress were self-imposed, and might be withdrawn at the pleasure of that body.

Capital Traction Company v. *Hof*, 174 U. S. 1, is a fair illustration, for it was there ruled, citing *Webster* v. *Reid*, 11 How. 437; *Callan* v. *Wilson*, 127 U. S. 550; *Thompson* v. *Utah*, 170 U. S. 343, that " it is beyond doubt, at the present day, that the provisions of the Constitution of the United States securing the right of trial by jury, whether in civil or in criminal cases, are applicable to the District of Columbia."

No reference whatever was made to section 34 of the act of February 21, 1871, 16 Stat. 419, c. 62, which, in providing for the election of a delegate for the District, closed with the words: " The person having the greatest number of legal votes shall be declared by the governor to be duly elected, and a certificate thereof shall be given accordingly ; and the Constitution and all laws of the United States, which are not locally inapplicable, shall have the same force and effect within the said District of Columbia as elsewhere within the United States."

Nor did the court in *Bauman* v. *Ross*, 167 U. S. 548, attribute the application of the Fifth Amendment to the act of Congress, although it was cited to another point.

The truth is that, as Judge Edmunds wrote, "the instances in which Congress has declared in statutes organizing territories, that the Constitution and laws should be in force there, are no evidence that they were not already there, for Congress and all legislative bodies have often made enactments that in effect merely declared existing law. In such cases they declare a preëxisting truth to ease the doubts of casuists." Cong. Rec. 56th Cong. 1st Sess. p. 3507.

In *Callan* v. *Wilson*, 127 U. S. 540, 550, which was a criminal prosecution in the District of Columbia, Mr. Justice Harlan, speaking for the court, said: "There is nothing in the history of the Constitution or of the original amendments to justify the assertion that the people of this District may be lawfully deprived of the benefit of any of the constitutional guarantees of life, liberty, and property—especially of the privilege of trial by jury in criminal cases." And further: "We cannot think that the people of this District have, in that regard, less rights than those accorded to the people of the territories of the United States."

In *Thompson* v. *Utah*, 170 U. S. 343, it was held that a statute of the State of Utah, providing for the trial of criminal cases other than capital, by a jury of eight, was invalid as applied on a trial for a crime committed before Utah was admitted; that it was not "competent for the State of Utah, upon its admission into the Union, to do in respect of Thompson's crime what the United States could not have done while Utah was a Territory;" and that an act of Congress providing for a trial by a jury of eight persons in the Territory of Utah would have been in conflict with the Constitution.

Article 6 of the Constitution ordains: "This Constitution, and the laws of the United States which shall be made in persuance thereof and all treaties made, or which shall be made under the authority of the United States, shall be the supreme law of the land."

And, as Mr. Justice Curtis observed in *United States* v. *Morris*,

1 Curtis, 23, 50, "nothing can be clearer than the intention to have the Constitution, laws, and treaties of the United States in equal force throughout every part of the territory of the United States, alike in all places, at all times."

But it is said that an opposite result will be reached if the opinion of Chief Justice Marshall in *American Insurance Company* v. *Canter*, 1 Pet. 511, be read "in connection with Art. III, secs. 1 and 2 of the Constitution, vesting 'the judicial power of the United States' in 'one Supreme Court, and in such inferior courts as the Congress may from time to time ordain and establish. The judges, both of the Supreme and inferior courts, shall hold their offices during good behaviour,'" etc. And it is argued : "As the only judicial power vested in Congress is to create courts whose judges shall hold their offices during good behaviour, it necessarily follows that, if Congress authorizes the creation of courts and the appointment of judges for a limited time, it must act independently of the Constitution, and upon territory which is not part of the United States within the meaning of the Constitution."

And further, that if the territories "be a part of the United States, it is difficult to see how Congress could create courts in such territories, except under the judicial clause of the Constitution."

By the ninth clause of section 8 of Article I, Congress is vested with power "to constitute tribunals inferior to the Supreme Court," while by section 1 of Article III the power is granted to it to establish inferior courts in which the judicial power of the government treated of in that article is vested.

That power was to be exerted over the controversies therein named, and did not relate to the general administration of justice in the territories, which was committed to courts established as part of the territorial government.

What the Chief Justice said was (p. 546) : "These courts, then, are not constitutional courts, in which the judicial power conferred by the Constitution on the general government can be deposited. They are incapable of receiving it. They are legislative courts, created in virtue of the general right of sovereignty which exists in the government, or in virtue of that clause which enables Congress to make all needful rules and regulations respecting the territory belonging to the United States. The jurisdiction with which they are invested is not a part of that judicial power which is defined in the third article of the Constitution, but is conferred by Congress, in the execution of those general powers which that body possesses over the territories of the United States."

The Chief Justice was dealing with the subject in view of the nature of the judicial department of the government and the distinction between Federal and state jurisdiction, and the conclusion was, to use the language of Mr. Justice Harlan in *McAllister* v. *United States*, 141 U. S. 174, " that courts in the territories, created under the plenary municipal authority that' Congress possesses over the territories of the United States, are not courts of the United States created under the authority conferred by that article."

But it did not therefore follow that the territories were not parts of the United States, and that the power of Congress, in general, over them, was unlimited; nor was there in any of the discussions on this subject the least intimation to that effect.

And this may justly be said of expressions in some other cases, supposed to give color to this doctrine of absolute dominion in dealing with civil rights.

In *Murphy* v. *Ramsey*, 114 U. S. 15, Mr. Justice Matthews said: " The personal and civil rights of the inhabitants of the territories are secured to them, as to other citizens, by the principles of constitutional liberty which restrain all the agencies of government, state and national. Their political rights are franchises, which they hold as privileges in the legislative discretion of the Congress of the United States."

In the *Mormon Church Case*, 136 U. S. 1, 44, Mr. Justice Bradley observed: " Doubtless Congress, in legislating for the territories, would be subject to those fundamental limitations in favor of personal rights which are formulated in the Constitution and its amendments; but these limitations would exist rather by inference and the general spirit of the Constitution from which Congress derives all its powers than by any express and direct application of its provisions."

That able judge was referring to the fact that the Constitution does not expressly declare that its prohibitions operate on the power to govern the territories, but because of the implication that an express provision to that effect might be essential, three members of the court were constrained to dissent, regarding it, as was said, " of vital consequence that absolute power should never be conceded as belonging under our system of government to any one of its departments."

What was ruled in *Murphy* v. *Ramsey* is that in places over which Congress has exclusive local jurisdiction its power over the political *status* is plenary.

Much discussion was had at the bar in respect to the citizenship of the inhabitants of Porto Rico, but we are not required to consider that subject at large in these cases. It will be time enough to seek a ford when, if ever, we are brought to the stream.

Yet although we are confined to the question of the validity of certain duties imposed after the organization of Porto Rico as a territory of the United States a few observations and some references to adjudged cases may well enough be added in view of the line of argument pursued in the concurring opinion.

In *American Insurance Company* v. *Canter*, 1 Pet. 511, 541— in which, by the way, the court did not accept the views of Mr. Justice Johnson in the Circuit Court or of Mr. Webster in argument—Chief Justice Marshall said : " The course which the argument has taken, will require, that, in deciding this question, the court should take into view the relation in which Florida stands to the United States. The Constitution confers absolutely on the government of the Union, the powers of making war, and of making treaties ; consequently, that government possesses the power of acquiring territory, either by conquest or by treaty. The usage of the world is, if a nation be not entirely subdued, to consider the holding of conquered territory as a mere military occupation, until its fate shall be determined at the treaty of peace. If it be ceded by the treaty, the acquisition is confirmed, and the ceded territory becomes a part of the nation to which it is annexed ; either on the terms stipulated in the treaty of cession, or on such as its new master shall impose.

On such transfer of territory, it has never been held, that the relations of the inhabitants with each other undergo any change. Their relations with their former sovereign are dissolved, and new relations are created between them, and the government which has acquired their territory. The same act which transfers their country, transfers the allegiance of those who remain in it; and the law, which may be denominated political, is necessarily changed, although that which regulates the intercourse, and general conduct of individuals, remains in force, until altered by the newly created power of the State. On the 2d of February, 1819, Spain ceded Florida to the United States. The sixth article of the treaty of cession contains the following provision: 'The inhabitants of the territories, which his Catholic Majesty cedes to the United States by this treaty, shall be incorporated in the Union of the United States, as soon as may be consistent with the principles of the Federal Constitution; and admitted to the enjoyment of the privileges, rights, and immunities of the citizens of the United States.' This treaty is the law of the land, and admits the inhabitants of Florida to the enjoyment of the privileges, rights, and immunities, of the citizens of the United States. It is unnecessary to inquire, whether this is not their condition, independent of stipulation. They do not, however, participate in political power; they do not share in the government, till Florida shall become a State. In the mean time, Florida continues to be a territory of the United States; governed by virtue of that clause in the Constitution, which empowers Congress 'to make all needful rules and regulations, respecting the territory, or other property belonging to the United States.' Perhaps the power of governing a territory belonging to the United States, which has not, by becoming a State, acquired the means of self-government, may result necessarily from the facts, that it is not within the jurisdiction of any particular State, and is within the power and jurisdiction of the United States. The right to govern may be the inevitable consequence of the right to acquire territory. Whichever may be the source, whence the power is derived, the possession of it is unquestioned."

FULLER, C. J., HARLAN, BREWER and PECKHAM, JJ., dissenting.

General Halleck, (Int. Law, 1st ed. chap. 83, § 14,) after quoting from Chief Justice Marshall, observed:

"This is now a well settled rule of the law of nations, and is universally admitted. Its provisions are clear and simple, and easily understood; but it is not so easy to distinguish between what are *political* and what are *municipal* laws, and to determine *when* and *how far* the constitution and laws of the conqueror change or replace those of the conquered. And in case the government of the new state is a constitutional government, of limited and divided powers, questions necessarily arise respecting the authority, which, in the absence of legislative action, can be exercised in the conquered territory after the cessation of war, and the conclusion of a treaty of peace. The determination of these questions depends upon the institutions and laws of the new sovereign, which, though conformable to the general rule of the law of nations, affect the construction and application of that rule to particular cases."

In *United States* v. *Percheman*, 7 Pet. 51, 87, the Chief Justice said:

"The people change their allegiance; their relation to their ancient sovereign is dissolved; but their relations to each other, and their rights of property, remain undisturbed. If this be the modern rule even in cases of conquest, who can doubt its application to the case of an amicable cession of territory? . . . The cession of a territory by its name from one sovereign to another, conveying the compound idea of surrendering at the same time the lands and the people who inhabit them, would be necessarily understood to pass the sovereignty only, and not to interfere with private property."

Again the court in *Pollard's Lessee* v. *Hagan*, 3 How. 212, 225 said:

"Every nation acquiring territory, by treaty or otherwise, must hold it subject to the constitution and laws of its own government, and not according to those of the government ceding it."

And in *Chicago, Rock Island & Pacific Railway Co.* v. *McGlinn*, 114 U. S. 546: "It is a general rule of public law, recognized and acted upon by the United States, that whenever political jurisdiction and legislative power over any territory are transferred from one nation or sovereign to another, the municipal laws of the country, that is, laws which are intended for the protection of private rights, continue in force until abrogated or changed by the new government or sovereign. By the cession public property passes from one government to the other, but private property remains as before, and with it those municipal laws which are designed to secure its peaceful use and enjoyment. As a matter of course, all laws, ordinances, and regulations in conflict with the political character, institutions, and constitution of the new government are at once displaced. Thus, upon a cession of political jurisdiction and legislative power—and the latter is involved in the former—to the United States, the laws of the country in support of an established religion, or abridging the freedom of the press, or authorizing cruel and unusual punishments, and the like, would at once cease to be of obligatory force without any declaration to that effect; and the laws of the country on other subjects would necessarily be superseded by existing laws of the new government upon the same matters. But with respect to other laws affecting the possession, use and transfer of property, and designed to secure good order and peace in the community, and promote its health and prosperity, which are strictly of a municipal character, the rule is general that a change of government leaves them in force until, by direct action of the new government, they are altered or repealed."

When a cession of territory to the United States is completed by the ratification of a treaty, it was stated in *Cross* v. *Harrison*, 16 How. 164, 198, that the land ceded becomes a part of the United States, and that as soon as it becomes so the territory is subject to the acts which were in force to regulate foreign commerce with the United States, after those had ceased which had been instituted for its regulation as a belligerent right; and the latter ceased after the ratification of the treaty. This statement was made by the Justice delivering the opinion as the result of the discussion and argument which he had already set forth. It was his summing up of what he supposed was decided on that subject in the case in which he was writing

The new master was, in the instance of Porto Rico, the United States, a constitutional government with limited powers, and the terms which the Constitution itself imposed, or which might be imposed in accordance with the Constitution, were the terms on which the new master took possession.

The power of the United States to acquire territory by conquest, by treaty, or by discovery and occupation, is not disputed, nor is the proposition that in all international relations, interests, and responsibilities the United States is a separate, independent, and sovereign nation; but it does not derive its powers from international law, which, though a part of our municipal law, is not a part of the organic law of the land. The source of national power in this country is the Constitution of the United States; and the government, as to our internal affairs, possesses no inherent sovereign power not derived from that instrument, and inconsistent with its letter and spirit.

Doubtless the subjects of the former sovereign are brought by the transfer under the protection of the acquiring power, and are so far forth impressed with its nationality, but it does not follow that they necessarily acquire the full *status* of citizens. The ninth article of the treaty ceding Porto Rico to the United States provided that Spanish subjects, natives of the Peninsula, residing in the ceded territory, might remain or remove, and in case they remained might preserve their allegiance to the crown of Spain by making a declaration of their decision to do so, "in default of which declaration they shall be held to have renounced it and to have adopted the nationality of the territory in which they reside."

The same article also contained this paragraph: "The civil rights and political *status* of the native inhabitants of the territories hereby ceded to the United States shall be determined by Congress." This was nothing more than a declaration of the accepted principles of international law applicable to the *status* of the Spanish subjects and of the native inhabitants. It did not assume that Congress could deprive the inhabitants of ceded territory of rights to which they might be entitled. The grant by Spain could not enlarge the powers of Congress, nor did it

purport to secure from the United States a guaranty of civil or
political privileges.

Indeed a treaty which undertook to take away what the Constitution secured or to enlarge the Federal jurisdiction would be
simply void.

"It need hardly be said that a treaty cannot change the Constitution or be held valid if it be in violation of that instrument.
This results from the nature and fundamental principles of our
government." *The Cherokee Tobacco*, 11 Wall. 616, 620.

So Mr. Justice Field in *Geofroy* v. *Riggs*, 133 U. S. 258, 267:
"The treaty power, as expressed in the Constitution, is in terms
unlimited except by those restraints which are found in that instrument against the action of the government or of its departments, and those arising from the nature of the government
itself and of that of the States. It would not be contended that
it extends so far as to authorize what the Constitution forbids,
or a change in the character of the government or in that of one
of the States, or a cession of any portion of the territory of the
latter, without its consent."

And it certainly cannot be admitted that the power of Congress to lay and collect taxes and duties can be curtailed by an
arrangement made with a foreign nation by the President and
two thirds of a quorum of the Senate. See 2 Tucker on the
Constitution, §§ 354, 355, 356.

In the language of Judge Cooley: "The Constitution itself
never yields to treaty or enactment; it neither changes with
time nor does it in theory bend to the force of circumstances.
It may be amended according to its own permission; but while
it stands it is 'a law for rulers and people, equally in war and
in peace, and covers with the shield of its protection all classes
of men, at all times and under all circumstances.' Its principles
cannot, therefore, be set aside in order to meet the supposed
necessities of great crises. 'No doctrine involving more perni-
cious consequences was ever invented by the wit of man than
that any of its provisions can be suspended during any of the
great exigencies of government.'"

I am not intimating in the least degree that any reason exists
for regarding this article to be unconstitutional, but even if it

were, the fact of the cession is a fact accomplished, and this court is concerned only with the question of the power of the government in laying duties in respect of commerce with the territory so ceded.

In the concurring opinion of Mr. Justice White, we find certain important propositions conceded, some of which are denied, or not admitted in the other. These are to the effect that " when an act of any department is challenged, because not warranted by the Constitution, the existence of the authority is to be ascertained by determining whether the power has been conferred by the Constitution, either in express terms or by lawful implication ;" that as every function of the government is derived from the Constitution, " that instrument is everywhere and at all times potential in so far as its provisions are applicable ;" that " wherever a power is given by the Constitution and there is a limitation imposed on the authority, such restriction operates upon and confines every action on the subject within its constitutional limits ;" that where conditions are brought about to which any particular provision of the Constitution applies, its controlling influence cannot be frustrated by the action of any or all of the departments of the government ; that the Constitution has conferred on Congress the right to create such municipal organizations as it may deem best for all the territories of the United States, but every applicable express limitation of the Constitution is in force, and even where there is no express command which applies, there may nevertheless be restrictions of so fundamental a nature that they cannot be transgressed though not expressed in so many words ; that every provision of the Constitution which is applicable to the territories is controlling therein, and all the limitations of the Constitution applicable to Congress in governing the territories necessarily limit its power ; that in the case of the territories, when a provision of the Constitution is invoked, the question is whether the provision relied on is applicable ; and that the power to lay and collect taxes, duties, imposts and excises, as well as the qualification of uniformity, restrains Congress from imposing an impost duty on goods coming into the United States from a territory which has been incorporated into and forms a part of the United States.

And it is said that the determination of whether a particular provision is applicable involves an inquiry into the situation of the territory and its relations to the United States, although it does not follow, when the Constitution has withheld all power over a given subject, that such an inquiry is necessary.

The inquiry is stated to be: "Had Porto Rico, at the time of the passage of the act in question, been incorporated into and become an integral part of the United States?" And the answer being given that it had not, it is held that the rule of uniformity was not applicable.

I submit that that is not the question in this case. The question is whether, when Congress has created a civil government for Porto Rico, has constituted its inhabitants a body politic, has given it a governor and other officers, a legislative assembly, and courts, with the right of appeal to this court, Congress can in the same act and in the exercise of the power conferred by the first clause of section eight, impose duties on the commerce between Porto Rico and the States and other territories in contravention of the rule of uniformity qualifying the power. If this can be done, it is because the power of Congress over commerce between the States and any of the territories is not restricted by the Constitution. This was the position taken by the Attorney General, with a candor and ability that did him great credit.

But that position is rejected, and the contention seems to be that if an organized and settled province of another sovereignty is acquired by the United States, Congress has the power to keep it, like a disembodied shade, in an intermediate state of ambiguous existence for an indefinite period; and, more than that, that after it has been called from that limbo, commerce with it is absolutely subject to the will of Congress, irrespective of constitutional provisions.

The accuracy of this view is supposed to be sustained by the act of 1856 in relation to the protection of citizens of the United States removing guano from unoccupied islands; but I am unable to see why the discharge by the United States of its undoubted duty to protect its citizens on *terra nullius*, whether temporarily engaged in catching and curing fish, or working mines, or taking away manure, furnishes support to the proposition that the power of Congress over the territories of the United States is unrestricted.

Great stress is thrown upon the word "incorporation," as if possessed of some occult meaning, but I take it that the act under consideration made Porto Rico, whatever its situation before, an organized territory of the United States. Being such, and the act undertaking to impose duties by virtue of clause one of section 8, how is it that the rule which qualifies the power does not apply to its exercise in respect of commerce with that territory? The power can only be exercised as prescribed, and even if the rule of uniformity could be treated as a mere regulation of the granted power, a suggestion to which I do not assent, the validity of these duties comes up directly and it is idle to discuss the distinction between a total want of power and a defective exercise of it.

The concurring opinion recognizes the fact that Congress, in dealing with the people of new territories or possessions, is bound to respect the fundamental guarantees of life, liberty, and property, but assumes that Congress is not bound, in those territories or possessions, to follow the rules of taxation prescribed by the Constitution. And yet the power to tax involves the power to destroy, and the levy of duties touches all our people in all places under the jurisdiction of the government.

The logical result is that Congress may prohibit commerce altogether between the States and territories, and may prescribe one rule of taxation in one territory, and a different rule in another.

That theory assumes that the Constitution created a government empowered to acquire countries throughout the world, to be governed by different rules than those obtaining in the original States and territories, and substitutes for the present system of republican government, a system of domination over distant provinces in the exercise of unrestricted power.

In our judgment, so much of the Porto Rican act as authorized the imposition of these duties is invalid, and plaintiffs were entitled to recover.

Some argument was made as to general consequences apprehended to flow from this result, but the language of the Constitution is too plain and unambiguous to permit its meaning to be thus influenced. There is nothing " in the literal construction so obviously absurd, or mischievous, or repugnant to the general spirit of the instrument, as to justify those who expound the Constitution " in giving it a construction not warranted by its words.

Briefs have been presented at this bar, purporting to be on behalf of certain industries, and eloquently setting forth the desirability that our government should possess the power to impose a tariff on the products of newly acquired territories so as to diminish or remove competition. That, however, furnishes no basis for judicial judgment, and if the producers of staples, in the existing States of this Union, believe the Constitution should be amended so as to reach that result, the instrument itself provides how such amendment can be accomplished. The people of all the States are entitled to a voice in the settlement of that subject.

Again, it is objected on behalf of the government that the possession of absolute power is essential to the acquisition of vast and distant territories, and that we should regard the situation as it is to-day rather than as it was a century ago. " We must look at the situation as comprehending a possibility—I do not say a probability, but a possibility—that the question might be as to the powers of this government in the acquisition of Egypt and the Soudan, or a section of Central Africa, or a spot in the Antarctic Circle, or a section of the Chinese Empire."

But it must be remembered that, as Marshall and Story declared, the Constitution was framed for ages to come, and that the sagacious men who framed it were well aware that a mighty future waited on their work. The rising sun to which Franklin referred at the close of the convention, they well knew, was that star of empire, whose course Berkeley had sung sixty years before.

They may not indeed have deliberately considered a triumphal progress of the nation, as such, around the earth, but, as Marshall wrote: "It is not enough to say, that this particular case was not in the mind of the convention, when the article was framed, nor of the American people, when it was adopted. It is necessary to go farther, and to say that, had this particular, case been suggested, the language would have been so varied, as to exclude it, or it would have been made a special exception."

This cannot be said, and, on the contrary, in order to the successful extension of our institutions, the reasonable presumption is that the limitations on the exertion of arbitrary power would have been made more rigorous.

After all, these arguments are merely political, and "political reasons have not the requisite certainty to afford rules of judicial interpretation."

Congress has power to make all laws which shall be necessary and proper for carrying into execution all the powers vested by the Constitution in the government of the United States, or in any department or officer thereof. If the end be legitimate and within the scope of the Constitution, then, to accomplish it, Congress may use "all means which are appropriate, which are plainly adapted to that end, which are not prohibited, but consistent with the letter and spirit of the Constitution."

The grave duty of determining whether an act of Congress does or does not comply with these requirements is only to be discharged by applying the well settled rules which govern the interpretation of fundamental law, unaffected by the theoretical opinions of individuals.

Tested by those rules our conviction is that the imposition of these duties cannot be sustained.

MR. JUSTICE HARLAN, dissenting.

I concur in the dissenting opinion of the Chief Justice. The grounds upon which he and Mr. Justice Brewer and Mr. Justice Peckham regard the Foraker act as unconstitutional in the particulars involved in this action meet my entire approval.

Those grounds need not be restated, nor is it necessary to re-examine the authorities cited by the Chief Justice. I agree in holding that Porto Rico—at least after the ratification of the treaty with Spain—became a part of the United States within the meaning of the section of the Constitution enumerating the powers of Congress and providing that "*all* duties, imposts and excises shall be uniform *throughout the United States.*"

In view, however, of the importance of the questions in this case, and of the consequences that will follow any conclusion reached by the court, I deem it appropriate—without redis-cussing the principal questions presented—to add some observa-tions suggested by certain passages in opinions just delivered in support of the judgment.

In one of those opinions it is said that "the Constitution was created by the people of the *United States,* as a union of *States,* to be governed solely by representatives of the *States;*" also, that "we find the Constitution speaking *only to States,* except in the territorial clause, which is absolute in its terms, and sug-gestive of no limitations upon the power of Congress in dealing with them." I am not sure that I correctly interpret these words. But if it is meant, as I assume it is meant, that, with the exception named, the Constitution was ordained by the States, and is addressed to and operates only on the States, I cannot accept that view.

In *Martin* v. *Hunter,* 1 Wheat. 304, 324, 326, 331, this court, speaking by Mr. Justice Story, said that "the Constitution of the United States was ordained and established, not by the States in their sovereign capacities, but emphatically, as the preamble of the Constitution declares, by the People of the United States."

In *McCulloch* v. *Maryland,* 4 Wheat. 316, 403–406, Chief Justice Marshall, speaking for this court, said : " The Govern-ment proceeds directly from the people; is 'ordained and es-tablished' in the name of the people; and is declared to be ordained, 'in order to form a more perfect union, establish jus-tice, ensure domestic tranquillity, and secure the blessings of liberty to themselves and their posterity.' The assent of the States, in their sovereign capacity, is implied in calling a Convention, and thus submitting that instrument to the people. But the people were at perfect liberty to accept or reject it; and their act was final. It required not the affirmance, and could not be negatived, by the state governments. The Constitution, when thus adopted, was of complete obligation, and bound the state sovereignties. . . . The Government of the Union, then, (whatever may be the influence of this fact on the case,) is, emphatically, and truly, a government of the people. In form and substance it emanates from them. Its powers are granted by them, and are to be exercised directly on them and for their benefit. This Government is acknowledged by all to be one of enumerated powers. . . . It is the Government of all; its powers are delegated by all; it represents all, and acts for all."

Although the States are constituent parts of the United States, the Government rests upon the authority of the people of the United States, and not on that of the States. Chief Justice Marshall, delivering the unanimous judgment of this court in *Cohens* v. *Virginia,* 6 Wheat. 264, 418, said : " That the United States form for many, and for most important purposes, a single nation, has not yet been denied. In war, we are one people. In making peace, we are one people. In all commercial regulations, we are one and the same people. In many other respects, the American people are one; and the government which is alone capable of controlling and managing their interests in all these respects is the Government of the Union. It is their Government, and in that character they have no other. America has chosen to be, in many respects and to many purposes, a nation; and for all these purposes her Government is complete; to all these objects it is competent. The people have declared that in the exercise of all powers given for those objects, it is supreme. It can, then, in effecting these objects, legitimately control all individuals or governments within the American territory."

In reference to the doctrine that the Constitution was established by and for the States as distinct political organizations, Mr. Webster said: " The Constitution itself in its very front refutes that. It declares that it is ordained and established by

the People of the United States. So far from saying that it is
established by the governments of the several States, it does
not even say that it is established by the people of the several
States. But it pronounces that it was established by the people of the United States in the aggregate. Doubtless, the people of the several States, taken collectively, constitute the people
of the United States. But it is in this their collective capacity,
it is as all the people of the United States, that they established
the Constitution."

In view of the adjudications of this court, I cannot assent to
the proposition, whether it be announced in express words or by
implication, that the National Government is a government of or
by the States in union, and that the prohibitions and limitations
of the Constitution are addressed only to the States. That is
but another form of saying that like the government created
by the Articles of Confederation, the present government is a
mere league of States, held together by compact between themselves; whereas, as this court has often declared, it is a government created by the People of the United States, with enumerated powers, and supreme over States and individuals, with
respect to certain objects, throughout the entire territory over
which its jurisdiction extends. If the National Government is,
in any sense, a compact, it is a compact between the People of
the United States among themselves as constituting in the aggregate the political community by whom the National Government was established. The Constitution speaks not simply to
the States in their organized capacities, but to all peoples, whether
of States or territories, who are subject to the authority of the
United States. *Martin* v. *Hunter*, 1 Wheat. 304, 327.

In the opinion to which I am referring it is also said that the
" practical interpretation put by Congress upon the Constitution
has been long continued and uniform to the effect that the Constitution is applicable to territories acquired by purchase or conquest only when and so far as Congress shall so direct;" that
while all power of government may be abused, the same may be
said of the power of the Government " under the Constitution as
well as outside of it;" that " if it once be conceded that we are
at liberty to acquire foreign territory, a presumption arises that

our power with respect to such territories is the same power which other nations have been accustomed to exercise with respect to territories acquired by them;" that "the liberality of Congress in legislating the Constitution into all our contiguous territories has undoubtedly fostered the impression that it went there by its own force, but there is nothing in the Constitution itself, and little in the interpretation put upon it, to confirm that impression;" that as the States could only delegate to Congress such powers as they themselves possessed, and as they had no power to acquire new territory, and therefore none to delegate in that connection, the logical inference is that "if Congress had power to acquire new territory, which is conceded, that power was not hampered by the constitutional provisions;" that if "we assume that the territorial clause of the Constitution was not intended to be restricted to such territory as the United States then possessed, there is nothing in the Constitution to indicate that the power of Congress in dealing with them was intended to be restricted by any of the other provisions;" and that "the executive and legislative departments of the Government have for more than a century interpreted this silence as precluding the idea that the Constitution attached to these territories as soon as acquired."

These are words of weighty import. They involve consequences of the most momentous character. I take leave to say that if the principles thus announced should ever receive the sanction of a majority of this court, a radical and mischievous change in our system of government will be the result. We will, in that event, pass from the era of constitutional liberty guarded and protected by a written constitution into an era of legislative absolutism.

Although from the foundation of the Government this court has held steadily to the view that the Government of the United States was one of enumerated powers, and that no one of its branches, nor all of its branches combined, could constitutionally exercise powers not granted, or which were not necessarily implied from those expressly granted, *Martin* v. *Hunter*, 1 Wheat. 304, 326, 331, we are now informed that Congress possesses powers *outside of the Constitution*, and may deal with new ter-

ritory, acquired by treaty or conquest, in the same manner *as
other nations have been accustomed to act with respect to territories acquired by them.* In my opinion, Congress has no existence and can exercise no authority outside of the Constitution.
Still less is it true that Congress can deal with new territories
just as other nations have done or may do with their new territories. This nation is under the control of a written constitution, the supreme law of the land and the only source of the
powers which our Government, or any branch or officer of it, may
exert at any time or at any place. Monarchical and despotic
governments, unrestrained by written constitutions, may do
with newly acquired territories what this Government may not
do consistently with our fundamental law. To say otherwise
is to concede that Congress may, by action taken outside of the
Constitution, engraft upon our republican institutions a colonial
system such as exists under monarchical governments. Surely
such a result was never contemplated by the fathers of the Constitution. If that instrument had contained a word suggesting
the possibility of a result of that character it would never have
been adopted by the People of the United States. The idea
that this country may acquire territories anywhere upon the
earth, by conquest or treaty, and hold them as mere colonies or
provinces—the people inhabiting them to enjoy only such rights
as Congress chooses to accord to them—is wholly inconsistent
with the spirit and genius as well as with the words of the Constitution.

   The idea prevails with some—indeed, it found expression in
arguments at the bar—that we have in this country substantially
or practically two national governments; one, to be maintained
under the Constitution, with all its restrictions; the other to be
maintained by Congress outside and independently of that instrument, by exercising such powers as other nations of the
earth are accustomed to exercise. It is one thing to give such
a latitudinarian construction to the Constitution as will bring
the exercise of power by Congress, upon a particular occasion
or upon a particular subject, within its provisions. It is quite a
different thing to say that Congress may, if it so elects, proceed
outside of the Constitution. The glory of our American system

of government is that it was created by a written constitution
which protects the people against the exercise of arbitrary, unlimited power, and the limits of which instrument may not be
passed by the government it created, or by any branch of it, or
even by the people who ordained it, except by amendment or
change of its provisions. "To what purpose," Chief Justice
Marshall said in *Marbury* v. *Madison*, 1 Cranch, 137, 176, "are
powers limited, and to what purpose is that limitation committed to writing, if these limits may, at any time, be passed
by those intended to be restrained? The distinction between a
government with limited and unlimited powers is abolished if
those limits do not confine the persons on whom they are imposed, and if acts prohibited and acts allowed are of equal obligation."

The wise men who framed the Constitution, and the patriotic
people who adopted it, were unwilling to depend for their safety
upon what, in the opinion referred to, is described as "certain
principles of natural justice inherent in Anglo-Saxon character
which need no expression in constitutions or statutes to give
them effect or to secure dependencies against legislation manifestly hostile to their real interests." They proceeded upon the
theory—the wisdom of which experience has vindicated—that
the only safe guaranty against governmental oppression was to
withhold or restrict the power to oppress. They well remembered that Anglo-Saxons across the ocean had attempted, in de-
fiance of law and justice, to trample upon the rights of Anglo-
Saxons on this continent and had sought, by military force, to
establish a government that could at will destroy the privileges
that inhere in liberty. They believed that the establishment
here of a government that could administer public affairs according to its will unrestrained by any fundamental law and
without regard to the inherent rights of freemen, would be ruinous to the liberties of the people by exposing them to the op-
pressions of arbitrary power. Hence, the Constitution enumer-
ates the powers which Congress and the other Departments
may exercise—leaving unimpaired, to the States or the People,
the powers not delegated to the National Government nor prohibited to the States. That instrument so expressly declares in

the Tenth Article of Amendment. It will be an evil day for American liberty if the theory of a government outside of the supreme law of the land finds lodgment in our constitutional jurisprudence. No higher duty rests upon this court than to exert its full authority to prevent all violation of the principles of the Constitution.

Again, it is said that Congress has assumed, in its past history, that the Constitution goes into territories acquired by purchase or conquest *only when and as it shall so direct*, and we are informed of the liberality of Congress in *legislating* the Constitution into all our contiguous territories. This is a view of the Constitution that may well cause surprise, if not alarm. Congress, as I have observed, has no existence except by virtue of the Constitution. It is the creature of the Constitution. It has no powers which that instrument has not granted, expressly or by necessary implication. I confess that I cannot grasp the thought that Congress which lives and moves and has its being in the Constitution and is consequently the mere creature of that instrument, can, at its pleasure, legislate or exclude its creator from territories which were acquired only by authority of the Constitution.

By the express words of the Constitution, every Senator and Representative is bound, by oath or affirmation, to regard it as the supreme law of the land. When the Constitutional Convention was in session there was much discussion as to the phraseology of the clause defining the supremacy of the Constitution, laws and treaties of the United States. At one stage of the proceedings the Convention adopted the following clause: "This Constitution, and the laws of the United States made in pursuance thereof, and all the treaties made under the authority of the United States, shall be the supreme law of the several *States* and of *their* citizens and inhabitants, and the judges of the several States shall be bound thereby in their decisions, anything in the constitutions or laws of the several States to the contrary notwithstanding." This clause was amended, on motion of Mr. Madison, by inserting after the words "all treaties made" the words "or which shall be made." If the clause, so amended, had been inserted in the Constitution as finally adopted, per-

haps there would have been some justification for saying that
the Constitution, laws and treaties of the United States constituted the supreme law only in the States, and that outside of
the States the will of Congress was supreme.   But the framers
of the Constitution saw the danger of such a provision, and put
into that instrument in place of the above clause the following :
" This Constitution, and the laws of the United States which
shall be made in pursuance thereof, and all treaties made, or
which shall be made, under the authority of the United States,
shall be *the supreme law of the land;* and the judges in every
State shall be bound thereby, anything in the constitution or
laws of any State to the contrary notwithstanding."   Meigs's
Growth of the Constitution, 284, 287.   That the Convention
struck out the words " the supreme law of the several States "
and inserted " the supreme law of the land," is a fact of no little
significance.   The " land " referred to manifestly embraced all
the peoples and all the territory, whether within or without the
States, over which the United States could exercise jurisdiction
or authority.

Further, it is admitted that *some* of the provisions of the Constitution do apply to Porto Rico and may be invoked as limit.
ing or restricting the authority of Congress, or for the protection
of the people of that island.   And it is said that there is a clear
distinction between such prohibitions " as go to the very root
of the power of Congress to act at all, irrespective of time or
place, and such as are operative only 'throughout the United
States' or among the several States."   In the enforcement of
this suggestion it is said in one of the opinions just delivered :
" Thus, when the Constitution declares that ' no bill of attainder
or *ex post facto* law shall be passed,' and that ' no title of nobility shall be granted by the United States,' it goes to the competency of Congress to pass a bill *of that description."*   I cannot accept this reasoning as consistent with the Constitution or
with sound rules of interpretation.   The express prohibition
upon the passage by Congress of bills of attainder, or of *ex post
facto* laws, or the granting of titles of nobility, goes no more
directly to the root of the power of Congress than does the express prohibition against the imposition by Congress of any

duty, impost or excise that is not uniform throughout the United States. The opposite theory, I take leave to say, is quite as extraordinary as that which assumes that Congress may exercise powers outside of the Constitution, and may, in its discretion, legislate that instrument into or out of a domestic territory of the United States.

In the opinion to which I have referred it is suggested that conditions may arise when the annexation of distant possessions may be desirable. "If," says that opinion, "those possessions are inhabited by alien races, differing from us in religion, customs, laws, methods of taxation and modes of thought, the administration of government and justice, according to Anglo-Saxon principles, may for a time be impossible; and the question at once arises whether large *concessions* ought not to be made for a time, that ultimately our own theories may be carried out, and the blessings of a free government under the Constitution extended to them. We decline to hold that there is anything in the Constitution to forbid such action." In my judgment, the Constitution does not sustain any such theory of our governmental system. Whether a particular race will or will not assimilate with our people, and whether they can or cannot with safety to our institutions be brought within the operation of the Constitution, is a matter to be thought of when it is proposed to acquire their territory by treaty. A mistake in the acquistion of territory, although such acquisition seemed at the time to be necessary, cannot be made the ground for violating the Constitution or refusing to give full effect to its provisions. The Constitution is not to be obeyed or disobeyed as the circumstances of a particular crisis in our history may suggest the one or the other course to be pursued. The People have decreed that it shall be the supreme law of the land at all times. When the acquisition of territory becomes complete, by cession, the Constitution necessarily becomes the supreme law of such new territory, and no power exists in any Department of the Government to make "concessions" that are inconsistent with its provisions. The authority to make such concessions implies the existence in Congress of power to declare that constitutional provisions may be ignored under special or embarrassing circumstances. No such dispensing power exists in any branch of our Government. The Constitution is supreme over every foot of territory, wherever situated, under the jurisdiction of the United States, and its full operation cannot be stayed by any branch of the Government in order to meet what some may suppose to be extraordinary emergencies. If the Constitution is in force in any territory, it is in force there for every purpose embraced by the objects for which the Government was ordained. Its authority cannot be displaced by concessions, even if it be true, as asserted in argument in some of these cases, that if the tariff act took effect in the Philippines of its own force, the inhabitants of Mandanao, who live on imported rice, would starve, because the import duty is many fold more than the ordinary cost of the grain to them. The meaning of the Constitution cannot depend upon accidental circumstances arising out of the products of other countries or of this country. We cannot violate the Constitution in order to serve particular interests in our own or in foreign lands. Even this court, with its tremendous power, must heed the mandate of the Constitution. No one in official station, to whatever department of the Government he belongs, can disobey its commands without violating the obligation of the oath he has taken. By whomsoever and wherever power is exercised in the name and under the authority of the United States, or of any branch of its Government, the validity or invalidity of that which is done must be determined by the Constitution.

In *DeLima* v. *Bidwell,* just decided, we have held that upon the ratification of the treaty with Spain, Porto Rico ceased to be a foreign country and became a domestic territory of the United States. We have said in that case that from 1803 to the present time there was not a shred of authority, except a *dictum* in one case, "for holding that a district ceded to and in possession of the United States remains for any purpose a foreign territory;" that territory so acquired cannot be "domestic for one purpose and foreign for another;" and that any judgment to the contrary would be "pure judicial legislation," for which there was no warrant in the Constitution or in the powers conferred upon this court. Although, as we have just decided,

Porto Rico ceased, after the ratification of the treaty with Spain, to be a foreign country within the meaning of the tariff act, and became a domestic country—"a territory of the United States"—it is said that if Congress so wills it may be controlled and governed outside of the Constitution and by the exertion of the powers which other nations have been accustomed to exercise with respect to territories acquired by them; in other words, we may solve the question of the power of Congress under the Constitution, by referring to the powers that may be exercised by other nations. I cannot assent to this view. I reject altogether the theory that Congress, in its discretion, can exclude the Constitution from a domestic territory of the United States, acquired, and which could only have been acquired, in virtue of the Constitution. I cannot agree that it is a domestic territory of the United States for the purpose of preventing the application of the tariff act imposing duties upon imports from foreign countries, but not a part of the United States for the purpose of enforcing the constitutional requirement that *all* duties, imposts and excises imposed by Congress "shall be uniform throughout the United States." How Porto Rico can be a domestic territory of the United States, as distinctly held in *De Lima* v. *Bidwell*, and yet, as is now held, not embraced by the words "throughout the United States," is more than I can understand.

We heard much in argument about the "expanding future of our country." It was said that the United States is to become what is called a "world power;" and that if this Government intends to keep abreast of the times and be equal to the great destiny that awaits the American people, it *must* be allowed to exert all the power that other nations are accustomed to exercise. My answer is, that the fathers never intended that the authority and influence of this nation should be exerted otherwise than in accordance with the Constitution. If our Government needs more power than is conferred upon it by the Constitution, that instrument provides the mode in which it may be amended and additional power thereby obtained. The People of the United States who ordained the Constitution never supposed that a change could be made in our system of government by mere judicial interpretation. They never contemplated any such juggling with the words of the Constitution as 'would authorize the courts to hold that the words " throughout the United States," in the taxing clause of the Constitution, do not embrace a domestic " territory of the United States " having a civil government established by the authority of the United States. This is a distinction which I am unable to make, and which I do not think ought to be made when we are endeavoring to ascertain the meaning of a great instrument of government.

There are other matters to which I desire to refer. In one of the opinions just delivered the case of *Neely* v. *Henkel*, 180 U. S. 119, is cited in support of the proposition that the provision of the Foraker act here involved was consistent with the Constitution. If the contrary had not been asserted I should have said that the judgment in that case did not have the slightest bearing on the question before us. The only inquiry there was whether Cuba was a foreign country or territory within the meaning not of the tariff act but of the act of June 6, 1900, 31 Stat. 656, c. 793. We held that it was a foreign country. We could not have held otherwise, because the United States, when recognizing the existence of war between this country and Spain, disclaimed "any disposition or intention to exercise sovereignty, jurisdiction or control over said island except for the pacification thereof," and asserted "its determination, when that is accomplished, to leave the government and control of the island to its people." We said: "While by the act of April 25, 1898, declaring war between this country and Spain, the president was directed and empowered to use our entire land and naval forces, as well as the militia of the several States to such an extent as was necessary, to carry such act into effect, that authorization was not for the purpose of making Cuba an integral part of the United States, but only for the purpose of compelling the relinquishment by Spain of its authority and government in that island and the withdrawal of its forces from Cuba and Cuban waters. The legislative and executive branches of the Government, by the joint resolution of April 20, 1898, expressly disclaimed any purpose to exercise sovereignty, juris-

diction or control over Cuba 'except for the pacification thereof,' and asserted the determination of the United States, that
object being accomplished, to leave the government and control
of Cuba to its own people.   All that has been done in relation
to Cuba has had that end in view, and, so far as this court is
informed by the public history of the relations of this country
with that island, nothing has been done inconsistent with the
declared object of the war with Spain.   Cuba is none the less
foreign territory, within the meaning of the act of Congress,
because it is under a Military Governor appointed by and representing the President in the work of assisting the inhabitants
of that island to establish a government of their own, under
which, as a free and independent people, they may control their
own affairs without interference by other nations.   The occupancy of the island by troops of the United States was the
necessary result of the war.   That result could not have been
avoided by the United States consistently with the principles
of international law or with its obligations to the people of
Cuba.   It is true that as between Spain and the United States
—indeed, as between the United States and all foreign nations
—Cuba, upon the cessation of hostilities with Spain and after
the Treaty of Paris was to be treated as if it were conquered
territory.   But as between the United States and Cuba, that
island is territory held in trust for the inhabitants of Cuba to
whom it rightfully belongs, and to whose exclusive control it
will be surrendered when a stable government shall have been
established by their voluntary action."   In answer to the suggestion that, under the modes of trial there adopted, Neely, if
taken to Cuba, would be denied the rights, privileges and immunities accorded by our Constitution to persons charged with
crime against the United States, we said that the constitutional
provisions referred to "have no relation to crimes committed
without the jurisdiction of the United States against the laws of
a foreign country."   What use can be made of that case in order
to prove that the Constitution is not in force in a territory of
the United States acquired by treaty, except as Congress may
provide, is more than I can perceive.

There is still another view taken of this case.   Conceding

that the National Government is one of enumerated powers to be exerted only for the limited objects defined in the Constitution, and that Congress has no power, except as given by that instrument either expressly or by necessary implication, it is yet said that a new territory, acquired by treaty or conquest, cannot become *incorporated* into the United States without the consent of Congress. What is meant by such incorporation we are not fully informed, nor are we instructed as to the precise mode in which it is to be accomplished. Of course, no territory can become a State in virtue of a treaty or without the consent of the legislative branch of the Government; for only Congress is given power by the Constitution to admit new States. But it is an entirely different question whether a domestic "territory of the United States," having an organized civil government, established by Congress, is not, for all purposes of government by the Nation, under the complete jurisdiction of the United States and therefore a part of, and incorporated into, the United States, subject to all the authority which the National Government may exert over any territory or people. If Porto Rico, although a territory of the United States, may be treated as if it were not a part of the United States, then New Mexico and Arizona may be treated as not parts of the United States, and subject to such legislation as Congress may choose to enact without any reference to the restrictions imposed by the Constitution. The admission that no power can be exercised under and by authority of the United States except in accordance with the Constitution is of no practical value whatever to constitutional liberty if, as soon as the admission is made—as quickly as the words expressing the thought can be uttered—the Constitution is so liberally interpretated as to produce the same results as those which flow from the theory that Congress may go outside of the Constitution in dealing with newly acquired territories, and give them the benefit of that instrument only when and as it shall direct.

Can it for a moment be doubted that the addition of Porto Rico to the territory of the United States in virtue of the treaty with Spain has been recognized by direct action upon the part of Congress? Has it not legislated in recognition of that treaty and appropriated the money which it required this country to pay?

If, by virtue of the ratification of the treaty with Spain, and the appropriation of the amount which that treaty required this country to pay, Porto Rico could not become a part of the United States so as to be embraced by the words "throughout the United States," did it not become "incorporated" into the United States when Congress passed the Foraker act? 31 Stat. 77, c. 191. What did that act do? It provided a civil government for Porto Rico, with legislative, executive and judicial departments; also, for the appointment by the President, by and with the advice and consent of the Senate of the United States, of a "governor, secretary, attorney general, treasurer, auditor, commissioner of the interior and a commissioner of education." §§ 17–25. It provided for an executive council, the members of which should be appointed by the President, by and with the advice and consent of the Senate. § 18. The governor was required to report all transactions of the government in Porto Rico to the President of the United States. § 17. Provision was made for the coins of the United States to take the place of Porto Rican coins. § 11. All laws enacted by the Porto Rican legislative assembly were required to be reported to the Congress of the United States, which reserved the power and authority to amend the same. § 31. But that was not all. Except as otherwise provided, and except also the internal revenue laws, the statutory laws of the United States, not locally inapplicable, are to have the same force and effect in Porto Rico as in the United States. § 14. A judicial department was established in Porto Rico, with a judge to be appointed by the President, by and with the advice and consent of the Senate. § 33. The court, so established, was to be known as the District Court of the United States for Porto Rico, from which writs of error and appeals were to be allowed to this court. § 34. All judicial process, it was provided, "shall run in the name of the United States of America, and the President of the United States." § 16. And yet it is said that Porto Rico was not "incorporated" by the Foraker act into the United States so as to be part of the United States within the meaning of the constitutional requirement that all duties, imposts and excises imposed by Congress shall be uniform "throughout the United States."

It would seem, according to the theories of some, that even if Porto Rico is in and of the United States for many important purposes, it is yet not a part of this country with the privilege of protesting against a rule of taxation which Congress is expressly forbidden by the Constitution from adopting as to any part of the "United States." And this result comes from the failure of Congress to use the word "incorporate" in the Foraker act, although by the same act all power exercised by the civil government in Porto Rico is by authority of the United States, and although this court has been given jurisdiction by writ of error or appeal to reëxamine the final judgments of the District Court of the United States established by Congress for that territory. Suppose Congress had passed this act: "*Be it enacted by the Senate and House of Representatives in Congress assembled*, That Porto Rico be and is hereby incorporated into the United States as a territory," would such a statute have enlarged the scope or effect of the Foraker act? Would such a statute have accomplished more than the Foraker act has done? Indeed, would not such legislation have been regarded as most extraordinary as well as unnecessary?

I am constrained to say that this idea of "incorporation" has some occult meaning which my mind does not apprehend. It is enveloped in some mystery which I am unable to unravel.

In my opinion Porto Rico became, at least after the ratification of the treaty with Spain, a part of and subject to the jurisdiction of the United States in respect of all its territory and people, and Congress could not thereafter impose any duty, impost or excise with respect to that island and its inhabitants, which departed from the rule of uniformity established by the Constitution.

HUUS *v.* NEW YORK AND PORTO RICO STEAMSHIP COMPANY.

CERTIFICATE FROM THE CIRCUIT COURT OF APPEALS FOR THE SECOND CIRCUIT.

No. 514. Argued January 11, 14, 1901.—Decided May 27, 1901.

Vessels engaged in trade between Porto Rican ports and ports of the United States are engaged in the coasting trade in the sense in which those words are used in the New York pilotage statutes; and steam vessels engaged in such trade are coastwise steam vessels under Revised Statutes, section 4444.

THIS was a libel filed in the District Court for the Southern District of New York to recover spoken pilotage upon the American built steamship Ponce, belonging to the defendant, a New York corporation.

The facts were that libellant, on June 25, 1900, offered his service as a Sandy Hook pilot to the master of the Ponce, then about entering the harbor of New York, her port of distination, from the port of San Juan, in the Island of Porto Rico. Libellant, who was a duly licensed Sandy Hook pilot, was the first and only one to offer his services. These services were declined by the master of the vessel, who was himself a licensed pilot for the harbor of New York under the laws of the United States. The steamship was at the time duly enrolled and licensed for the coasting trade under the laws of the United States, and was engaged in trade between Porto Rico and New York. The libel was dismissed by the District Court, 105 Fed. Rep. 74, an appeal taken to the Circuit Court of Appeals, which certified to this court the following questions of law, concerning which it desired instructions :

"1. Since the proclamation of the treaty of peace between the United States and the Kingdom of Spain, and the passage of the act of Congress entitled ' An act temporarily to provide revenues and civil government for Porto Rico, and for other purposes,' (approved April 12, 1900,) do Porto Rican ports remain foreign ports in the sense in which those words are used in the statutes of the State of New York regulating pilotage?

"2. Are vessels engaged in trade between Porto Rican ports and ports of the United States engaged in the coasting trade in the sense in which those words are used in the statutes of the State of New York regulating pilotage?

"3. Are steam vessels engaged in trade between Porto Rican ports and ports of the United States coastwise steam vessels in the sense in which those words are used in section 4444 of the Revised Statutes of the United States?"

Mr. William Lindsay for appellant.

W. F. Kingsbury Curtis for appellee.   *Mr. William Edmond Curtis* was on his brief.

MR. JUSTICE BROWN, after stating the case, delivered the opinion of the court.

Conceding it to be within the power of Congress to assume control of and regulate the whole system of pilotage, as applied to vessels engaged in foreign or interstate commerce, it has for obvious reasons left to the several States the power to legislate upon this subject, and to prescribe rules for the licensing and government of pilots, the collection of their fees, and such other incidental matters as the nature of their services in the particular localities may require. The power to do this was recognized by this court in *Cooley* v. *Board of Wardens,* 12 How. 299, though it was subsequently said to be subject to such restrictions as Congress might see fit to impose. *Spraigue* v. *Thompson,* 118 U. S 90.

By Rev Stat. sec. 4235, it is expressly enacted that "until further provision is made by Congress, all pilots in the bays, inlets, rivers, harbors, and ports of the United States shall continue to be regulated in conformity with the existing laws of the States respectively wherein such pilots may be," subject, however, to a prohibition (sec. 4237) against "any discrimination in the rate of pilotage or half-pilotage between vessels sailing between the ports of one State and vessels sailing between the ports of different States, or any discrimination against vessels propelled in whole or in part by steam;" and to a further restriction (sec. 4401) that "all coastwise seagoing vessels . . . shall be subject to the navigation laws of the United States, . . . and that every coastwise seagoing steam vessel subject to the navigation laws of the United States, and to the rules and regulations aforesaid, not sailing under register, shall, when under way, except on the high seas, be under the control and direction of pilots licensed by the inspectors of steamboats." To further effectuate its control over coastwise seagoing vessels, it is provided by sec. 4444 that "no State or municipal government shall impose upon pilots of steam vessels any obligation to procure a state or other license in addition to that issued by the United States. . . . Nor shall any pilot charges be levied by any such authority upon any steamer piloted as provided by this title," . . . although "nothing in this title shall be construed to annul or affect any regulation established by the laws of any State requiring vessels entering or leaving a port in any such State, *other than coastwise steam vessels*, to take a pilot duly licensed or authorized by the laws of such State, or of a State situated upon the waters of such State."

The general object of these provisions seems to be to license pilots upon steam vessels engaged in the *coastwise* or interior commerce of the country, and at the same time, to leave to the States the regulation of pilots upon all vessels engaged in *foreign* commerce.

This view was evidently accepted by the legislature of New York, which, in section 2119 of the Consolidated Act of 1882, declares that "no master of any vessel navigated under a coasting license and employed in the coasting trade by way of Sandy Hook, shall be required to employ a licensed pilot when entering or departing from the harbor of New York;" but reserving its own control of vessels engaged in the foreign trade by enacting further in the same section that "all masters of foreign vessels and vessels from a *foreign port*, and all vessels sailing under register bound to or from the port of New York or by the way of Sandy Hook, shall take a licensed pilot, or, in case of refusing to take such pilot, shall himself, owners or consignees, pay the said pilotage as if one had been employed, and such pilotage shall be paid to the pilot first speaking or offering his services as pilot to such vessel," with a final proviso that "this section shall not apply to vessels propelled wholly or in part by steam, owned or belonging to citizens of the United States, and licensed and engaged in the coasting trade."

As the statement of facts connected with the question certified shows that the Ponce was an American built steamship, sailing from New York, belonging to a New York corporation, enrolled and licensed for the coasting trade, navigated by a master duly licensed to act as pilot in the bay and harbor of New York, under the laws of the United States, and was engaged in trade between the Island of Porto Rico and the port of New York, the only question remaining to be considered is whether she was a *coastwise seagoing steam vessel* under Rev. Stat. sec. 4401, and actually employed in the coasting trade by way of Sandy Hook under sec. 2111 of the New York Consolidation Act.

Under the commercial and navigation laws of the United States merchant vessels are divisible into two classes: First, vessels registered pursuant to Rev. Stat. sec. 4131. These must be wholly owned, commanded and officered by citizens of the United States, and are alone entitled to engage in foreign trade; and, second, vessels enrolled and licensed for the coasting trade or fisheries. Rev. Stat. sec. 4311. These may not engage in foreign trade under penalty of forfeiture. Section 4337. This class of vessels is also engaged in navigation upon the Great Lakes and the interior waters of the country— in other words, they are engaged in domestic instead of foreign trade.

The words "coasting trade," as distinguishing this class of vessels, seem to have been selected because at that time all the domestic commerce of the country was either interior com-

merce, or coastwise, between ports upon the Atlantic or Pacific
coasts, or upon islands so near thereto, and belonging to the several States, as properly to constitute a part of the coast. Strictly
speaking Porto Rico is not such an island, as it is not only situated some hundreds of miles from the nearest port on the Atlantic
coast, but had never belonged to the United States, or any of
the States composing the Union. At the same time trade with
that island is properly a part of the domestic trade of the country since the treaty of annexation, and is so recognized by the
Porto Rican or Foraker act. By section 9 the Commissioner
of Navigation is required to "make such regulations . . .
as he may deem expedient for the nationalization of all vessels
owned by the inhabitants of Porto Rico on April 11, 1899,
. . . and for the admission of the same to all the benefits
of the coasting trade of the United States; and the coasting
trade between Porto Rico and the United States shall be regulated in accordance with the provisions of law applicable to
such trade between any two great coasting districts of the United States." By this act it was evidently intended, not only
to nationalize all Porto Rican vessels as vessels of the United
States, and to admit them to the benefits of their coasting
trade, but to place Porto Rico substantially upon the coast of
the United States, and vessels engaged in trade between that
island and the continent, as engaged in the coasting trade.
This was the view taken by the executive officers of the government in issuing an enrollment and license to the Ponce, to
be employed in carrying on the coasting trade, instead of treating her as a vessel engaged in foreign trade.

That the words "coasting trade" are not intended to be
strictly limited to trade between ports in adjoining districts is
also evident from Rev. Stat. sec. 4358, wherein it is enacted that
"the coasting trade between the territory ceded to the United
States by the Emperor of Russia, and any other portion of the
United States, shall be regulated in accordance with the provisions of law applicable to such trade between any two great districts." These great districts were, for the more convenient
regulation of the coasting trade, divided by the act of March 2,
1819, 3 Stat. 492, c. 48, as amended by the act of May 7, 1822,

3 Stat. 684; Rev. Stat. sec. 4348, as follows: "The first to include all the collection districts on the seacoast and navigable rivers between the eastern limits of the United States and the southern limits of Georgia; the second to include all the collection districts on the seacoast and navigable rivers between the river Perdido and the Rio Grande; and the third to include all the collection districts on the seacoast and navigable rivers between the southern limits of Georgia and the river Perdido." A provision similar to that for the admission of the Territory of Alaska was also adopted in the act to provide a government for the Territory of Hawaii, (31 Stat. 141, sec. 98,) which provides that all vessels carrying Hawaiian registers on August 12, 1888, and owned by citizens of the United States or citizens of Hawaii, "shall be entitled to be registered as American vessels, . . . and the coasting trade between the islands aforesaid and any other portion of the United States shall be regulated in accordance with the provisions of law applicable to such trade between any two great coasting districts."

This use of the words " coasting trade " indicates very clearly that the words were intended to include the domestic trade of the United States upon other than interior waters. The District Court was correct in holding that the Ponce was engaged in the coasting trade, and that the New York pilotage laws did not apply to her.

The second and third questions are therefore answered in the affirmative. An answer to the first question becomes unnecessary.

CARSON *v.* BROCKTON SEWERAGE COMMISSION.

ERROR TO THE SUPREME JUDICIAL COURT OF MASSACHUSETTS.

No. 249. Argued April 18, 1901.—Decided May 27, 1901.

Whether the construction of a public sewer by assessments upon adjoining property entitles the owners of such property to the free use of such sewer, or only to the right to a free entrance to it of their particular sewers, is a question of local policy.

Notwithstanding that such sewer was built by assessments upon the property benefited, it is competent for the legislature to require persons making use of it to pay a reasonable sum for such use.

Where an ordinance fixes the charges that shall be paid for the use of a common sewer, no notice is required to be given to the property owners of an assessment for that purpose.

THIS was a petition to the justices of the Supreme Judicial Court for the county of Suffolk, for a writ of certiorari to the Board of Sewer Commissioners of the city of Brockton, directing them to bring up certain proceedings connected with the assessment of taxes upon petitioner's land to the amount of $42.53, for the maintenance and operation of a public sewer, and for an order quashing the proceedings.

The petitioner alleged the assessment to be illegal and void:

1. Because the city ordinance does not provide for notice to or hearing of persons whose estates are affected thereby, in violation of the state constitution;

2. Because the method of computing the sewer charges is unreasonable and disproportionate;

3. Because petitioner, having already paid for the sewers connected with his land, cannot be compelled to pay a special tax for the maintenance and operation of sewers from which he receives no special benefit;

4. Because such tax or sewer rental is in violation of the Fourteenth Amendment to the Federal Constitution;

5. Because such tax is permissible only when founded upon peculiar and special benefits to the property so taxed, and then only to the amount of such benefits;

6. Because lands assessed for the construction of sewers cannot be said to receive an additional and special and peculiar benefit from the general oversight and operation of the same.

By an act of the legislature of Massachusetts, passed May 6, 1892, c. 245, "to give greater power to cities and towns in relation to the construction of sewers," it was enacted as follows:

"SEC. 1. The city council of any city except Boston, or a town, in which common sewers are laid under the provisions of sections one, two and three of chapter fifty of the Public Statutes, or a system of sewerage is adopted under the provisions of section seven of said chapter, may by vote establish just and equitable annual charges or rents for the use of such sewers, to be paid by every person who enters his particular sewer into the common sewer, and may change the same from time to time. Such charges shall constitute a lien upon the real estate using such common sewer, to be collected in the same manner as taxes upon real estate, or in an action of contract in the name of such city or town. Sums of money so received may be applied to the payment of the cost of maintenance and repairs of such sewers or of any debt contracted for sewer purposes."

Pursuant to this authority the city council of Brockton, on August 23, 1894, adopted an ordinance, of which the following is the material provision:

"SEC. 4. Every person or owner of an estate who enters his particular sewer into a common sewer shall pay for the use of such sewer an annual rental determined upon the basis of water service, as follows: For unmetered water service, eight dollars; for metered water service, thirty cents per 1000 gallons of sewerage delivered to the sewer, the quantity so delivered to be determined by the meter readings taken by the water commissioners, but the annual charge shall in no case be less than eight dollars, it being provided, however, that in cases where said commissioners shall deem the same to be equitable, a discount may be made, such discount to be determined by said commissioners and approved by the mayor and aldermen; and it being further provided that any such person or owner may place at his own expense a water meter, which shall be approved by the said commissioners, to measure the amount of water which does not enter the sewer.

"Such charges shall be collected quarterly and shall constitute a lien upon the real estate using the sewer, to be collected in the same manner as taxes upon real estate or in an action of contract in the name of the city of Brockton."

The petition was denied, and the petitioner sued out this writ of error.

Mr. William H. Carson, in person for plaintiff in error.

No appearance for defendants in error.

MR. JUSTICE BROWN, after stating the case, delivered the opinion of the court.

This case involves the single question whether a municipal ordinance, making an annual assessment upon property owners for the use of a common sewer, infringes upon any provision of the Constitution of the United States.

The Supreme Judicial Court of Massachusetts held that the petitioner received a special benefit in the use of the sewer for which he might be charged; that the city, by building the sewer and receiving a part of its cost from the petitioner, did not bind itself that the sewer should be maintained forever, or that the petitioner should be at liberty to use it free of further expense; that the charge for using it was a benefit distinct from that originally conferred by building it; that there was no charge unless the sewer were used; that the only questions were whether petitioner's sewer entered the common sewer, and what amount of sewage was delivered to it; and that, if the petitioner wished to be heard on either of these facts, he could resort to the courts; that the city counsel had a right to fix the charges without notice to the parties interested, unless, under the pretence of fixing an equitable rate, the ordinance should do what amounted to the taking or destruction of property.

The ordinance imposes an annual rental of eight dollars for

unmetered water service, and for metered water service thirty cents per thousand gallons of sewage delivered to the sewer —the quantity to be so delivered to be determined by the meter readings—with the privilege to the commissioners of making a discount when equitable. As the Supreme Judicial Court held that the municipality had power to adopt this ordinance under the public statutes of the Commonwealth, and that such statutes were no violation of the state constitution, we are concerned only with the question whether the petitioner was thereby deprived of his property without due process of law, or denied the equal protection of the laws within the Fourteenth Amendment.

The validity of the legislative act is assailed upon the ground that no notice was required to be given to the property owner, nor provision made for a hearing, and that the authority given to the city council of Brockton to change the rate of sewerage charges and assessments from time to time manifested an intention on the part of the legislature to assess such property without regard to benefits. There is no doubt that, when land is proposed to be taken and devoted to the public service, or any serious burden is laid upon it, the owner of the land must be given an opportunity to be heard with respect to the necessity of the taking, and the compensation to be paid by the city. *Davidson* v. *New Orleans*, 96 U. S. 97; *Palmer* v. *McMahon*, 133 U. S. 660; *Stuart* v. *Palmer*, 74 N. Y. 183, subsequently reëxamined in this court in *Spencer* v. *Merchant*, 125 U. S. 345.

Obviously these cases have no application to an ordinance which fixes beforehand the price to be paid for certain privileges, and leaves it optional with the taxpayer to avail himself of such privileges or not. As well might it be insisted that an ordinance which fixes water rates, proportioned to the amount furnished, is void, because no notice is required to be given before such rate is fixed, or the taxpayer is assessed his proportionate charge under the ordinance. Where the use of such privilege is left optional with the taxpayer by his election to avail himself of it or not, he contracts with the city to pay the rental fixed by its ordinance, if he elect to use it. In such case there is no room for the question of notice. Where notice will avail nothing, no notice is required. *Reclamation District* v. *Phillips*, 108 California, 306; *Amery* v. *Keokuk*, 72 Iowa, 701; *Commonwealth* v. *Lehigh Valley Railroad Co.*, 129 Penn. St. 429.

Thus in *Hagar* v. *Reclamation District*, 111 U. S. 701, it was said by Mr. Justice Field (p. 708): "Undoubtedly where life and liberty are involved, due process requires that there be a regular course of judicial proceedings, which imply that the party to be affected shall have notice and an opportunity to be heard. So, also, where title or possession is involved. But where the taking of property is in the enforcement of a tax, the proceeding is necessarily less formal, and whether notice to him is at all necessary may depend upon the character of the tax, and the manner in which the amount is determinable. . . . Of the different kind of taxes which the State may impose, there is a vast number of which, from their nature, no notice can be given to the taxpayer, nor would notice be of any possible advantage to him. Such as poll taxes, license taxes, (not dependent upon the extent of his business,) and, generally, specific taxes on things or persons or occupations. In such cases, the legislature, in authorizing the tax, fixes its amount, and that is the end of the matter." See also *Parsons* v. *District of Columbia*, 170 U. S. 45. Under the circumstances of this case no notice was necessary.

Similar considerations apply to the defence that petitioner has been, or is about to be, deprived of his property without due process of law. But of what property has he been deprived? None whatever. There has not been, nor is there anything to indicate there ever will be, any taking of his property within the meaning of the law. Assuming that the imposition of a burden which manifestly belongs to the public, upon private property, constitutes a deprivation of such property within the meaning of the Fourteenth Amendment, there is nothing of the kind involved in this case. There is not even compulsory taxation of the property. The act of the legislature (chap. 245, act of 1892) merely provides that the city council "may by vote establish just and equitable annual charges or rents for the use of such sewers, to be paid by every person who enters his particular sewer into the common sewer, and may change the same from time to time." The municipal ordinance fixes the annual rentals, determinable upon a certain basis of water service, with a provision that the commissioners may make an equitable discount from such rates at their discretion. This was all there was to it. The lot owner could use the sewer or not, as he chose. If he used it, he paid the rental fixed by the ordinance. If he made no use of it, he paid nothing. There is no element of deprivation here or even of taxation, but one of contract, into which the lot owner might or might not enter. There is no allegation in the petition that the petitioner was required by the board of health to discharge into the public sewer. There is no allegation that the particular charges fixed by the commissioners are unreasonable, only that the *method* is unreasonable, that is, that *any* charge is unreasonable.

The stress of petitioner's argument appears to be laid upon the proposition that his property having been once assessed for the construction of the common sewer, he has a right to the free use of such sewer forever afterwards, and that the expense of its maintenance must be raised by general taxation and not by special assessment. This, however, is a question of state policy. It was for the legislature to say whether the construction of the sewer entitled the adjoining property owners to the free use of it, or only to the right to a free entrance to it of their particular sewers. As held by the Supreme Judicial Court, there can be no doubt that the adjoining property owners did receive a special benefit in being permitted to discharge their private sewers into it. The amount of such benefit was, under the statutes of the Commonwealth, determinable by the city council, which fixed upon a certain rate for unmetered service, and a certain other rate per thousand gallons of sewage discharged for metered service. We have held in the recent case of *Parsons* v. *District of Columbia*, 170 U. S. 45, that it was competent for the legislative power to assess the amount of benefit specially received by abutting property, and so long as such amount is not grossly excessive, or out of all proportion to the benefit received, there is no reason to complain, particularly if, as held by the Supreme Judicial Court in this case, the question of connecting with the public sewer be left optional with the property owner.

The case is somewhat analogous to that of *Sands* v. *Manistee River Improvement Co.*, 123 U. S. 288, wherein we held that the exaction of tolls, under a state statute, for the use of an improved national waterway, is not within the prohibition of the due process of the law clause of the Constitution. Said Mr. Justice Field (p. 293): "The tolls exacted from the defendant are merely compensation for benefits conferred, by which the floating of his logs down the stream was facilitated. . . . Tolls are the compensation for the use of another's property or of improvements made by him, and their amount is determined by the cost of the property or of the improvements, and consideration of the returns which such values or expenditures should yield. The legislature, acting upon information received, may prescribe, at once, the tolls to be charged, but ordinarily it leaves their amount to be fixed by officers or boards appointed for that purpose."

It is true that in *Sears* v. *Street Commissioners of Boston*, 173 Mass. 350, decided in May, 1899, construing a similar statute applicable to the city of Boston, the Supreme Judicial Court made a decision which it is difficult to reconcile with its opinion in the case under consideration, and held that "where lands have paid assessments for special benefits from the construction of all sewers by whose operation they are affected, it cannot be said that they receive an additional special and peculiar benefit from the general oversight and operation of the sewers of Boston, such as to subject them to a second special assessment. Expenses of this kind should be made the subject of general taxation," citing a number of cases in support of this proposition, none of which appear to be in point. *Hammett* v. *Philadelphia*, 65 Penn. St. 146, was a case of widening and repaving a public street; *Washington Avenue*, 69 Penn. St. 352, one of compelling the owners of farm lands lying within one mile on each side of a public highway to pay for grading, macadamizing and improving it, by an assessment upon their lands by the acre; *Appeal of Williamsport*, 41 Atlantic Rep. 476, one of reconstructing a sewer originally built by the city; *Erie* v. *Russell*, 148 Penn. St. 384, a similar case, except that the sewer was originally built by local assessments; *Dietz* v. *City of Neenah*, 91 Wisconsin, 422, a question of want of notice; *Dyar* v. *Farmington*, 70 Maine, 515, one of assessment for building a railroad; *Hanscom* v. *Omaha*, 11 Nebraska, 37, one of the extent to which property was benefited by constructing a sewer. It needs no argument to show that these cases had no pertinence. The question of notice or want of notice was also considered in the *Sears* case, but the court did not decide that question, intimating, however, an opinion somewhat adverse to the validity of the statute upon this ground.

We are not required, however, to reconcile these cases. It is sufficient that the Supreme Judicial Court held that this case was "free from the elements which in *Sears* v. *Street Commissioners* led to the conclusion that the petitioner was assessed without regard to the benefits received by him." Notwithstanding the former case, we think the court was correct in holding in this case that the petitioner and other property owners whose lots abutted on this public sewer did receive a benefit not common to the inhabitants of the city generally, in being permitted to discharge into it the contents of their private sewers, that the amount of such benefit was determinable by the city council, and that in its action there was nothing violative of the Federal Constitution. It was properly said by Chief Justice Holmes in this connection: "No one denies that it was a special benefit to the petitioner to have the sewer built in front of his land. That benefit was the probability that the sewer would be available for use in the future; but the city by building it and receiving a part of the cost from the petitioner did not impliedly bind itself or the general taxes that the sewer should be maintained forever, and that the petitioner should be at liberty to use it free of further expense. If building the sewer was a special benefit, keeping the sewer in condition for use by such further expenditure as was necessary was a further special benefit to such as used it."

The judgment of the Supreme Judicial Court is therefore

Affirmed.

HOMER RAMSDELL TRANSPORTATION COMPANY *v.* LA COMPAGNIE GÉNÉRALE TRANSATLANTIQUE.

CERTIFICATE FROM THE CIRCUIT COURT OF APPEALS FOR THE SECOND CIRCUIT.

No. 168. Argued March 6, 1901.—Decided May 27, 1901.

The statutes of New York impose compulsory pilotage on foreign vessels inward and outward bound to and from the port of New York by way of Sandy Hook.

In an action at common law the shipowner is not liable for injuries inflicted exclusively by negligence of a pilot accepted by a vessel compulsorily.

THIS was an action at law, brought by the Homer Ramsdell Transportation Company, a corporation of New York, against the Compagnie Générale Transatlantique, a corporation of the Republic of France, to recover damages caused by the defendant's steamship, the Bretagne, striking and injuring the plaintiff's pier in New York harbor.

The answer alleged, among other things, "that at the time of the said collision the said steamship La Bretagne was in the command, and her movements and navigation entirely under the orders and direction, of a pilot, duly licensed under, and compulsorily imposed upon the defendant by the authority of the State of New York; and that the regular officers and crew of the said steamship in the service of the defendant had no part in the navigation of the said steamer except to carry out or execute the orders of the said pilot, which they did promptly and efficiently in every particular."

The case was referred by the Circuit Court of the United States for the Southern District of New York to Hon. William G. Choate, who reported in favor of the defendant, and filed an opinion published in 63 Fed. Rep. 848. That court gave judgment on his report for the defendant; and the plaintiff appealed to the Circuit Court of Appeals for the Second Circuit, which certified to this court, together with the pleadings, the judgment of the Circuit Court, and the report and opinion of the referee, the following statement of facts and questions of law:

"The defendant in error is a foreign corporation, owning and plying a regular line of steamers between Havre and New York. On the morning of December 10, 1892, one of the defendant's steamers, La Bretagne, while outward bound from the port of New York to Havre by way of Sandy Hook, with cargo and passengers, struck the plaintiff's pier, damaging it to the amount of upwards of thirteen thousand dollars. The said vessel, at the time she left her pier, was in all respects seaworthy and properly manned, equipped and supplied, and her owner exercised due diligence to make her so. She had on board a Sandy Hook pilot, duly licensed under and pursuant to the laws of the State of New York, and was navigated under his direction up to the time of said collision, and all his orders were promptly and efficiently obeyed and carried out by the master, officers and crew of said steamship. The said collision and the damage resulting therefrom were caused solely by the negligence and want of skill and care on the part of the said pilot, and not by any want of skill or negligence on the part of the master, other officers, or crew of the said steamship.

"Certain questions of law arise in the cause concerning which the court desires the instructions of the Supreme Court for its proper decision, and which are as follows:

"First. Whether the provisions of chapter 467 of the laws of New York passed June 28, 1853, as amended by chapter 196 of the laws of said State passed April 11, 1854; chapter 243 of the laws of the said State passed April 3, 1857; chapter 930 of the laws of the said State passed May 16, 1867, and chapter 548 of the laws of said State passed May 2, 1870, and consolidated into sections 2093 to 2123, inclusive, of chapter 410 of the laws of said State passed July 1, 1882, impose compulsory pilotage on foreign vessels inward and outward bound to and from the port of New York by way of Sandy Hook, in view of the decisions of the New York Court of Appeals.

"Second. Whether in an action at common law the shipowner is liable for injuries inflicted exclusively by negligence of a pilot accepted by a vessel compulsorily."

Mr. William H. Harris for plaintiff in error.

Mr. Edward K. Jones for defendant in error.

MR. JUSTICE GRAY, after stating the case, delivered the opinion of the court.

The question whether the statutes of the State of New York impose compulsory pilotage on foreign vessels inward and outward bound to and from the port of New York by way of Sandy Hook depends, as both counsel admit, upon the true construction of the provisions which are copied in the margin.[1]

---

[1] The statute of 1854, c. 196, § 2, reënacted in the statute of 1882, c. 410, § 2100, provides that the commissioners of pilots "shall have the power to regulate the stationing of pilot boats for the purpose of receiving pilots from outward bound vessels, may alter or amend any existing regulations of pilots, and make and duly promulgate and enforce new rules or regulations, not inconsistent with the laws of this State, or of the United States, which shall be binding and effectual upon all pilots licensed by them, and upon all parties employing such pilots. They may declare and enforce forfeitures of pilotage upon any mismanagement or neglect of duty by the pilots licensed by them; they may declare and impose and collect fines and penalties not exceeding two hundred and fifty dollars for each offence, to prevent any of the pilots licensed by them from combining injuriously with each other, or with other persons, and to prevent any person licensed by them from acting as a pilot during his suspension, or after his license may be revoked; and the said commissioners may establish and enforce all other needful rules and regulations for the conduct and government of the pilots licensed by them, and the parties employing them; and they may enforce and receive accounts of all moneys collected for pilotage by the pilots licensed by them, and may impose and collect from such pilots a sum not exceeding three per cent on the amount thereof, to defray their necessary expenses, including clerk hire and office rent."

By the statute of 1867, c. 930, also reënacted in the statute of 1882, c. 410, § 2100, "Any pilot bringing in a vessel from sea shall, by himself or one of his boat's company, be entitled to pilot her to sea when she next leaves the port, unless in the mean time a complaint for misconduct or incapacity shall have been made against such pilot, or one of his boat's company, and proved before the board of commissioners of pilots; provided, however, that if the owner of any vessel shall desire to change such pilot, then the said commissioners may assign any other pilot in the same pilot boat to pilot said vessel to sea."

The statute of 1857, c. 243, reënacted in the statute of 1882, c. 410, § 2119, after providing how the master of a vessel sailing under a coasting license to or from the port of New York by the way of Sandy Hook, "desirous of piloting his own vessel," may obtain a license for such purpose from the commissioners

---

The statute of 1857, c. 243, reënacted in the statute of 1882, c. 410, § 2119, provides as follows: "If the master of any vessel above one hundred and fifty and not exceeding three hundred tons burthen, and owned by a citizen of the United States, and sailing under a coasting license to or from the port of New York by the way of Sandy Hook, shall be desirous of piloting his own vessel, he shall first obtain a license for such purpose from the commissioners of pilots, who are hereby authorized and required to grant the same, if such master shall after an examination had by said commissioners be deemed competent; which said license shall be and continue in force one year from the date thereof, or until the determination of any voyage during which the license may expire. For such license, the master to whom it shall be granted shall pay to the said commissioners four cents per ton. All masters of foreign vessels and vessels from a foreign port, and all vessels sailing under register, bound to or from the port of New York by the way of Sandy Hook, shall take a licensed pilot, or, in case of refusal to take such pilot, shall himself, owners or consignees, pay the said pilotage as if one had been employed, and such pilotage shall be paid to the pilot first speaking or offering his services as pilot to such vessel. Any person not holding a license as pilot under this act, or under the laws of the State of New Jersey, who shall pilot, or offer to pilot, any ship or vessel to or from the port of New York by the way of Sandy Hook, except such as are exempt by virtue of this act, or any master or person on board a steam-tug or towboat, who shall tow such vessel or vessels, shall be deemed guilty of a misdemeanor, and, on conviction, shall be punished by a fine not exceeding one hundred dollars or imprisonment not exceeding sixty days; and all persons employing a person to act as pilot, not holding a license under this act, or under the laws of the State of New Jersey, shall forfeit and pay to the board of commissioners of pilots the sum of one hundred dollars."

By the statute of 1854, c. 196, § 5, reënacted in the statute of 1882, c. 410, § 2120, "Any person not holding a license as pilot under this act, or under the laws of the State of New Jersey, who shall pilot or offer to pilot any ship or vessel to or from the port of New York by the way of Sandy Hook, shall be deemed guilty of a misdemeanor, and, on conviction, shall be punished by a fine not exceeding one hundred dollars or imprisonment not exceeding sixty days; and all persons employing a person to act as pilot not holding a license under this act, or under the laws of the State of New Jersey, shall forfeit and pay to the board of commissioners of pilots the sum of one hundred dollars."

of pilots, provides that every master of a foreign vessel bound
to or from the port of New York by the way of Sandy Hook
"shall take a licensed pilot, or, in case of refusal to take such
pilot, shall himself, owners or consignees, pay the said pilotage
as if one had been employed, and such pilotage shall be paid to
the pilot first speaking or offering his services as pilot to such
vessel." It then goes on to provide that "any person not holding a license as pilot under this act," or under the laws of New
Jersey, who shall pilot any vessel to or from the port of New
York by the way of Sandy Hook, shall be punished by fine or
imprisonment, and that " all persons employing a person to act
as pilot, and not holding a license under this act," or under the
laws of New Jersey, shall pay a fine.

By these provisions, not only is the master of a foreign vessel
required to take a licensed pilot, or, in case of refusal to take
such pilot, required to pay pilotage to the pilot first offering his
services; but the subsequent provision as to any " person not
holding a license under this act," construed in connection with
the previous provision as to licensing the master of a coasting
vessel as its pilot, evidently includes the master of a foreign
vessel, and subjects him to fine or imprisonment if he pilots his
own vessel.

The requirement to take a licensed pilot or pay pilotage, together with the penalty imposed on a master who pilots his
own foreign vessel, clearly impose compulsory pilotage. And
it was held by this court in *The China*, (1868) 7 Wall. 53, that
the statute of 1857 imposed such pilotage.

The statute of 1867, c. 930, reënacted in the statute of 1882,
c. 410, § 2100, enacts that a pilot bringing in a vessel from sea,
may by himself or one of his boat's company, pilot her to sea
when she next leaves the port; provided that if the owner shall
desire to change the pilot, the commissioners of pilots may assign another one of the same pilot boat. But the right of the
owner to object to one pilot does not make the selection of another by the commissioners a voluntary act of his.

The cases in the New York Court of Appeals, cited by the
plaintiff, do not affect this question. In *Brown* v. *Ellworth*,
(1875) 60 N. Y. 249, the only point decided was that a pilot

licensed by the law of New Jersey could not recover pilotage under the statute of New York. And in *Gillespie* v. *Zittlosen*, (1875) 60 N. Y. 449, the only point decided was that the pilot first offering his services could not recover pilotage if the master took another licensed pilot.

The answer to the first question certified must therefore be that the statutes of New York do impose compulsory pilotage on foreign vessels inward and outward bound to and from the port of New York by the way of Sandy Hook.

This action is at common law. It is not, and, being for damages inflicted on land, could not be, in admiralty. *The Plymouth*, (1865) 3 Wall. 20.

At common law, no action can be maintained against the owner of a vessel for the fault of a compulsory pilot.

In *Carruthers* v. *Sydebotham*, (1815) 4 M. & S. 77, 85, Lord Ellenborough, in holding that the act of the pilot was not the act of the master or mariners or owner of the ship, said: "Now to make the pilot the representative of the master, and consequently to exempt the underwriter from liability for his acts, it must first be shown that there is a privity between the pilot and the master, so that the one may be considered as the representative or agent of the other. But does the master appoint the pilot? Certainly not. The regulations of the general pilot act impose a penalty upon the master of every ship which shall be piloted by any other person then a pilot duly licensed, within any limits for which pilots are lawfully appointed. And there is an exception of such places for which pilots are not appointed. But if the master cannot navigate without a pilot except under a penalty, is he not under the compulsion of law to take a pilot? And if so, is it just that he should be answerable for the misconduct of a person whose appointment the provisions of the law have taken out of his hands, placing the ship in the hands and under the conduct of the pilot? The consequence is, that there is no privity between them."

In *Attorney General* v. *Case*, (1816) 3 Price, 302, 322, in the Court of Exchequer, the master of the vessel whose owners were held liable, as the court said, "was not compellable, at that time, in any way, either under the penalty of double the wages, or of paying even the single wages, to have any pilot on board. It was his own act to have him; and it can be only in the case of such an officer having been forced upon them, and without his own election, that the responsibility of the owner can possibly be discharged."

In *The Maria*, (1839) 1 W. Rob. 95, 106, Dr. Lushington, on a full review of those cases, held that upon general principles, and independently of the express provisions in the English statutes, the compulsory taking of a pilot relieved the owner from all responsibility for his acts.

In *Lucey* v. *Ingram*, (1840) 6 M. & W. 302, 315, Baron Parke, delivering the judgment of the Court of Exchequer, spoke of the exemption of the master who was compelled to take a pilot, from liability by the common law independent of statute, as follows: "It may, indeed, be admitted, that in many of the cases, the judges, in giving their judgments, refer to the obligation of the master to take a pilot, as the ground on which his irresponsibility is founded; and no doubt that is the foundation, and probably the only foundation, on which it can rest independently of the statutes; but the language of the exempting clause in the last pilot act certainly carries the doctrine further, and it may well be conceived that this extension of the common-law doctrine was not accidental, but intentional. The object of the legislature, in establishing pilots, has been to secure, as far as possible, protection to life and property, by supplying a class of men better qualified than ordinary mariners to take charge of ships in places where, from local causes, navigation is attended with more than common difficulty. To effect this object, it has in general been made the duty of the master of every ship, on arriving at any of the places in question, to take a pilot on board, and to give up to him the navigation of the vessel. The master, however well qualified to conduct the ship himself, is bound under a penalty in a great measure to divest himself of its control, and to give up the charge to the pilot. As a necessary consequence, the master and owners are exempted from responsibility for acts resulting from the mismanagement of the pilot." He then proceeded to consider the extension of the exemption by statute, which has no bearing on this case.

In *The Halley*, (1868) L. R. 2 P. C. 193, 201, the Judicial Committee of the Privy Council agreed with Sir Robert Phillimore in the same case in the Court of Admiralty, L. R. 2 Ad. & Ec. 3, " in his statement of the common law of England, with respect to the liability of the owner of a vessel for injuries occasioned by the unskillful navigation of his vessel, while under the control of a pilot, whom the owner was compelled to take on board, and in whose selection he had no voice; and that this law holds that the responsibility of the owner for the acts of his servant is founded upon the presumption that the owner chooses his servant and gives him orders which he is bound to obey, and that the acts of the servant, so far as the interests of third persons are concerned, must always be considered as the acts of the owner."

There is no occasion to refer further to the English cases in admiralty, because in England it is held that the ship is not responsible in admiralty, where the owner would not be at common law, differing in this respect from our own decisions. *The China*, 7 Wall. 53; *Ralli v. Troop*, (1894) 157 U. S. 386, 402, 420; *The John G. Stevens*, (1898) 170 U. S. 113, 120–122; *The Barnstable*, (1901) 181 U. S. 464.

In *The China*, affirming the decision of the Circuit Court in admiralty, the liability of a vessel *in rem* for a collision from the fault of a compulsory pilot was put upon the maritime law, the court saying: " The maritime law as to the position and powers of the master, and the responsibility of the vessel, is not derived from the civil law of master and servant, nor from the common law." " According to the admiralty law, the collision impresses upon the wrongdoing vessel a maritime lien. This the vessel carries with it into whosesoever hands it may come. It is inchoate at the moment of the wrong, and must be perfected by subsequent proceedings." " The proposition of the appellants would blot out this important feature of the maritime code, and greatly impair the efficacy of the system. The appellees are seeking the fruit of their lien." 7 Wall. 68.

Such was the view of that case taken by the whole court in *Ralli v. Troop*, in which the majority of the judges said of it: " That decision proceeded, not upon any authority or agency of the pilot, derived from the civil law of master and servant, or from the common law, as the representative of the owners of the ship and cargo;" "but upon a distinct principle of the maritime law, namely, that the vessel in whosesoever hands she lawfully is, is herself considered as the wrongdoer liable for the tort, and subject to a maritime lien for the damages." 157 U. S. 402. And the dissenting judges said that in *The China* "this court held, contrary to the English, but conformably to the continental authorities, that a vessel was liable for the consequences of a collision through the negligence of a pilot taken compulsorily on board, although it was admitted that, if the action had been at common law against the owner, and probably also *in personam* in admiralty, there could have been no recovery, as a compulsory pilot is in no sense the agent or servant of the owner." 157 U. S. 423.

In none of the cases in which actions at law have been maintained against the owner of a ship for the fault of a pilot was the owner compelled to employ the pilot.

In *Bussy* v. *Donaldson,* (1800) 4 Dall. 194, in the Supreme Court of Pennsylvania, an action on the case was brought against the owner of a ship for damages by collision; and the defence that the ship " was in the charge of a public pilot of the port (a person not the choice, nor the voluntary agent, of the owner) when the injury was committed," was overruled. But the statute of Pennsylvania, cited in that case, simply provided that the pilot first offering himself to any inward bound ship should be entitled to take charge of her; and that, if the master of any ship should refuse or neglect to take a pilot, the master, owner or consignee, should forfeit and pay a sum equal to half pilotage, to the use of the society for the relief of distressed and decayed pilots, their widows and children. Penn. Stat. April 11, 1793, §§ 8, 10; 3 Dall. Laws, 424, 426. The subsequent pilot laws of Pennsylvania have made similar provisions. *Cooley* v. *Board of Wardens,* (1851) 12 How. 299. And the Supreme Court of Pennsylvania has held that they did not make the employment of a pilot compulsory, saying: " The legislature have wisely decided not to compel the owners to supply one, but have permitted them, if they please, to compound by paying half pilotage, for the benevolent and beneficial purpose of relieving distressed and decayed pilots, their widows and children. The act sets out an inducement to avail themselves of their services, but does not compel them to do so." *Flanigen* v. *Washington Ins. Co.*, (1847) 7 Penn. St. 306, 312. And see *The Creole*, (1853) 2 Wall. Jr. 485, 516, 517.

So in *Williamson* v. *Price*, (1826) 4 Martin (N. S.) 399, the Supreme Court of Louisiana maintained an action for a collision by a vessel "at the time under the care and consequently the control of a licensed pilot." But the statutes of Louisiana, likewise, only provided that "if the master of any ship or vessel coming to the port of New Orleans shall refuse to receive on board and employ a pilot, the master or owner of such ship or vessel shall pay to such pilot, who shall have offered to go on board and take charge of the pilotage of the vessel, half pilotage." Law of Territory of Orleans of March 31, 1805, § 17, p. 140; Louisiana Rev. Stat. 1853, p. 457, § 17; Rev. Stat. 1856, pp. 403, 404, §§ 9, 19. And this court has held that those statutes are not compulsory. *The Merrimac*, (1871) 14 Wall. 199, 203.

In *Yates* v. *Brown*, (1829) 8 Pick. 22, in the Supreme Judicial Court of Massachusetts, in which the owners of a vessel were held liable for a collision by the fault of a pilot, it is only stated that he was duly authorized to pilot the ship, that he held his commission under the executive authority of the Commonwealth, and that the owners had selected him for this service. And in Massachusetts, as has been observed by its court, "the statute does not make it incumbent on the master of a vessel, subject to pilotage, to receive a pilot, if he chooses to navigate her himself," although it makes him and the owner liable to pay full pilotage fees if a pilot offers his services and they are refused. *Martin* v. *Hilton*, (1845) 9 Met. 371, 373.

In *Denison* v. *Seymour*, (1832) 9 Wend. 1, in the Supreme Court of New York, the taking of a pilot was not compulsory, and the court said: "The officer here called the pilot is not the same as the pilot recognized in the laws regulating foreign commerce."

In *Atlee* v. *Packet Co.*, (1874) 21 Wall. 389, which was a suit *in personam* in the admiralty, where the owners of a vessel were held liable for the fault of a pilot, it does not appear that they acted under compulsion in appointing him, and the question of their liability for his acts was not discussed.

In *Sherlock* v. *Alling*, (1876) 93 U. S. 99, the case came to this court on writ of error from the Supreme Court of the State of Indiana, and therefore none but Federal questions were within the jurisdiction of this court; and the only questions decided, or which could have been decided, were that an act of Indiana making any person liable for the death of another caused by his wrongful act or omission was not, as applied to a tort committed on navigable waters within the State, an encroachment on the commercial powers of Congress; and that an act of Congress making the master and owners of a vessel liable for injuries to passengers under certain circumstances afforded no defence to the action.

The liability of the owner at common law for the act of a pilot on his vessel is well stated by Mr. Justice Story in his Treatise on Agency, (2d ed.) § 456*a:* "The master of a ship, and the owner also, is liable for any injury done by the negligence of the crew employed in the ship. The same doctrine will apply to the case of a pilot, employed by the master or owner, by whose negligence any injury happens to a third person or his property; as, for example, by a collision with another ship, occasioned by his negligence. And it will make no difference in the case, that the pilot, if any is employed, is required to be a licensed pilot; provided the master is at liberty to take a pilot, or not, at his pleasure; for, in such a case, the master acts voluntary, although he is necessarily required to select from a particular class. On the other hand, if it is compulsive upon the master to take a pilot, and, *a fortiori*, if he is bound to do so under a penalty, then, and in such case, neither he, nor the owner, will be liable for injuries occasioned by the negligence of a pilot; for, in such a case, the pilot cannot be deemed properly the servant of the master or the owner, but is forced upon them, and the maxim, *Qui facit per aliam facit per se*, does not apply."

The answer to the second question must therefore be that in

an action at common law the shipowner is not liable for injuries inflicted exclusively by negligence of a pilot accepted by a vessel compulsorily.

Answer to the first question in the affirmative; to the second in the negative.

LAKE STREET ELEVATED RAILROAD COMPANY *v.* FARMERS' LOAN AND TRUST COMPANY.

ERROR TO THE SUPREME COURT OF THE STATE OF ILLINOIS.

No. 669. Submitted May 13, 1901.—Decided May 27, 1901

The action of the Supreme Court of Illinois in this case on April 17, 1901, was a full compliance with the mandate of this court in this case, 177 U. S. 51.

THE case is stated in the opinion of the court.

Mr. Herbert B. Turner and *Mr. William Berry* for the motion to dismiss.

Mr. Clarence A. Knight opposing.

MR. JUSTICE SHIRAS delivered the opinion of the court.

When this cause was before us at October term, 1899, it was determined that the jurisdiction of the Circuit Court of the United States for the Northern District of Illinois had attached, as respected the Lake Street Elevated Railroad Company and its property, before the institution, by the Lake Street Elevated Railroad Company, in the Superior Court of Cook County, Illinois, of a suit involving the same parties and questions as those in the Federal court; and, accordingly, it was held that the decree of injunction granted by the Superior Court and affirmed by the Appellate Court and by the Supreme Court of Illinois, enjoining and restraining the Farmers' Loan and Trust Company from proceeding with its suit in the Circuit Court of the United States, had been improperly granted; and thereupon the judgment of the Supreme Court was reversed, and the cause remanded to that court for further proceedings not inconsistent with the opinion of this court. 177 U. S. 51, 62.

In pursuance of the mandate and in conformity with the opinion of this court, the Supreme Court of Illinois, on April 17, 1901, reversed and set aside the judgment of the Appellate Court and the injunction decree of the Superior Court.

This action of the Supreme Court of Illinois was a full compliance with the mandate of this court.

But it is now complained that the Supreme Court went further, and beyond our mandate, in directing the Superior Court to dismiss the bill; and this writ of error was sued out asking us to supervise and reverse the action of the Supreme Court in that respect.

But the Supreme Court, in directing a dismissal of the bill, was in the exercise of its own jurisdiction over the cause pending in the Superior Court of Cook County. Whether it should order that court to suspend action until the Federal court had exhausted its jurisdiction or to dismiss the bill, leaving the parties to abide by the decree of the court whose jurisdiction had first attached, was for the Supreme Court of Illinois to determine, and as such action in nowise involved any Federal question this court has no jurisdiction to review it.

It cannot be said that, by ordering the dismissal of the bill, the Supreme Court of Illinois passed upon Federal questions involved in the litigation in such a sense as to give this court jurisdiction to review its decree. The record of the case when here before discloses that, so far as Federal rights were concerned, they were asserted by the defendants in the Superior Court, and hence the dismissal of the bill, if it affected such Federal rights at all, was not a decision *against* the parties invoking them, which alone would give us jurisdiction.

The writ of error is

Dismissed.

REAGAN *v.* UNITED STATES.

APPEAL FROM THE COURT OF CLAIMS.

No. 239.   Argued April 15, 1901.—Decided May 27, 1901.

In 1896, commissioners appointed by judges of the United States Court in the Indian Territory were inferior officers, not holding their offices for life, or by any fixed tenure, but subject to removal by the appointing power.

Commissioners appointed by that court prior to the act of March 1, 1895, were entitled to reappointment under that act, but were removable at pleasure unless at that date, or at the date of removal, causes for removal were prescribed by law.

As no causes for removal had been prescribed by law at the date of the removal of claimant in 1896, he was subject to removal by the judge of his district, and the action of that judge in removing him was not open to review in an action for salary.

APPELLANT filed his petition in the Court of Claims, October 13, 1897, and an amended petition October 27, 1899, seeking to recover salary as United States Commissioner in the Indian Territory, at the rate of $1500 per annum, from February 1, 1896, to September 30, 1899, aggregating $5375.

The findings of fact and conclusion of law were as follows:

"I. The claimant was, on the 25th day of April, 1893, appointed by the United States court for the Indian Territory United States commissioner within said Territory, under the provisions of section 39 of an act of Congress approved May 2, 1890, chapter 182, (1st Suppl. Rev. Stat. 737,) and upon the 1st day of March, 1895, the claimant was one of the present commissioners, then holding office under an existing appointment.   On April 17, 1895, the following order was entered of record in the United States court in the Indian Territory, Southern District:

"'It appearing from the records of this court that the said William R. Reagan was a duly appointed, qualified and acting commissioner for the United States court for the Third Judicial Division of the Indian Territory, located at Chickasha, on the 1st day of March, 1895, it is hereby ordered that in accordance with the act of Congress approved March 1, 1895, the said William R. Reagan be, and he is hereby, continued in office, and the bond hereinbefore recited be, and the same is, in all things approved and confirmed.   C. B. KILGORE, *Judge.*' "

" II. He continuously performed the duties and received the salary of said office until the 31st day of January in the year 1896, when the following letter was entered upon the records of the United States court in the Indian Territory, in the Southern District, by the Hon. Constantine B. Kilgore, judge of said court :

" ' IN CHAMBERS,
" ' ARDMORE, INDIAN TERRITORY, *January 31st*, 1896.
" ' HON. WILLIAM R. REAGAN, United States commissioner for the fourth commissioner's district in and for the southern district of the Indian Territory.
" ' *Sir :* I feel it my duty to declare the office of commissioner in that district vacant and to notify you that you are no longer United States commissioner for that district, and your successor will be named at once.
" ' There are many reasons which I could assign for my action in this behalf, but I will only suggest one now, that is, your age and the infirmities incident thereto render you, in my judgment, in many respects unfit for the office.
" ' Very respectfully, your obedient servant,
" ' C. B. KILGORE,
" ' *Judge U. S. Dist. Court, S. Dist.*'

" The letter was not sent to the claimant or served upon him. No other statement of cause was made. The claimant was given no notice of any charge against him. No hearing was allowed the claimant and no opportunity to submit proof in his defence.

" III. The claimant protested that said letter was insufficient to effect his removal, and duly served such protest upon the Hon. Constantine B. Kilgore, judge of said court.

" IV. On February 10, 1896, one John R. Williams, who had

been designated by said judge as United States commissioner in the claimant's place, came to claimaint's office with two armed deputy marshals, and, presenting his order of appointment, demanded possession of the dockets, books, and papers belonging to claimant's office as United States commissioner.

" V. The order of appointment of said Williams is as follows :

" ' IN CHAMBERS,
" ' ARDMORE, INDIAN TERRITORY, *January 31st,* 1896.

" ' John R. Williams, a resident of Ryan, Southern District of Indian Territory, is hereby appointed United States commissioner in and for the Fourth District of the Southern District of the Indian Territory.

" ' Said appointment to take effect at once.

" ' It is further ordered that said commissioner shall reside at Ryan, and that he shall hold court at Ryan and at the town of Duncan in said district until further ordered, the time to be divided so as to dispose of the business at both points, which time shall be determined upon hereafter.

" ' C. B. KILGORE,
" ' *Judge U. S. Ct., So. Dist.*'

" VI. The claimant protested and refused to recognize said Williams as his successor in said office, excepting so far as he was compelled thereto by the exercise of superior force on the part of the deputy marshals aforesaid and said Williams. Thereupon the claimant and said Williams joined in the following instrument of writing:

" ' DUNCAN, INDIAN TERRITORY, }
    Southern District.   }

" ' This instrument of writing witnesseth:

" ' That whereas C. B. Kilgore, judge of the United States court for the Southern District of the Indian Territory, on the 31st day of January, A. D. 1896, made, and caused to be entered upon the docket of his court at Ardmore, Indian Territory, an order declaring my office of United States Commissioner for the Ryan division of said district vacant; and at the same time appointing John R. Williams to be my successor in said office,

and the said Reagan having appealed to the courts of the United
States from said order, on the ground that said order is contrary
to the law:

"'Now, therefore, it is agreed by and between the parties hereto that said Reagan will turn over and surrender the dockets,
books, and papers belonging to said office under protest, and
that said Williams receives the same with the understanding
that said Reagan yields no rights by so doing that he would
otherwise have.

"Witness our hands this 10th day of February, A. D. 1896.

"'JNO. R. WILLIAMS.

"'WM. R. REAGAN.'

"VII. The claimant received a salary of $1500 per annum
up to the 3d day of February, 1896, but since that date has
not been paid said salary or any part thereof.

"VIII. Claimant took no other or further action to assert
his claim to said office or to obtain a reversal of the action of
Judge Kilgore until the institution of this proceeding.

"IX. From the 3d day of February, 1896, until the 7th day
of October, 1897, John R. Williams, who was appointed by
Judge Kilgore to said office in claimant's stead, exercised said
office and was paid the salary thereof. On said date one Horace
M. Wolverton was appointed as the successor of said John R.
Williams by Hon. Hosea Townsend, United States judge for
said district, and since that time has exercised said office and
has been paid the salary thereof.

"X. From the 3d day of February, 1896, until the commencement of this action, the disbursing clerk of the Department of
Justice paid to the persons who succeeded claimant to said office
the salary of said office in the absence of any notice on the part
of claimant that he claimed to be lawfully entitled to said office
and the salary thereof, or any claim or demand on the part of
claimant for the payment to him of such salary for said period
of time or any part thereof.

" Conclusion of Law.

"Upon the foregoing findings of fact, the court decide, as a
conclusion of law, that the petition be dismissed."

Judgment was thereupon rendered dismissing the petition, and the case was brought to this court by appeal. The opinion below is reported 35 C. Cl. 90.

Mr. William B. King for appellant.

Mr. Assistant Attorney General Pradt for appellee.

MR. CHIEF JUSTICE FULLER delivered the opinion of the court.

Section 39 of the act of May 2, 1890, 26 Stat. 98, c. 182, provided:

"That the United States court in the Indian Territory shall have all the powers of the United States Circuit Courts or Circuit Court judges to appoint commissioners within said Indian Territory, who shall be learned in the law, and shall be known as United States commissioners; but not exceeding three commissioners shall be appointed for any one division, and such commissioners when appointed shall have, within the district to be designated in the order appointing them, all the powers of commissioners of Circuit Courts of the United States.

"They shall be *ex officio* notaries public, and shall have power to solemnize marriages.

"The provisions of chapter ninety-one of the said laws of Arkansas, regulating the jurisdiction and procedure before justices of the peace, are hereby extended over the Indian Territory; and said commissioners shall exercise all the powers conferred by the laws of Arkansas upon justices of the peace within their districts; but they shall have no jurisdiction to try any cause where the value of the thing or the amount in controversy exceeds one hundred dollars."

The act of March 1, 1895, 28 Stat. 693, c. 145, provided for additional judges of the court, and by section 4:

"That each judge of said court shall have the powers conferred by law upon the United States Circuit Courts to appoint commissioners within the district in which he presides, who, at the time of their appointment, shall be duly enrolled attorneys of some court of record of the United States or of some State,

and shall be competent and of good standing, and shall be
known as United States commissioners, but not exceeding six
commissioners shall be appointed for any district hereinbefore
constituted:

" *Provided,* That the present commissioners shall be included
in that number and shall hold office under their existing appointments, subject to removal by the judge of the district
where said commissioners reside, for causes prescribed by law.
The judge for each district may fix the place where, or the
time when, each commissioner shall hold his regular terms of
court.

"The order appointing such commissioners shall be in writing and shall be spread upon the records of one of the courts of
the district for which they are appointed; and such order shall
designate, by metes and bounds, the portion of the district for
which they are appointed. They shall have all the powers of
commissioners of the Circuit Courts of the United States.

"They shall be *ex officio* notaries public and *ex officio* justices of the peace within and for the portion of the district for
which they are appointed, and shall have the power as such to
solemnize marriages."

Appellant was appointed a commissioner April 25, 1893, and
was such on March 1, 1895. In view of the proviso he was
continued in office until January 31, 1896, when he was removed by the judge of the district where he resided, and another person appointed.

He now contends that the removal was void, because the
cause assigned for the action of the judge was not a "cause
prescribed by law," and because he was given no notice of any
charge against him, and no hearing, contrary to the statute.

The commissioners appointed by the judges of the United
States Court in the Indian Territory are inferior officers, not
holding their offices for life, or by any fixed tenure, and they
fall within the settled rule that the power of removal is incident
to the power of appointment. *Ex parte Hennen,* 13 Pet. 230,
258; *Parsons* v. *United States,* 167 U. S. 324. But it is assumed that because of the language of the proviso, commissioners appointed by the court prior to March 1, 1895, formed an

exceptional class from commissioners appointed by the judges of that court after that date, and hold office until they are removed for causes prescribed by existing law, or until Congress passes a law defining such causes. The latter view may be rejected at once, for the words " causes prescribed by law," manifestly relate to causes prescribed when the act was approved, or at least when the removal was made. Not only is there nothing here to give them any other meaning, but it cannot be presumed that Congress intended to forbid the exercise by the judges of their power in the matter of these appointments in the instance of these particular commissioners, or to provide that they should hold office during life, or until Congress should specify causes subjecting them to removal, while all other commissioners were removable at the will of the power appointing them.

The proviso was enacted apparently out of abundant caution lest the legislation in respect of the United States Court in the Indian Territory might operate in itself to turn the then commissioner out of office, and if Congress had intended in addition that they should hold office free from the rule applicable to others, we think that the intention would have been plainly expressed.

The inquiry is therefore whether there were any causes of removal prescribed by law, March 1, 1895, or at the time of the removal. If there were, then the rule would apply that where causes of removal are specified by constitution or statute, as also where the term of office is for a fixed period, notice and hearing are essential. If there were not, the appointing power could remove at pleasure or for such cause as it deemed sufficient.

The suggestion that the proviso refers to such causes as courts might recognize as just will not do, for " prescribed by law " is prescribed by legislative act, and removal for cause, when causes are not defined nor removal for cause provided for, is a matter of discretion and not reviewable.

It does not appear that any causes for removal of these court officers were ever affirmatively specified by Congress; but it is said that Congress had prescribed such causes by the adoption in the Indian Territory of certain laws of Arkansas. By section thirty-one of the act of May 2, 1890, some of those laws were put in force in the Indian Territory, and by section thirty-nine the commissioners were authorized to exercise all the powers conferred by the laws of Arkansas on justices of the peace within their districts, and the provisions of chapter ninety-one of those laws regulating the jurisdiction of and procedure before justices of the peace were extended to that Territory. By the act of March 1, 1895, these were reënacted, and chapters forty-five and forty-six of Mansfield's Digest, treating of criminal law and criminal procedure, were also put in force there.

The argument is that the effect of these provisions was to put the commissioners in the place of justices of the peace in Arkansas, and that consequently the causes prescribed by law for the removal of justices of the peace must be taken as prescribed by law as causes for the removal of commissioners.

In our opinion this conclusion does not follow. In order to clothe the commissioners with the powers pertaining to justices of the peace, this was conveniently accomplished by reference, but that did not convert these officers of the United States Court in the Indian Territory into justices of the peace or change the relations between them and the judges of that court. Justices of the peace in Arkansas by state constitution and laws hold office for two years, and cannot be removed except for cause, and on notice and hearing. The commissioners hold office neither for life, nor for any specified time, and are within the rule which treats the power of removal as incident to the power of appointment, unless otherwise provided. By chapters forty-five and forty-six, justices of the peace on conviction of the offences enumerated are removable from office, but these necessarily do not include all causes which might render the removal of commissioners necessary or advisable. Congress did not provide for the removal of commissioners for the causes for which justices of the peace might be removed, and if this were to be ruled otherwise by construction, the effect would be to hold the commissioners in office for life unless some of those specially enumerated causes became applicable to them.

We agree with the Court of Claims that this would be a most unreasonable construction and would restrict the power of removal in a manner which there is nothing in the case to indicate could have been contemplated by Congress.

If causes of removal had been prescribed by law before the removal of appellant that would have presented a different question, but as there were then none such, the proviso did not operate to take him out of the rule expounded in *Ex parte Hennen*, and the mere fact that in that particular this part of the proviso was inoperative as to him did not change the result.

Judgment affirmed.

SIMON *v.* CRAFT.

ERROR TO THE SUPREME COURT OF THE STATE OF ALABAMA.

No. 191.   Argued March 12, 1901.—Decided May 27, 1901.

The essential elements of due process of law are notice and opportunity to defend, and in determining whether such rights are denied, the court is governed by the substance of things and not by mere form.

A person charged with being of unsound mind is not denied due process of law by being refused an opportunity to defend, when, in fact, actual notice was served upon him of the proceedings, and when, if he had chosen to do so, he was at liberty to make such defences as he deemed advisable.

The due process clause in the Fourteenth Amendment to the Constitution does not necessitate that the proceedings in a state court should be by a particular mode, but only that there shall be a regular course of proceedings, in which notice is given of the claim asserted, and an opportunity afforded to defend against it.

This court accepts as conclusive the ruling of the Supreme Court of Alabama that the jury which passed upon the lunacy proceeding considered in this case was a lawful jury, that the petition was in compliance with the statute, and that the asserted omissions in the recitals in the verdict and order thereon were at best but mere irregularities which did not render void the order of the state court, appointing a guardian.

THIS is a writ of error to review a judgment of the Supreme

Court of Alabama affirming a judgment in favor of John N. Craft, the defendant in error herein. The judgment thus affirmed was entered by a lower state tribunal upon a verdict rendered on the second trial of an action in ejectment, wherein Jetta Simon, plaintiff in error herein, was plaintiff.

In brief, the facts are as follows: In 1889, Jetta Simon, a widow, resided in Mobile, Alabama, with several minor children. She lived at that time in a house of which she was the owner, being the real estate affected by the action of ejectment heretofore referred to. On January 30, 1889, Ralph G. Richard filed in the probate court of Mobile County, Alabama, a petition for an inquisition of lunacy as to Mrs. Simon. In this petition it was represented that Richard was a friend of Mrs. Simon and of her family; that she was of the age of forty-nine years, a resident of Mobile, of unsound mind and incapable of governing herself or of conducting and managing her affairs. Upon this petition an order was entered for a hearing on February 6, 1889, and that a jury " be drawn, as the law directs, for the trial of this issue." The order also provided that a writ issue to the sheriff, " requiring him to take the said Jetta Simon, so that he have her in this court to be presented at said trial, if consistent with the health and safety of said Simon." The writ issued. Therein was stated the substance of the allegations of the petition, and that the order had been entered appointing February 6, 1889, " for hearing said petition and for the due trial thereof." The command of the writ was that—

" If it be consistent with the health and safety of said Jetta Simon, you are hereby required to take her body, so that you may have her in said court, to be present at said trial, and before the jury then to be empanelled to make said inquisition.

"And have you then and there this writ with your return thereon as to how you have executed the same."

The writ was duly returned with the following endorsement:

" Received January 31st, 1889, and on the same day I executed the within writ of arrest by taking into my custody the within-named Jetta Simon and handing her a copy of said writ, and as it is inconsistent with the health or safety of the within-named Jetta Simon to have her present at the place of trial, and on the advice of Dr. H. P. Hirshfield, a physician, whose certificate is hereto attached, she is not brought before the honorable court.

"W. H. HOLCOMBE, *Sheriff.*

"Mobile, February 5th, 1889. By WM. H. SHEFFIELD, *D. S.*"

The certificate referred to reads as follows:

"MOBILE, ALA., *Jan. 30th*, 1889.

"To the Sheriff of Mobile County, Ala.:

"I, H. P. Hirshfield, a regular physician, practicing in Mobile County, Ala., hereby certify that I am acquainted with Mrs. Jetta Simon, and have examined her condition on yesterday and find that she is a person of unsound mind, and it would not be consistent with her health or safety to have her present in court in any matter now pending.

"H. P. HIRSHFIELD, M. D."

One Vaughan was appointed by the probate court the guardian *ad litem* of Mrs. Simon "in the matter of the petition to inquire into her lunacy." The appointment was accepted, and the guardian filed in said proceeding an answer averring "that he wholly denies all the matters and things stated and contained in said petition, and requires strict proof to be made thereof according to law." Thereupon a hearing was had before a jury, who returned a verdict that Mrs. Simon was "of unsound mind." The probate court then entered the following order or decree:

"JETTA SIMON, Lunatic.

"STATE OF ALABAMA, } Probate Court of said County,
    Mobile County. }

"February 6th, 1889.

"This being the day appointed, by reference to an entry thereof made upon the minutes of the court on the 30th of January, 1889, for the hearing of the petition of Ralph G. Richard, filed, alleging the lunacy of the said Jetta Simon and praying an inquisition thereof, and it being shown that it would not be consistent with the health and safety of said lunatic to bring her into court at this time, and it appearing that due process had been served upon said lunatic notifying her of this proceeding, now comes the said Richard and a jury of good and lawful men, who reside in the county of Mobile, and who, having been summoned, to wit, John Pollock, Jr., and eleven others, who, having heard the evidence, the arguments of counsel, and the charge of the court in the premises, and being first duly tried, empanelled and sworn well and truly to make inquisition of the facts alleged in said petition and a true verdict to render according to the evidence, upon their oath say, 'We, the jury, find Mrs. Jetta Simon to be of unsound mind.'

"It is ordered, adjudged and decreed by the court that said petition and all other proceedings thereon, together with the aforesaid verdict of said jury declaring the said Jetta Simon a lunatic, be recorded."

Subsequently, on February 11, 1889, Richard was duly appointed guardian of the estate of Mrs. Simon, and regular proceedings were had by which, under authority of the court, a sale of the real estate in question was ordered to be made for the payment of the debts of Mrs. Simon and for the support and maintenance of her family. Such sale was had in May, 1889, when Henry J. Simon became the purchaser, who sold the property to John N. Craft, defendant in error herein. In September, 1895, more than six years after the sale to Simon, the action in ejectment heretofore referred to was instituted against one Brown, a tenant of Craft. Craft, as landlord, was subsequently substituted in the stead of Brown. Upon a second trial of the issues joined, the defendant Craft, among other evidence, introduced the record of the proceedings in the probate court upon the inquisition of lunacy, to which reference has already been made, and the record of the subsequent proceedings resulting in the sale to Henry J. Simon. Objection to the introduction of such records was made upon specified grounds, all of which are stated in the margin.[1] The objec-

---

[1] 1st. In that there was no process issued notifying Jetty Simon to be present at the trial of the inquest of lunacy that was held.

2d. In that no provision was made in or by said proceedings whereby said Jetty Simon might be present at the inquest of lunacy that was held.

3d. In that the writ of arrest issued for the body of Jetty Simon was

tions were overruled and the record allowed to be read in evidence, to which action of the court exception was duly taken. The approval by the Supreme Court of Alabama of this ruling is what is here complained of.

The opinion of the Supreme Court of Alabama reversing the judgment entered on a verdict in favor of Mrs. Simon rendered at the first trial of the action of ejectment is contained in 118 Alabama, 625. The judgment entered in favor of Craft upon the second trial was affirmed upon the authority of the previous opinion.

Mr. Harry T. Smith for plaintiff in error. *Mr. Gregory L. Smith* was on his brief.

Mr. H. Pillans for defendant in error. *Mr. D. P. Bestor* was on his brief.

MR. JUSTICE WHITE, after making the foregoing statement, delivered the opinion of the court.

By subdivision 6 of section 787 of the Civil Code of Alabama of 1886 courts of probate in that State are vested with original jurisdiction over the appointment and removal of guardians for minors and persons of unsound mind. Pertinent provisions of

conditional in form and conferred upon the sheriff the power to determine whether it should be executed or not.

4th. In that the writ of arrest left it to the judgment of the sheriff whether the said Jetty Simon should be allowed to appear at the trial of the inquest of lunacy.

5th. In that the writ of arrest authorized the sheriff to restrain Jetty Simon of her liberty and deprive her of the opportunity to be heard at the inquest of lunacy.

6th. In that the sheriff's return shows that under the writ of arrest he restrained Jetty Simon of her liberty and did not permit her to be present at the trial of the inquest of lunacy.

7th. Because the statute under which Jetty Simon was restrained of her liberty and deprived of her property is in conflict with article V of the amendments to the Constitution of the United States, which provides, "Nor be deprived of life, liberty, or property without due process of law," and in conflict with article XIV of the amendments to said Constitution:

the statutes of Alabama relating to the mode of appointment of guardians of persons of unsound mind, contained in said Civil Code, are excerpted in the margin.[1]

In the proceedings to inquire into the sanity of Mrs. Simon the writ which issued to the sheriff was evidently based upon the following clause of section 2393 of the Civil Code of 1886:

1*a*. In that it authorizes a citizen to be deprived of his or her liberty without due process of law.

2*a*. In that it authorizes a citizen to be deprived of his or her property without due process of law.

8th. Because said proceedings in the probate court are irrelevant and immaterial to any issue in the cause.

[1] Sundry sections of Part 2, Title 5, Chapter 4, of the Civil Code of Alabama of 1886, pp. 535, *et seq.* :

"2390 (2753, 2754). Appointment.—The court of probate has authority, and it is a duty to appoint guardians for persons of unsound mind residing in the county, having an estate, real or personal, and of persons of unsound mind residing without the State, having within the county property requiring the care of a guardian, under the limitations, and in the mode hereinafter prescribed.

"2391. Guardian not appointed until after inquisition.—A guardian for a person alleged to be of unsound mind, residing in the county, must not be appointed until an inquisition has been had and taken as hereinafter directed.

"2392 (2757). Inquisition; proceedings.—Upon the petition of any of the relatives or friends of any person alleged to be of unsound mind, setting forth the facts and name, sex, age and residence of such person, accompanied by an affidavit that the petitioner believes the facts therein stated to be true, the court of probate of the county in which such person alleged to be of unsound mind resides, must appoint a day, not more than ten days from the presentment of such petition, for the hearing thereof.

"2393 (2758). Jury summoned; writ of arrest.—The judge of probate must issue a writ directed to the sheriff, commanding him to summon twelve disinterested persons of the neighborhood for the trial thereof, and also issue subpœnas for witnesses, as the parties may require, returnable to the time of trial; he must also issue a writ directed to the sheriff to take the person alleged to be of unsound mind, and, if consistent with his health or safety, have him present at the place of trial.

"2394 (2759). Oaths of jurors; vacancies filled.—At the time set for the trial, if good cause be not shown for continuance, the jury must be impanelled and sworn well and truly to make inquisition of the facts alleged in the petition, and a true verdict render according to the evidence. If any of the jurors are excused from serving, fail to attend, or are set aside for any cause, their places may be supplied from the bystanders.

"2393. The judge of probate . . . must also issue a writ directed to the sheriff to take the person alleged to be of unsound mind, and, if consistent with his health or safety, have him present at the place of trial."

The invalidity of the proceedings in the inquisition of lunacy which formed the basis of the subsequent proceedings for the sale of the property of Mrs. Simon is in substance predicated on the contention that the writ directed to the sheriff authorized that official to determine whether it was consistent with the health and safety of Mrs. Simon to be present at the trial of the question of her sanity; that the sheriff decided this question against her, and she was detained in custody and not allowed to be present at the hearing on the inquisition. The latter claim, however, is founded upon the return endorsed by the sheriff on the writ directed to him. At the trial below there

---

"2395 (2760). On verdict of insanity, papers filed, and guardian appointed.—If the jury find by their verdict that the facts alleged in the petition are true, and that such person is of unsound mind, the court must cause the petition and all the proceedings thereon to be recorded, and appoint a suitable guardian of such person.

　*　　*　　*　　*　　*　　*　　*　　*

"2396 (2761). Proceedings when person of unsound mind is confined in asylum.—If the person alleged to have been of unsound mind is a resident of the county, and is at the time of the application confined in an hospital or asylum within or without the State, the inquisition may be had or taken without notice to him, but on the filing of the application the court must appoint a guardian *ad litem* to represent and defend for him; it is the duty of such guardian by answer to put in issue the facts stated in the application, and to employ counsel at the expense of such person of unsound mind to appear and defend.

"2397 (2804). Application for revocation of guardianship.—At any time after the inquisition the person ascertained to be of unsound mind, by himself or by next friend, may apply in writing to the court of probate for a revocation of the proceedings against him, and of the letters of guardianship; the application to be accompanied by the certificate in writing of two physicians or of two other competent persons, stating that after examination of such person they believe him to be of sound mind.

"2398 (2804). Proceedings on application.—On the filing of such application the court must appoint a day for the hearing thereof, not more than ten days thereafter, and the guardian and the person at whose instance the inquisition was had and taken must be cited to appear and show cause against it.

was no offer to prove, by any form of evidence, that Mrs. Simon was in fact of sound mind when the proceedings in lunacy were instituted, or that she desired to attend, and was prevented from attending, the hearing, or was refused opportunity to consult with and employ counsel to represent her. The entire case is thus solely based on the inferences which are deduced, as stated, from the face of the return of the sheriff. And upon the assumptions thus made it is contended that the statute as well as the proceedings thereunder were violative of the clause of the Fourteenth Amendment to the Constitution of the United States, which forbids depriving any one of life, liberty or property without due process of law.

It is not seriously questioned that the Alabama statute provided that notice should be given to one proceeded against as being of unsound mind of the contemplated trial of the question of his or her sanity. Indeed, it would seem that it was not urged before the Supreme Court of Alabama that the statutes

---

" 2399 (2805, 2806). Contest of application.—If the guardian or person at whose instance the inquisition was had and taken appear and deny the allegations of the application, the court must appoint a day for the trial of such contest, not more than ten days thereafter, and must cause a jury to be summoned for the trial thereof, and the like proceedings must be had as upon the original inquisition; or if there be no contest of the allegations of the application, and the court is satisfied of the truth thereof, a decree must be entered revoking the proceedings on the inquisition and the guardianship, and declaring that the ward must be restored to the custody and management of his estate.

" 2400 (2807). Judgment on contest; costs thereof.—If on the trial of the contest, the jury find the facts stated in the application to be true, the court must enter a decree revoking the proceedings on the inquisition and the guardianship, and declaring that the ward must be restored to the custody and management of his estate, and must adjudge the costs as is just and equitable, but if the verdict of the jury negatives the facts stated in the application, a judgment of dismissal at the cost of the applicant or of the next friend must be entered.

" 2401 (2808). Revocation on application of guardian.—If, at any time after his appointment, the guardian becomes satisfied that the ward has been restored to sanity, and is capable of managing his estate, and the judge of probate is of opinion, from the proof and the facts stated, that such representation is correct, he must make an order that the guardian be discharged, and that the estate of the ward be restored to him.

of that State failed to provide for notice, and that court assumed
in its opinion that no question of that character was presented.
As a matter of fact, a copy of the writ which issued and which
embodied a notice of the date of the hearing of the proceedings
in lunacy is shown by the record to have been actually served
on Mrs. Simon.   As early as 1870 the Supreme Court of Alabama in *Fore* v. *Fore,* 44 Alabama, 478, 483, held that the service of the writ upon a supposed lunatic was the notice required
by the statute and brought the defendant into court, and that
if he failed to avail of such matters of defence as he might have,
he must suffer the effect of his failure to do so.

We excerpt in the margin the portion of the opinion of the
Supreme Court of Alabama which dealt with the objection
that Mrs. Simon was deprived of opportunity to be heard.[1]

The contention now urged is that notice imports an opportunity to defend, and that the return of the sheriff conclusively
established that Mrs. Simon was taken into custody and was
hence prevented by the sheriff from attending the inquest or
defending through counsel if she wished to do so in consequence
of the notice which she received.   It seems, however, manifest
—as it is fairly to be inferred the state court interpreted the

---

[1] "The second ground of objection is that the appellee had no opportunity
to be heard at the inquisition.   This objection is based upon the character
and wording of the writ directed to the sheriff.   The provision of the
statute is that the judge must 'issue a writ, directed to the sheriff, to take
the person alleged to be of unsound mind, and, if consistent with his health
or safety, have him present at the place of trial.'   The writ that issued,
after setting out the facts averred in the petition, proceeded: 'Now, therefore, if it be consistent with the health and safety of said Jetta Simon, you
are hereby required to take her body so that you may have her in said
court,' etc.   The statute is that the sheriff be directed to take her body,
and, if consistent with health, etc.   By the statute it is made the duty of
the sheriff to take the body without condition, and, if consistent with health
and safety, to have her present at the trial.   The writ issued, directed to the
sheriff, 'if consistent with health and safety, to take her body,' etc.   The return of the sheriff shows that the writ was executed in accordance with the
statute.   It is: 'I executed the within writ of arrest by taking into my custody the within-named Jetta Simon and handing a copy of said writ, and as
it is inconsistent with the health or safety . . . to have her at the place
of trial . . . she is not brought before the court.'   Technically the
writ of the judge was not accurately correct.   Its meaning, however, is

statute—that the purpose in the command of the writ, "to take the person alleged to be of unsound mind, and, if consistent with his health or safety, have him present at the place of trial," was to enforce the attendance of the alleged *non compos*, rather than to authorize a restraint upon the attendance of such person at the hearing. In other words, that the detention authorized was simply such as would be necessary to enable the sheriff to perform the absolute duty imposed upon him by law of bringing the person before the court, if in the judgment of that officer such person was in a fit condition to attend, and hence it cannot be presumed, in the absence of all proof or allegation to that effect, that the sheriff in the discharge of this duty, after serving the writ upon the alleged lunatic, exerted his power of detention for the purpose of preventing her attendance at the hearing, or of restraining her from availing herself of any and every opportunity to defend which she might desire to resort to, or which she was capable of exerting. The essential elements of due process of law are notice and opportunity to defend. In determining whether such rights were denied we are governed by the substance of things and not by mere form. *Louisville & Nashville Railroad Co.* v. *Schmidt*, 177 U. S. 230. We cannot, then, even on the assumption that Mrs. Simon was of sound mind and fit to attend the hearing, hold that she was denied due process of law by being refused an opportunity to defend, when, in fact, actual notice was served upon her of the proceedings, and when, as we construe the statute, if she had chosen to do so, she was at liberty to make such defence as she deemed advisable. The view we take of the statute was evidently the one adopted by the judge of the probate court, where the proceedings in lunacy were heard, since that court, upon the return of the sheriff, and the failure of the alleged lunatic to appear, either in person or by counsel, in order to protect her interests, entered an order appointing a guardian *ad litem* "in the matter of the petition to inquire into her lunacy;" and

---

evident. The sheriff's return was complete and regular in every respect. We do not doubt she was brought into the court in the manner prescribed by statute, and that she was subject to its jurisdiction. The second objection cannot be sustained." 118 Alabama, 636.

an answer was filed by such guardian denying all the matters and things stated and contained in the petition, and requiring strict proof to be made thereof according to law.

It is also urged as establishing the nullity of the appointment of a guardian of the estate of Mrs. Simon that the proceedings failed to constitute due process of law, because: 1, they were special and statutory, and the petition failed to state sufficient jurisdictional facts: 2, a jury was not empanelled as provided by law: and 3, there was no finding in the verdict of the jury or the order entered thereon ascertaining and determining all the facts claimed to be essential to confer jurisdiction to appoint a guardian. But the due process clause of the Fourteenth Amendment does not necessitate that the proceedings in a state court should be by a particular mode, but only that there shall be a regular course of proceedings in which notice is given of the claim asserted and an opportunity afforded to defend against it. *Louisville & Nashville Railroad Co.* v. *Schmidt*, 177 U. S. 230, 236, and cases cited. If the essential requisites of full notice and an opportunity to defend were present, this court will accept the interpretation given by the state court as to the regularity under the state statute of the practice pursued in the particular case. Tested by these principles, we accept as conclusive the ruling of the Supreme Court of Alabama that the jury which passed on the issues in the lunacy proceeding was a lawful jury, that the petition was in compliance with the statute, and that the asserted omissions in the recitals in the verdict and order thereon were at best but mere irregularities which did not render void the order of the state court, appointing a guardian of Mrs. Simon's estate.

Judgment affirmed.

PIRIE v. CHICAGO TITLE AND TRUST COMPANY.

APPEAL FROM THE CIRCUIT COURT OF APPEALS FOR THE SEVENTH CIRCUIT.

No. 391. Argued January 18, 21, 1901. — Decided May 27, 1901.

Frank Brothers were adjudged bankrupts in February, 1899. For a long time prior to that Pirie & Co. had dealt with them, selling them merchandise. Within four months prior to the adjudication of bankruptcy Pirie & Co. received from them $1336.79, leaving a balance still due and unpaid of $3093.98. When this payment was made Frank Brothers were hopelessly insolvent to the knowledge of Frank Brothers, but Pirie & Co. and their agents had no knowledge of it, and had no reasonable cause to believe that the bankrupts, by such payment intended to give a preference, nor did they intend to do so. Pirie & Co. proved their claim against the estate, and received a dividend thereon, which they still hold.

The provisions in the Bankrupt Act of July 1, 1898, c. 541, § 60, that "a person shall be deemed to have given a preference if, being insolvent, he has procured or suffered a judgment to be entered against himself in favor of any person, or made a transfer of any of his property, and the effect of the enforcement of such judgment or transfer will be to enable any one of his creditors to obtain a greater percentage of his debt than any other of such creditors, of the same class," means that a transfer of property includes the giving or conveying anything of value, anything which has debt paying or debt securing power; and money is property.

If the person receiving such preference did not have cause to believe that it was intended, he may keep the property transferred, but, if it be only a partial discharge of his debt cannot prove the balance.

When the purpose of a prior law is continued, its words usually are, and an omission of the words imply an omission of the purpose.

The object of a bankrupt act is, so far as creditors are concerned, to secure equality of distribution, among all, of the property of the bankrupt.

Subdivision c of section 60 of the bankrupt act is applicable to the cases arising under subdivision b, and allows a set-off, which might not be otherwise allowed.

In proceedings in bankruptcy in the matter of Frank Brothers, bankrupts, in the District Court for the Northern District of Illinois, the appellants filed a claim for goods, wares and merchandise, sold and delivered to said bankrupt firm for the sum of $3093.98. The claim was allowed, and subsequently a dividend of fifteen per cent was paid thereon.

On the 31st of August, 1899, the appellee, the Chicago Title and Trust Company, filed a petition for a reconsideration of the claim and its rejection on the ground that Carson, Pirie, Scott & Company had within four months prior to the filing of the petition in bankruptcy received from the bankrupts large sums of money as preferences, which preferences had not been surrendered. The recovery of the dividend paid was also prayed for.

To the petition, Carson, Pirie, Scott & Company made the following answer:

"They admit that they have collected in the usual and ordinary course of their business, from said bankrupts, Frank Brothers, within four (4) months prior to the filing of the petition in bankruptcy, the sum of one thousand three hundred and thirty-six and $\frac{79}{100}$ dollars ($1336.79).

"Further answering, Carson, Pirie, Scott & Company say, that they did not know, or have reason to believe, that the said Frank Brothers were insolvent at the time the payments were made, nor did they have reasonable cause to believe that such payments were made with any intent to give them a preference, nor did said Frank Brothers intend the payments so made to be preferences."

The matter came up before Frank L. Wean, referee, and he substantially found the facts, from the stipulation of the parties, as hereinafter stated in the findings of the Circuit Court of Appeals, and that the payments constituted a preference. He adjudged therefore that the claim be reconsidered and rejected, and the dividend paid thereon be given up. On review the District Court also found the facts as the referee found them, and on the 9th of May, 1900, made and entered an order, the conclusion of which was as follows:

"It is therefore ordered, adjudged and decreed that said claim of said Carson, Pirie, Scott & Company, heretofore filed herein and allowed, should be reconsidered.

"That said claim of Carson, Pirie, Scott & Company should be rejected and expunged.

"That said Carson, Pirie, Scott & Company forthwith pay

to the trustee herein the amount of the dividend heretofore paid to them by the trustee herein, to wit, the sum of $464.10."

Carson & Company excepted, and subsequently took an appeal to the Circuit Court of Appeals, which court affirmed the order of the District Court, upon its opinion in *Columbus Electric Co.* v. *Worden, Trustee, In re Fort Wayne Electric Corporation*, 99 Fed. Rep. 400. The case was then brought here.

The findings and facts and conclusions of law of the Circuit Court of Appeals are as follows:

"*First.* That on February 11, 1899, August Frank, Joseph Frank and Louis Frank, trading as Frank Brothers, were duly adjudged bankrupts.

"*Second.* That for a long time prior thereto appellants carried on dealings with the said bankrupt firm—said dealings consisting of a sale by said appellants to said Frank Brothers of goods, wares and merchandise amounting to the total sum of $4403.77.

"*Third.* That said appellants in the regular and ordinary course of business, and within four months prior to the adjudication in bankruptcy herein, did collect and receive from said bankrupts as partial payment of said account for such goods, wares and merchandise so sold and delivered to said Frank Brothers, the sum of $1336.79, leaving a balance due, owing and unpaid, amounting to $3093.98.

"*Fourth.* That at the time this payment was made said Frank Brothers were wholly and hopelessly insolvent to the knowledge of said Frank Brothers, and that when said payments were made and at the time of the adjudication in bankruptcy of the bankrupts herein, the assets of said bankrupts did not exceed the sum of $125,000, while their liabilities exceeded $500,000.

"*Fifth.* That at the time of the payment above set forth neither said appellants nor any of their agents had knowledge of the insolvency of said Frank Brothers, or had reasonable cause to believe that said Frank Brothers were insolvent, and that when said payment was made said appellants did not have reasonable cause to believe that said bankrupts by said payment intended thereby to give a preference. Nor did said bankrupts by said payments intend thereby to give a preference.

"*Sixth.* That at or about the time of the first meeting of the creditors herein, to wit, on March 17, 1899, said appellants duly filed a claim herein against said bankrupts' estate for their balance of said claim for goods, wares and merchandise sold by them to the bankrupts, as aforesaid—said balance amounting to the sum of $3093.98, and that at or about the time of the said first meeting of creditors herein said claim was duly allowed at the sum last above set forth; that thereafter, and on the 28th day of April, 1899, a dividend of 15 per cent upon all claims which were allowed against said bankrupts' estate was duly declared by the referee herein, and that said dividend was paid to the various creditors who had proved their claims, including appellants'; that the amount of the dividend paid to appellants was $464.10, which money appellants still retain, no part thereof having been repaid or returned to the trustee herein or anybody acting on behalf of said trustee.

"*Seventh.* That at the time of the allowance of said claim and the declaration of said dividend and the payment thereof, the trustee was not aware of the fact that said appellants had received any preference on their claim and demand against said bankrupts.

"*Eighth.* That said appellants have refused to surrender to the trustee the amount of the payment made to them by said bankrupts above set forth as a condition of the allowance of their said claim, and have by their counsel declared that it is the intention of said claimants to retain the full amount of said payment so made to them by said bankrupts, and not to surrender the same.

"*Ninth.* That the appellee, Chicago Title and Trust Company, trustee, which had been duly appointed trustee of the bankrupt estate of said Frank Brothers, filed its petition praying that the claim of appellants against the bankrupts' estate be reconsidered and rejected, and that said appellants be ordered and required to repay to the trustee the amount of the dividend on the said claims theretofore paid to appellants, the grounds of said petition being that said appellants had within four months prior to the adjudication in bankruptcy of said bankrupts received large sums of money as preferences, which preferences said appellants had not surrendered; that said appellants appeared in said proceedings and answered said petition.

"That the referee upon the evidence presented before him decided that the said payment made by the bankrupts to said appellants constituted a preference, and that by reason of said preferences the appellants' claim should be reconsidered and rejected, and that appellants should repay to appellee the amount of the dividend on appellants' said claim theretofore paid by appellee to them, the sum of $464.10; that upon appellants' application and upon the certification of the questions presented to the United States District Court for the Northern District of Illinois, the decree of the referee was confirmed, and an order in the District Court was entered in accordance with the referee's said report, from which order an appeal was taken to this court.

"Upon the foregoing facts this court makes the following conclusions of law:

"*First.* That the payment made by appellants to the bankrupts at the time and in the manner above shown constitutes a preference, and that by reason of the failure and refusal of said appellants to surrender said preferences they were not entitled to prove their claim against the bankrupts' estate.

"*Second.* That the District Court had the power and authority to order, require and compel appellants to repay to the trustee the amount of the dividend received by appellants."

Mr. Henry Ach and *Mr. A. J. Pflaum* for appellants. *Mr. George Packard*, *Mr. Joseph M. Rothschild*, and *Mr. S. O. Levinson* were on their brief.

Mr. Eli B. Felsenthal and *Mr. Herman Frank* for appellee.

MR. JUSTICE MCKENNA, after stating the case as above, delivered the opinion of the court.

The question presented by this record is whether payments in money made by an insolvent debtor to a creditor, the debtor not intending to give a preference, and the creditor not having rea-

sonable cause to believe a preference was intended, did nevertheless constitute a preference within the meaning of the bankrupt
act of 1898, and were required to be surrendered as a condition
of proving the balance of the debt or other claims of the creditor.

The solution of the question depends primarily upon the in-
trepretation of subdivisions " a " and " b," section 60, of the law
of July 1, 1898, c. 541, and certain related sections. Subdivision " a " of section 60 is as follows:

" Preferred Creditors.—a. A person shall be deemed to have
given a preference if, being insolvent, he has procured or suffered a judgment to be entered against himself in favor of any
person, or made a transfer of any of his property, and the effect
of the enforcement of such judgment or transfer will be to enable any one of his creditors to obtain a greater percentage of
his debt than any other of such creditors of the same class."

It will be observed that payments in money are not expressly
mentioned. Transfers of property are, and one of the contentions of appellants is that by " transfers of property," payments
in money are not intended. The contention is easily disposed
of. It is answered by the definitions contained in section 1.
It is there provided that " ' transfer ' shall include the sale and
every other and different mode of disposing of or parting with
property or the possession of property, absolute or conditional,
as a payment, pledge, mortgage, gift or security." It seems
necessarily to mean that a transfer of property includes the
giving or conveying anything of value—anything which has
debt paying or debt securing power.

We are not unaware that a distinction between money and
other property is sometimes made, but it would be anomalous
in the extreme that in a statute which is concerned with the
obligations of debtors and the prevention of preferences to creditors, the readiest and most potent instrumentality to give a
preference should have been omitted. Money is certainly property, whether we regard any of its forms or any of its theories.
It may be composed of a precious metal, and hence valuable of
itself, gaining little or no addition of value from the attributes
which give it its ready exchangeability and currency. And its

other forms are immediately convertible into the same precious metal, and even without such conversion have, at times, even greater commercial efficacy than it. It would be very strange indeed if such forms of property, with all their sanctions and powers, should be excluded from the statute, and the representatives of private debts which we denominate by the general term "securities" should be included. We certainly cannot so declare upon one meaning of the word "transfer." If the word itself permitted such declaration, which we do not admit, the definition in the statute forbids it. "Transfer" is defined to be not only the sale of property, but "every other mode of disposing or parting with property." All technicality and narrowness of meaning is precluded. The word is used in its most comprehensive sense, and is intended to include every means and manner by which property can pass from the ownership and possession of another, and by which the result forbidden by the statute may be accomplished—a preference enabling a creditor "to obtain a greater percentage of his debt than any other creditors of the same class."

But it is said "that Congress in passing the law had in mind the distinction between the payments of money and the transferring of property; otherwise they indulged in tautology" in subdivision (*d*). By that it is provided: "If a debtor shall, directly or indirectly, in contemplation of the filing of a petition by or against him, *pay money* or *transfer property* to an attorney and counsellor at law, solicitor in equity, or proctor in admiralty, for services to be rendered, the transaction shall be reexamined by the court on petition of the trustee or any creditor, and shall only be held valid to the extent of a reasonable amount to be determined by the court, and the excess may be recovered by the trustee for the benefit of the estate."

That all the words of a statute should, if possible, be given effect we concede, but tautology sometimes occurs. Is there not an example in subdivision (*e*) of section 67 (which, by the way, and notwithstanding, is relied on by the appellants)? It provides that "all conveyances, *transfers*, assignments, or incumbrances of his property, or any part thereof, made or given by a person adjudged a bankrupt," in fraud of creditors, shall be null and void as to them.

Manifest tautology, but certainly not used to detract from the definition of "transfer" in section 1, or to exclude application of that section in proper cases. Conveyances, assignments and incumbrances of property are but modes of its absolute or conditional disposition (transfer), as payment of money is a mode of its disposition (transfer), and there was a particular expression of each mode on account of the primary purpose to be secured in each case—the purpose being, in 60 (*d*), to control payments to attorneys; in 67 (*e*), the purpose being to prohibit the disposition of property by the debtor to persons other than creditors in fraud of the act.

But, construing transfers of property to include payments of money, it is nevertheless urged, that not only must the act and state of mind of the giving debtor be considered, but the act and state of mind of the receiving creditor must be considered. It is not enough that an advantage in fact be given, but to make it a preference "the person receiving it or to be benefited thereby, or his agent acting therein, shall have had reasonable cause to believe that it was intended, thereby to give a preference." In other words, it is contended that the quoted words should be read into subdivision (*a*) from subdivision (*b*), and the necessity of doing so is claimed to be established by other sections of the statute. The other sections are inserted in the margin.[1]

---

[1] SEC. 60 *c*. If a creditor has been preferred, and afterwards in good faith gives the debtor further credit without security of any kind for property which becomes a part of the debtor's estate, the amount of such new credit remaining unpaid at the time of the adjudication in bankruptcy may be set off against the amount which would otherwise be recoverable from him.

SEC. 3. Acts of Bankruptcy.—*a*. Acts of bankruptcy by a person shall consist of his having (1) conveyed, transferred, concealed, or removed, or permitted to be concealed or removed, any part of his property with intent to hinder, delay or defraud his creditors, or any of them; or (2) transferred, while insolvent, any portion of his property to one or more of his creditors with intent to prefer such creditors over his other creditors.

SEC. 3 *b*. A petition may be filed against a person who is insolvent and who has committed an act of bankruptcy within four months after the commission of such act. Such time shall not expire until four months after (1) the date of the recording or registering of the transfer or assignment when the act consists in having made a transfer of any of his property

Section 60 (*b*) is as follows:

" If a bankrupt shall have given a preference within four months before the filing of a petition, or after the filing of the petition and before the adjudication, and the person receiving it, or to be benefited thereby, or his agent acting therein, shall have had reasonable cause to believe that it was intended thereby to give a preference, it shall be voidable by the trustee, and he may recover the property or its value from such person."

Subdivisions (*a*) and (*b*) are concerned with a preference given by a debtor to his creditor. Subdivision (*a*) defines what shall constitute it, and subdivision (*b*) states a consequence of it— gives a remedy against it. The former defines it to be a transfer of property which will enable him to whom the transfer is made to obtain a greater percentage of his debt than other creditors. The latter provides a consequence to be that the transfer may be avoided by the trustee and the property or its value recovered, provided, however, that the preference was given four months before the filing of the petition in bankruptcy or before the adjudication, and the creditor had reason to believe a preference was intended. So far, so clear. If the con-

---

with intent to hinder, delay or defraud his creditors or for the purpose of giving a preference as hereinbefore provided, or a general assignment for the benefit of his creditors, if by law such recording or registering is required or permitted, or if it is not, from the date when the beneficiary takes notorious, exclusive, or continuous possession of the property unless the petitioning creditors have received actual notice of such transfer or assignment.

SEC. 67 *d.* Liens given or accepted in good faith and not in contemplation of or in fraud upon this act, and for a present consideration, which have been recorded according to law, if record thereof was necessary in order to impart notice, shall not be affected by this act.

SEC. 68. Set-offs and Counter Claims.—*a.* In all cases of mutual debts or mutual credits between the estate of a bankrupt and a creditor the account shall be stated, and one debt shall be set-off against the other, and the balance only shall be allowed or paid.

b. A set-off or counter claim shall not be allowed in favor of any debtor of the bankrupt which (1) is not provable against the estate; or (2) was purchased by or transferred to him after the filing of the petition, or within four months before such filing, with a view to such use and with knowledge or notice that such bankrupt was insolvent, or had committed an act of bankruptcy.

ditions mentioned exist, the preference may be avoided. But if the person receiving the preference did not have cause to believe it was intended, what then? It follows that the condition being absent, its effect will be absent. In other words, he may keep the property transferred to him, whether it be a complete or partial discharge of his debt. But if only a partial discharge, may he prove the balance of his debt or other debts?

Section 57 (*g*) provides for such case. "The claims of creditors," it provides, "who have received preferences shall not be allowed unless such creditors have surrendered their preferences."

There is certainly no ambiguity so far. What a preference is, is plain. What the effect of it is, if taken under the conditions mentioned, is equally plain. So taken, it may be recovered back. If not so taken, it may be kept or surrendered. Unless surrendered, he who received it cannot prove his debt or other debts. His election is between keeping the preference and surrendering it. That is the favor of the law to his innocence, but, aiming to secure equality between him and other creditors, can the law indulge farther? He may have been paid something—maybe a greater percentage than other creditors can be. That is his advantage, and he may keep it. If paid a less percentage he can obtain as much as other creditors by surrendering the payment, and an equality of distribution of the assets of the bankrupt is assured. The effect is equitable, and that it was intended is supported by prior legislation.

The bankrupt act of 1867 had provisions against preferences. Secs. 23 and 35; secs. 5084 and 5128, Rev. Stat. They could be recovered and had to be surrendered to enable the creditor to prove his debt, but the law was careful to express upon what condition in each case. They could be recovered back if the creditor had "reasonable cause to believe" the debtor was insolvent, and they were given "in fraud of the provisions of the act." Sec. 5128, Rev. Stat. They had to be surrendered if received under like condition. Section 5084, Rev. Stat., provided that "any person who . . . has accepted any preference *having reasonable cause to believe that the same was made or given by the debtor* contrary to any provision of the act,

March 2, 1867, chap. 176, . . . shall not prove the debt or claim on account of which the preference is given, nor shall he receive any dividend therefrom until he shall first surrender to the assignee the property, money, benefit or advantage received by him under such preference."

The words in italics are omitted from the act of 1898. Was the omission without purpose? The omission of a condition is certainly not the same thing as the expression of a condition. Was it left out in words to be put back by construction? Taken from the certainty given by prior use and prior decisions and committed to doubt and controversy? There is a presumption against it. When the purpose of a prior law is continued usually its words are, and an omission of the words implied an omission of the purpose. This rule we lately applied in *Bardes* v. *First National Bank of Hawarden*, 178 U. S. 524. In that case, in determining whether the jurisdiction of the Circuit and District Courts of the United States was concurrent with the state courts in certain suits at law and equity between the assignee in bankruptcy and the adverse claimant of property of the bankrupt, the statutes of 1841 and 1867 were compared with that of 1898, and from the omission from the latter of certain provisions of the former statutes it was decided that such jurisdiction did not exist. It was said by the court, speaking by Mr. Justice Gray: "We find it impossible to infer that when Congress, in framing the act of 1898, entirely omitted any similar provision, and substituted the restricted provisions of section 23, it intended that either of those courts should retain the jurisdiction which it had under the obsolete provision of the earlier acts."

We might rest the discussion here, but counsel have ably urged against our interpretation of the statute considerations which should be noticed. They assert its incorrectness because: (1) That the provisions of 57 (*g*) which deny allowance to the claims of creditors unless such creditors surrender the preferences they have received, are penal and should be strictly construed. Being penal, it is contended, there should be a guilty intent to incur their punishment. (2) Of the defectiveness of 60 (*a*) and the necessity of explaining it and enlarging it by other provisions. (3) Of the consequences of the construction—consequences which are declared to be anamolous and even absurd.

1. We cannot concur in the view that 57 (g) is a penal requirement. It is hardly necessary to assert that the object of a bankrupt act, so far as creditors are concerned, is to secure equality of distribution among them of the property of the bankrupt—not among some of the creditors, but among all of them. Such object could not be secured if there were no provisions against preferences—no provisions for defeating their purpose. And it is no reflection on the statute that it does not do so entirely. It allows complete payments, and counsel has seen and urges what seems to be inequitable in that—the giving favor to the diligence which secured it, and strongly argues that if complete payments may be retained without penalty, why not partial payments; if diligence (and diligence is made a great deal of in the argument) is favored in the one case, why not in the other? The view is too narrow and partial. Comparing such creditors, there may be inequality, but, considering other creditors, what shall be said? Some thought must be had of them, and considering them—indulgent creditors as well as diligent creditors—an attempt to secure the best remedies and results in the circumstances was, no doubt, the aim of the legislature. An advantage may be left with the preferred creditor. As we have already said, if the preference exceed the share of the bankrupt's estate which the creditor would be entitled to, he may keep the preference. If it be less, he may surrender it and share equally with the other creditors. If the purposes of the statute are to be considered, this is certainly not punishment but benefit. If it is discrimination at all, it is discrimination against the other creditors.

2. Undoubtedly all the sections of the act must be construed together as means to effect its purpose, and some of its sections are closely related. It does not follow, however, that each section should not be given the meaning its language conveys, if clear and consistent. It does not follow that because the terms of a section are defined elsewhere, or the consequences of its provisions are expressed elsewhere, it becomes a nullity or is

defective. Not that we may not " travel outside," to use counsel's expression, of any section if it be necessary to travel
outside. We may travel outside for some things, not necessarily for all things. The argument is, you must travel outside of subdivision (a) for a time within which the preference
must have been given, and four months are selected in analogy
to subdivision (b) and of section 3 (b). This may be conceded,
and the meaning of subdivision (a) would not be otherwise altered. There would still remain a clear definition of a preference.

The argument is strong which is urged to support a four
months' limitation, but it can be argued in opposition that subdivision (a) needs no explanation from other parts of the statute " in order to obtain a time limit on the question of preference." It can be argued that subdivision (a) gives such limit
in the existence of insolvency. But we are not required to decide either way on this record. A time limit is entirely independent of the belief of the creditor or of the belief which may
be attributed to him—entirely independent of his right to a
greater proportion of the bankrupt's property than other creditors. It is urged, however, that a time limit—whether of four
months or extending indefinitely before the filing of the petition in bankruptcy having no limit but the statute of limitations
—differently affects the creditor receiving the preference, and
the difference should be considered in construing the statute.
It is pointed out that insolvency has a different meaning under
the act of 1898 than it had under the act of 1867. Under the
latter, the debtor was insolvent when he was unable to pay his
debts in the ordinary course of business. Under the former,
when the aggregate of his property at a fair valuation is insufficient to pay his debts, and, it is said, this being practically impossible to ascertain on account of the uncertainty of its fac-
tors, therefore a time limit to a preference is necessary, and
also that there should be a guilty knowledge on the part of the
creditor of the guilty intent upon the part of the debtor. There
are two weaknesses in the argument. It ascribes a penal character to section 57 (g), and regards the requirement of the surrender of the preference as a condition of proving debts as a

punishment and not a provision to secure equality among creditors. On this we have sufficiently commented. The other weakness in the argument is that it exaggerates the difference between the definitions of insolvency and overlooks an advantage to the creditor in the definition contained in the act of 1898. Inability to pay debts in the ordinary course of business usually accompanies an insufficiency of assets. It may not, of course. At times a debtor's property, though amply sufficient in value to discharge all his obligations, may not be convertible without sacrifice into that form by which payments may be made. The law regards that possibility. In this there is indulgence to the debtor, and through him to preferred creditors. But the discussion need not be extended. The law has made its definition of insolvency, whatever the effect may be, and has determined by that definition consequences not only to the debtor but to his creditors and to purchasers of his property.

3. It is but one rule of construction that the consequences of a statute may be considered in construing its meaning. The rule may be counterpoised by other rules; it may be prevailed over by that one which requires the intent of the statute to be looked for in its words. Where they are clear and involve no absurdity, they are its only expositors. It is not contended that the provisions which we are considering are not clearly expressed and adequate to convey a definite meaning. It is true, it is urged that the word preference imports the conscious participation of the creditor and debtor in the same intent. We cannot concur in that view, and we are brought to the consequences of the construction which we have put upon section 60. It is denominated absurd by appellants. What is the test of absurdity? The contradiction of reason, it may be said, and to make an immediate application to legislation, the contradiction of the reason which grows out of the subject matter of the legislation and the purpose of the legislators. But all legislation is not simple nor its consequences obvious or to be controlled, even if obvious. Whether there should be any legislation at all and its extent and form may be matters of dispute. Its consequences may be viewed with favor or with alarm;

some regretted but accepted as inevitable—accepted as the shadow side of the good. In such situation it is for the legislature to determine, and it is very certain that the judiciary should not refuse to execute that determination from its view of some consequence which (to use the thought and nearly the words of Chief Justice Marshall) may have been contemplated and appreciated when the act was passed, and considered as overbalanced by the particular advantages the act was calculated to produce. *United States* v. *Fisher*, 2 Cranch, 358, 389. Therefore the sound rule expressed in *Sturgis* v. *Crowninshield*, 4 Wheat. 202: "It would be dangerous in the extreme, to infer from extrinsic circumstances, that a case for which the words of an instrument expressly provide, shall be exempted from its operation. Where words conflict with each other, where the different clauses of an instrument bear upon each other, and would be inconsistent unless the natural and common import of words be varied, construction becomes necessary, and a departure from the obvious meaning of words is justifiable. But, if in any case, the plain meaning of a provision, not contradicted by any other provision in the same instrument, is to be disregarded, because we believe the framers of that instrument could not intend what they say, it must be one in which the absurdity and injustice of applying the provision to the case, would be so monstrous that all mankind would, without hesitation, unite in rejecting the application."

So in *United States* v. *Goldenberg*, 168 U. S. 95, 103, where Mr. Justice Brewer, answering the argument based on the consequences of an act of Congress against the meaning expressed by its words, said:

"No mere omission, no mere failure to provide for contingencies, which it may seem wise to have specifically provided for, justify any judicial addition to the language of the statute. In the case at bar the omission to make specific provision for the time of payment does not offend the moral sense; *Holy Trinity Church* v. *United States*, 143 U. S. 457; it involves no injustice, oppression or absurdity, *United States* v. *Kirby*, 7 Wall. 482; *McKee* v. *United States*, 164 U. S. 287; there is no overwhelming necessity for applying in the one clause the same limitation of time which is provided in the other. *Non constat* but that Congress believed it had sufficiently provided for payment by other legislation in reference to retaining possession until payment or security therefor; or that it failed to appreciate the advantages which counsel insists will enure to the importer in case payment does not equally with protest follow within ten days from the action of the collector; or that, appreciating fully those advantages, it was not unwilling that he should enjoy them."

Let us apply these principles to the present case. The consequences of the construction of the Circuit Court of Appeals is said to be that it will "harass and embarrass the business of the country," and the specification is that any payment to a creditor may become a preference and the alternative forced upon him of giving it up or losing the right to prove his claim or claims against his debtor's estate. That consequence does not seem to us very formidable even in the instance of payments to private bankers by their depositors as illustrated by counsel or, as also illustrated if the payments should be distributed as gifts to relatives, or to endow universities, and cannot be obtained to be surrendered. Granting that such situation may be produced, is it anything after all but putting the creditor to an election of comparative and debatable courses where some loss must occur, whichever be taken? Business life has many such examples, and a law which has that consequence in seeking equality among creditors is certainly not absurd in even the loosest and most inconsiderate meanings of the word. Other illustrations are used which present the same situation or depend upon it—that is, the election which a preferred creditor is forced to make in order to prove his debts. A trader is insolvent and owes $100,000. His assets are $75,000. He owes $50,000 to A and B; the other $50,000 to other letters of the alphabet. He makes payments to the latter in order to prefer them, and then goes into bankruptcy. A and B having nonpreferred, hence provable claims, elect a trustee. What of the other creditors? Counsel having full control of the imaginary situation makes them ignorant of the debtor's affairs, and therefore unwilling to risk a division with A and B. That it is possible for such

ignorance and doubt to exist may be conceded, but it does not
occur to us how either can reasonably continue for the time
debts may be proved against the estate under the disclosures
required of the bankrupt by the statute, and the information
obtained by the trustee of the estate in its administration.

But is said a debtor may even make money by going into
voluntary bankruptcy, and the result is worked out by circumstances carefully imagined to that end, combined with, as
absolutely necessary to the result, the ignorance and timidity
of creditors. The illustration is that, suppose a bankrupt has
made partial payments to every one of his creditors within four
months preceding bankruptcy; that his assets at the time of
the filing of the petition amounted to $50,000, and his liabil-
ities to $100,000. Hesitating in this extraordinary situation
to surrender their payments—no creditor being tempted by
$50,000—the conclusion is confidently advanced that "if the
construction of the court below is sound, there are no creditors
who have provable claims against the bankrupt." And the
query is put, who gets the $50,000? The implied answer is,
that the bankrupt gets them, and the result is easily pronounced absurd. It is an absurdity which the "construction
of the court below" is not responsible for. What a court would
do with such a scheme as a fraud upon the act, we are not called
upon to say. We may well doubt if a scheme of that kind will
ever come up for decision. We find it impossible to conceive
a case in which $50,000, or, indeed, any surplus, would not be
an inducement to some creditor to add it, or some portion of it,
to the payment of his claim.

It is further contended "that to constitute a preference under
the bankruptcy act within either 57 (*g*) or 60 (*a*), at least *the
intent on the part of the bankrupt to prefer must be* present."
In support of this it is said that an act of bankruptcy consists
under section 3 (2) of a transfer by a debtor while insolvent of
any portion of his property to one or more of his creditors, *with
intent to* prefer such creditors over other creditors, and in such
a case a petition in involuntary insolvency may be filed against
him. Section 3 *b*. It is hence deduced, reading those provisions
with section 60 (*a*), that preferences under the latter must be

taken with the intent declared in the former, because it is not reasonable to assume that Congress intended that there could be preferences which were not acts of bankruptcy. The claim overlooks the fact that the language of section 3 (2) implies a difference between a preference and the intent with which it is given, and besides confounds the different purposes of the sections and their different conditions. It was for Congress to decide whether the consequences to a debtor of being forced into bankruptcy so far transcended the consequences to a creditor by a surrender of his preference, as to make the former depend upon an intent to offend the provision of the statute and the latter not so depend. And we see nothing unreasonable in the distinction or purpose. Nor does the contention of appellants find support in the provisions of the act of 1867, and the cases of *Mays* v. *Fritton*, 20 Wall. 414, and *Wilson* v. *City Bank*, 17 Wall. 473. In that act there was a careful expression of the intent of the debtor (section 5021, Rev. Stat.) and as careful an expression of the state of mind of the preferred creditor. Secs. 5084, 5128.

Nor again do we find anything which militates against our conclusion in subdivision "*c*" of section 60. That subdivision is applicable to the cases arising under "*b*," and allows a set-off which otherwise might not be allowed.

The interpretation of the statute which we have given has also been given by the Circuit Court of Appeals of the Ninth Circuit, in a well considered opinion by Circuit Judge Morrow, in the matter of *Fixen, Bankrupt*, 102 Fed. Rep. 295.

The second assignment of error is that the court erred in compelling the appellants to repay the amount of dividends received by them. Error is asserted because of the provision of subdivision "*b*" of section 23. The whole section is as follows:

"Jurisdiction of the United States and state Courts.—*a*. The United States Circuit Courts shall have jurisdiction of all controversies at law and in equity, as distinguished from proceedings in bankruptcy, between trustees as such and adverse claimants, concerning the property acquired or claimed by the trustees, in the same manner and to the same extent only as though bankruptcy proceedings had not been instituted and such controversies had been between the bankrupts and such adverse claimants.

"*b.* Suits by the trustee shall only be brought or prosecuted in the courts where the bankrupt, whose estate is being administered by such trustee, might have brought or prosecuted them if proceedings in bankruptcy had not been instituted unless by consent of the proposed defendant.

"*c.* The United States courts shall have concurrent jurisdiction with the courts of bankruptcy, within their respective territorial limits, of the offences enumerated in this act."

The proceedings we are reviewing were not a suit within the meaning of that section, and the order of the court requiring the repayment of the dividend was properly and legally made.

Judgment affirmed.

The CHIEF JUSTICE, Mr. Justice SHIRAS, Mr. Justice WHITE and Mr. Justice PECKHAM dissented.

———————

UNITED STATES *ex rel.* QUEEN *v.* ALVEY.

ORIGINAL.—PETITION FOR MANDAMUS.

No. 17. Original. Argued February 25, 1901.—Decided May 27, 1901.

Under the circumstances set forth in its opinion this court thinks that the rule respecting appeals to the Court of Appeals of the District of Columbia must receive the interpretation here which was given to it by the Court of Appeals.

UPON filing the petition for mandamus a rule was issued and served. The respondents have replied thereto. The question presented is the interpretation of a rule of the Court of Appeals of the District of Columbia hereinafter set out.

The case of petitioners as presented by their petition is substantially as follows: Marcella Jarboe, a widow, died without issue in the District of Columbia, on the 28th day of March,

1899, aged 88 years. The petitioners were her heirs at law. After her death a paper writing, purporting to be her will, dated February 24, 1892, and two other paper writings purporting to be codicils, dated respectively October 20, 1892, and February 15, 1898, were offered for probate by William Myer Lewin, executor, in the Supreme Court of the District of Columbia, holding a special term for orphans' court business, as her last will and testament.

The relators filed caveats to the probate of the will traversing the due execution of the papers as a will and alleging incapacity, undue influence and fraud. Upon the issue thus formed testimony was taken, and at its close the court instructed the jury to render a verdict for the will and codicils. Exception was duly made, and subsequently, on May 10, 1900, a motion for new trial was made and overruled, and an order was passed admitting the will and codicils to probate and directing letters testamentary to issue. An appeal was allowed to the Court of Appeals of the District, and a bond fixed for costs, not to operate as a supersedeas. The bond was duly approved, and filed May 17, 1900.

On July 2, 1900, the trial justice extended the time for filing the transcript forty days from the expiration of the time then limited. The transcript, however, was not filed within the extended time, and Mr. Justice Cole again extended it to October 15, 1900.

The transcript was filed October 9, 1900, but not until after appellees had given notice of a motion to docket and dismiss under the rule. When the motion came on to be heard it was abandoned, and by leave of the court a motion to dismiss was substituted. It was granted October 19, 1900, and the appeal dismissed with costs. This petition was then filed. The rule, the interpretation of which is involved, is as follows :

" All cases, the records and transcripts of which shall be received by the clerk of this court before the last twenty days of the term, shall be considered for trial in the course of that term ; but such cases shall be placed on the docket in the order of time in which the records or transcripts shall be received ; and if received within twenty days of the next succeeding term, either party shall be entitled to a continuance; but when an appeal is entered in the court below which shall operate as a supersedeas of the judgment, order or decree appealed from, or when there has been a special order or appeal bond for the stay or supersedeas of the judgment, decree or order appealed from, in all such cases it shall be the duty of the appellant, within forty days from the time of the appeal entered and perfected in the court below (unless such time for special and sufficient cause be extended by the court below, or the judge thereof by whom the judgment, decree, or order may have been rendered, such time to be definite and fixed), to produce and file with the clerk of this court a transcript of the record of such cause; and if he shall fail to file the transcript within the time limited therefor the appellee shall be allowed to file a copy or transcript of the record with the clerk of this court, and the cause shall stand for trial in the like manner as if the transcript had been filed by the appellant in due time; or the said appellee may, on producing a certificate from the clerk of the court below, stating the cause, and that an appeal has been entered, and the date thereof, and that the judgment, decree or order appealed from is stayed or superseded by bond or otherwise, have the said appeal docketed and dismissed; or, *in any and all cases*, the appellee may, after the time limited for filing the transcript in this court by the appellant, and his or her default in respect thereto, upon producing a certificate showing the entry of appeal and the date thereof, have said appeal docketed and dismissed; and in no case shall the appellant be entitled to docket a case and file the record after said appeal shall have been docketed and dismissed under this rule, unless by special order of the court, upon satisfactory reason shown."

The answer of the respondents alleged the promulgation of the rule in pursuance of the act of Congress creating the court, and that under the same act on the 29th of September, 1894, the court amended the rules in several respects and promulgated them as amended. The amendment consisted in the insertion of the words "*in any and all cases*" for the words "*in any case*," in numbered rule XV.

Mr. Walter D. Davidge and *Mr. Walter D. Davidge, Jr.,* for petitioners.

Mr. A. S. Worthington for respondents. *Mr. Charles L. Frailey* was on his brief.

Mr. Justice McKenna, after stating the case as above, delivered the opinion of the court.

By the act of Congress of February 9, 1893, which established a Court of Appeals for the District of Columbia, it was provided—

"That any party aggrieved by any final order, judgment or decree of the Supreme Court of the District of Columbia, or of any justice thereof, may appeal therefrom to the Court of Appeals hereby created; and upon such appeal the Court of Appeals shall review such order, judgment or decree, and affirm, reverse or modify the same as shall be just."

And it was also provided—

"That said Court of Appeals shall establish by rule of court such terms in the court in each year as to it may seem necessary: Provided, however, that there shall be at least three terms in each year, and it shall make such rules and regulations as may be necessary and proper for the transaction of its business and the taking of appeals to said court. And said Court of Appeals shall have power to prescribe what part or parts of the proceedings in the court below shall constitute the record on appeal and the form of bills of exception, and to require that the original papers shall be sent to it instead of copies thereof, and generally to regulate all matters relating to appeals, whether in the court below or in said Court of Appeals."

Under this provision the rule set out in the return of the respondents was established and amended. The question now is as to the interpretation of the rule. It will be observed that the rule states that "when an appeal is entered in the court below which shall operate as a supersedeas of the judgment, order or decree appealed from, or when there has been a special order or appeal bond for the stay or supersedeas of the judg-

ment, decree or order appealed from, in all such cases it shall
be the duty of the appellant, within forty days from the time
of the appeal entered and perfected in the court below, (unless
such time for special and sufficient cause be extended by the
court below, or the judge thereof by whom the judgment, decree
or order may have been rendered, such time to be definite and
fixed,) to produce and file with the clerk of this court a transcript of the record of such cause."

The contention of the parties turns on this provision. Is it to
be interpreted independently or in connection with and as receiving meaning from the subsequent provision commencing
with the words " *in any and all cases ?* " Or, in other words, is
the rule to be applied differently when the appeal operates as a
supersedeas from what it does when the appeal does not so
operate? The appeal of relators did not so operate, and the
relators contend that their cause " was not of the class of cases
to which the rule relates," and therefore no rule or authority
imposed on them the duty of filing the transcript within the
forty days, but that their case falls under that part of the rule
which provides for filing the record in cases where there was
no supersedeas or stay. " It does not enlarge in any manner,"
counsel say, " the cases specified in the former part of the rule,
and to which the duty of filing within forty days is confined."
The Court of Appeals held otherwise, and declares in its reply,
which is very circumstantial, that the rule even as originally
framed was intended to have a different meaning from that
which relators put upon it, but upon doubts arising it was
amended to remove the doubts, and "in all cases, whether
there had been a supersedeas or not, to fix a period of time
within which the transcript should be filed in the Court of Appeals (subject to the authority given by the rule itself, to the
court below or a judge thereof, to extend the time). Otherwise
there would have been no provision at all for cases in which
there should be no supersedeas."

The answer also states—

" The rule as so understood and construed by the respondents
has been enforced in every case in which it has been brought
to the attention of the respondents. So far as they know no

case has arisen since September 29, 1894, in which the transcript has not been filed within forty days from the time of the appeal entered and perfected in the court below, except where the time has been extended in accordance with the rule, by an order made by a judge of the court below before the expiration of the time limited by the rule or by a previous order. In the case of the *District of Columbia* v. *Humphrey*, 11 App. D. C. 68, the appeal was dismissed solely because the transcript was not filed in the Court of Appeals within the forty days prescribed by the rule in question, and without reference to whether the appeal operated as a supersedeas. The opinion of the Court of Appeals in that case was published among the regular reports of that court in 1898."

Under these circumstances we are of the opinion that the rule must receive the interpretation which was given it by the Court of Appeals.

Rule discharged.

CLEWS *v.* JAMIESON.

CERTIORARI TO THE CIRCUIT COURT OF APPEALS FOR THE SEVENTH CIRCUIT.

No. 245.   Argued April 17, 18, 1901.—Decided May 27, 1901.

As the governing committee of the stock exchange had no personal interest in the fund in question in this suit, which was placed in its possession in the trust and confidence that it would see that the purposes of the deposit were fulfilled, and that the moneys were paid out only in accordance with the terms of the trust under which it was deposited, there can be no question that the fund became thereby a trust fund in the possession of the governing committee, and the disposition of which, in accordance with the trust, they were called upon to secure. The committee occupied, from the time of the deposit of the funds, a fiduciary relation towards the parties depositing it, and became a trustee of the fund, charged with the duty of seeing that it was applied in conformity with the provisions creating it.

The jurisdiction of the court below was plainly established, because, under the circumstances, the complainant had no adequate and full remedy at law.

It plainly appears in this case from the pleadings that the sales and purchases of stock were in fact made subject to the rules of the stock exchange, and all the transactions regarding the sales and purchases must
be regarded as having taken place with direct reference and subject to
those rules.

A principal can adopt and ratify an unauthorized act of his agent, who in
fact is assuming to act in his behalf, although not disclosing his agency
to others, and when it is so ratified, it is as if the principal had given an
original authority to that effect, and the ratification relates back to the
time of the act which is ratified.

A contract which is, on its face one of sale, with a provision for future delivery is valid, and the burden of proving that it is invalid, as being a
cover for the settlement of differences, rests with the party making the
assertion.

There is nothing in these contracts which shows that they were gaming
contracts, and in violation of the statutes of Illinois; and there is no
evidence that they were entered into pursuant to any understanding
whatever that they should be fulfilled by payments of the difference between the contract and the market price at the time set for delivery.

The sales were made subject to the rules of the exchange, but those rules
do not assume to exclude the jurisdiction of the courts, or to provide an
exclusive remedy which the parties must follow.

The complainants were justified in the course which they pursued, and the
price at which the stock sold was a fair basis upon which to determine
the amount of damages.

THE petitioners and complainants, being residents of the State
and city of New York, commenced this suit in equity in the
United States Circuit Court for the Northern District of Illinois
against certain of the defendants composing the governing
committee of the Chicago Stock Exchange, to recover funds
deposited with them, in trust, and also to recover damages
against other defendants composing the firm of Jamieson &
Company, brokers belonging to the exchange, alleged to have
been sustained by the complainants by a violation by those defendants of their contract to purchase and pay for certain stock
sold them by the complainants. Still other defendants composed the firm of Schwartz & Company, the brokers who effected
the sales of the stock for the complainants, no recovery being
sought against them. All of the defendants were residents of
the State of Illinois. The Circuit Court after a hearing gave
judgment for a dismissal of the bill for want of any privity of
contract between complainants and defendants, Jamieson &

Company, against whom a money recovery was sought. On appeal the Circuit Court of Appeals for the Seventh Circuit affirmed the judgment of dismissal, and in the opinion discussed only the question whether or not the contract sued on was a gaming one and in violation of the statute of Illinois on that subject, sections 130 and 131 of the Criminal Code hereinafter set forth. It held that the contract violated those sections and that the bill was properly dismissed for want of equity, and it therefore affirmed the decree of dismissal. The complainants thereupon petitioned this court for a writ of certiorari, which was granted, and the case brought here.

No important question arises upon the pleadings, with the exception that it was set up by way of defence that the complainants had an adequate remedy at law, and the facts upon which the defence is rested are sufficiently adverted to in the opinion. The pleadings admit that the sales and purchases of stock were all made subject to the rules of the exchange. The case was referred to a master to take testimony and to report the same to the court with his conclusions thereon, and it was subsequently brought to a hearing upon the master's report and the testimony taken before him and upon a stipulation as to facts, entered into between the parties. The facts reported by the master are, among others, the following:

There has existed in the city of Chicago since the year 1882 a voluntary association known as the Chicago Stock Exchange, composed of brokers having places of business in the vicinity of the exchange, and who are elected to membership therein in accordance with the provisions of the constitution and by-laws; the association is governed by a governing committee composed of the president of the exchange *ex officio,* and twenty-four members, and every member is required to sign the constitution and by-laws, or assent thereto in writing, and obligate himself to abide thereby and by the rules theretofore or thereafter to be adopted.

Article 17 of the constitution provides as follows:

"SEC. 1. No fictitious sales shall be made. Any member contravening this section shall, upon conviction, be suspended by the governing committee.

"SEC. 2. Any member who shall make fictitious or trifling bids or offers or who shall offer to buy or sell any stock or security other than government bonds at a less variation than one eighth of one per cent shall, upon conviction, be subject to suspension, or such other penalty as the governing committee shall impose."

Article 29 is as follows:

"Any member of this exchange who is interested in or associated with, or whose office is connected directly or indirectly by wire or other method or contrivance, with any organization, firm or individual engaged in the business of dealing in differences or quotations on the fluctuations in the market price of any commodity or security without a *bona fide* purchase or sale of said commodity or security in a regular market or exchange, shall, on conviction thereof, be deemed to have committed an act or acts detrimental to the interest and welfare of the exchange."

Articles 16 and 17 of the by-laws read as follows:

"ARTICLE XVI.

"SEC. 1. In any contract either party may call at any time during the continuance of the same for a deposit of ten dollars per share upon the par value of the securities bought and sold; and whenever the market price of the securities shall change so as to reduce the margin of said deposit, either way below the ten dollars, either party may call for a deposit sufficient to restore the margin to ten dollars, and this may be repeated as often as the margin may be so reduced. In all cases where deposits are called they shall be made within one banking hour from the time of such call.

"SEC. 2. In case either party shall fail to comply with the demand for a deposit in accordance with the provisions of this article, the party calling, after having given due notice, may report the default to an officer of the exchange, who shall repurchase or resell the security forthwith in the exchange, and any difference that may accrue shall be paid over to the party entitled thereto. The notice above referred to shall be either personal or shall be left in writing at the office of the party to be notified, or in case he has no office, then by public announcement whenever the exchange may be in session.

*   *   *   *   *   *   *   *

"Article XVII.

"Should any member neglect to fulfill his contract on the day it becomes due, the party or parties contracting with him shall, after giving notice as required by section 2 of the preceding article, employ an officer of the board to close the same forthwith in the exchange by purchase or sale as the case may require, unless the price of settlement has been agreed upon by the contracting parties. In case of a failure of a creditor to close the contract as above, the price shall be fixed by the price current at the time such contract ought to have been closed under the rule. In all cases where an officer may be directed to buy or sell securities under this rule, the name of the member defaulting, as well as that of the member giving the order, shall be announced. No order for the purchase or sale of securities under this rule shall be executed unless made out in writing over the signature of the party giving the order, who shall state the reason therefor; and it shall be the duty of the officer who executes the order to indorse thereon the name of the purchaser or seller, the price and the hour at which the contract is closed, and hand the same to the secretary of the board, who shall within twenty-four hours ascertain whether the party for whose account the order was given has paid the difference, if any, arising from the transaction; if not, the secretary shall report the default to the president. The duty devolved upon the officers of the exchange under this rule shall be performed without charge. No party shall be permitted to supply offers to buy or sell securities closed for his account under the rule; and when a contract is closed under this rule, any action of the defaulter, direct or indirect, by which the prompt fulfillment of such contract is delayed, hindered or evaded, to the detriment of the other contracting party, shall subject the offending party to suspension for not less than thirty days in the discretion of the governing committee, by a vote of two thirds of the members present at the meeting.

When contracts are closed out under the rule, any member supplying the bid or offer, and not duly receiving or delivering the stock, as the case may be, renders himself liable to prosecution under this article. Should any stock thus sold not be delivered until the next day, the contract shall continue, but the defaulting party shall not be liable to pay such damage as may be assessed by the arbitration committee. The same rules as to notice, time and places that govern defaults in other contracts shall apply to borrowed securities, which on non-delivery or receipt, must be borrowed or loaned in open market, except in case of actual default in receiving or delivering after notice to close the loan; then the same are to be bought or sold, as the case may be, for account of the defaulter in the manner provided in this article."

The rules of the clearing house in regard to buying or selling for " the account" (under which these transactions were had) read as follows:

"·*Clearing House Rules.*

"Sec. 1. Under the following regulations transactions may be made for 'the account' in any securities listed for that purpose dealt in at the exchange.

"Sec. 2. Deliveries of cash, stock or transactions for 'the account' shall be made on the last day of each month. Provided, however, should the last day of any month occur on a holiday, or on a day when the exchange is closed for business, then in that case deliveries shall be made on the first business day preceding.

"Sec. 3. All purchases and sales 'for the account' shall be entered upon the blanks furnished by the manager sealed for that purpose, and said blanks properly filled out, balanced, accompanied by a proof-sheet, and signed, must be delivered to said manager before 9:45 A. M. It shall be the duty of the manager to compare and examine the statements rendered, and to report, should any errors be found, to the parties making such errors before 12 M., by written notice, which must be called for at the manager's office. Parties in error must at once proceed to adjust the same and correct their statements. All balances due from members as shown by the statements shall be paid by certified check drawn to the order of the bank designed for that purpose, and delivered to the manager before 10:15 A. M., the same day, except on Saturdays, when the balance must be paid before 9:45 A. M.

"Sec. 4. On balances due to members, as shown by the statements, a draft for the amount, payable to their own order, shall be drawn upon the bank designated for that purpose, and delivered to the manager before 10:20 A. M. (except on Saturday). The manager shall cause said draft, if correct, to be accepted by said bank and returned to the parties entitled thereto at the manager's office.

"Sec. 5. At or before 9:45 A. M. parties who have not borrowed or loaned their stock balances for 'the account' shall extend said balances on their statements at the closing bid price, designated as short or long, and shall request the manager, in writing, to borrow or loan said stock balances for their account and risk at the closing bid price. Notice that such loans have been made and the names of the parties thereto will be delivered at the manager's office on or before 2 P. M. Loans made by the manager are for one day only, unless renewed between members.

"Sec. 6. Stock balances as shown by the statements rendered for cash settling days must be delivered and paid for at the closing bid price of the previous day, as per manager's notices, before 1:30 P. M. Provided, however, if satisfactory evidence is shown the clearing house committee that the cash stock is on hand in New York or in transit for Chicago, three days' grace will be given the seller to make the delivery with interest, failing in which the clearing house committee shall cause to be purchased for account of delinquent said stock in whichever market in their judgment seems best; and the party so failing to deliver shall be held responsible for all loss or damage arising therefrom, but when a failure to receive or deliver occurs, nothing in these notifications shall be construed to relieve the last contracting parties to the transaction from the liabilities to each other.

"Sec. 7. Whenever a member fails to pay the balance due on his statement by 10:15 A. M. (except on Saturday), the manager shall notify the presiding officer of the exchange, whose duty it shall be to forthwith cause the stock balance, as shown by the statement of the delinquent, to be bought in or sold out under the rules, as the case may be, and assess the party in interest on the statement *pro rata*. In case any member owes an additional amount caused by errors, disputes or assessments, said amount shall be paid within one hour from the time of notification of the same, otherwise the party will be considered as having failed, and be treated accordingly.

"SEC. 8. Whenever a member is unable to meet his contracts or transactions made for 'the account,' he shall make a statement of his transactions, to be audited that day, and deliver it to the manager or presiding officer of the exchange.

"SEC. 9. The manager or any assistants employed by him in the manager's office are positively prohibited from receiving any securities or currency or any other evidences of value, except the checks and drafts hereinbefore mentioned in these rules.

"SEC. 10. The same rules as to notice, time and place that govern defaults in other contracts shall apply to transactions for 'the account.'

"SEC. 11. Neither the exchange nor any of its members (except those making the errors), the manager or any assistants employed by him, shall be responsible for any errors made in the statements to the manager, but the errors must be settled and adjusted at once between the members making said errors when notified by the manager to do so. The manager shall report any neglect or refusal to comply with these rules to the presiding officer of the exchange.

"SEC. 12. The margin to be deposited on stocks trading in clearing house, selling at $100 or over per share, shall be $10 per share, and on all stocks selling under $100 per share, the margin shall be $5 per share.

"SEC. 13. Margins deposited on trades in the clearing house shall be considered as a margin or as a part of same under section 1 of article XVI of the by-laws of the stock exchange. All such margins to be deposited in the clearing house.

"SEC. 14. The clearing of trades and money is not completed

until the trades and substitutions are all made and notice posted to that effect by the manager of the clearing house.

"SEC. 15. The brokers [shall] have the party they may trade with or party received from the clearing house on the substitution of the day before, in case of any failures between the hours the sheets are put in the clearing house, 9:45 A. M., and the time the notice is posted that the substitutions are ready for that day.

"SEC. 16. In the event of the announcement of the failure of any member to meet his contract, all stock bought in, on or sold out for him as 'account' stock shall be settled outside of the clearing house, and only such stocks as appear on the substitution sheet of the day of the failure shall be allowed to clear on the clearing house sheet of the following morning.

"SEC. 17. When any member fails to execute any contracts required of him by the clearing house, the margin checks deposited by such member for the protection of other members contracting with him through the clearing house, shall be held first for that special purpose, and after satisfying the claims of such members to the extent of the margin rule of the clearing house, the balance, if any, shall be held for a period not exceeding ten days, as a trust fund for a *pro rata* distribution among other creditors, who are members of the Chicago Stock Exchange."

The master further reported the facts relating to the sales in dispute as follows:

"That such rules and by-laws being in force, complainants, on the 16th day of July, 1896, wired their brokers, Schwartz, Dupee & Company, as follows:

"'Sell 500 Diamond Match at 220½ for account;' which was done; that later on the same day said brokers wired complainants as follows:

"'Sold 500 Diamond Match at 221⅛ for the account.'

"That on the 20th day of July, 1896, said brokers received telegram from complainants as follows:

"'Sell 200 Diamond Match at 221 for the account at opening of market.'

"That later on the same day said brokers wired complainants as follows:

"'Sold 200 Diamond Match at 221½ for the account.'

"That on the 25th day of July, 1896, complainants wired said brokers as follows:

"'Change the Diamond Match over to August account at 2½%. If you cannot do it let us know at once.'

"That shortly after on the same day complainants wired said brokers as follows:

"'You sent us the difference this morning at 2¼; at what difference can you do it now?'

"That later on the same day complainants wired said brokers as follows:

"'Change the 500 at 2 cents or better.'

"That afterwards, and about 12 o'clock on the same day, said brokers wired complainants as follows:

"'Bought Diamond Match 227 for the account; sold 500, 229 account 2d.'

"That on July 27th complainants wired said brokers as follows:

"'Change 200 more Diamond Match 2% or better.'

"Later on the same day said brokers wired complainants as follows:

"'We changed the 200 Match at 2¼ difference. Will give you price later.'

"And shortly afterwards on the same day said brokers wired complainants as follows:

"'Bought 200 Match 226¾, account; sold 200 second account 229.'

"That these purchases on July account balanced the sales on July account and left the brokers with sales made for complainants of 700 shares of the stock of the Diamond Match Company for the August account; that on the 3d day of August the clearing department of said stock exchange sent to Schwartz, Dupee & Company, and Jamieson & Company, clearing house sheets as follows."

Here follow copies of the sheets; that of Schwartz & Company showed that all trades on their sheet had been settled, with the exceptions therein stated, among which were 1150 shares of Diamond Match Company's stock at $222, for which

Jamieson & Company had been substituted as buyers; the notice to Jamieson & Company from the clearing department contained a like statement, showing that Jamieson & Company had bought 1150 shares of Diamond Match stock at $222, Schwartz & Company being substituted as sellers.

The 1150 shares of Diamond Match stock at $222 were made up in part of 700 shares sold by Schwartz & Company upon August account for the complainants, and the substitution of Jamieson & Company for the parties to whom such 700 shares had been originally sold was made by the clearing department of said stock exchange according to its uniform custom.

The master also found that Jamieson & Company had settled with Schwartz & Company for 450 shares of the 1150 shares referred to between those parties on the clearing house sheet of August 3, 1896, but that such settlement did not include the 700 shares in question in this case.

The master further found as to the manner of making sales " for the account:"

"That the method of doing business on said exchange is as follows: At ten o'clock there is an official call at which the secretary and manager call all the stocks, bonds and securities on the official printed list, and as this call progresses, any member wishing to buy or sell, bids thereon, and the record is made of the transaction; after which there is an irregular call, which closes at half-past one, when the manager of the clearing house announces the clearing house or settlement price for the day, which are the closing prices on the exchange for the respective stocks and securities; that the manager then substitutes trades and sends out cards to all buying or selling on account for the current month, or for the next month; that on the 25th of the · month and thereafter until the second day before the end of the month, two calls are made, one for the current month and one for the next ensuing month, and this is done to allow those who wish to do so to change their accounts over to the next month. That this substitution was made by the clearing department by a system somewhat similar to that employed by the clearing house for banks, that is, that where a broker has purchased and sold during the day the same amount of the same kind of stocks or bonds, his account is balanced by the clearing department, and all margins deposited by such broker may be withdrawn; that when sales and purchases are made by different brokers, one buying and the other selling, the same kind of stocks or bonds, a substitution is made by the manager of the clearing department, by which it appears that the broker selling has sold such stock, not to the person to whom it was originally sold, but to a person or persons other than those to whom such sales were originally made, and who originally bought of some one else, and that a broker purchasing stock has purchased from some broker other than the broker from whom he originally purchased the same; for instance, if A had sold 100 shares of stock to X, and B has bought the same amount of the same stock from Y, and X and Y's accounts are balanced by other transactions, the substitution would make it appear that A had sold 100 shares to B, and B had bought 100 shares from A, and the names of the parties with whom the original transactions had actually been made by A and B would not appear on the clearing house sheet; that in the transactions on said exchange it is then customary for the parties thus substituted and brought into the relation of buyer and seller with each other by the manager, to assent to the new relations thus formed, and to confirm the transactions as thus adjusted by the manager, and to put up the margins required by the rules, unless margins are already on deposit in the exchange, in which cases they are transferred by the manager to the new account.

"IV. That being advised of the substitution on account, as aforesaid, said Schwartz, Dupee & Company, and said Jamieson & Company on said 3d day of August, 1896, exchanged trading cards with each other, on which appears the following:

"'CHICAGO, Aug. 3, 1896.

"'M. Jamieson & Co.:

"'We hereby confirm sales made by us for the account to-day under the rules of the Chicago Stock Exchange, also substitution trades.

Am't.	Kind of property.	Price.
	Substitution trades—Sold.	
1150	Match	222
	Difference { Collect	
	{ Pay	287 50
(Signed)		" ' Schwartz, B. & Co.'

" ' Chicago, Aug. 3, 1896.

" ' M. Schwartz:

" ' We hereby confirm purchases made by us for the account to-day, under the rules of the Chicago Stock Exchange, also substitution trades.

Am't.	Kind of property.	Price.
	Substitution trades—Bought.	
1150	D. Match	222
	Difference { Collect	
	{ Pay	287 50
(Signed)		" ' Jamieson & Co.'

" That these cards were handed by the parties receiving them to the clearing house department, so that it appeared at the close of business on said 3d day of August, by the clearing sheet, that Schwartz, Dupee & Company had sold to Jamieson & Company on account, for August, 1150 shares of the stock of the Diamond Match Company, 700 shares of which are the stock in controversy in this case, delivery of which under rule 2 of the clearing house was to be made on the last day of August, 1896; that Schwartz, Dupee & Company and Jamieson & Company each deposited with the said clearing house, seven thousand dollars, as margins on said 700 shares of stock, which amount is still held by the said stock exchange, in trust.

" V. That on the 3d day of August, 1896, the governing committee, of which defendant Jamieson was then *ex officio* president, by virtue of his being then president of said exchange, held a meeting at which the following resolution was adopted, the said defendant Jamieson voting in favor of its adoption:

" ' *Resolved*, That the exchange adjourn on Tuesday morning, the 4th instant, and remain closed pending further action by this committee.'

" That pursuant to said action, said exchange did not open

on said August 4, or thereafter, until the 5th day of November, 1896.

"VI. That on the 31st day of August, 1896, Schwartz, Dupee & Company tendered to Jamieson & Company ten certificates of the stock of the Diamond Match Company for one hundred shares each, and three like certificates 50 shares each, making 1150 shares of said stock, which said Jamieson & Company examined and refused to receive."

On September 9, 1896, Schwartz & Company wrote the following letter to Jamieson & Company:

"CHICAGO, September 9, 1896.

"Messrs. Jamieson & Co., No. 187 Dearborn street, Chicago, Illinois.

"Dear Sirs: On August 31, 1896, we tendered you seven hundred (700) shares of Diamond Match stock in settlement of sales made by us. The sales made were 500 shares July 25th, and 200 shares July 27th, 1896, you being substituted through the clearing house of the Chicago Stock Exchange August 3, 1896, as the purchaser of the said stock.

"This is to notify you that said sales were made by us as agents for Henry Clews & Co. of New York, who may rightfully take any proceedings to enforce the contracts for said sales and who are authorized to make settlements therefor.

"Very truly yours,

"SCHWARTZ, DUPEE & Co."

(This tender of the 700 shares was part of the total tender made to Jamieson & Company on the 1150 shares sold them.)

The complainants on the next day (the 10th of September) gave Jamieson & Company notice in writing of their intention to sell 700 shares of Diamond Match Company stock, at public sale, to the highest bidder, and named the place and time, and that they would hold Jamieson & Company responsible for any loss on the sale on account of the contracts.

It was further admitted "that Schwartz, Dupee & Company have no claim whatever of any kind or character against the fourteen thousand dollars, seven thousand dollars of which was respectively contributed by Schwartz, Dupee & Company and

Jamieson & Company to the clearing house of the Chicago Stock Exchange." And it was testified that the $7000 deposited by Schwartz & Company were for the account of complainants, in whom is the real interest in such fund.

The stipulation as to facts signed by the parties for the purpose of the trial contained long and detailed statements of the actions of Jamieson & Company and Schwartz & Company in relation to all purchases and sales by them of Diamond Match Company's stock, for both July and August accounts, whether between themselves directly or not, and the stipulation ended with this statement:

"That the transactions heretobefore set out in this stipulation of purchase and sale of Schwartz, Dupee & Company, Jamieson & Company and the other brokers whose names are stated, with the exception of those transactions which are marked as substitutions, were had by the brokers on behalf of different clients or principals whom they represented, and those transactions, so far as the different principals are concerned, were not settled or canceled by any of the substitutions, nor by any of the settlements between the brokers, except so far as where one client or principal of a broker was, through such broker, both a purchaser and a seller.

"In other words, the settlements by substitutions or otherwise through the clearing house were merely settlements between the members of the stock exchange, and were not settlements or cancellations of the contracts between the principals whom the brokers represented and the brokers themselves, except where the same broker had both purchased and sold for the same client."

It was also admitted that when complainants gave their orders to sell and at the time that they were executed by Schwartz & Company the latter did not have in their hands any stock of the Diamond Match Company belonging to the complainants, nor did Schwartz & Company at any time thereafter have in their hands any of the stock of that company, which was the property of the complainants; that the 1150 shares of capital stock of the Diamond Match Company tendered to Jamieson & Company by Schwartz & Company in behalf of complainants, on August 31, 1896, were not the property of the complainants, nor any part thereof; that the 700 shares of stock alleged to have been sold in the bill of complaint, on September 22, 1896, were not delivered to the alleged purchaser after the sale, but were delivered to J. W. Conley, a member of the firm of Schwartz & Company by the individual who conducted the sale on behalf of the complainants, for safekeeping by Conley. The stock tendered belonged to Schwartz & Company, who tendered it on behalf and for the benefit of complainants.

The various facts set forth in the stipulation form a somewhat complicated mass of detail, and, when taken in connection with the oral evidence and the findings of the master, it is not clear that they are all perfectly consistent.

Upon the hearing before the master, Mr. Joseph R. Wilkins, the secretary and chairman of the Chicago Stock Exchange and manager of the clearing house, was called as a witness in behalf of the complainants. After giving a statement of the manner in which business was done on the exchange in relation to sales for "the account," he testified that the expressions in the telegrams from the complainants to the brokers, in which the word "difference" occurred, did not mean the difference between the then market price of the stock and the contract price, but meant the charges made for carrying the stock for the customer until the next delivery day. The price for this service differs from day to day, and is matter of agreement for each transaction; it is in effect the interest charged by the individual who carries the stock, on the amount necessary to carry it until the next delivery day. The rate of interest differs, of course, according to the demand, and is matter of agreement between the parties. The charge bears no relation whatever to the difference between the market price and the contract price of the stock. He also testified that a sale for "the account" on any day up to the 25th of the month means a sale of the stock which has to be delivered and paid for and taken at the end of the month. In other words, an actual delivery of the stock is contemplated by such a contract, and if a change from that delivery day to the next delivery day, thirty days thereafter, is asked for, it will depend upon the agreement of the parties upon what terms it shall be made. He also said that a sale for "the account" under the rules of the exchange assumed that there might be changes or substitutions of names during the period between the sale and the delivery day, and this happened by reason of the clearing house custom, under which all the sheets showing the transactions of the brokers in the sales and purchases in a given stock during the day were examined in the clearing house and the sheets balanced, so that at the end it appears there are a certain number of shares sold and the same number bought, and if the sheets do not balance the work stops and does not go on until a balance is made. After the balance is arrived at the substitution of names takes place, and the tickets or cards are sent to the brokers who are "long" and "short" of the stock respectively, and they then send to each other cards confirming the sale each day, and the cash deposit with the committee is added to by the one side and taken from by the other, according to the fluctuation of the stock, so that the full amount of deposit is kept at all times with the committee until the transaction is closed.

In regard to the fluctuations of price from day to day and the manner in which a party selling or buying at a certain price finally obtains or pays it on delivery day, although the original purchaser may have substituted another name at a different price, the witness explained that such original price was realized by means of the margin in the hands of the committee, which was added to daily by the party against whom the price of the stock turned, and drawn from by the party in whose favor it turned, so that, taking such payments and adding the price the stock actually sold for on the delivery day, the party selling or purchasing obtains his original selling or purchase price, which results in a loss or a gain, as the price of the stock on delivery day is higher or lower than the original contract price.

Mr. Henry D. Estabrook for petitioners. *Mr. Frank O. Lowden* and *Mr. Herbert J. Davis* were on his brief.

Mr. John H. Hamline and *Mr. Horace Kent Tenney* for

respondents. *Mr. Frank H. Scott* and *Mr. Frank E. Lord*
were on *Mr. Hamline's* brief. *Mr. Samuel P. McConnell*,
Mr. M. Lester Coffeen, *Mr. Charles F. Harding* and *Mr. James
H. Wilkerson* were on *Mr. Tenney's* brief.

Mr. Justice Peckham, after making the above statement of
facts, delivered the opinion of the court.

It is contended that there is an adequate and complete remedy at law for any liability that may arise by reason of the
transactions above set forth, and that therefore the bill was
properly dismissed and the decree of dismissal should be affirmed by this court.

It is undisputed that the defendants, the governing committee of the stock exchange, have in their hands the sum of
$14,000, the absolute title to which they do not claim. That
sum was deposited with them by Schwartz & Company and
Jamieson & Company, each depositing one half, for the purpose
of thereby securing the performance of the contract entered into
by those parties, and which sum was only to be taken from the
possession of the governing committee for the purpose of ful-
filling the condition upon which its deposit with the committee
was made. As that committee had no personal interest in or
title to the fund and it was placed in its possession in the trust
and confidence that it would see that the purposes of the deposit
were fulfilled and the moneys paid out only in accordance with
the terms of the trust under which it was deposited, there can
be no question that the fund thereby became a trust fund in
the possession of the governing committee and the disposition
of which in accordance with the trust those members were
called upon to secure. The complainants claim that pursuant
to the conditions of the trust they are entitled to the money
deposited with the committee. It is shown that the money
deposited by Schwartz & Company was deposited by them for
and in behalf of the complainants, and Schwartz & Company
lay no claim to the fund or any portion of it. Complainants
demanded from the committee the payment of the whole fund
to them on the ground that they were entitled to such payment

by the terms of the trust, and because of the violation of the contract by Jamieson & Company, to secure which the latter deposited $7000 of the fund in question. The committee has refused to pay over any portion of this fund to complainants, although it lays no claim to it, or any portion of it, on its own behalf. There is a dispute in regard to the right of the complainants to any portion of this fund, and a refusal on the part of the committee to pay it over to them. By reason of the facts, the committee occupied, from the time of the deposit of the funds, a fiduciary relation towards the parties depositing it, and it became a trustee of the fund charged with the duty of seeing that it was applied in conformity with the provisions creating it.

Pomeroy in his work on Equity Jurisprudence, second edition, instances, among other equitable estates and interests which come within the jurisdiction of a court of equity, those of trusts. In volume one, at section 151, he says: "The whole system fell within the exclusive jurisdiction of chancery; the doctrine of trusts became and continues to be the most efficient instrument in the hands of a chancellor for maintaining justice, good faith, and good conscience; and it has been extended so as to embrace not only lands, but chattels, funds of every kind, things in action, and moneys."

All possible trusts, whether express or implied, are within the jurisdiction of the chancellor. In this case the committee, as trustee, was charged with the performance of some active and substantial duty in respect to the management and payment of the funds in its hands, and it was its duty to see that the objects of its creation were properly accomplished. The fact that the relief demanded is a recovery of money only is not important in deciding the question as to the jurisdiction of equity. The remedies which such a court may give "depend upon the nature and object of the trust; sometimes they are specific in their character, and of a kind which the law courts cannot administer, but often they are of the same general kind as those obtained in legal actions, being mere recoveries of money. A court of equity will always, by its decree, declare the rights, interest or estate of the *cestui que trust*, and will compel the

trustee to do all the specific acts required of him by the terms
of the trust. It often happens that the *final* relief, to be obtained by the *cestui que trust* consists in the recovery of money.
This remedy the courts of equity will always decree when
necessary, whether it is confined to the payment of a single
specific sum, or involves an accounting by the trustee for all
that he has done in pursuance of the trust, and a distribution
of the trust moneys among all the beneficiaries who are entitled to share therein." 1 Pom. Eq. Jur. sec. 158.

In cases where the equity doctrine of trusts has been extended so as to embrace other relations of a fiduciary kind,
while it may not be said that a court of equity possesses exclusive jurisdiction, yet it is well settled that in such case there
is so much of the trust character between the parties so situated
that the jurisdiction of equity, though not exclusive, is acknowledged. 1 Pom. Eq. Jur. sec. 157.

In *Foley* v. *Hill*, 2 H. L. Cas. 28, a question arose over that
sort of relation which exists between a banker and his depos-
itor, and it was held to be merely that of debtor and creditor.
The court added however that, as between principal and factor,
an equitable jurisdiction attached, because the latter partook of
the character of a trustee, and that "so it is with regard to an
agent dealing with any property. . . . And though he is
not a trustee according to the strict technical meaning of the
word, he is *quasi* a trustee for that particular transaction," and,
therefore, equity has jurisdiction.

In *Marvin* v. *Brooks*, 94 N. Y. 71, it was held that an agent
who had been entrusted with his principal's money to be expended for a specific purpose might be required to account in
equity, and that upon such an accounting the burden was upon
him to show that his trust duties had been performed and the
manner of their performance. The jurisdiction was placed
upon the ground of a fiduciary or trust relation, and it was held
that a court of equity had jurisdiction over trusts and those
fiduciary relations which partake of that character, and in such
cases the right to an accounting is well established; but it was
held that the existence of a bare agency was not sufficient. It

must be an agency coupled with some distinct duty on the part of the agent in relation to funds or some specific property.

In 2 Story's Eq. Jur. (12th ed.) it is stated, at section 975*a*, that in general a trustee is suable in equity in regard to any matters touching the trust.

In *Oelrichs* v. *Spain*, 15 Wall. 211, 228, the court remarked that there being an element of trust in the case, that element, wherever it existed, always confers jurisdiction in equity.

That the governing committee could file a bill of interpleader against the complainants and the other defendants, alleging that each claimed the fund, or some portion thereof, and ask the court to determine which of the parties was entitled to the same, furnishes no reason for excluding the jurisdiction of equity in this case.

It may be somewhat doubtful whether an action against these defendants could be maintained at law, the contract not being originally between Schwartz & Company and Jamieson & Company, but only becoming so by way of substitution under the rules of the clearing house, and the relief sought being different between the two sets of defendants, Jamieson & Company and the members of the governing committee of the stock exchange. The maintenance of this suit enables the whole question between all the parties to be determined therein, and prevents the necessity of any action at law or other proceeding in the courts for the purpose of determining the ultimate and final rights of all the parties to this suit. Such relief cannot be obtained in any one action at law.

Upon all the facts we think that the jurisdiction of the court was plainly established, because under the circumstances the complainants had no adequate and full remedy at law.

We are then brought to the question decided by the Circuit Court, which held that there was no privity of contract between the complainants and Jamieson & Company. Aside from the general rule that a party sending an order to a broker doing business in an established market or trade for a transaction in that trade, thereby confers upon the broker authority to deal according to any well-settled usage in such trade or market, *Bibb* v. *Allen*, 149 U. S. 481, 489, it plainly appears in this case from the pleadings that the sales and purchases of stock were in fact made subject to the rules of the exchange, the complainants alleging in their bill that such was the fact, while the defendants Jamieson & Company in their answer make a like claim.

All the transactions regarding the sales and purchases of the various shares of stock mentioned in this case must, therefore, be regarded as having taken place with direct reference and subject to those rules.

The Circuit Court did not question that upon the facts stated a contract came into existence whereby primarily Schwartz & Company were obliged to sell to Jamieson & Company 700 shares of the stock named at the price of $222 per share, and it found no difficulty in holding that the undisclosed principals of either of these parties were entitled to step into the places of these respective brokers, and in their own name and for their own benefit insist upon the enforcement of the contract according to its terms; that under the rules of the exchange each of the brokers bound himself to the other broker and the principals whom the other broker represented to carry out the terms of the contract, but the court held that the evidence disclosed that Schwartz & Company were only clothed with the authority to sell the stock at $229, and that their principals, the complainants herein, were not bound by a sale at any figure less than that sum, and that neither Schwartz & Company nor any persons with whom that firm had contracted could have compelled the complainants to deliver the stock at a price less than $229. As the fact appeared that the contract between the respective brokers was for a sale at $222, the defendants Jamieson & Company, even under the substitution provided for by the rules of the stock exchange, could not hold complainants as principals of the contract for a sale at that price, and the court held that for want of mutuality the complainants are in no position to hold those defendants; that there was no identity of contract between the one the complainants authorized and the one entered into between the brokers, and the fact that the complainants now choose to accept it is of no consequence, the legal fact remained that they are not so bound, and, not being so bound, the defendants Jamieson & Company on their part are not legally bound.

In this case, although the brokers on the exchange acted in their own name, yet in fact each acted for undisclosed principals. In regard to 700 shares Schwartz & Company acted for the complainants, and in regard to 450 shares they acted in behalf of other clients. If the contract had been for the sale and purchase of these shares at $229, there would have been no difficulty in the case upon the principle adopted by the Circuit Court. The bar to a recovery lay in the alleged fact that the sale was without authority, although really procured by Schwartz & Company while acting as agents of the complainants.

A principal can adopt and ratify an unauthorized act of his agent who in fact is assuming to act in his behalf, although not disclosing his agency to others, and when it is so ratified it is as if the principal had given an original authority to that effect and the ratification relates back to the time of the act which is ratified. He must disavow the act of his agent within a reasonable time after the fact has come to his knowledge, or he will be deemed to have ratified it. Bringing a suit upon the contract of his agent which was unauthorized at the time and in excess of the authority conferred upon the agent is a ratification of the unauthorized act; and it is no answer to the ratification that prior to its taking place the principal is not bound, and hence there is no right on the part of the other party to enforce as against him the unauthorized act of his agent. These principles are well known, and may be found laid down in the following text books and authorities: Story on Agency, 9th ed. sec. 90, note 7; secs. 248, 251 and 251*a*, and note; secs. 258, 259; Livermore on Agency, page 44; Dunlap's Paley on Agency, 4th Am. ed. marginal page 324, note; *Lucena* v. *Craufurd*, 1 Taunton, 325, 334, 336; *Routh* v. *Thompson*, 13 East, 274, 283; *Hagedorn* v. *Oliverson*, 2 Maule & Selw. 485; *Fleckner* v. *Bank of United States*, 8 Wheat. 338, 363; *Law* v. *Cross*, 1 Black, 533, 539, citing *Hoyt* v. *Thompson*, 19 N. Y. 207, 218, 219; *Cooke* v. *Tullis*, 18 Wall. 332, 338.

Therefore if in fact the sale at $222 had been unauthorized

on the part of Schwartz & Company, the subsequent ratification of their unauthorized act by the complainants was the same as a precedent authority to them.   The failure of the complainants to repudiate the action of their agents in the sale immediately after it was reported to them would operate as a ratification. They not only failed to repudiate, but actually approved the action, and notified the defendants Jamieson & Company that the sales made by Schwartz & Company to the extent of 700 shares of stock had been made for them, and that they should hold Jamieson & Company liable upon the contract and for any damage caused by its violation.

It is argued, however, on the part of complainants that there was no unauthorized action by Schwartz & Company, and in proof thereof an explanation is given and an argument made founded thereon in relation to the peculiar facts which attend the sale and purchase of stock on "the account" on the floor of the stock exchange at Chicago.   The very term itself imports, as is stated and as the evidence shows, a sale of stock to be delivered at a future time, and under the rules of the exchange that time means the last day of the month in which the sale or purchase is made.

Under these same rules, when an agreement to sell for future delivery is effected, each party places a margin in the hands of the governing committee for the purpose of securing the performance of the contract, and, as is set forth in the foregoing statement of facts, this sum is kept intact in the hands of the committee until the final closing of the transaction, and upon a sale for "the account" the fluctuation in the price of the stock is provided for by payment into the fund upon the part of the one against whom the price of the stock has turned, and by drawing out of that same fund by the party in whose favor the price was, and so at the delivery day, whatever the price may be, the party selling gets the market price of the stock on that day, and the difference between that and the contract price he has received by payments into the fund in the hands of the governing committee by the other party and his withdrawal of the same sums, making in that way the contract price of the stock. Hence, it is argued, on the part of complainants that the sale at $222 was entirely proper, and in accordance with the previous authority given complainants' agents, because the difference between $229 and $222 complainants' agents had already received by a draft drawn upon the fund in the hands of the governing committee. This is upon the assumption that there had been a margin put up by the parties to the sales on the July account in accordance with the rules, which had been carried over to the August account, and that into this deposit the money had been paid as the stock dropped from July 25 to August 3, and Schwartz & Company had drawn the same out.

If this plainly appeared in the testimony, the findings or the stipulation of the parties, it would be an answer to the contention that the act of Schwartz & Company in selling at $222 was unauthorized. It is, however, answered on the part of Jamieson & Company that there is no evidence that this fund had been drawn from and paid into by the respective parties, and hence there is no basis of fact appearing in the record upon which the argument can rest. Counsel allege that the statements on the part of the complainants are at variance with the conceded facts in the case. They say in the first place that the bill itself avers that this deposit was made when the contract of August 3 for 1150 shares, was entered into, and that the answers of the governing committee and of Jamieson & Company expressly state that the deposit was made on that day. If this fund were not created until August 3, it could not have been drawn from by the agents of complainants in the July previous, and so it would be impossible for the complainants to have received moneys from that fund prior to that date. Although the rules of the stock exchange require the deposit of these margins, and in cases where a sale for "the account" has been changed from one month to the following, the rules and the practice of the exchange require that the deposit on the old account shall be transferred to the new, yet still it is said that the rules or practice requiring such deposit cannot supply the place of evidence of a fact when the pleadings expressly state the opposite.

It seems to us quite evident, after a perusal of the whole record and from the manner in which the case was tried, that it was assumed that a deposit of the moneys for the first July sales was made and that such deposit remained and went over into the new account of and for August delivery, although such assumed fact may be inconsistent with the allegation in the pleadings in regard to the date of the deposit, which was alleged to be August 3. There is perhaps this technical inconsistency, yet assuming it to be as claimed on the part of counsel for Jamieson & Company, it does not touch the fact that the complainants ratified the action of their agents, Schwartz & Company, in selling at $222.

Aside from these questions, however, it is claimed on the part of Jamieson & Company that the record shows there never was any privity of contract between these parties, complainants on the one side, and Jamieson & Company on the other, because there were contracts on the part of Schwartz & Company for other dealers in the same stock, and that such contracts were not closed on August 3. Their claim is, even assuming that on August 3, Schwartz & Company contracted to sell to Jamieson & Company 1150 shares of stock at $222, deliverable August 31, the record shows that the complainants were not alone the interested parties to that contract. It is averred that 700 shares of the 1150 shares sold by Schwartz & Company to Jamieson & Company, on August 3, were for the account of the complainants, but it also appears that of the 1150 shares, 450 were sold for the account of others. These latter shares have, however, been settled for between the respective brokers. We are not concerned with the terms of the settlement or any admission or liabilities resulting therefrom, but the fact of such settlement eliminates all questions in regard to those 450 shares and leaves the 700 shares remaining, which were the shares sold by Schwartz & Company as agents for the complainants. The fact that there were in this sale of August 3 other shares than the 700, and that in regard to those others some had been sold originally by Schwartz & Company to other and different brokers than Jamieson & Company, will not prevent the contract as to the 700 shares from being enforced by complainants against Jamieson & Company, although but for such settlement there might have been some embarrassment in maintaining a suit against the latter for a portion only of the total shares sold them, while the other portion was represented by different clients of Schwartz & Company. The splitting up of the contract into two or more claims in behalf of different principals of Schwartz & Company and bringing different suits by the different principals against Jamieson & Company, on the single contract, might be in violation of the general rule refusing to recognize such right, but where all other claims have been settled and there remains but the one demand against the defendants, the objection does not apply, and we see no reason why the complainants may not take advantage of the contract made by their agents and enforce the same against Jamieson & Company.

Selling "for the account" is not an invention of the Chicago Stock Exchange. It has been practiced upon the London and the New York and other stock exchanges for many years, and the general rules governing it are much the same on all of them. Thus it is said in Dos Passos on Stock Brokers and Stock Exchanges, page 276, as follows:

"It also appears in accordance with the usages of the stock exchange that the broker may, in executing the order of a client, enter into a contract for the specific amount of stock ordered to be bought or sold, or may include such order with others he may have received in a contract for the entire quantity, or in quantities at his convenience.

"Neither in stock exchange contracts is there any real appropriation to any particular client of any particular stock in any transaction entered into with the jobber. Each transaction only forms an item in an account with that jobber, or, more correctly, with the house generally—that is to say, specific delivery or acceptance of that amount of stock is not necessarily made; but the transaction is liable to be balanced at any time during that account by a counter transaction by the same broker on behalf of the same or any client, or even on his own behalf, so that the balance only of all purchases and sales of that particular stock made by the broker in the house generally is to be finally accepted or delivered by him, and this through the instrumentality of the clearing house and the system of tickets."

The rules of the Chicago exchange clearly contemplate and provide for a substitution of names between the selling and the delivery days, and each party is kept secured by the margin originally put up, which is added to and taken from as the stock fluctuates in price from day to day. Hence it may be that the parties buying or selling may by virtue of this rule be liable to different principals represented in one original contract between the brokers. Whatever the rules or practice of the exchange may be, it is of course plain that no principal can be held to the performance of a contract which he never made, authorized or ratified. The stipulation made between the parties relating to this matter, while not entirely plain, might affect the right to maintain this action but for the fact that all other claims were settled, leaving only the controversy regarding the 700 shares to be disposed of between these parties. Upon the facts before us we think there was sufficient privity of contract between them to sustain this suit.

The view taken by the Circuit Court of Appeals in regard to this case was that the contracts were void as being in violation of the terms of the Illinois statute, sections 130 and 131, which are set forth in the margin.[1] It is a very far-reaching decision,

---

[1] Sec. 130. Whoever contracts to have or give to himself or another the option to sell or buy, at a future time, any grain or other commodity, stock of any railroad or other company, or gold, or forestalls the market by spreading false rumors to influence the price of commodities therein, or corners the market, or attempts to do so in relation to any such commodities, shall be fined not less than $10 nor more than $1000 or confined in the county jail not exceeding one year, or both; and all contracts made in violation of this section shall be considered gambling contracts, and shall be void.

Sec. 131. All promises, notes, bills, bonds, covenants, contracts, agreements, judgments, mortgages, or other securities or conveyances made, given, granted, drawn or entered into, or executed by any person whatsoever, where the whole or any part of the consideration thereof shall be for any money, property or other valuable thing, won by any gaming or playing at cards, dice or other game or games, or by betting on the side or hands of any person gaming, or by wager or bet upon any race, fight, pastime, sport, lot, chance, casualty, election or unknown or contingent event whatever, or for the reimbursing or paying any money or property knowingly lent or advanced at the time or place of such play or bet, to any person or persons so gaming or betting, or that shall, during such play or betting, so play or bet, shall be null and void and of no effect.

and if followed would invalidate most transactions of every stock exchange in the country "for the account." We are unable to agree with the opinion of the court on this question.

"The generally accepted doctrine in this country is, as stated by Mr. Benjamin, that a contract for the sale of goods to be delivered at a future day is valid, even though the seller has not the goods, nor any other means of getting them than to go into the market and buy them; but such a contract is only valid when the parties really intend and agree that the goods are to be delivered by the seller and the price to be paid by the buyer; and, if under guise of such a contract, the real intent be merely to speculate in the rise or fall of prices, and the goods are not to be delivered, but one party is to pay the other the difference between the contract price and the market price of the goods at the date fixed for executing the contract, then the whole transaction constitutes nothing more than a wager, and is null and void."

This quotation with the doctrine therein stated is approved in *Irwin* v. *Williar*, 110 U. S. 499, 508.

As a sale for future delivery is not on its face void, but is a perfectly legal and valid contract, it must be shown by him who attacks it that it was not intended to deliver the article sold, and that nothing but the difference between the contract and the market price was to be paid by the parties to the contract. And the fact that at the time of making a contract for future delivery the party binding himself to sell has not the goods in his possession and has no means of obtaining them for delivery, otherwise than by purchasing them after the contract is made, does not invalidate the contract. *Hibblewhite* v. *McMorine*, 5 M. & W. 462. Parke, Alderson and Maule, barons, before whom the case was heard, were unanimously of this opinion.

In order to invalidate a contract as a wagering one, both parties must intend that instead of the delivery of the article there shall be a mere payment of the difference between the contract and the market price. *Pearce* v. *Rice*, 142 U. S. 28; *Pickering* v. *Cease*, 79 Illinois, 328. In the latter case it was stated:

"Agreements for the future delivery of grain, or any other commodity, are not prohibited by the common law, nor by any statute of the State, nor by any policy adopted for the protection of the public. What the law does prohibit, and what is deemed detrimental to the general welfare, is speculating in differences in market values. The alleged contracts for August and September come within this definition. No grain was ever bought and paid for, nor do we think it was ever expected any would be called for, nor that any would have been delivered had demand been made. What were these but 'optional contracts,' in the most objectionable sense; that is, the seller had the privilege of delivering or not delivering, and the buyer the privilege of calling or not calling for the grain, just as they chose. On the maturity of the contracts, they were to be filled by adjusting the differences in the market values. Being in the nature of gambling transactions, the law will tolerate no such contracts."

And in *Pearce* v. *Rice*, 142 U. S. 28, 40, it was remarked:

"But the evidence before us is overwhelming to the effect that the real object of the arrangement between Hooker & Company and Foote was, not to contract for the actual delivery, in the future, of grain or other commodities—which contracts would not have been illegal (*Pickering* v. *Cease*, 79 Illinois, 328, 330)—but merely to speculate upon the rise and fall in prices, with an explicit understanding, from the outset, that the property apparently contracted for was not to be delivered, and that the transactions were to be closed only by the payment of the differences between the contract price and the market price at the time fixed for the execution of the contract."

A contract which is on its face one of sale with a provision for future delivery, being valid, the burden of proving that it is invalid, as being a mere cover for the settlement of "differences," rests with the party making the assertion. A defence of the illegality of the contract was pleaded by the defendant in *Cothran* v. *Ellis*, 125 Illinois, 496. In speaking of the burden of proof the court (at page 506) said:

"The facts alleged in the defendant's pleas, and put in issue by the plaintiff's traverse, are the only controverted facts in this case, and the *onus probandi* was upon the defendant. If the latter had offered no evidence at all, it would not have been necessary for the plaintiff to offer any, for the jury are always bound to find the facts against the party having the burden of proof, if he offers no evidence in support of the issues."

In *Irwin* v. *Williar*, 110 U. S. 499, 507, the trial judge in substance charged the jury that the burden of showing that the parties were carrying on a wagering contract and were not engaged in legitimate trade or speculation rests upon the defendant. Contracts for the future delivery of merchandise or stock are not void, whether such property is in existence in the hands of the seller or to be subsequently acquired. On their face these transactions are legal, and the law does not, in the absence of proof, presume that the parties are gambling. The proof must show that there was a mutual understanding that the transaction was to be a mere settlement of differences; in other words, a mere wagering contract. This charge was approved by this court, and the principle was again approved in *Bibb* v. *Allen*, 149 U. S. *supra*.

Taking the contracts in this case as evidenced by the various telegrams passing between the complainants and their agents, Schwartz & Company, and having in mind the manner in which the business was in fact transacted, we are unable to find any evidence upon which to base a holding that the contracts came within the statutes of Illinois on the subject of gaming. There was no proof that there was a mutual understanding that the transactions were to be settled by a mere payment of "differences," and that there was to be no delivery, nor, in our judgment could any inference to that effect be legitimately drawn from the undisputed facts. In the first place it is proper to consider the rules of the stock exchange where the business was done. We find that article 17 of the constitution provides in section 1, "that no fictitious sale shall be made. Any member contravening this section shall upon conviction be suspended by the governing committee." Article 29 prohibits any member of the exchange from being interested in or associated with any organization engaged in the business of dealing in differences or quotations on the fluctuations in the market price of any commodity or security, without a *bona fide* purchase or sale of said commodity or security in a regular market or exchange. These two rules provide on their face that no sale for mere collection of differences is allowed ; that every sale must be one in good faith for the delivery, either present or future, of the article sold. Sales "for the account" under the rules are made upon the basis of an intended actual delivery of the stock at the time when due. The evidence upon this point is undisputed.

A contract for the mere settlement of differences is a violation of the rules of the organization under which these brokers were doing business. Neither the rules of the exchange nor those of the clearing house set forth in the foregoing statement provide for these wagering contracts. Some of them provide for the course to be pursued where a member fails to fulfill his contract. They do not provide as a means for the fulfillment of such contract the payment of "differences," but point out a course which the party claiming the fulfillment may pursue as against the party who violates the contract. Rule 17 treats the party failing to fulfill as a defaulter, and his name as a defaulter is announced. Sections 1 and 2 of article 16 provide for the failure of either party to keep up his margin, and the failure is described as a default. To say that such rules afford strong ground to infer an understanding between the parties doing business subject to them—that their contract was not one of actual sale, but merely one to speculate upon "differences" —is, in our opinion, to presume an illegal contract against its plain terms, and without any sound basis for the presumption. Thus, if an individual agreeing to purchase and pay for certain stock at a future date fails or refuses to perform his contract, the stock is sold under the rule, the price received and the difference between the price at which it sold and the contract price he is held answerable for. That would be his legal liability, in any event, and we cannot agree that the rules made for the case of a violation of contract provide or were intended to provide a means for its fulfillment. In case of a violation, the rules merely afford an expeditious means of ascertaining the amount of the damages. Of course, we do not say that these rules actually prevent gambling on the exchange. It is possible, if not probable, that gambling may be and is in fact carried on there, but it must be in violation of and not pursuant to the rules.

Recurring, then, to the terms of these contracts, there is nothing therein which shows that they were gaming contracts, and hence in violation of the Illinois statute. They were plain directions to sell certain named stock for "the account," the meaning of which was that the stock was to be sold for actual delivery on the next delivery day, being the last day of the month. Such a direction presumes the intention to deliver the stock at the time named upon the receipt of the purchase price thereof as agreed upon at the time of the sale. There is no presumption opposed to this view in the absence of any evidence upon which it can rest. The fact that at the time of the sale the complainants did not own any of the stock cannot support the presumption, because it is perfectly valid to make such a sale, and an illegal intent accompanying the performance of a perfectly legal act cannot be presumed. The subsequent telegrams directing the changing of the delivery time from the July to the August account, after inquiring in regard to the difference upon which such change could be effected, furnish no evidence of any illegal intention in connection either with the original or the changed contracts.

The "difference," as explained by the testimony set out in the foregoing statement, related to the charges to be made for carrying the stock from the July to the August delivery day, and did not relate to the payment of any difference between the contract price and market price of the stock. A direction to change the 500 shares from the July account to the August account would mean, as Mr. Wilkins, the manager of the stock exchange, testified, that the party who had agreed to sell 500 shares of stock, deliverable in July, did not wish to deliver on that day, and the direction to change to the August account meant that the agents were to buy in that number of shares and sell them out again for the August account, keeping "short" the same amount of stock and making the difference in that case of $2.50 a share, or $250 on every 100 shares of stock, for carrying it for another month, and this charge was the interest which the party would have to pay to him who was on the other side of the market and who would carry it to the next delivery day, 30 days thereafter.

There is nothing in the whole transaction from which it can be reasonably said that at the time when the original July order to sell was given there was any intention to do otherwise than make delivery of the stock at the July settlement day, and a delivery must have been then made by the very terms of the contract, as also under the rules of the exchange, unless there might thereafter be a change of that agreement by postponing the delivery to the August account. If there were no such subsequent agreement, then the delivery must have been made in July, but the seller might, in order to make it, enter into another agreement with some one else to take it off his hands upon such terms as might be agreed upon. There is absolutely no evidence that these contracts were entered into pursuant to any understanding whatever that they should be fulfilled by payments of the difference between the contract and the market price at the time set for delivery. To hold otherwise would entirely prevent any dealing in stocks for "the account," including of course a case where for any reason the delivery day should be changed from the one originally intended to another and a future day.

To uphold the rulings of the Circuit Court of Appeals herein the cases of *Pickering* v. *Cease*, 79 Illinois, 328; *Lyon* v. *Culbertson*, 83 Illinois, 33; *Tenney* v. *Foote*, 95 Illinois, 99; *Pearce* v. *Foote*, 113 Illinois, 228; *Cothran* v. *Ellis*, 125 Illinois, 496; *Schneider* v. *Turner*, 130 Illinois, 28, and *Soby* v. *The People*, 134 Illinois, 66, have been cited. We have examined them all, and are unable to see that they justify the ruling herein.

These cases hold these various propositions:

(1) That "option contracts" to sell or deliver grain or other commodity, or railroad or other stock, which contracts are intended to be settled by payment of differences at the settling date, are invalid. 79, 83, 113 and 125 Illinois, *supra*.

(2) A contract to have or give to himself an option to sell or buy at a future time any grain, etc., subjects the party to fine or imprisonment, and all contracts made in violation of the stat-

ute are gambling contracts and void under section 130, Criminal Code, and all notes or securities, part of the consideration of which is money, etc., won by wager upon an unknown or contingent event, as described in section 131 of the code, are also void. 95 and 113 Illinois, *supra.*

(3) An "option contract" to sell or buy at a future time grain or other commodity or stock, etc., is void under the Illinois statute, even though a settlement by differences was not contemplated. 130 Illinois, *supra.*

(4) The keeper of a shop or office where dealing is carried on in stock, etc., on margins, without any intention of delivering articles bought or sold, is guilty of an offence under the Illinois act of 1887. 134 Illinois, *supra.*

The cases of *Pearce* v. *Rice*, 142 U. S. *supra*, and *Irwin* v. *Williar*, 110 U. S. *supra*, are referred to in some of these cases as holding that dealings in differences, where the contract provides therefor, are void.

These Illinois cases, it will be seen upon examination, do not touch the case before us, which is a contract for future delivery, where there is no evidence that such delivery was not contemplated and a settlement by payment of differences only intended. The "option contracts" spoken of in those cases are explained in the cases themselves to mean what is commonly called "puts and calls," where there is no obligation on the part of the person to sell or to buy, and that class of contracts is the class covered by the statute. There is nothing in the evidence in this record that seems to us to afford any reasonable ground for holding that the contract in this case was on its face illegal as in violation of any statute in Illinois, or that, while valid on its face, the contract was really a guise under which to enable the parties to gamble on the differences in the price of stock sold and bought.

The further objection that these contracts having been made with reference to the rules of the exchange, the parties must in pursuing a remedy be confined to that which the rules provide, to the exclusion of the jurisdiction of ordinary courts of justice, we do not regard as well taken.

The sales were made subject to the rules referred to, but so far as regards a remedy for their violation, those rules provide a means by which parties may seek and obtain relief in accordance with their terms. They do not assume to exclude the jurisdiction of the courts, or, in other words, they do not assume to provide an exclusive remedy which the parties must necessarily follow, and which they have no right to refuse to follow without violating such rules, and thereby violating their contract. Any rule which would exclude the jurisdiction of the courts over contracts or transactions such as are here shown would not be enforced in a legal tribunal.

It is also objected that the means taken to obtain a price for the stock after a tender thereof had been refused by Jamieson & Company were inadequate for that purpose, if not fraudulent, and that, hence, there is no proof properly before the court as to the value of the stock on August 31, when it was tendered, or September 22, when it was sold, and it is also contended that there was no fair sale, but a mere sham, colorable in itself and fraudulent as against the defendants Jamieson & Company; that the only price of the stocks contemplated in the contracts at the time they were entered into and in case of a violation thereof, was the price to be fixed by the stock exchange by actual sales on the delivery days, and that as the exchange was closed from August 3 until November 5 following, no means existed by which that price could be ascertained.

We think the course pursued by the complainants was a proper one. On August 31, the exchange being closed, Schwartz & Company, acting in behalf of the complainants, tendered to Jamieson & Company 1150 shares of the stock in question, 700 of which included the shares sold by them for the complainants. This tender was refused. It is objected that the stock did not belong to the complainants when tender thereof was made to Jamieson & Company. That was not material. Their agents, Schwartz & Company, who did own the stock, made tender of it to Jamieson & Company and demanded the contract price in payment thereof. If that price had been paid and the delivery of the stock made to Jamieson & Company it would have been a good delivery. They would have had the title to the stock as against every one, Schwartz & Company included. It was a matter, therefore, of no importance that the complainants at the time this stock was tendered did not have the legal title to it. Under these circumstances, what could the complainants or their agents, Schwartz & Company, do? A tender of the stock had been made and had been refused. The stock exchange was closed by order of its governing committee, and Jamieson had voted in favor of its closing. Were there no means by which the value of the stock at or about this time could be ascertained while the stock exchange was closed? We think there were, and we also think that the course pursued by the complainants was a proper and appropriate one.

Accordingly Jamieson & Company were notified that the stock would be sold to the highest bidder at a time and place mentioned, and that they would be held responsible for any loss that might result from their refusal to take and pay for the stock as agreed upon. They were also informed at or about that time that the sales made by Schwartz & Company had been made for complainants as to 700 of such shares. On the day named the stock was put up for sale, and it is not an important fact that it did not belong to the complainants. It was stock over which they had control, and it was offered for sale on the part of the complainants with the approval and assent of its owners, and if it had been bought by any individual at the sale other than the one who did bid it in such purchaser would have obtained a good title to the stock on payment of the price bid. Wide publicity had been given on the part of the complainants of the time when and the place where this sale would occur, and the highest bid was made by an individual who was a member of the firm of Schwartz & Company, but there were many other people there who had the right, and, as it appears, were urged to bid, and there was neither fraud nor deception in the fact that a bid was made by a member of the firm as stated. The price at which the bidding closed was fixed after a chance for full and open competition upon the part of all who were present, and although the complainants entered into some arrangement with their agents by which the latter produced the stock and offered it for sale on account of and for the complainants, yet no injurious effect upon the transaction was thereby caused, and it in no way injured Jamieson & Company. That the bid was a fair indication of what was then regarded as the value of the stock, we think admits of very little question. When the exchange opened in November the stock sold at $130, and continued near that figure for some time.

Under all the facts in the case we think the complainants were justified in the course they pursued, and that the price at which the stock sold was a fair basis upon which to determine the amount of damages sustained by the complainant by reason of the refusal of Jamieson & Company to fulfill their contract of purchase.

For these reasons the decrees of the Circuit Court of Appeals and the Circuit Court must be reversed, and the case remanded to the latter court for such further proceedings therein as are not inconsistent with the opinion of this court, and it is so ordered.

MR. JUSTICE HARLAN, dissenting.

I dissent from the opinion and judgment in this case upon the ground stated by the Circuit Court of Appeals, namely, that the transactions involved in this litigation constituted gambling in "differences," in violation of the statute of Illinois.

CALHOUN GOLD MINING COMPANY *v.* AJAX GOLD MINING COMPANY.

ERROR TO THE SUPREME COURT OF THE STATE OF COLORADO.

No. 195.   Argued March 13, 14, 1901.—Decided May 27, 1901.

The rights conferred upon the locators of mining locations by Rev. Stat. § 2322, are not subject to the right of way expressed in § 2323, and are not limited by § 2336.

As to § 2336, by giving to the oldest or prior location, where veins unite, all ore or mineral within the space of intersection, and the vein below the point of union, the prior location takes no more, notwithstanding that § 2322 gives to such prior location the exclusive right of possession and enjoyment of all the surface included within the limits of the location, and of all veins, lodes and ledges throughout their entire depth, the top or apex of which lies inside of such surface lines extended downward, vertically.   *Held* that § 2336 does not conflict with § 2332, but supplements it.

A locator is not confined to the vein upon which he based his location, and upon which the discovery was made.

A patent is not simply a grant for the vein, but a location gives to the locator something more than the right to the vein which is the subject of the location.

Patents are proof of the discovery.   They relate back to the location of the claims, and cannot be collaterally attacked.

THE case is stated in the opinion of the court.

Mr. W. E. So Relle for plaintiff in error.

Mr. Joseph C. Helm for defendant in error.   *Mr. Ernest A. Colburn* and *Mr. Charles H. Dudley* were on his brief.

MR. JUSTICE McKENNA delivered the opinion of the court.

This action was brought in one of the District Courts of the State of Colorado by the defendant in error to recover damages from plaintiff in error for certain trespasses on, and to restrain it from removing ore from ground claimed to be within the boundaries of, the mining claims of defendant in error. The answer of plaintiff in error justified the trespasses and asserted a right to the ore by reason of the ownership of another mining claim and the ownership of a certain tunnel site.

The rights of the parties are based on and their determination hence involves the construction of the following sections of the Revised Statutes of the United States, empowering the location of mining claims:

" SEC. 2322. The locators of all mining locations heretofore made or which shall hereafter be made, on any mineral vein, lode, or ledge, situated on the public domain, their heirs and assigns, where no adverse claim exists on the tenth of May, eighteen hundred and seventy-two, so long as they comply with the laws of the United States, and with state, territorial and local regulations not in conflict with the laws of the United States governing their possessory title, shall have the exclusive right of possession and enjoyment of all the surface included within the lines of their locations, and of all veins, lodes and ledges throughout their entire depth, the top or apex of which lies inside of such surface lines extended downward vertically, although such veins, lodes, or ledges may so far depart from a perpendicular in their course downward as to extend outside the vertical side lines of such surface locations. But their right of possession to such outside parts of such veins or ledges shall be confined to such portions thereof as lie between vertical planes drawn downward as above described, through the end lines of their locations, so continued in their own direction that such planes will intersect such exterior parts of such veins or ledges. And nothing in this section shall authorize the locator or possessor of a vein or lode which extends in its downward course beyond the vertical lines of his claim to enter upon the surface of a claim owned or possessed by another.

" SEC. 2323. Where a tunnel is run for the development of a vein or lode, or for the discovery of mines, the owners of such tunnel shall have the right of possession of all veins or lodes within three thousand feet from the face of such tunnel on the lines thereof, not previously known to exist, discovered in such tunnel, to the same extent as if discovered from the surface;

and location on the line of such tunnel of veins or lodes not appearing on the surface, made by other parties after the commencement of the tunnel, and while the same is being prosecuted with reasonable diligence, shall be invalid; but failure to prosecute the work on the tunnel for six months shall be considered as an abandonment of the right to all undiscovered veins on the line of such tunnel."

"Sec. 2336. Where two or more veins intersect or cross each other, priority of title shall govern, and such prior location shall be entitled to all ore or mineral contained within the space of intersection; but the subsequent location shall have the right of way through the space of intersection for the purposes of the convenient working of the mine. And where two or more veins unite, the oldest or prior location shall take the vein below the point of union, including all the space of intersection."

The especial controversy is whether the rights conferred by section 2322 are subject to the right of way expressed in section 2323 and limited by section 2336. Or, in other words, as to the latter section, whether by giving to the oldest or prior location, where veins unite, " all ore or mineral contained within the space of intersection," and "the vein below the point of union," the prior location takes no more, notwithstanding that section 2322 gives to such prior location "the exclusive right of possession and enjoyment of all the surface included within the lines" of the location, "and of all veins, lodes and ledges throughout their entire depth, the top or apex of which lies inside of such surface lines extended downward vertically."

The defendant in error denied such effect to sections 2323 and 2336, and brought this suit, as we have said, against plaintiff in error for damages and to restrain plaintiff in error from removing ore claimed to be within the boundaries of the claims of defendant in error, to which ore defendant in error claimed to be entitled by virtue of section 2322. The judgment of the lower court sustained the claim of the defendant in error and damages were awarded it, and the plaintiff in error was enjoined from further prosecuting work. An appeal was taken to the Supreme Court of the State and the judgment was affirmed. Thereupon this writ of error was allowed.

The annexed plat exhibits the relative location of the respective properties of the parties. The Champion location was dropped from the case. There is no controversy as to the validity of the respective locations, none as to the tunnel site or of the steps necessary to preserve it. Indeed, the facts are all stipulated, and that the respective locations are evidenced by patents, the defendant in error being the owner of the Monarch and the Mammoth Pearl, and the plaintiff in error the owner of the Victor Consolidated and the tunnel site. The facts are stated by the Supreme Court of the State as follows:

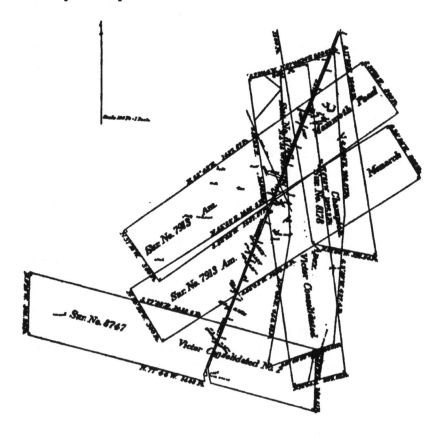

"That each of appellee's claims was located prior to either the lode claim or tunnel site of appellant; that the receiver's receipt on each of the claims of appellee issued prior to the location of the tunnel site and prior to the issuance of receiver's receipt on the Victor Consolidated; that the patents upon the lode claims of appellee issued prior to the patent on the lode claim of appellant; that the patent to the apex issued prior to the location of the tunnel site and on the Mammoth Pearl and Monarch subsequent to such location; that the vein of the Victor Consolidated was discovered and located from the surface, was not known to exist prior to such discovery, extends throughout the entire length of that claim, and on its strike crosses each of the veins of the claims of appellee upon which they were respectively discovered and located; that the tunnel cuts numerous blind veins underneath the surface of the claims of appellee, which do not appear upon the surface and were not known to exist prior to the location of the tunnel; that the vein of the Victor Consolidated was cut in this tunnel underneath the claims of appellee and ore of the value of four hundred dollars removed therefrom. It also appears that the patents upon the lode claims of appellee embrace the conflict with the Victor Consolidated without any reservation as to either surface or veins, and in this respect conform to the receiver's receipts upon such claims; that the patent on the Victor Consolidated excludes the surface in conflict with the claims of appellee and all veins having their apex within such conflict, which are the same exceptions contained in the receiver's receipt for that claim; that the portal to the Ithaca tunnel site was at the date of its location on public domain; that work thereon was prosecuted diligently, and that the location of such tunnel was in all respects regular; that all necessary steps were taken by appellant to locate the blind veins cut in such tunnel, which are in controversy in this case; that the record titles of the claims of appellee are vested in it, and the record titles of the Victor Consolidated, the Ithaca tunnel site, the blind veins discovered therein underneath the claims of appellee, are vested in the appellant. The record discloses that appellant offered testimony tending to prove that at the date of the location of its tunnel site mineral in place had not been discovered on the Monarch and Mammoth Pearl lode-claims."

The assignments of error present the following propositions

which it is stipulated the case involves and to which the decision may be directed:

" First. Whether or not the Ithaca tunnel (the tunnel claimed by plaintiff in error) is entitled to a right of way through defendant in error's lode claims.

" Second. Whether or not plaintiff in error has acquired by virtue of said tunnel site location the ownership and right to the possession of the blind veins cut therein, to wit, veins or lodes not appearing on the surface and not known to exist prior to the date of location of said tunnel site.

" Third. Whether or not plaintiff in error is the owner and entitled to the ore contained in the vein of its Victor Consolidated claim, within the surface boundaries and across lode claims of defendant in error.

" Fourth. Whether or not plaintiff in error should have been allowed to introduce evidence for the purpose of showing that there was no discovery of mineral in place on the Monarch and Mammoth Pearl claims of defendant in error prior to the location of said tunnel site."

The third proposition involves the relation of sections 2322 and 2336. It is first discussed by plaintiff in error, and is given the most prominence in the argument and we, therefore, give it precedence in the order of discussion. It presents for the first time in this court the rights of a junior location of a cross vein within the side lines of a senior location under section 2336. Prior to the decision by the Supreme Court of Colorado in the case at bar that court had decided that the junior location was "entitled to all of the ore found on his vein within the side lines of the senior location, except at the space of intersection of the two veins." *Branagan* v. *Dulaney*, (1885) 8 Col. 408; *Lee* v. *Stahl*, (1886) 9 Col. 208; *Morgenson* v. *Middlesex M. & M. Co.*, (1887) 11 Col. 176; *Lee* v. *Stahl*, 13 Col. 174. In *Coffee* v. *Emigh*, (1890) 15 Col. 184, it was held that the rule laid down in the foregoing cases had become established law. The claims of the plaintiff in error were located after the decisions, and it is contended that the rule laid down by them became a rule of property in the State, and it is earnestly urged that to reverse the rule now would take from plaintiff in error that which it "had reason to believe was a vested right in the Victor Consolidated vein."

There are serious objections to accepting that consequence as determinative of our judgment. We might by doing so confirm titles in Colorado but we might disturb them elsewhere. The statute construed is a Federal one, being a law not only for Colorado but for all of the mining States, and, therefore, a rule for all, not a rule for one, must be declared. Besides what consideration should have been given to prior cases, the Supreme Court of the State was better able to judge than we are. It may be that the repose of titles in the State was best effected by the reversal of the prior cases. At any rate, a Federal statute has more than a local application, and until construed by this court cannot be said to have an established meaning. The necessity of this is illustrated, if it need illustration, from the different view taken of sections 2322 and 2336 in California, Arizona and Montana, from that taken in the prior Colorado cases. The Supreme Courts respectively of those States and that Territory have adjudged a superiority of right to the cross veins to be in the senior location. Manifestly, on account of this difference if for no other, this court must interpret the sections independently of local considerations. And in doing so we do not find in the sections much ambiguity so far as the issue raised by the record is concerned; indeed, not even much necessity for explanation. Section 2336 does not conflict with section 2322, but supplements it. Section 2336 imposes a servitude upon the senior location, but does not otherwise affect the exclusive rights given the senior location. It gives a right of way to the junior location. To what extent, however, there may be some ambiguity; whether only through the space of the intersection of the veins, as held by the Supreme Courts of California, Arizona and Montana, or through the space of intersection of the claims, as held by the Supreme Court of Colorado in the case at bar. It is not necessary to determine between these views. One of them is certainly correct, and therefore the contention of the plaintiff in error is not correct, and more than that it is not necessary to decide on this record. A complete interpretation of the sections would, of course, determine between those views, but on that determination other rights than those submitted for judgment may be passed upon, and we prefer therefore to reserve our opinion.

There was some contrariety of views in the cases on other points. There was discussion as to whether veins cross on their strike or their dip, and it was held that they could cross on both strike and dip, but as to the exact application of section 2336 to either there was some disagreement.

The Supreme Court of Arizona said: "Congress had in mind, at the time of the enactment of the law of 1872, that, as mining rights then stood, A's lode might legally cross B's lode on the strike, and whether on the dip or not, makes no difference; and section 2336 was designed to define the rights of A and B in the space of intersection." *Watervale Mining Co.* v. *Leach*, 33 Pac. Rep. 418, 424.

The Supreme Court of California held in *Wilhelm* v. *Silvester*, 101 California, 358, that the provisions of the section could readily be construed as intending to protect the rights of old ledge locations. And speaking of veins intersecting on their dip, said, "moreover, there is strong reason for thinking that such an intersection was the very one in mind of Congress when it passed section 2336; for in that section, and speaking of the same subject, it says that 'where two or more veins *unite*, the oldest or prior location shall take the vein *below* the point of union,' and if the other kind of intersection (on the strike) was in the minds of the legislators at that time they would not have used the word 'below;' for 'below' would not apply at all to a union on the strike of two veins, such as the appellant's rights depend on in the case at bar." But the Chief Justice of the State, concurring in the result, observed:

"I think, however, that too much is conceded, both in the opinion of the court and in the argument of counsel for respondent, in assuming that the provisions of § 2336 cannot be applied to locations made since the passage of the mining law of 1872 on veins which intersect upon their strike, without bringing it in conflict with the plain terms of § 2322. This wholly unwarranted assumption has been the source of all the trouble and difficulty which the land office and some of the state courts have encountered in their attempts to construe provisions of a statute which are in perfect harmony, but which have been erroneously supposed to be inconsistent."

The Supreme Court of Colorado concurred in the conclusions of the courts of Arizona and California, and expressed its own view as follows:

" Our conclusion is that the provisions of § 2336 apply to locations made under the act of 1872, as well as before; p. 616, refer to the intersection or crossing of veins either upon their strike or dip; that the space of intersection in determining the ownership of ore within such space, means either intersection of veins or conflicting claims, according to the facts in each particular case, and grants a right of way to the junior claimant for the convenient working of his mine through such space upon the veins (underneath the surface) which he owns or controls outside of that space. This construction renders the two sections entirely harmonious, gives effect to every clause and part of each, and in so far as § 2336 regulates or in any manner provides for rights as between conflicting claims, it applies only to intersections consistent with all the provisions of § 2322."

See for the views of the Supreme Court of Montana, *Pardee v. Murray*, 4 Montana, 234.

2. The other assignments of error relate to rights claimed by plaintiff in error by the location of the tunnel site, and present the questions whether such location gave to the plaintiff in error the following rights: of way through the lode claims of the defendant in error; of possession of the blind veins cut by the tunnel underneath the claims of the defendant in error.

The plaintiff in error asserts the right of way for its tunnel under section 2323 by implication, and from that implication, and the rule it contends for as to cross veins, deduces its right to all of the blind veins. The contention as to cross veins we have answered, and the deduction as to blind veins is not justified. The section contemplates that tunnels may be run for the development of veins or lodes, for the discovery of mines, gives a right of possession of such veins or lodes, if not previously known to exist, and makes locations on the surface after the commencement of the tunnel invalid. There is no implication of a displacement of surface locations made before the commencement of the tunnel. Indeed, there is a necessary implication of their preservation. And there can be no implication of a conflict with the rights given by section 2322. The exclusiveness of those rights we have declared. The tunnel can only be run in subordination to them. How else can section 2322 be given effect? There are no exceptions to its language. The locators "of any mineral veins, lode or ledge" are given not only "an exclusive right of possession and enjoyment" of all the surface included within the lines of their locations, but "*of all veins, lodes and ledges throughout their entire depth*, the top or apex of which lies inside of such surface lines extended downward vertically." A locator therefore is not confined to the vein upon which he based his location and upon which the discovery was made. "All veins or lodes having their apices within the planes of the surface lines extended downward are his, and possession of the surface is possession of all such veins or lodes within the prescribed limitations." The Laws of Mines and Mining, page 442, by Barringer & Adams.

Under the old law the miner "located the *lode*. Under the new (the act of 1872) he must locate a piece of land containing the top or apex of the lode. While the vein is still the principal thing, in that it is for the sake of the vein that the location is made, the location must be of a piece of land including the top or apex of the vein. If he make such a location, containing the top or apex of his discovered lode, he will be entitled to all other lodes having their tops or apices within their surface boundaries." Lindsay on Mines, sec. 71.

And this court said, speaking by Mr. Justice Brewer, in *Campbell* v. *Ellet*, 167 U. S. 116 :

"But the patent is not simply a grant of the vein, for, as stated in the section, 'a patent for any land claimed and located for valuable deposits may be obtained in the following manner.' It must also be noticed that section 2322, in respect to locators, gives them the exclusive right of possession and enjoyment of all the surface within the lines of their locations, and all veins, lodes and ledges, the tops or apices of which are inside such lines. So that a location gives to the locator something more

than the right to the vein which is the occasion of the location."
See also *Del Monte Mining & Milling Co.* v. *Last Chance Mining & Milling Co.*, 171 U. S. 55.

The only condition is that the veins shall apex within the surface lines. It is not competent for us to add any other condition. Blind veins are not excepted, and we cannot except them. They are included in the description, "all veins" and belong to the surface location..

3. The same reasoning disposes of the claim of plaintiff in error to the right of way for its tunnel through the ground of defendant in error, so far as the right of way is based on the statutes of the United States. So far as it is based on the statutes of Colorado it is disposed of by their interpretation by the Supreme Court of Colorado, and, expressing it, the court said:

"It is contended by counsel for appellant that, under sec. 2338, Rev. Stat. U. S. and sec. 3141, Mills' Ann. Stat.; 59 Pac. Rep. 607, 617, it is entitled to such right. The first of these sections provides that, in the absence of necessary legislation by Congress, the legislature of a State may provide rules for working mines involving easements, drainage, and other necessary means to their complete development, and that these conditions shall be fully expressed in the patent. The section of Mills referred to provides that a tunnel claim located in accordance with its provisions shall have the right of way through lodes which may lie in its course; but it will be observed that this section only refers to tunnels located for the purpose of discovery, and if any of its provisions are still in force, which appears to be doubted in *Ellet* v. *Campbell, supra*, they can have no application to the case at bar, because the section of the Revised Statutes only provides for easements for the development of mines, and the section of Mills relied upon does not attempt to confer any such rights, but is limited to the one purpose of discovery. In this respect it has been clearly superseded by the act of Congress, so that, if appellant is entitled to the right claimed, it must attach by virtue of some provision of this act."

4. An assignment of error is based upon an offer of plaintiff in error to prove, that at the time of the location of the Ithaca tunnel site, no ore had been discovered in two of the patented

claims of the defendant in error, to wit, the Monarch and the Mammoth Pearl. The ruling was right. The patents were proof of the discovery and related back to the date of the locations of the claims. The patents could not be collaterally attacked. This has been decided so often that a citation of cases is unnecessary.

Judgment affirmed.

DISTRICT OF COLUMBIA *v.* TALTY.

APPEAL FROM THE COURT OF CLAIMS.

No. 288. Argued April 12, 15, 1901.—Decided May 27, 1901.

In this case this court holds, (1) that it was not error in the court below to try the case on the amended petition; (2) that the report to the Government of a person employed by the Attorney General in this case was properly rejected as evidence; (3) that there was no error in the rulings of the court below.

THE case is stated in the opinion of the court.

Mr. Robert A. Howard for appellant. *Mr. Assistant Attorney General Pradt* was on his brief.

Mr. V. B. Edwards for appellee.

MR. JUSTICE MCKENNA delivered the opinion of the court.

This action was brought in the Court of Claims under the act of Congress approved June 16, 1880, 21 Stat. 284, c. 243, entitled "An act to provide for the settlement of all outstanding claims against the District of Columbia, and conferring jurisdiction on the Court of Claims to hear the same, and for other purposes."

The purpose of the action was to recover compensation for work done and materials furnished under certain contracts entered into between the District of Columbia and appellee. The court found that there was due appellee, on March 18, 1876, the sum of $4180.44, and entered judgment for that amount on February 20, 1899, from which judgment the District took this appeal.

The original petition was filed on December 15, 1880, issue was joined, and the case referred to a referee. Two reports were made by him, but no trial was had thereon. On the 22d of June, 1897, on motion of claimant (appellee), the District consenting, the case was referred to Frank W. Hackett, Esq., to take and state the account between the parties. On the 28th of July, 1897, the claimant, by leave of the court, filed an amended petition, in which he alleged that he entered into a contract with the board of public works of the District on the 7th of August, 1873, numbered 826, for the improvement of certain streets, and for which he was to receive the prices established and paid by the board for work of similar character. The contract was extended, respectively, on the 17th of September, 1875, and the 3d of December, 1875, to embrace work on other streets. The contract and the extensions filed by the Commissioners were referred to. It was alleged that the work was done to the satisfaction of the District, and was duly measured and certified to by the engineers of the District, and that the work done amounted to the sum of $49,323.54; amount paid thereon, $49,033.91; leaving due and payable January 15, 1876, the sum of $289.63. The measurements returned by the Commissioners of the District were referred to.

It was also alleged as follows:

"That under the provisions of the new contracts, called extensions, the claimant performed a large amount of work on K street, between Third and Seventh northeast, which was duly accepted by the District of Columbia, and certified measurements issued at the written contract rates, one measurement for $10,504.60, which was audited by the board of audit, and one measurement for $2570.30. This last measurement was not audited by the board of audit, not having reached said board of audit prior to the abolition of said board, and the said measurements, amounting to the sum of $13,074.90, remain due and unpaid, less the sum of $9184.45 paid on account thereof in partial measurements, leaving a balance due on the work done under the extensions of said contract the sum of $3890.45, due and payable March 18, 1876.

"(See measurement of March 18, 1876. Returned by Commissioners of the District of Columbia.)

" The claimant therefore demands judgment against the District of Columbia in the sum of four thousand one hundred and eighty dollars and eight cents ($4180.08), as a debt against the District of Columbia, due and payable as follows . $289.63, January 15, 1876, and $3890.45, March 18, 1876, and such other sums as your petitioner shall prove to be due to him from the District of Columbia, but which your petitioner cannot at present specifically state, for want of records not in his possession."

There are a number of assignments of error, but, to quote counsel for the District—

" The errors insisted upon by the defendant, the District of Columbia, all arise from, and may be said to be included in, the failure to try the case originally brought, and the impossibility, on account of the loss and destruction of all the records in the case, to state an account between the claimant and the District."

It is, therefore, also insisted by the District that claimant (appellee) " should have endeavored to substitute a petition for the one lost," and he not having done so, no trial could have been had.   It was on this assumption that the District requested the court to find (and error is now assigned because the court did not find) that the original petition was based on other contracts than contract No. 826 and its extensions; that the case was referred to Daniel Donovan, who made two reports, and that all of the papers, including the original petition, contracts, (except contract No. 826 and the two extensions thereof,) vouchers and report of referee have been lost; that Donovan died without making a report under the second reference, and that after the reference to Hackett the petition was amended by leave of the court; that Donovan reported certain excess of payments amounting to $1377.03 under contract No. 826, and was made by allowing for work at " board rates," instead of contract rates; than the greater part of the settlement made under contract No. 828, and all settlements under the extensions of the contract, were made at "board rates."

It was not error in the court to try the case on the amended petition. It was filed without objection being made, but it would have been no error even if objection had been made. It finally rested the right of recovery upon contract No. 828 and its extension. That contract and its extensions were relied on in the original petition. There was, therefore, only a limitation of the action, not a change of it. If the District had any rights or defences on account of the other contracts, such rights and defences could have been set up or established by evidence. We said in *United States* v. *Burns*, 12 Wall. 246-254, that "the Court of Claims, in deciding upon the rights of claimants, is not bound by any special rule of pleading."

The District further contends that the proof of the claimant was defective, and did not justify the report of 'the referee and the judgment of the court upon it; and also contends that what is claimed to be a report of Donovan should have been received as evidence. Its rejection was certainly not error. Treating it as a report, it was not acted on in any way; besides, it was in no sense a report. It is claimed to be, it is true, but it was found in a letter written by Donovan, in pursuance of his employment at $10 per day, by the Attorney General of the United States, to Mr. Brannigan, one of the assistant attorneys of the Department of Justice. In other words, what is claimed to have been found out and reported by Donovan, as attorney against the claim, is urged as evidence against the claim. The paper is as follows:

"WASHINGTON, D. C., June 25, 1891.

" Felix Brannigan, Esq.,

"Assistant Attorney, Department of Justice.

"Sir: In compliance with your request and verbal instructions, and under my appointment by the Attorney General of the United States of November 11, 1890, I have made a careful, thorough and searching investigation of all records, vouchers and other papers pertaining to every of the cases relating to the District of Columbia now pending in the United States

Court of Claims under the act of June 16, 1880, and beg to submit the following report as the result of such research:

*　　*　　*　　*　　*　　*　　*　　*

"Stephen Talty
　　　　　v.　　　　} No. 335.　Referred list (p. 36).
The District of Columbia.

"This case involved the examination and stating of accounts under seven separate and distinct contracts. It was referred to me as referee by the Court of Claims, and was reported under a rule of said court, heretofore referred to in Murray's case No. 90.

Said report shows that claimant is entitled to recover
　　from the defendant under his several contracts the
　　sum of...................................... $1814 79
It further shows that claimant is indebted to defend-
　　ant, by reason of overpayment by the board of
　　audit, in the amount of........................ 989 71

Thus leaving a net balance due claimant.......... $825 08

Attached to said report and forming part thereof is the following set-off:

File No. 22. 61.5 square yards of cobblestone pave-
　　ment relaid, allowed at 37 cents per yard in lieu of
　　30 cents, the contract rate.　Excess, 7 cents per yard
　　on 61.5 yards............................... $4 30
8237.98 cubic yards grading, allowed at 40 cents per
　　yard in lieu of 30 cents, the contract rate.　Excess,
　　10 cents per yard on 8237.98 yards............. 823 79
784.20 cubic yards rock excavation, allowed at $1.00
　　per yard in lieu of 30 cents per yard, the contract
　　rate.　Excess, 70 cents per yard on 784.20 yards.. 548 94

　　　　Total.................................. $1377 03

"A more thorough and exhaustive examination of the records which constitute my former report in this case convince me that the counterclaim therein reported is both erroneous and unjust, for the reason that M street northwest, wherein the alleged excessive allowances were made claimant, was in fact, at the time he did the work, an old graveled street, and also that rock was encountered in part of it.   It also appears that nearly the whole of the work was done under the Commissioners of the District of Columbia, and that the prices certified by the engineer and paid Talty were what are commonly known as ' board rates,' and that said allowances were not made under any mistake of fact, but that they were the prices universally paid to other contractors doing similar work at the time.

" I am therefore of the opinion that my former report should be amended by striking out the set-off therein stated, and finding a balance due Talty of $825.08, with interest from March 1, 1876.   All of which is respectfully submitted.

(Signed)   " DAN DONOVAN.

" Correct copy.   A. McKENZIE,

" *Acting Auditor, District of Columbia.*"

The objections to the reports of the referee are untenable. It is impossible, however, to quote the reports without unduly extending this opinion.   It is enough to say they were stated to have been founded upon depositions of witnesses and " original sheets of measurements taken from the field book of the engineer measuring the work."   They exhibited the measurements and the quantity of material in tabulated form.   Other papers were used and figures taken from original books in the possession of the District.   The referee reported :

" It was agreed at the hearing that the papers heretofore filed in this case have disappeared.   Search in the office of the attorney for the District, and at the house of Mr. Donovan, former referee, has failed to discover anything of these papers.   It is not charged that their disappearance is due to any fault of the claimant.

" In these circumstances the referee is satisfied to rely upon the memoranda in the sheets just referred to."

The report also set out contract No. 828 and its extension, and an itemized account of the work done and materials furnished, certified by the assistant engineer of the District.

It is not necessary to set out at length the objections to the report and those to the rulings of the court in refusing certain findings. We have examined and considered them and are of the opinion that there was no error in the rulings of the court, and the judgment is

Affirmed.

RUSSELL *v.* UNITED STATES.

APPEAL FROM THE COURT OF CLAIMS.

No. 242. Argued April 16, 1901.—Decided May 27, 1901.

This was an action at law against the United States upon an alleged implied contract to pay for the use of a patented invention belonging to the plaintiffs in error, in rifles used by the Government which had been purchased under contract from a Norwegian Company. It was conceded that a contract must be established in order to entitle appellants to recover, as the Court of Claims has no jurisdiction of demands against the United States founded on torts. *Held,* that on the facts proved in this case no such contract was proved against the United States, and that if the petitioners have suffered injury, it has been through the infringement of their patent, and not by a breach of contract.

THIS was an action for $100,000, brought in the Court of Claims by the appellants, upon an implied contract, asserted to have arisen from the use by the United States of Krag-Jorgensen rifles, which rifles contained, it is claimed, certain features, which were the invention of Russell, one of the appellants. The United States demurred to the petition, and the demurrer was sustained.

The facts as presented by the petition are as follows: That on or about August 3, 1880, letters patent No. 230,823, for certain new and useful improvements in firearms, were granted to Russell, and that he and Livermore are now the owners of such invention.

That pursuant to an advertisement by a board of officers convened under the act of Congress, approved February 24, 1881, to select a magazine rifle for the service of the United States, Russell submitted to said board an operative magazine rifle made in accordance with his letters patent, and on or about December 16, 1890, submitted to another board of officers, convened for like purpose, the same rifle. The officers made reports on the rifle, which reports, it is alleged, may be found in certain Congressional documents designated by number and of the session of Congress of whose records they constitute a part.

On the 15th of September, 1892, a second board recommended the adoption of the magazine rifle presented to it by the Krag-Jorgensen Gevaerkompagni of Christiania, Norway, and the rifle was provisionally adopted by the War Department for the use of the United States Army. The rifle is termed in the petition " Army rifle."

The petition recites a correspondence between Russell and the Chief of Ordnance of the United States Army, giving its substance, which may be omitted, as the letters are hereafter set out in full.

It is also alleged that on June 7, 1893, the Krag-Jorgensen Gevaerkompagni and the United States, represented by Brigadier General D. W. Flagler, United States Army, Chief of Ordnance, under the direction and by the authority of the Secretary of War, entered into a contract, whereby that company granted to the United States the right to manufacture an unlimited number of said "Army rifles." As much of the contract as we consider important is hereinafter set out.

That the United States did proceed to manufacture said "Army rifles," and introduce them for use in the United States Army, and since January 1, 1894, commenced to account, and has ever since accounted, to the Krag-Jorgensen Company for royalties, at the rate named in the contract, and paid certain sums on account thereof. The company failed to furnish an indemnifying bond, but the United States, with consent of the company, withheld a certain amount of the royalties, which aggregated on or about June 16, 1895, the sum of $25,000. The company then gave a bond with sureties, and the said sum was paid to it. The bond was conditioned as follows:

" That whereas the Krag-Jorgensen Gevaerkompagni of Christiania, Norway, has, on the seventh day of June, 1893, entered into a contract with the United States, represented by Brigadier General D. W. Flagler, Chief of Ordnance, for granting unto the United States full rights to manufacture an unlimited number of the Krag Jorgensen magazine firearms, for the military service of the United States, under the American patents Nos. 429,811, of June 10, 1890, and 492,212, of February 21, 1893, granted to O. W. J. Krag and E. Jorgensen, during the lifetime of the said patents, and by the said contract covenanted to indemnify the United States, and all persons acting under them, for all liability on account of any patent rights granted by the United States which may affect the right to manufacture therein contracted for, and further covenanted and agreed to furnish, before the payment of any royalties by the United States, a good and sufficient bond in the penal sum of twenty-five thousand dollars, to protect and defend the United States against all suits and claims by any and all persons for infringement on their inventions in the manufacture of said arms, and to pay all judgments that may be obtained against the United States for the same:

"Now, therefore, if the said Krag-Jorgensen Gevaerkompagni shall and will in all respects indemnify the United States, and all persons acting under them, for all liability on account of any patent rights granted by the United States which may affect the right to manufacture granted by said contract, and shall and will fully protect and defend the United States against all suits and claims by any and all persons for infringement of their inventions in the manufacture of said arms, and pay all judgments that may be obtained against the United States, or any officer or agent thereof for the same, then the above obligation shall be void and of no effect; otherwise to remain in full force and virtue."

It is alleged that the United States manufactured and used upwards of 75,000 "Army rifles" containing Russell's invention, and derived a profit thereby of $1 on each rifle.

The petition concluded as follows:

"By reason of the foregoing facts the claimants say:

"That neither the said contract, entered into by the United States and the Krag-Jorgensen Gevaerkompagni (Exhibit L) nor the said bond of indemnity delivered by the Krag-Jorgensen Gevaerkompagni to the United States, did provide the claimants with a remedy against the said Krag-Jorgensen Gevaerkompagni, for the use made by the United States of the claimants' said patented invention in accordance with the first alternative proposed by the Ordnance Department in its said letter to the claimant Russell, bearing date November 18, 1892 (Exhibit B).

"That there is to be implied from the use by the United States of the claimants' said patented invention, as hereinbefore related, a contract, whereby the United States agreed to pay to the claimants reasonable compensation for the same, and whereby the amount of such compensation was to be ascertained by means of a suit to be brought by the claimants in this court, in accordance with the second alternative proposed in the said letter (Exhibit B), and that the sum of $100,000 would be reasonable compensation for the said use, and that the United States has failed to pay the claimants the said sum of $100,000, or any sum or sums whatsoever for or on account of the said use, although duly requested thereunto.

"The claimants are the only persons owning or interested in the claim above set forth, and no assignment or transfer of the said claim or of any part thereof or interest therein has been made. The claimants are justly entitled to receive and recover from the United States the sum of one hundred thousand dollars ($100,000), after allowing all just credits and offsets. The claimants have always borne true allegiance to the government of the United States, and have not in any way aided, abetted or given encouragement to rebellion against the said government, and they believe the facts hereinabove stated to be true.

"Wherefore the claimants pray for judgment against the United States in the sum of one hundred thousand dollars ($100,000), and for such further relief as this honorable court may be entitled to grant, both at law and in equity, in the premises."

The following is the correspondence:

"EXHIBIT A.

(Copy.)

"WASHINGTON, D. C., November 16, 1892.

"To the Chief of Ordnance, U. S. Army.

"Sir: In the interest of Major Wm. R. Livermore, U. S. Army, and myself, I have the honor to invite attention to claims 22, 28 and 29 in U. S. patent No. 230,823, owned by us, as we believe their provisions to be infringed in the construction of the Krag-Jorgensen magazine gun lately adopted by the War Department, the points of resemblance being in the connection between the magazine and the receiver.

" In considering the allowance to inventors, we would request that our claims for these vital points of construction be regarded.

" Very respectfully, your obedient servant,

(Signed) " A. H. Russell,
Capt. of Ordnance, U. S. Army.

" Exhibit B.

" 5839.

" Ordnance Office, War Department,

" Washington, November 18, 1892.

"Capt. A. H. Russell, Ordnance Department, U. S. A., cor. 15th St. and N. Y. Ave., Washington, D. C.

" Sir: In reference to your letter of the 16th instant claiming the use of your patent right in the Krag-Jorgensen gun, lately adopted by this department for trial, which has been received and placed on file, I am instructed by the Chief of Ordnance to inform you that the business arrangements with the Krag-Jorgensen Company for the manufacture of this arm have not yet been completed.

" On the one hand, that company may agree to indemnify the United States on account of any patent rights granted by the United States which may affect the manufacture of the guns, in which case your recourse would be to communicate directly with the company.

" On the other hand, should the government proceed to manufacture the arms without such arrangement, your course will be to bring a suit against the government in the Court of Claims after manufacture has progressed.

" Respectfully, (Signed) C. W. Whipple,
" Capt. Ord. Dept., U. S. A.

"EXHIBIT C.

"WASHINGTON, D. C., December 9, 1892.

"To the Chief of Ordnance, U. S. Army.

"Sir: In reference to my letter to the Chief of Ordnance of November 16, 1892, and to the answer of November 18th from the Ordnance Office in reply thereto, concerning the claims 22, 28 and 29 in the U. S. patent to me, No. 230,823·(a copy of the specifications of which was enclosed in my. letter), I desire to say that I could practically have no remedy for infringement of my patent against the Krag-Jorgensen Company, as they have not, that I am aware of, any property in this country; and also that I presume it would be more satisfactory to the United States, as it certainly would be to me, to have whatever may be justly due to me on my patents allowed without litigation.

"I therefore hope that the Ordnance Office will bear my letter of November 16th, and this letter, in mind, and allow me a hearing before any business arrangement with the Krag-Jorgensen Company is closed.

"Very respectfully, your obedient servant,

(Signed)　　　"A. H. RUSSELL,

"*Capt. of Ordnance, U. S. Army.*

"EXHIBIT D.

"Ordnance Office, War Department,

"WASHINGTON, December 19, 1892.

"Capt. A. H. Russell, World's Columbian Exposition, Chicago, Ill.

"Sir: Referring to your letters of the 16th ult. and the 9th inst., on the subject of infringement of your patent in the manufacture of the Krag-Jorgensen magazine firearm, I am instructed by the Chief of Ordnance to inform you that in a letter received from the Commissioner of Patents dated 15th inst. he states that the invention of H. I. Krag and Erik Jorgensen for improvement in machine firearms has been examined and the invention has been found patentable in view of the state of the art, but that other applications are pending which appear to conflict in subject-matter ; therefore the application of Krag and Jorgensen will be withheld from issue until that question is settled definitely.

"I am also instructed to inform you that in a letter to the Commissioner of Patents, dated 16th inst., the Chief of Ordnance transmitted to him copies of your above-mentioned letters of 16th ult. and 9th inst.

"Should you desire further presentation of your patent, it is suggested that you communicate direct with the Commissioner of Patents.

"Respectfully,   (Signed)   CHARLES SHALER,
"*Capt. Ord. Dept. U. S. A.*

"EXHIBIT E.

"1429 New York Ave.,
"WASHINGTON, D. C., February 6, 1893.

"To the Commissioner of Patents.

"Sir: In an official communication from the Chief of Ordnance U. S. A., dated December 19, 1892, it is suggested that I 'communicate direct with the Commissioner of Patents' in regard to the following matter.

"I presented to the Ordnance Office the claim that the gun recommended for adoption by the U. S. Army, known as the Krag-Jorgensen gun, infringed claims 22, 28 and 29 of my patent No. 230,823, dated August 3, 1880, and I ask that the government do justice by me in case of using such device.

"The Chief of Ordnance states that there were claims now pending in the Patent Office, and referred me to you, with the information that copies of my letters had been sent you December 16, 1892.

"Copies of these letters are enclosed, with copies of replies from the Ordnance Office.

"I have the honor to inquire what further action should be taken by me.

"Very respectfully,   (Signed)   A. H. RUSSELL.

"EXHIBIT F.

" Department of the Interior, United States Patent Office,

"WASHINGTON, D. C., February 14, 1893.

" Capt. A. H. Russell, U. S. A., 1429 New York avenue, Washington, D. C.

" Sir: I have your letter of the 6th instant, and in reply you are advised that it is not seen how the Patent Office has any jurisdiction in the matter concerning which you write. Questions of infringement can be determined only by the courts.

" Very respectfully,   (Signed)   W. E. SIMONDS,
" 12,509 Div. A–1893.   *Commissioner.*

"EXHIBIT G.

" CHICAGO, ILL., June 30, 1893.

"To the Chief of Ordnance, U. S. Army, Washington, D. C.

" Sir: In reference to correspondence regarding infringement of my patent No. 230,823 by the manufacture of the Krag-Jorgensen magazine rifle recommended by the magazine gun board, I have the honor to state that I communicated direct with the Commissioner of Patents, as suggested in the letter of December 19, 1892, from the Ordnance Office, and was told that the Patent Office had no jurisdiction. The patent infringed is one of long standing, and no claim is made that the Krag-Jorgensen patents infringe, but it is claimed that the construction of the gun embodying those patents does infringe my patent of 1880.

"I therefore respectfully renew the request contained in my letter of December 9, 1892, that the Ordnance Office will ' allow me a hearing before any business arrangement with the Krag-Jorgensen Company is closed.'

" Very respectfully, your obedient servant,
             (Signed)   " A. H. RUSSELL,
                    " *Capt. of Ord., U. S. Army.*

"EXHIBIT H.

"(1st indorsement.)

" Ordnance Office, Washington, July 7, 1893.

" Respectfully returned to Capt. A. H. Russell, Government

Building, Jackson Park, Chicago, Ill., with information that a statement of this case should be made in writing for file at this office, and for future reference, as the case stated cannot be determined by the Ordnance. Office.

"The agreement with the Krag-Jorgensen people is such that they are required to guarantee the United States against all damages for infringement.

(Signed) "CHARLES SHALER,
"*Acting Chief of Ordnance.*

"EXHIBIT I.

"CHICAGO, ILL., November 22, 1893.

"To the Chief of Ordnance, U. S. Army, Washington, D. C.

"Sir: In response to your communication of July 7, 1893, in the 1st indorsement on my letter of June 30, 1893, Ordnance Office files 3515 of 1893, I hereby respectfully state my position with regard to the Krag-Jorgensen rifle construction recently adopted for the U. S. Army, and now in process of manufacture at the Springfield Armory.

"I do not claim to be the inventor of all the mechanism of said gun. There are probably several important and meritorious inventions involved. The grant of a patent by the Patent Office on some features, however, does not authorize the making or using of other features covered by other patentees, as I am informed, and as seems a reasonable construction of law.

"The specific features used in said gun and claimed by me, and believed to be covered by my U. S. patent No. 230,823 of August 3, 1880, are embraced in the 22d, 28th and possibly in the 29th claims of said patent. A free description of the general features of my invention as used in the Krag-Jorgensen gun would be 'a magazine feeding into the side of the receiver under a bridge, with the entrance way under the bridge narrowed at the rear of the receiver, so as to permit but a slight projection of the side of the cartridge into the receiver when the bolt is drawn back, but with a wider opening further forward, so that as the cartridge moves forward impelled by the bolt, it will find a full-width passage under the bridge, through which it passes into the receiver.'

" This bridge and magazine opening is described in my patent, referred to, page 6, lines 100 *et seq.*, as follows:

" ' In filling the chamber of the magazine the gate is forced downward as the cartridges are filled in, leaving ample space to insert cartridges between the top edge of the magazine wall and a bridge or top part, M, and thus supply the magazine. When relieved of the downward pressure the gate ascends far enough to prevent egress of cartridges in any other way than sidewise from the magazine mouth into the receiver B' beneath the bridge.

" ' The bridge M is of peculiar formation on its under and inner surface, and is at one side of the longitudinal center of the barrel and breech-bolt housing. (See particularly Figs. 9, 10, 11, 12, 13 and 14, where is represented the manner of curving or recessing the bridge so as to admit cartridges to the receiver and guide and control their movements as supplied to the receiver from the magazine mouth, and thence conducted to firing chamber by the thrust of the breech bolt A^2.)

" ' Supposing the breech bolt to be retraced and about to be advanced, the operation of supplying and seating a cartridge from the magazine is as follows, reference being had to the last referred to figures and to Figs. 3, 4 and 6, ignoring for the present the hinged gate N : The topmost cartridge is elevated by the pusher against the bridge, so that its flange projects partially, but very slightly, into the path of travel of the breech-bolt head, (see dotted lines, Fig. 11,) and in which position it is prevented from accidental inward movement by a slight ridge or swell, n, (see Figs. 4 and 11,) at the rear portion of the edge or wall of the opening in the receiver bottom, with which the magazine communicates, and by a similar ridge or downward swell, n', on the bridge.

" ' The point or nose of the cartridge is guided past the spring l and into the firing chamber b, by the flaring way or incline m as the bolt advances.

" ' Before the cartridge has been moved forward far enough by the advance of the bolt to jam or bind crosswise, the flange will have been moved to a point where the two ridges n n' begin to slope respectively downward and inward gradually to the plane of the bottom or lowest part of the receiver, and upward and inward, thus allowing the flange to pass toward its place in the receiver. The bridge is also cut away on a curve or incline m^2, forward and inward from its lower edge to m', so that as the front of the cartridge is entering the chamber the rear is being gradually brought into line therewith.

" 'About the time the flange has been advanced to the point indicated by m', Fig. 9, the curved recess or incline m^2 of the inner and under side of the bridge, which had previously served to gradually admit the inward passage of the flange, and which at this point terminates in a swell or ridge similar to n', now serves to prevent its escape or outward movement and to direct it into the proper position to be pushed home and firmly seated by the bolt, as already described. . . . The bridge M is shown as formed with or attached to the magazine. It may, however, obviously be formed with the shoe or breech of the gun partly over and at one side of the receiver chamber.'

" In the main, this description applies to the Krag-Jorgensen gun quite as well as to my own gun, and my opinion is confirmed by that of experts that this part of my construction has been adopted in that arm.

" I am informed that my rights under my patent depend on the claims therein, and my belief is sustained by expert opinion that the Krag-Jorgensen gun under construction at Springfield infringes the 22d claim of my patent, which reads as follows:

" ' 22. The combination of the shoe chamber or receiver, the bridge at the side and top of the receiver, and the magazine chamber having an inlet to receive the cartridges inserted downward outside and beneath the bridge, and a mouth to conduct them beneath the bridge into and at the side of the receiver, substantially as hereinbefore set forth.'

" Taking the elements of this claim separately, it will be seen that the Krag-Jorgensen gun has 'the shoe or receiver.' All bolt guns and many others have it. The Krag gun is, however, one of a few to have 'the bridge at the side and top of the receiver.' It also has 'the magazine chamber having an inlet to receive the cartridges inserted downward outside and beneath the bridge.' It is true that the Krag Jorgensen gun shows the loading opening in a different place from that illustrated in my patent, but my claim is not limited as to the exact location of the loading opening, and the whole tenor of my patent is against the theory that I am limited in this claim to the precise construction shown in my drawing. Further, I am informed that the rule of law is that a patent claim, if valid, covers its mechanical equivalent—that is, other devices operating in a similar way to a like result, which is certainly the case in this instance. My claim further specifies 'a mouth by which to conduct them' (the cartridges) 'beneath the bridge into and at the side of the receiver, substantially as hereinbefore set forth.' This language, like the rest of the claim, applies quite as well to the Krag-Jorgensen gun as to the one invented and constructed by me.

"It thus seems demonstrable that the Krag-Jorgensen gun, having adopted part of my invention, has also adopted that part covered by claim 22.

"Claim 28 in my specified patent is as follows:

"'28. The combination, substantially as hereinbefore set forth, of the reciprocating breech bolt, the shoe chamber or receiver, the magazine chamber having a mouth, as described, the cartridge-supplying mechanism, and the bridge M, having the ridge or swell n', and otherwise curved or recessed, substantially in the manner and for the purpose set forth.'

"The Krag-Jorgensen gun has 'the reciprocating breech bolt, the shoe chamber or receiver, the magazine chamber having a mouth;' it has 'cartridge-supplying mechanism;' and 'the bridge having the ridge or swell' below the main part of the bridge and extending forward precisely in the same manner and for the same purpose as the swell n' of my patent, and is otherwise curved or recessed substantially in the manner set forth in my patent. I therefore feel warranted in the belief that my claim 28 covers the Krag-Jorgensen construction, and is infringed thereby.

"Claim 29 of my patent is as follows:

"'29. The bridge M, located relatively to the receiver and mouth of the magazine essentially as shown and described, and having the rear ridge n', and the curved or inclined surfaces m'

and m^2, substantially as and for the purpose hereinbefore set forth.'

"The construction of the Krag-Jorgensen gun substantially conforms to the terms of this claim also, yet I am not quite certain that the bridge in that gun has a projection the equivalent of that described as m^2 in my patent. I presume this question can only be determined by careful expert examination.

"I, therefore, base my claim for compensation on the infringement of my claims 22 and 28, and the probable infringement of claim 29, in my said patent No. 230,823, of August 3, 1880, by the Krag-Jorgensen construction.

"I am not fully informed as to the terms of the contract between the United States and the owners of the Krag-Jorgensen patent. Assuming that the owners of said patent are in ignorance of my rights in the premises, I respectfully request that a copy of this communication may be sent to the said parties, and a duplicate of this paper is forwarded for that purpose, with an extra copy of my patent to go with the duplicate. I further request that I may be furnished with the name and address of the responsible parties representing the Krag-Jorgensen interest.

"I am aware that in the event of a suit in equity, the alleged infringing parties have a statutory right to challenge the validity of my patent, and, to avoid litigation, I am willing to go further than to make a mere statement of the *prima facie* case as above, and show to infringing parties or their experts that my claims are well within my rights, provided I am met by these parties in a fair spirit, and with a desire to make a just compensation when my title to the property is shown.

"My first official notice to the Ordnance Department of this infringement is dated November 16, 1892, (Ordnance Office file 5839 of 1892,) but a gun presenting the special features here mentioned was submitted by me to the board of magazine guns, convened by General Orders 31, H. Q. A., March 21, 1881, and it is described in the report of that board. It is now in my possession subject to examination.

"Very respectfully, your obedient servant,

(Signed) "A. H. RUSSELL."

"EXHIBIT K.

(Copy.)

"Ordnance Office, War Department,

"WASHINGTON, December 1, 1898.

"Capt. A. H. Russell, Ordnance Department, U. S. A. Government Building, Jackson Park, Chicago, Ill.

"Sir: I am instructed by the Chief of Ordnance to acknowledge the receipt of your letter of the 22d inst., and to inform you that the terms of the contract between the United States and the Krag-Jorgensen Company contains a clause to the following effect:

"'The said party of the first part shall indemnify the United States and all persons acting under them for all liability on account of any patent rights granted by the United States which may affect the right to manufacture herein contracted for.'

"You have requested that a copy of your communication and a copy of your patent should be forwarded by this office to the company, and for that purpose you have forwarded duplicates of your letter and of the patent specifications. It is considered best that you should forward these communications direct; they are, therefore, returned to you for the purpose. The address of the contracting company is 'The Krag-Jorgensen Gewehr Kompagnie, Christiania, Norway.' The other papers, enclosures to Ordnance Office file 3515, containing letter and copies of patent specifications, are filed in this office for future reference.

"Your attention is again invited to the statement of the first indorsement on that file, which states that the case 'cannot be determined by the Ordnance Department.'

"Respectfully,   (Signed)   CHARLES SHALER,
                                    "*Capt., Ord. Dept., U. S. A.*"

The parts of the contract between the United States and the Krag-Jorgensen Company, which are relevant to the question presented in this case, are as follows:

"It is further stipulated and agreed that before any royalties are paid by the United States, the Krag-Jorgensen Gevarkompagni shall furnish a good and sufficient bond in the penal sum of twenty-five thousand dollars ($25,000.00), to protect and defend the United States against all suits and claims by any and all persons for infringement of their inventions in the manufacture of these arms, and to pay all judgments that may be obtained against the United States for the same.

"2d. All the — herein contracted for shall be delivered by the said party of the first part.

"3d. The said party of the first part shall indemnify the United States and all persons acting under them for all liability on account of any patent rights granted by the United States which may affect the right to manufacture herein contracted for."

Mr. James H. Hayden for appellants.　*Mr. Joseph K. Mc-Cammon* was on his brief.

Mr. Charles C. Binney for appellee.　*Mr. Assistant Attorney General Pradt* was on his brief.

MR. JUSTICE McKENNA, after making the above statement of the case, delivered the opinion of the court.

It is conceded that a contract must be established to entitle appellants to recover, and, it is contended, that one is established by the correspondence between the Ordnance Department and Russell in regard to the use of the "Army rifle," which, it is claimed, contained features of Russell's invention. That is, not an express contract is claimed, but an implied contract is claimed. This court has held that under the act of March 3, 1887, 24 Stat. 505, c. 359, defining claims of which the Court of Claims had jurisdiction, the court had no jurisdiction of demands against the United States founded on torts. *Schillinger* v. *United States,* 155 U. S. 163; *United States* v. *Berdan Fire-Arms Co.,* 156 U. S. 552. In other words, to give the Court of Claims jurisdiction the demand sued on must be founded on a convention between the parties—" a coming together of minds." That there was such "coming together of minds" is asserted in the case at bar, and *United States* v. *Palmer,* 128 U. S. 262, is cited to sustain the assertion. That case was considered and commented on in *Schillinger* v. *United States, supra*, and it was said to be " an action to recover for the authorized use of a patent by the government, and these observations in the opinion are pertinent : ' This is not a claim for an infringement, but a claim of compensation for an authorized use—two things totally distinct in the law, as distinct as trespass on lands is from use and occupation under a lease. The first sentence in the original opinion of the court below strikes the keynote of the argument on this point. It is as follows : " The claimant in this case invited the government to adopt his patented infantry equipments, and the government did so. It is conceded on both sides that there was no infringement of the claimant's patent, and that whatever the government did was done with the consent of the patentee and under his implied license." We think that an implied contract for compensation'fairly arose under the license to use, and the actual use, little or much, that ensued thereon.' "

The facts of the case fully supported the remarks of the court. The petitioner Palmer was the inventor, patentee and owner of improvement of infantry equipments. They were submitted to a board of officers appointed to consider and report upon the subject of proper equipment for infantry soldiers. The board recommended Palmer's invention. The recommendation was approved by the General of the Army and the Secretary of War, and the invention was manufactured by the government and used.

McKeever v. *United States*, 14 C. Cl. 396, affirmed on appeal by this court, rested on the same facts as the *Palmer* case, the only difference being that McKeever's invention was a cartridge box. There was a recommendation by the board, and the manufacture and use of the cartridge box by the government.

But there is a wide difference between the facts in those cases and the facts in the case at bar. The rifle of the petitioners was not adopted by the board ; the Krag-Jorgensen rifle was. The contention is, however, that the latter rifle contained some of the features of petitioners' invention, and that by adopting it the Ordnance Department conceded that fact and the rights of petitioners to compensation. We are unable to draw tnat conclusion from the correspondence, conceding the power of the Ordnance Department to make the concessions.

The first letter of Captain Russell " invites attention to claims 22, 28 and 29 " of his patent, and expresses a belief that " the Krag-Jorgensen magazine gun *lately adopted by the War Department*" infringed them " in the connection between the magazine and the receiver." The letter concluded as follows: "In considering the allowance for inventions we would request that our claims for these vital points of construction be regarded." A somewhat vague request. However, the letter was replied to (November 18), and he was told that " the business arrangements with the Krag-Jorgensen Company for the manufacture of this arm have not yet been completed," and it is represented to him that the company may agree to indemnify the United States, in which case his "recourse would be to communicate directly with the company." Or if the government should proceed to manufacture the arms without such arrangement, his course would be "to bring suit against the government in the Court of Claims after manufacture has progressed." Of what and on account of what was he to communicate to the Krag-Jorgensen Company, and on account of what was he to bring suit against the government ? On account of an implied contract which had arisen or would arise between him and the United States? Certainly not but on account of an infringement of his invention which might arise. And this was his interpretation, for he writes on the 9th of December that he " could practically have no remedy for infringement of any patent against the Krag-Jorgensen Company, as they have not, that I am aware of, any property in the United States." He requested a hearing before " any business arrangement with the Krag-Jorgensen Company " should be closed.

In reply to that letter he was told by the Ordnance Department that his letters had been referred to the Commissioner of Patents, and that the Commissioner " states that the invention of H. I. Krag and Erik Jorgensen for improvement in magazine firearms has been examined, and the invention has been found patentable." He is then requested, in "further presentation" of his patent, to " communicate direct with the Commissioner of Patents." He did so, and was informed that it was not seen how the Patent Office had any jurisdiction in the matter. "Questions of infringement," he was told, "can be determined only by the courts." Letter February 14, 1893. Waiting until June 30, he informs the Ordnance Department of the reply of the Commissioner of Patents, claimed again the Krag-Jorgensen to be an infringement of his patent and repeated the request of December 9, 1892, that the Ordnance Office allow him "a hearing before any business arrangement with the Krag-Jorgensen Company" be closed. On July 7 that letter was returned to Captain Russell with the endorsement, " that a statement of the case be made in writing for file at this office, and for future reference, as the case stated cannot be determined by the Ordnance Office. The agreement with the Krag-Jorgensen people is such that they are required to guarantee the United States against all damages for infringement."

In answer to this letter Captain Russell's letter of November 22, 1893, (Ex. I,) was written. It need not be reproduced at length. It described his invention and wherein the Krag-Jorgensen rifle infringed that invention, and stated that he based his claim "for compensation on the infringement" of his claims 22 and 28 "and the probable infringement of claim 29" of his patent No. 230,823, of August 3, 1800, " by the Krag-Jorgensen construction." The letter concluded as follows:

"I am not fully informed as to the terms of the contract between the United States and the owners of the Krag-Jorgensen patent. Assuming that the owners of said patent are in ignorance of my rights in the premises, I respectfully request that a copy of this communication may be sent to said parties, and a duplicate of this paper is forwarded for that purpose, with an extra copy of my patent to go with the duplicate. I further request that I may be furnished with the name and address of the responsible parties representing the Krag-Jorgensen interest.

" I am aware that in the event of a suit in equity the alleged infringing parties have a statutory right to challenge the validity of my patent, and, to avoid litigation, I am willing to go further than to make a mere statement of the *prima facie* case as above, and show to the infringing parties or their experts that my claims are well within my rights, provided I am met by these parties in a fair spirit and with a desire to make just compensation when my title to the property is shown.

"My first official notice to the Ordnance Department of this infringement is dated November 16, 1892, (Ordnance Office file 5839 of 1892,) but a gun presenting the special features here mentioned was submitted by me to the board on magazine guns, convened by General Orders 31, H. Q. A., March 21, 1881, and it is described in the report of that board. It is now in my possession subject to examination."

He received the following reply, which seems decisive against the contention of petitioners that there was a concession of their patented rights and implied contract to compensate petitioners:

"I am instructed by the Chief of Ordnance to acknowledge the receipt of your letter of the 22d instant, and to inform you that the terms of the contract between the United States and the Krag-Jorgensen Company contain a clause to the following effect:

"'The said party of the first part shall indemnify the United States and all persons acting under them for all liability on account of any patent rights granted by the United States which may affect the right to manufacture herein contracted for.

"'You have requested that a copy of your communication and a copy of your patent should be forwarded by this office to the company, and for that purpose you have forwarded duplicates of your letter and of the patent specifications. It is considered best that you should forward these communications direct; they are, therefore, returned to you for the purpose.'

"The address of the contracting party is 'The Krag-Jorgensen Geivehr-Kompagnie, Christiania, Norway.' The other papers, enclosures to Ordnance Office file 3515, containing letter and copies of patent specifications, are filed in this office for future reference.

"Your attention is again invited to the statement of the first endorsement on that file, which states that the case 'cannot be determined by the Ordnance Department.'"

Not only is the foregoing letter closing the correspondence decisive against petitioners, but we can discern nothing which tends to support their contention and claim. It was not deemed necessary even to grant his request for a hearing. His rifle was not adopted; another was. There was no concession of his rights. He was told twice that his case could not be determined by the Ordnance Department. There was probably, however, no thought of an arbitrary invasion of his rights. The Ordnance Office sought the opinion of the Commissioner of Patents, and was informed that the Krag-Jorgensen improvement in machine firearms had been examined, and the invention had been found patentable in view of the state of the art. The patent of petitioners was part of the state of the art. The opinion of the Commissioner, of course, was not necessarily conclusive. As he himself said, "Questions of infringement belong to the courts." And because such questions are for the courts the Ordnance Office, no doubt, took indemnity from the Krag-Jorgensen Company, not in concession of petitioners' claim, but for protection against it, if protection should be necessary, and whether it would be or not the Ordnance Office very naturally resolved not to determine. The prudence which takes a bond against a claim cannot be said to constitute or raise a contract in favor of the claim—cannot be said to have intended to create the liability which was meant to be forestalled. Indeed, the Ordnance Office twice wrote Captain Russell that his case could not be determined by it. No contract therefore based on the action of that office can be claimed. If petitioners have suffered injury it has been through the infringement of their patent, not by a breach of contract, and for the redress of an infringement the Court of Claims has no jurisdiction. This doctrine may be technical. If the United States was a person, on the facts of this record (assuming, of course, the petition to be true) it could be sued as upon an implied contract, but it is the prerogative of a sovereign not to be sued at all without its consent or upon such causes of action as it chooses. It has not chosen to be sued in an action sounding in tort this court has declared, as we have seen.

Judgment affirmed.

MR. JUSTICE HARLAN did not participate in this case.

Mr. Justice Shiras, Mr. Justice White and Mr. Justice Peckham dissented.

LANTRY v. WALLACE.

ERROR TO THE CIRCUIT COURT OF APPEALS FOR THE EIGHTH CIRCUIT.

No. 180. Argued March 11, 1901.—Decided May 27, 1901.

This action was brought by the receiver of a national bank under Rev. Stat. § 5151, providing that share holders of every such association shall be held individually responsible, equally and ratably, and not one for another, for all contracts, debts and engagements of such association, to the amount of their stock therein, at the par value thereof, in addition to the amount invested in such shares.

Assuming that the defendant became a shareholder in a national bank in consequence of fraudulent representations of the bank's officers, two questions are presented for determination: 1, Whether such representations, relied upon by defendant, constituted a defence in this action, brought by the receiver only for the purpose of enforcing the individual liability imposed by § 5151, Rev. Stat., upon shareholders of national banking associations? which question is answered in the negative; and, 2, Can the defendant, because of frauds of the bank whereby he was induced to become a purchaser of its stock, have a judgment against the receiver, on a counterclaim for money paid by him for stock, to be satisfied out of the bank's assets and funds in his control and possession? which question is also answered in the negative.

The present action is at law, its object being to enforce a liability created by statute for the benefit of creditors who have demands against the bank of which the plaintiff is receiver. If the defendant was entitled, under the facts stated, to a rescission of his contract of purchase, and to a cancellation of his stock certificate, and, consequently, to be relieved from responsibility as a shareholder of the bank, he could obtain such relief only by a suit in equity to which the bank and the receiver were parties.

Whether a decree based upon the facts set forth in the answer, even if established in a suit in equity, would be consistent with sound principle, or with the statute regulating the affairs of national banks, and securing the rights of creditors, is a question upon which this court does not express an opinion.

The purchase of this stock by the bank under the circumstances was ultra vires, but that did not render the purchase void.

THE case is stated in the opinion of the court.

Mr. C. N. Sterry for plaintiffs in error. *Mr. Eugene Hagen* and *Mr. I. E. Lambert* were on his brief.

Mr. William C. Cochran for defendant in error. *Mr. J. McD. Trimble* and *Mr. W. H. Wallace* were on his brief.

MR. JUSTICE HARLAN delivered the opinion of the court.

This action was brought by the receiver of the Missouri National Bank of Kansas City, Missouri, under section 5151 of the Revised Statutes, providing that the shareholders of every national banking association shall be held individually responsible, equally and ratably, and not one for another, for all contracts, debts and engagements of such association, to the amount of their stock therein, at the par value thereof, in addition to the amount invested in such shares.

The case was determined in the Circuit Court upon demurrer to the answer and cross-petition of the defendant Lantry, and the action of the court in sustaining the demurrer and giving judgment for the plaintiff was affirmed in the Circuit Court of Appeals, Judge Thayer delivering the opinion of the court. 97 Fed. Rep. 865. Judge Sanborn dissented for the reasons set forth in his dissenting opinion in *Scott* v. *Latimer*, 89 Fed. Rep. 843, 857–862; 60 U. S. App. 720, 743–751, which was the case recently decided by this court under the title of *Scott* v. *Deweese*, 181 U. S. 202.

The petition set forth the appointment by the Comptroller of the Currency on the 3d day of December, 1896, of the plaintiff Wallace as receiver of the bank. It alleged that at the time of the bank's failure the defendant was the owner and holder of two hundred shares of its stock of the par value of $100 each; that on the 30th of July, 1897, it appearing to the satisfaction of the Comptroller that it was necessary to enforce the individual liability of stockholders, as prescribed by sections 5151 and 5234 of the Revised Statutes of the United States, that officer made an assessment upon shareholders for $250,000 to be paid by them ratably on or before the 30th of August, 1897; and that he had made demand upon stockholders for

$100 upon each share of capital stock held and owned by them respectively at the time of the failure of the bank.

The defendant made two separate defences. In setting out the first defence he denied that he was then or had ever been the owner of the two hundred shares of the stock referred to in the petition otherwise than as set forth by him.

The case made by the answer of the defendant was substantially as follows:

A short time prior to April 18, 1896, D. V. Rieger, president of the bank, and who had been such from its organization, solicited the defendant and one Calvin Hood to purchase some shares of the stock of the bank and become stockholders. He persistently urged upon them that it was the desire of the bank to have them own its stock and their names connected with it, as they were men of means and had a large business acquaintance in the State of Kansas, and their connection with the bank would be of benefit to it by attracting and securing a large amount of Kansas business otherwise not obtainable.

In the preliminary negotiations for the purchase of the stock Rieger represented that the bank was in a sound, healthy financial condition, free from debts, earning large profits, and paying dividends, and that he was ready and willing to submit to them a detailed statement showing its financial condition.

In consequence of his statements the defendant and Hood called upon Rieger at the banking house of the bank with a view of investigating its condition.

During such preliminary negotiations Rieger continued to act as president of and for the bank, and all statements made by him during the negotiations were made in his capacity as president of the bank, with its knowledge, consent and authority.

The defendant and Hood informed Rieger that they had been induced by him to investigate the condition of the bank with a view of purchasing some of its shares, and they called on him for a full and complete history and detailed account of its business and financial condition. He at once promised to submit to them a faithful statement and history of the bank from its organization, and agreed to submit such statement to any expert bank examiner they might select, if they desired him to do so.

The defendant together with Hood then entered upon such investigation which covered a period of several weeks—Rieger representing at the time that the bank was originally organized with a capital stock of $500,000, all of which had been actually paid for by the subscribers thereto and the money deposited as required by law; that some time in July, 1893, on account of the extreme stringency in money matters and panics, the bank suspended, but upon full investigation by the Comptroller of the Currency, and a full report of the national bank examiner submitted to that officer, it was permitted to resume business; that the Comptroller required the bank to reduce its capital stock to the extent of $250,000, to cover any loss it might have sustained previous to that time; and that this left outstanding the sum of $250,000 of the capital stock, all of which had been actually issued and paid for by the shareholders of the bank at that time.

Rieger submitted to the defendant a report by the Comptroller in support of his statements, which was in words and figures as follows: "This bank [referring to the Missouri National Bank of Kansas City] suspended on the 17th inst., because of the run on the part of its depositors. There was nothing in its condition to warrant this run or occasion suspicion as to its insolvency. It seems to have been prudently managed and its resources are unusually free from items of questionable value, there being no bad debts. The bank is solvent and should be permitted to resume."

He also submitted to defendant a bulletin issued by the Comptroller, dated July 28, 1893, which was in words as follows: "The Missouri National Bank of Kansas City, Missouri, having complied with the conditions imposed by the Comptroller of the Currency, and its capital stock being unimpaired, has this day been permitted to reopen its doors for business. The bank opens with plenty of money on hand, and is wholly solvent and safe."

He represented to the defendant that the statements contained in the above report and bulletin were absolutely true and correct, and that the bank had been frequently examined after it resumed business in July, 1893, by bank examiners and experts, who uniformly and truthfully reported the bank in good, healthy and prosperous condition, and entirely free from bad loans or unsecured paper; that all the paper held by the bank was fresh, clean paper and well secured, and that the interest on its securities had been promptly paid, and there was not among its assets a single item or piece of paper that had not been secured or kept alive as provided by the banking laws of the United States. A list of the securities and assets of the bank was submitted to the defendant by Rieger, he stating that each and every item on the list was worth its face value and was fully secured, and that the bank was and had been since its organization doing a large and profitable business, accumulating a large surplus and paying an annual dividend of six per cent to the stockholders.

After defendant received the above statements from Rieger he secured the services of two expert bank examiners to whom Rieger made the same statements and representations about the assets and condition of the bank, thereby inducing them to believe such statements to be true and to so report to the defendant. Relying upon such representations, defendant agreed with Rieger, as president, to purchase certain shares of its capital stock, which the latter represented was the property of the bank, and which he represented had been theretofore acquired in a lawful way, and paid cash therefor, receiving from him " a certificate of stock dated the 18th day of April, 1896, and numbered 611, and purporting to be issued by the bank and under its seal as of that date, and that this certificate represented the two hundred shares of stock which the defendant supposed he was purchasing, being the same stock mentioned and described in plaintiff's petition, for this defendant never at any time nor upon any occasion purchased or attempted to purchase any other stock of said bank for himself, and never at any time received for himself any certificate of stock purporting to be a certificate for stock of said bank other than the said certificate numbered 611." The bank received the money so paid by the defendant for the stock, namely, the sum of $20,000, and the same was for the use and benefit of the bank.

The defendant continued to be the holder of such certificate of stock, without the knowledge or the means of knowing that the bank was insolvent, until on or about the 2d day of December, 1896. From the time of its delivery to him the bank was apparently doing a very large and prosperous business, having a daily average of about $1,500,000 deposits, and apparently on an average in good bills receivable about $1,300,000. But owing to the vast number of books and the complicated system of bookkeeping kept by the bank, and the artful manner in which its insolvency was and had been secreted by its officers, no one except the officers having knowledge of its condition could or would have supposed from any investigation made within any reasonable limit of time that the bank was insolvent, or that any bills receivable were fictitious, fraudulent or dead paper, or that any of the representations made by the president of the bank were false and untrue, or that it was the purchaser of a large amount of its own stock.

On or about December 2, 1896, there were rumors of the insolvency of the bank. Immediately after learning of such rumors the defendant began to make the most diligent efforts to ascertain the cause thereof, and to ascertain its actual condition. During such investigation, which was only a day prior to the bank's suspension, its president, cashier and other officers persistently insisted that it was in a good, healthy financial condition, and was perfectly solvent, as they had represented it to be. But the defendant was unable to ascertain at that time any information showing the real and actual condition of the bank beyond the representations of its officers.

On the 3d of December, 1896, the receiver of the bank appointed by the Comptroller of the Currency took the actual and exclusive control and possession of the bank and its assets, books, papers and records, and excluded every one, including the officers of the bank and the defendant from the right or opportunity of making any inspection of such books, records, papers or assets. Although repeatedly requested to give to the defendant and his associates an opportunity of making a careful and complete investigation of the affairs and condition of the bank after the same had passed into the possession of the plaintiff as such receiver, the latter persistently refused and denied such request, and after the bank was placed in the hands of the receiver the defendant called upon him and demanded access to the books, papers and documents of the bank, there being no other source from which he could ascertain the real history of the bank and its business transactions or its real condition. He frequently called upon the receiver and asked him for information as to its real history and actual condition, but the only information he could get from that officer was that the bank was solvent and under his management would pay all of its debts, liabilities and obligations without making any assessment or call upon the stockholders. The receiver continued to make the statement that the bank was in a solvent condition, up to the time the order of assessment was issued by the Comptroller as alleged in the petition. By reason of the statements and representations made by the receiver the defendant was led to believe and did believe that the bank was solvent and would pay all its obligations, and that its embarrassments were due to the complication of business matters and would only be temporary.

The defendant also alleged in his answer that on or about September 1, 1897, upon repeated requests, the Comptroller gave to defendant and his associate Hood permission to inspect the assets, books and records of the bank, and to secure all information possible as to its actual condition, and also permission to see the reports of the examinations made under the direction of the Comptroller and the receiver; and that thereupon for the first time after the appointment of the receiver the defendant was permitted to and did make a careful and thorough examination into the actual condition and affairs of the bank; that as a result of that investigation defendant for the first time ascertained the actual condition of the bank and its affairs, and for the first time learned that the representations made by Rieger about its condition were knowingly false, fraudulent and untrue, and were made by him as president of the bank with full knowledge of the bank and the directors and the managers thereof that they were false, fraudulent and untrue and for the purpose of inducing the defendant and Hood to invest in the capital stock of the bank;

That at no time after its organization in 1891 was the bank solvent or able to meet its debts; that no part of its original capital stock was ever actually paid for, as required by the banking laws, nor was any part of the reduced capital stock of $250,000 ever paid for; that the stock had in many instances been issued to irresponsible parties and worthless notes taken therefor; that at the time the bank was represented by Rieger to be in good, sound financial condition it had on hand and included as part of its good assets fraudulent, fictitious and worthless paper greatly in excess of its capital stock; that there was at the time fraudulently concealed and covered up, in paper represented to be good, $50,000 of Rieger's personal indebtedness absolutely worthless; and that the bank never from its organization earned a dividend, but had paid out on dividends over $70,000 in order to conceal and cover up its actual condition; that Rieger when he made the above representations knew the stock to be absolutely worthless and that the capital stock had never been properly issued or paid for; that a large portion of the stock was claimed to be owned by the bank, and that the bank had on hand notes and paper that were fraudulent, fictitious and worthless in a greater amount than its entire capital stock; and that with full knowledge of its absolutely insolvent condition he represented the bank to be in good condition in order to induce defendant to purchase the two hundred shares of the capital stock and defraud him of the $20,000; that at the time of such purchase the defendant believed the statements made to him to be absolutely true in every particular, and was governed by them in such belief; but that such stock was not at any time of any value whatever;

That by reason of the facts above stated and set forth no consideration was ever received by the defendant from the bank for the payment of the $20,000 at the time of the purchase of said stock;

That immediately after the investigation permitted by the Comptroller, and on or about the 27th of October, 1897, he called at the banking house of the bank, where its affairs were being adjusted by the receiver, but finding only the receiver in possession and custody of the bank, its assets, books, records and affairs, and being unable to find any officer of the bank, he tendered to the receiver said certificate of stock numbered 611, above referred to, for cancellation, notifying and informing him that because of the fraud and deceit that had been practiced upon him, he disaffirmed the contract of purchase or pretended purchase of stock, and demanded that the receiver receive the certificate, cancel it and repay to the defendant the sum of $20,000 paid by him as above stated, or such proportionate part thereof as he would be entitled to receive as a creditor of the bank for that amount; but such tender and demand the receiver refused to accept or accede to; and,

That, from the time the bank went into the hands of the receiver until the filing of the answer, there had not been any officer of the bank living or residing in Missouri upon whom any service of summons or other process could be had in any suit that might have been commenced against the bank in Missouri, save and except only the receiver as representing the bank, and that since December 3, 1896, the bank had no usual place of business whatever in Missouri where process could be left or served upon any person conducting the bank's business, save and except as process might be served upon the plaintiff as such receiver.

For a second and further defence the defendant alleged that prior to and during the negotiations with him and Hood, the president, cashier and other officers of the bank, knowing well its insolvency, purchased or pretended to purchase from stockholders or alleged stockholders, with the funds of the bank, shares of its outstanding capital stock in order to prevent exposure by the stockholders owning such shares of its actual condition and the threatened throwing of such shares on the market at prices that would advertise the bank's insolvency; that in order to prevent an open and apparent violation of law, as well as to deceive the public and the Comptroller of the Currency and his associates and employés, the officers of the bank engaged in this transaction and in transactions of buying or attempting to buy such shares of its own stock so outstanding, would cause the certificates of stock purchased or pretended to be purchased from the parties holding the same to be endorsed by those to whom the certificates were issued, either in blank or in the names of the president, cashier or some one of the clerks or other parties connected with the bank, and would then procure a delivery of such endorsed certificate, paying for the same with money belonging to the bank, or by surrendering notes held by it against such parties for such stock, or by payment of money and surrender of notes; that then to account for the funds so used, the parties to whom the certificates would be assigned, or whom the bank pretended were the owners of them, would make a promissory note or notes to the bank for the amount of money used in the purchase of such stock, such note or notes being payable to the bank and unsecured except as the certificates of stock were issued to secure the same; that all stock purchased or attempted to be purchased by the officers of the bank was paid for out of its funds and notes taken from its officers, agents and servants to the bank to represent the funds so used;

That each and every one of the persons engaged in this transaction and who executed any or all of the notes referred to was at the time and ever since had been absolutely and hopelessly insolvent, to the full knowledge of every officer of the bank engaged in these transactions; that such pretended or attempted purchase of shares of stock was made by the bank directly with the owners or holders thereof and with the people in whose names the stock stood, to the knowledge of each one of the persons owning or holding the stock or in whose name it stood, and that none of the transactions concerning the negotiations for the pretended purchase of stock was made to or with the owners, holders or the persons in whose names it stood by any employés in whose names the certificates were taken or to whom the certificates were delivered in blank, except in the case of the president and cashier, who in such negotiations and pretended purchases were, to the knowledge of those with whom they dealt, acting for and on account and in the name of the bank;

That in some instances, and perhaps all, the certificates of stock surrendered to the bank were cancelled, and new certificates issued to irresponsible persons, who were to hold the same for the use and benefit of the bank; that owing to the fact that the entire history of these transactions, so far as it appeared in writing, was and is contained in the books, records and papers of the bank in the sole custody of the receiver, the defendant was unable to give a more detailed statement and history of the transactions, or to state from whom all the purchases were made, or to whom certificates were assigned, or by whom held, or to whom they might have been transferred;

That none of said stock was taken or purchased or procured by the bank to prevent any losses or loss upon debts previously contracted in good faith or purchased in any way authorized by law, but the same was purchased by the bank with its funds for the purpose of preventing the stock from being sold in open market, and to prevent any investigation being made as to the actual condition of the bank by the parties owning the same; that none of the parties to whom new certificates of stock were issued have paid anything for it, nor did they pay or cause to be paid the notes executed to the bank, nor did they intend to pay the notes when they were executed, because they were executed with the fraudulent purpose of concealing the stock purchased by the bank;

That at the time of the negotiations for the purchase of said stock, and at the purchase thereof, the bank had purchased with its funds, in the manner set forth, about $80,000 of the $250,000 of the reduced capital stock of the bank;

That during said negotiations with the president and other officers of the bank by the defendant and Hood, and at the time of the purchase of said stock, the president and other officers of the bank represented to them that all of its capital stock had been subscribed for and issued to actual purchasers in good faith, and was then held and owned by such parties as stockholders of the bank, except an amount of the capital stock which the bank then had on hand which had been taken in by it to prevent a loss on indebtedness previously contracted in good faith, and had been so taken without violating the banking laws of the United States; that defendant believing those statements purchased of the bank two hundred shares of its capital stock of the par value of $20,000, which sum he paid therefor, and a certificate was issued to him by the bank; and,

That at the time of the purchase or attempted purchase by him of said stock for which the certificate was issued, the president and other officers of the bank, in order to have its books show correctly the amount of the outstanding stock, caused some or all of the parties who held certificates of stock in their names that had been purchased for and on account of the bank to surrender to the bank enough of such certificates for cancellation, so that the certificate issued to the defendant could be issued therefor and in the place thereof, and immediately upon the surrender of such certificates to the officers of the bank, and without the knowledge of the defendant, the certificates were cancelled by the bank to an amount sufficient to enable it to issue the certificate so received by defendant; and that the parties who held said stock never at any time received the purchase money paid by the defendant for it, but the same was retained and kept by the bank. Wherefore defendant demanded that the action be dismissed.

The defendant also filed a cross-petition and counterclaim, incorporating therein by reference all the allegations of his first and second defence. He alleged that by reason of the facts stated and the fraud and deceit practiced upon him by the bank and its officers, he has been damaged in the sum of $20,000, with interest from April 18, 1896. He further alleged that he had presented such claim to the receiver for allowance as a claim against the bank, and that the same had been rejected and refused by the receiver. He therefore prayed judgment against the bank for the above sum with interest, and asked that the same be paid ratably by the receiver out of the assets and funds of the bank in his control and possession.

We have given a full statement of the averments of the defendant's pleadings because in an attempt to condense them something might be omitted that was deemed by the plaintiff in error essential to his case, and because the questions presented for consideration may be regarded as important.

Assuming that the defendant became a shareholder in consequence of the fraudulent representations of the bank's officers, as set forth in the answer and cross-petition or counter-claim, two principal questions are presented for determination : 1. Whether such representations, relied upon by the defendant, constituted a defence in the present action brought by the receiver only for the purpose of enforcing the individual liability imposed by section 5151 of the Revised Statutes upon the shareholders of national banking associations. 2. Can the defendant, because of the frauds of the bank whereby he was induced to become a purchaser of its stock, have a judgment against the receiver on the counterclaim in this action for the money paid by him for stock, to be satisfied out of the bank's assets and funds in his control and possession ?

The present action is beyond question one at law. Its object is to enforce a liability created by statute for the benefit of creditors who have demands against the bank of which the plaintiff is receiver. The defendant stood upon the books of the bank as a shareholder at the time it was placed in the hands of the receiver and he was accorded the privileges appertaining to that position. He claims exemption from the responsibility attaching to him, under the statute, as a shareholder upon the ground that in consequence of the frauds practiced upon him he was entitled to disaffirm, and that he had upon due notice to the receiver disaffirmed, the contract under which he purchased the stock in question. He seeks to have the certificate received by him treated as cancelled. Clearly such a defence is of an equitable nature, and could not be recognized and sustained except in some proceeding to which the bank, at least, was a party. If the defendant was entitled, under the facts stated, to a rescission of his contract of purchase, and to a cancellation of his stock certificate, and consequently to be relieved from all responsibility as a shareholder of the bank, he could obtain such a relief only by a suit in equity to which the bank and the receiver were parties.

The defendant alleges that he tendered to the receiver the certificate of stock received by him for cancellation, notifying and informing the latter that, because of the fraud and deceit practised upon him by which he was induced to purchase or attempt to purchase the stock represented by the certificate, he disaffirmed the contract of purchase or pretended purchase of the stock, and demanded that the receiver receive the certificate and cancel it and repay the sum of twenty thousand dollars paid by him, or such proportionate part thereof as he would be entitled to receive as a creditor of the bank for that amount, which tender and demand the receiver refused to accept or accede to. Such tender was an idle ceremony and added nothing to the rights of the defendant; for the receiver had no power to accept or cancel the certificate or to relieve the defendant from the responsibility attaching to him as one appearing upon the books of the bank as a shareholder and to whom had been accorded by the bank the privileges of a shareholder. His duty was to take charge of the assets of the bank and to enforce such assessment upon shareholders as was made by the Comptroller in virtue of the statute.

Nor could the bank, after its suspension and the appointment of a receiver, have assumed to discharge the defendant from any liability attaching to him as a shareholder. Upon the failure of the bank the rights of creditors attached and could not be affected by anything that the bank or its officers might, after such failure, have done or omitted to do. In *Earle* v. *Pennsylvania,* 178 U. S. 449, 455, we held that when a national bank suspends and is placed in the hands of a receiver the entire control and administration of its assets are committed to the receiver and the comptroller, subject to whatever rights of priority, if any, may have been previously acquired by proceedings lawfully instituted against the bank before its suspension. So that the only way in which the defendant could have effectively raised the question of his liability as a shareholder, arising from frauds committed by the bank or its officers before its suspension whereby he was induced to become a shareholder, was by a suit in equity against the bank and the receiver. Instead of pursuing that course, he sought by interposing an equitable defence to defeat this action at law brought by the receiver under the statute. That cannot be done, because under the Constitution of the United States the distinction between law and equity is recognized, so that in actions at law in a Circuit

Court of the United States equitable defences are not permitted. So, also "if the defendant," this court has said, " have equitable grounds for relief against the plaintiff, he must seek to enforce them by a separate suit in equity." *Northern Pacifo Railroad* v. *Paine,* 119 U. S. 561, 563. See also *Bennett* v. *Butterworth*, 11 How. 669; *Thompson* v. *Railroad Companies*, 6 Wall. 134; *Scott* v. *Neely*, 140 U. S. 106; *Scott* v. *Armstrong*, 146 U. S. 499, 512.

We must not be understood as expressing any opinion upon the question whether the defendant could have been discharged from liability as a shareholder if the facts stated in his answer by way of defence had been established in a separate suit in equity. Whether a decree based upon the facts set forth in the answer, even if established in a suit in equity brought against the bank and the receiver after the appointment of a receiver, would be consistent with sound principle or with the statute . regulating the affairs of national banks and securing the rights of creditors, is a question upon which we do not now express an opinion. We mean at this time only to adjudge that the facts set forth in the answer present grounds of relief which cannot be made available by way of defence in this action at law, and if sufficient to protect the defendant against the liability attaching to him as a shareholder, must be alleged and proved in a suit in equity to which the bank and the receiver are made parties.

Some of the observations made in *Scott* v. *Deweese*, 181 U. S. 202, are quite applicable to the present case. That was an action at law to enforce the individual liability imposed by section 5151 of the Revised Statutes. The defendant in that case sought to escape such liability upon the ground, in part, that he had been induced by false representations of the bank's officers to accept a certificate for a certain amount of its increased capital stock. No suit had been instituted to cancel the certificate or to rescind the subscription of stock. The court said: "The present suit is primarily in the interest of creditors of the bank. It is based upon a statute designed not only for their protection but to give confidence to all dealing with national banks in respect to their contracts, debts and engagements, as well as to stockholders generally. If the subscriber became a shareholder in consequence of frauds practiced upon him by others, whether they be officers of the bank or officers of the Government, he must look to them for such redress as the law authorizes, and is estopped, as against creditors, to deny that he is a shareholder, within the meaning of section 5151, if at the time the rights of creditors accrued he occupied and was accorded the rights appertaining to that position." Whether the defendant in that case could have been relieved from liability as a shareholder and had his subscription of stock cancelled, if he had in good faith and in due time before the suspension of the bank instituted proceedings to obtain relief, was not decided.

The defendant, however, contends that the present suit is not embraced by the rule just announced because, he insists, the purchase by the bank of its stock—which he was induced thereafter by its fraud to purchase from it—was not simply voidable but was absolutely *void;* consequently, the sale to him of such stock was void and he did not by his purchase and by taking a certificate of stock become a shareholder within the meaning of section 5151.

It is true that the statute declares that no national bank shall be the purchaser or holder of any of its own shares of capital stock. Rev. Stat. § 5201. But will a violation of this provision by the bank relieve from liability one who holds a certificate of its stock and enjoys the right of a shareholder?

The statute forbids a national bank to lend money upon real estate as security. Rev. Stat. § 5137. Nevertheless, this court has frequently held that the borrower cannot escape liability for the repayment of the money so borrowed, nor dispute the right of the bank to enforce the security taken in violation of the statute; that it was for the Government and not for the borrower to complain of the bank's departure from the rule prescribed by statute. *Scott* v. *Deweese,* 181 U. S. 202, and authorities there cited.

In *National Bank* v. *Stewart,* 107 U. S. 676, 677, it appeared that a bank had loaned money on the security of its shares of stock held by the borrower. The debt not having been paid, the bank sold the stock and applied the proceeds to the payment of an equal amount of the debt. The stockholder then sued the bank to recover the value of the stock, relying on section 5201 of the Revised Statutes forbidding a national bank to make any lóan or discount on the security of the shares of its own capital stock. The trial court held that as the statute forbade the bank to accept its own shares of stock as security for money loaned, the plaintiff was entitled to recover. The judgment was reversed by this court, which held that the statute imposed no penalty, either on the bank or borrower, if a loan was made in violation of its provisions; and that if the prohibition could be urged against the validity of the transaction by any one except the Government, it could only be done while the security was subsisting in the hands of the bank.

So in *Scott* v. *Deweese* above cited, which involved a construction of section 5205 providing that no increase of a bank's capital stock shall be valid until the whole amount of such increase shall have been paid in, and until the Comptroller certifies that the amount of the increase has been duly paid in as part of the capital of the association. This court said: "The statute does not, in terms, make *void* a subscription or certificate of stock based upon increased capital stock actually paid in, simply because the whole amount of any proposed or authorized increase has not in fact been paid into the bank. . . . That the bank, after obtaining authority to increase its capital, issued certificates of stock without the knowledge or approval of the Comptroller and proceeded to do business upon the basis of such increase before the whole amount of the proposed increase of capital had been paid in, was a matter between it and the Government under whose laws it was organized, and did not render void subscriptions or certificates of stock based upon capital actually paid in, nor have the effect to relieve a shareholder, who had become such by paying into the bank the amount subscribed by him, from the individual liability imposed by section 5151."

In view of these decisions it cannot be held that the purchase by the bank of its own shares of stock was void. It was of course a matter of which the Government by its officers could take cognizance; and it may be that it was a matter of which stockholders, having an interest in the proper administration of the affairs of the bank, could complain in a proceeding instituted by them to restrain the bank from violating the statute. But, when the violation of the statute has occurred, it is not a matter of which a shareholder can complain in order that he may be relieved from the liability attaching to him as a shareholder and which the receiver seeks to enforce under the orders of the Comptroller. In the present case Judge Thayer, delivering the opinion of the Circuit Court of Appeals, well said: " In considering the second defence which was interposed by the defendant, it is important to bear in mind that the two hundred shares of stock which he purchased from the bank was not void stock, but was stock which, according to the averments of the answer, had once been issued to other persons and had been reacquired by the bank by purchasing it from such other persons to prevent them from throwing it on the market at ruinous prices. It is necessary to infer from the averments of the answer that this stock had once passed the scrutiny of the Comptroller, and had been outstanding and had been held by other persons since the organization of the bank in the year 1891. The purchase of this stock by the bank under the circumstances disclosed by the answer was doubtless *ultra vires,* but the purchase in question did not render the stock void. In purchasing it the bank made an unlawful use of its funds, for which the officers concerned in the transaction could have been held responsible, as for any other unlawful act, if the corporation had sustained damage; but in point of fact, by the sale of the stock to the defendant that portion of its capital which had been dissipated by the purchase was restored by the resale, and no loss seems to have been incurred. We are at a loss to understand how this transaction on the part of the bank can operate to relieve the defendant from his liability as a stockholder in a suit brought by the receiver to recover a stock assessment which was levied solely for the benefit of corporate creditors. The sale of the stock to the defendant after the bank had purchased the same was not unlawful, since it operated to restore that part of the capital that had been retired, and to that extent repaired the wrong which might otherwise have been done to the bank's creditors." 97 Fed. Rep. 865, 868.

It only remains to inquire whether, in any view of the case, the cross-petition or counterclaim can be sustained. We think not. The receiver sued in this case for the benefit of creditors who, it must be assumed upon this record, knew nothing of the circumstances under which the defendant became a shareholder. They trusted the bank and those who appeared on the list of shareholders required to be kept by section 5210 of the Revised Statutes, which list, that section declares, "shall be subject to the inspection of all the shareholders and creditors of the association." Referring to that section, this court, in *Pauly* v. *State Loan & Trust Co.*, 165 U. S. 606, 621, 622, said: "Manifestly, one, if not the principal, object of this requirement, was to give creditors of the association, as well as state authorities, information as to the shareholders upon whom, if the association becomes insolvent, will rest the individual liability for its contracts, debts and engagements." *Pullman* v. *Upton*, 96 U. S. 328, 330, 331; *National Bank* v. *Case*, 99 U. S. 628, 631. "It is true that one who does not in fact invest his money in such shares, but who, although receiving them simply as collateral security for debts or obligations, holds himself out in the books of the association as the true owner, may be treated as the owner, and therefore liable to assessment, when the association becomes insolvent and goes into the hands of a receiver. But this is upon the ground that by allowing his name to appear upon the stock list as owner he represents that he is such owner; and he will not be permitted, after the bank fails and' when an assessment is made, to assume any other position as against creditors. If, as between creditors and the person assessed, the latter is not held bound by that representation, the list of stockholders required to be kept for the inspection of creditors and others would lose most of its value."

We perceive no ground whatever upon which the defendant can have a judgment upon his cross-petition or counterclaim against the receiver. That officer had nothing to do with the fraudulent transactions of the bank prior to its suspension. His duty was to take charge of its assets, and have them admin-

istered according to the rights of parties existing at the time of
such suspension. Whether, if the defendant claimed a judgment against the bank or its officers for the alleged fraud or
deceit of the latter officers, he could participate in the distribution of the proceeds of the stock assessment until all the contract obligations of the bank had been met, was not decided
by the Circuit Court of Appeals. That question was wisely
reserved for decision when it should arise and become necessary
to be decided. It was deemed by that court only necessary to
adjudge that the receiver was entitled to a judgment against
the defendant, and that the latter was not entitled in this action
to a judgment against the receiver on account of frauds committed by the bank or its officers. In that view we concur.

Perceiving no error of law in the record, the judgment below is

Affirmed.

HOOD *v.* WALLACE.

ERROR TO THE CIRCUIT COURT OF APPEALS FOR THE EIGHTH CIRCUIT.

No. 179.   Argued March 11, 1901.—Decided May 27, 1901.

Lantry v. *Wallace, ante* 536, followed.

THE case is stated in the opinion.

The counsel were the same as in *Lantry* v. *Wallace,* and the
two cases were argued together.

MR. JUSTICE HARLAN delivered the opinion of the court.

The pleadings in this case are the same as in *Lantry* v. *Wallace,* just decided. The demurrer to the answer and cross-petition of Hood was sustained in an elaborate opinion by Judge
Phillips, holding the Circuit Court. 89 Fed. Rep. 11. The

judgment in that court was affirmed in the Circuit Court of Appeals. *Lantry* v. *Wallace*, 97 Fed. Rep. 865.

For the reasons stated in the opinion just rendered in *Lantry's* case, the judgment in this case is

Affirmed.

COMMERCIAL BANK *v.* CHAMBERS.

ERROR TO THE SUPREME COURT OF THE STATE OF UTAH.

No. 270. Argued and submitted April 26, 1901.—Decided May 27, 1901.

As the constitution of Utah distinguished between stock and credits in determining the amount of property of a national bank subject to taxation, shares of stock were not credits, and resident and non-resident shareholders were not entitled to deduct *bona fide* indebtedness from their shares of stock.

The assessed value of real estate owned by a bank in other States than that in which the bank is located, is not to be deducted in determining the amount of assessable property of the bank, unless authorized by the laws of the State in which the bank is situated.

THE plaintiff in error is a national banking association, doing business at Ogden City, Weber County, Utah. The action below was brought by the bank to enjoin the collection of the alleged illegal portion of certain taxes levied against its shareholders for the year 1898.

Certain provisions of the constitution and laws of Utah which are claimed to be pertinent to the controversy are excerpted in the margin.[1]

-----

[1] Provisions of the constitution of Utah relied on by plaintiff in error (Rev. Stat. Utah, 1898, p. 61):

"ARTICLE XIII, SEC. 2. [WHAT PROPERTY TAXABLE. DEFINITIONS. REVENUE.]—All property in the State, not exempt under the laws of the United States, or under this constitution, shall be taxed in proportion to its value, to be ascertained as provided by law. The word property, as used in this article, is hereby declared to include moneys, credits, bonds, stocks, franchises and all matters and things (real, personal and mixed) cap-

The substance of the complaint was that although the assessor in valuing the shares of stock of the bank deducted the

---

able of private ownership; but this shall not be so construed as to authorize the taxation of the stocks of any company or corporation, when the property of such company or corporation represented by such stock, has been taxed. . . ."

"Sec. 3. [Legislature to provide uniform tax. Exemptions.]— The legislature shall provide by law a uniform and equal rate of assessment and taxation on all property in the State, according to its value in money, and shall prescribe by general law such regulations as shall secure a just valuation for taxation of all property; so that every person and corporation shall pay a tax in proportion to the value of his, her or its property: *Provided*, That a deduction of debts from credits may be authorized. . . ."

Provisions of the Revised Statutes of Utah relied on by plaintiff in error (Rev. Stat. Utah, 1898, pp. 579, 581):

"2506. All taxable property must be assessed at its full cash value. . . ."

"2507. Bank Stock. Verified Statement.— The stockholders in every bank or banking association organized under the authority of this State or of the United States must be assessed and taxed on the value of their shares of stock therein in the county, town, city or district where such bank or banking association is located, and not elsewhere, whether such stockholders reside in such place or not. To aid the assessor in determining the value of such shares of stock, the cashier or other accounting officer of every such bank must furnish a verified statement to the assessor, showing the amount and number of shares of the capital stock of each bank, the amount of its surplus or reserve fund or undivided profits, amount of investments in real estate, which real estate must be assessed to said bank and taxed as other real estate, and the names and places of residence of its stockholders, together with the number of shares held by each.

"2508, Id. Deductions.—In the assessment of the shares of stock mentioned in the next preceding section each stockholder must be allowed all the deductions and exemptions allowed by law in assessing the value of other taxable personal property owned by individual citizens of this State, and the assessment and taxation must not be at a greater rate than is made or assessed upon other moneyed capital in the hands of individual citizens of this State.

"2509, Id.—In making such assessment there must also be deducted from the value of such shares such sum as is in the same proportion to such value as the assessed value of the real estate of such bank or banking association in which such shares are held bears to the whole amount of the capital stock, surplus, reserve and undivided profits of such bank or banking association.

*   *   *   *   *   *   *   *

"2518. What debts deductible from credits.—In making up the

proportionate amount of the assessed value of the real estate of complainant situated in the State of Utah, he neglected and refused to deduct the value of real estate owned by the bank situated without such State, and also refused to allow to certain non-resident stockholders deductions from the valuations of their shares of stock to the amount of their *bona fide* debts, though allowing deductions of that kind in favor of resident shareholders. Having tendered to the defendant what it claimed to be the lawful amount of the tax due from it, the bank brought this action to enjoin any attempt to collect the full amount of the tax as laid, and to compel acceptance of the sum which had been tendered. The trial court decided in favor of the bank. On appeal, however, the Supreme Court of the State held that the bank was not entitled to the relief prayed, and reversed the judgment in its favor with costs. Error was prosecuted to the judgment of reversal, and the cause is now in this court for review. 61 Pac. Rep. 560.

---

amount of credits which any person is required to list he will be entitled to deduct from the gross amount of such credits the amount of all *bona fide* debts owing by him, but no acknowledgment of indebtedness not founded on actual consideration, and no such acknowledgment made for the purpose of being so deducted, must be considered a debt within the intent of this section; and no person is entitled to a deduction on account of an obligation of any kind given to an insurance company for the premium of insurance, nor on account of any unpaid subscription to any institution or society, nor on account of a subscription to or instalment payable on, the capital stock of any company or corporation; and no liability of any person or persons, company or corporation, as surety for another must be deducted; and no other liability of any person or persons, company or corporation, on any bond or undertaking must be deducted; and no deduction must be made in any case unless the party claiming such deduction discloses to the assessor, under oath, the name or names of the persons to whom such party is indebted, and the amount of such indebtedness to each, and also that such indebtedness is not barred by the statute of limitations, or, in case such indebtedness is so barred, acknowledges such indebtedness in writing, duly subscribed. No debt is to be deducted unless the statement shows the amount of such debt, as stated under oath in the aggregate. Whenever one member of a firm or one of the proper officers of a corporation has made a statement showing the property of the firm or corporation, another member of the firm or another officer need not include such property in the statement made by him; but this statement must show

Mr. Abbot R. Heywood for plaintiff in error submitted on his brief.

Mr. James N. Kimball for defendant in error. *Mr. George Halverson* was on his brief.

MR. JUSTICE WHITE, after making the foregoing statement, delivered the opinion of the court.

It is urged that " by the action of the taxing officer and the Supreme Court of Utah the shareholders of the Commercial National Bank of Ogden were treated contrary to the provisions of section 5219 of the Revised Statutes of the United States; and, further, that they were denied the equal protection of the laws." Subsidiarily, it is contended first, that the assessor erroneously refused to deduct the *bona fide* debts of non-resident shareholders from the value of their shares of stock, contrary to the provisions of the laws of Utah and the requirements of said section 5219 of the Revised Statutes of the United States (excerpted in the margin[1]), and, second, that the bank was entitled to a deduction from the assessed valuation of the stock, not only of the value of its real estate situated in

the name of the person or officer who made the statement in which such property is included. The fact that such statement is not required, or that a person has not made such statement under oath, or otherwise, does not relieve the property from taxation."

[1] Section 5219, Revised Statutes of the United States:

"SEC. 5219. Nothing herein shall prevent all the shares in any association from being included in the valuation of the personal property of the owner or holder of such shares, in assessing taxes imposed by authority of the State within which the association is located; but the legislature of each State may determine and direct the manner and place of taxing all the shares of national banking associations located within the State, subject only to the two restrictions, that the taxation shall not be at a greater rate than is assessed upon other moneyed capital in the hands of individual citizens of such State, and that the shares of any national banking association owned by non-residents of any State shall be taxed in the city or town where the bank is located, and not elsewhere. Nothing herein shall be construed to exempt the real property of associations from either state, county or municipal taxes, to the same extent, according to its value, as other real property is taxed."

Utah, but the value of real estate situated outside of the limits
of the State.

We will first consider the contention respecting the failure
to deduct *bona fide* debts from the value of the stock of non-
resident shareholders. The Supreme Court of Utah, referring
to the provisions of the constitution of Utah, noted in the margin,
of the statement of facts preceding this opinion, held that as the
constitution of the State distinguished between stock and credits
and authorized only a deduction of debts from credits, shares
of stock were not credits, and both resident and non-resident
shareholders were not entitled to deduct *bona fide* indebtedness
from the value of their shares of stock. This construction of
the statute is binding on this court. *First National Bank of
Garnett* v. *Ayers*, 160 U. S. 660, 664; *First National Bank
of Aberdeen* v. *Chehalis County*, 166 U. S. 440, 444. The claim
of the benefit of the provisions of section 5219 of the Revised
Statutes of the United States is unavailing, for the reason that
there was neither averment nor proof of facts taking the case
out of the operation of recent decisions of this court. Those
decisions held that the term "moneyed capital," as employed
in section 5219 of the Revised Statutes, forbidding greater taxation of shareholders of national banks than is imposed on other
moneyed capital, does not include capital which does not come
into competition with the business of national banks, and that
it must be satisfactorily made to appear by the proof that
the moneyed capital claimed to be given an unjust advantage
is of the character just stated. *First National Bank of Wel-
lington* v. *Chapman*, 173 U. S. 205, 219, and cases cited.

There is obviously no merit in the further contention that re-
versible error was committed because of the refusal to deduct
from the value of the shares of stock of the bank the assessed
value of real estate owned by the bank, situated in other States
than Utah. There was no proof that such a deduction was authorized by the laws of Utah in valuing shares of stock of other
than national banking associations. On the contrary, the Supreme Court of Utah, from an examination of the several constitutional and statutory provisions respecting the subject of taxation in Utah, concluded that the only deductions which were

authorized in the assessment of the shares of stock of national banks or other corporations organized and doing business in the State, were deductions from the value of the shares of the value of real estate situate in Utah. Manifestly, the purpose was to prevent double taxation by the State, a tax on the real estate as such and a further tax thereon by a tax on the stock to the extent that such real estate entered into the value of the stock. As the national banking law, however, permits the taxation of shares of stock of a national bank in the State where the bank is domiciled, the State of domicil is of course entitled to collect taxes upon the full value of such shares of stock. While real estate of a bank situated outside of the State of domicil is taxed in the State of its situs, yet the value of such real estate necessarily enters into and is considered in estimating the value of the shares of stock, and to deduct the value of the real estate would, to the extent of such deduction, reduce the real value of the shares, without a compensatory equivalent. These views and those expressed by the Supreme Court of Utah accord with the doctrine enunciated in *Dwight* v. *Boston*, 12 Allen, 316, 332, and *American Coal Co.* v. *County Commissioners*, 59 Maryland, 185, 193. In the latter case the principle was thus expressed (p. 194):

"The true criterion, as fixed by the statute, is the true value of the stock, without reference to the question where, or in what manner or nature of property or security, the capital stock may be invested. Whether that be invested in real estate, or other property beyond the jurisdiction of this State, the latter having control over the shares and their true value, the peculiar nature and value of the investment of the capital stock of the corporation, beyond the limits of the State, can form no proper subject for specific deduction or abatement from the true value of the shares of stock, when presented to be assessed for purposes of taxation. It is exclusively with the shares of stock, and their true value, as representing the entire corporate assets, that the tax commissioner has to deal, and not with the nature and locality of the investment of the capital stock of the corporation, except as to the real estate of the company situate within this State."

As the shares of stock were taxed as other similar property
in Utah and no discrimination was occasioned, we can perceive
no ground for concluding that the refusal to deduct the value of
the real estate in question constituted either a violation of section 5219, Revised Statutes, or a denial of the equal protection
of the laws.

                                        Judgment affirmed.

                         ————————

FULLER *v.* UNITED STATES.

ORIGINAL APPLICATION FOR MANDAMUS.

No. 7.   Argued and submitted April 15, 1901.—Decided May 27, 1901.

The court below, of original jurisdiction in this case, had authority, upon
   newly discovered evidence, to grant to the railway company a new trial,
   after the final decision of this case at law in that court.
It was competent for Congress to confer upon such court, established under
   the authority of the United States, the power to grant a new trial in an
   action at law upon grounds discovered after the expiration of the term at
   which the verdict or decision was rendered.
The statute does not declare that the right to apply for a new trial upon
   newly discovered evidence after the term shall be any the less when the
   original term is superseded; nor that a new trial of an action at law shall
   not be applied for or granted, while the case is pending in the appellate
   court.
The statute of Arkansas in question is applicable only to actions and proceedings at law in the courts of that Territory, as distinguished from
   suits or proceedings in equity; and as application under that statute,
   within the time prescribed, for a new trial in an action at law, upon
   grounds discovered after the term at which the verdict or decision was
   rendered, was a matter of right, which did not require leave of any
   court.

THIS was a motion for leave to file a petition of mandamus,
and a petition therefor.

Mr. Richard B. Shepard, Mr. Harrison O. Shepard and *Mr.
William T. Hutchings,* for petitioner, submitted on their brief.

Mr. James Hagerman for respondent. *Mr. Clifford L. Jackson* and *Mr. Joseph M. Bryson* were on his brief.

MR. JUSTICE HARLAN delivered the opinion of the court.

Orange Fuller, assignee of Butler Brothers, brought an action on the 23d day of January, 1892, in the United States Court in the Indian Territory against the Missouri, Kansas and Texas Railway Company to recover the damages alleged to have been sustained in consequence of the negligence of the defendant resulting in the destruction of certain property of Butler Brothers by fire.

On the 1st day of May, 1894, the venue of the case was changed to the second judicial division of the Territory, now the Central District, and the result of a trial before the court and a jury was a verdict and judgment in favor of the plaintiff for $8500.

The judgment was superseded, and the cause was taken to the United States Circuit Court of Appeals for the Eighth Circuit, where the record was filed April 3, 1895. In that court the judgment was affirmed on the 30th day of December, 1895. 72 Fed. Rep. 467.

The judgment of affirmance was superseded, and the case was brought to this court upon writ of error sued out by the railway company, the transcript of record being filed here on March 10, 1896. In this court the judgment of the Circuit Court of Appeals was affirmed January 3, 1898. 168 U. S. 707. Our mandate was issued March 3, 1898, and filed in the United States Court in the Indian Territory on July 22, 1898.

On the 20th day of April, 1896, while the case was pending in this court, the railway company filed in the United States Court in the Indian Territory a petition for rehearing upon the ground of newly discovered evidence. Subsequently, at different dates, amended petitions were filed by the company for a new trial. To those amended petitions answers were made, and it was objected that the court was without jurisdiction or authority to grant a new trial, and that it could not consider the alleged newly-discovered evidence.

On the 15th of January, 1900, after the filing in the court of original jurisdiction of the mandate of this court, Judge Clayton of that court granted the application of the railway company and made an order for a new trial.

It should be here stated that in that court there were other cases of a like character with the present case—all growing out of the fire on account of which the present action was brought. One of those cases was *Missouri, Kansas & Texas Railway Company* v. *Wilder.* In that case the United States Court of Appeals for the Indian Territory adjudged that the plaintiff was not entitled to recover. 53 S. W. Rep. 490.

In the order granting a new trial in the present case it was stated: " Now at this day comes the above named defendant, and in support of the fourth amended petition for a new trial in this case files certified copy of opinion of the United States Court of Appeals for the Indian Territory, in the case of *Missouri, Kansas & Texas Railway Company, Appellant,* v. *William L. Wilder, Appellee,* and the court having fully considered the said amended petition for a new trial in this case, which was heretofore continued by agreement between counsel to this the December, 1899, term of this court, together with the evidence thereon, and the briefs of counsel filed both in support and in opposition to said amended petition for new trial, and the court having been fully advised in the premises finds that the original petition for new trial was filed on the twentieth day of April, 1896, in accordance with section 5155 of Mansfield's Digest of the Statutes of Arkansas, extended over the Indian Territory by act of Congress, May 2, 1890, and that summons was duly issued and served upon the plaintiff as required by said statute, and that the said plaintiff has duly entered his appearance in these proceedings, and filed answer to the original petition for new trial, as well as to the different amendments thereto, as such amendments have been based upon such additional evidence as the defendant alleges was discovered subsequent to the filing of the original petition for new trial and amended petitions for new trial respectively, and the court further finds that the evidence fully sustains the said petition for new trial and that under the statute hereinbefore referred to, and in view of the opinion of the United States Court of Appeals for the Indian Territory in the case of the *Missouri, Kansas & Texas Railway Company, Appellant,* v. *William L. Wilder, Appellee,* a companion case to this, that said petition should be sustained. Wherefore the court orders that the said petition be sustained and a new trial be granted."

After setting out the above, the return made by Judge Clayton to the rule herein continued: "Respondent would respectfully further show that it appeared upon the hearing of the petition for new trial in the *Orange Fuller* case that the defendant railway company had used all possible diligence to discover the actual origin of the fire upon which that suit was founded prior to the time of the trial of that case, and that it was to a large extent prevented from so doing by the strong influence which plaintiffs in said case exerted over the minds of the people in the community where the fire occurred, and also by the action of one of the plaintiffs who it was shown urged several of the witnesses to conceal what information they had with reference to the origin of the fire, and that after the trial of the *Orange Fuller* case in the District Court the said railway company continued its efforts to discover the real origin of the fire, and as a result of its efforts it produced reliable new witnesses who proved by newly discovered and strong circumstantial evidence that the fire was set out through accident by one Dole Baugh, and also proved by the admission of the said Dole Baugh, made while the fire was burning, that he had so set it out, and it was this evidence which largely induced the United States Court of Appeals for the Indian Territory to render their opinion of reversal heretofore cited in the said *Wilder* case as well as induced the plaintiffs in the companion cases to dismiss their cases, and that not only the judges of the District Court and the United States Court of Appeals for the Indian Territory, but the plaintiffs in the above suits, became satisfied that these suits were all matters of great injustice and wrong, and satisfied this respondent that the original judgment in the *Orange Fuller* case was also a great injustice and wrongfully obtained and had the actual truth been fully known and not purposely concealed that the judgment in the *Orange Fuller* case would not have been affirmed by the United States Circuit Court of Appeals for the Eighth Circuit or this honorable court, and your respondent further realized that a greater injustice and wrong would be done by permitting the plaintiffs in this said *Orange Fuller* case to recover when the truth had been suppressed, and when all other plaintiffs were prevented from recovering because of the fact that the whole truth had come to light."

The present proceeding is an application for leave to file a petition for a mandamus commanding the judge of the United States Court in the Indian Territory to set aside the above order of January 15, 1900, granting a new trial, and to execute the mandate of this court.

The question presented is whether the court of original jurisdiction had authority upon newly discovered evidence to grant to the railway company a new trial after the final decision in this court.

By the act of Congress of May 2, 1890, providing a temporary government for the Territory of Oklahoma and enlarging the jurisdiction of the United States Court in the Indian Territory and for other purposes, it was declared:

"§ 31. That certain general laws of the State of Arkansas in force at the close of the session of the general assembly of that State of 1883, as published in 1884 in the volume known as Mansfield's Digest of the Statutes of Arkansas, which are not locally inapplicable or in conflict with this act or with any law of Congress, relating to the subjects specially mentioned in this section, are hereby extended over and put in force in the Indian Territory until Congress shall otherwise provide, that is to say, the provisions of the said general statutes of Arkansas relating . . . to pleadings and practice, chapter 119." 26 Stat. 81, 94, c. 182.

In Mansfield's Digest of the Statutes of Arkansas, c. 119, will be found the following sections under the head of Pleadings and Practice:

"§ 5151. A new trial is a reëxamination in the same court of an issue of fact after a verdict by a jury or a decision by the court. The former verdict or decision may be vacated and a new trial granted, on the application of the party aggrieved, for any of the following causes, affecting materially the substantial rights of such party: . . . *Second.* Misconduct of the jury or prevailing party. . . . *Seventh.* Newly disc :- ered evidence, material for the party applying, which he could not, with reasonable diligence, have discovered and produced at the trial."

"§ 5153. The application for a new trial must be made at the term the verdict or decision is rendered, and, except for the cause mentioned in subdivision 7 of section 5151, shall be within three days after the verdict or decision was rendered, unless unavoidably prevented.

"§ 5154. The application must be made by motion, upon written grounds, filed at the time of making the motion. The grounds mentioned in the second, third and seventh subdivisions of section 5151 must be sustained by affidavits showing their truth, and may be controverted by affidavits.

"§ 5155. Where grounds for new trial are discovered after the term at which the verdict or decision was rendered, the application may be made by petition filed with the clerk not later than the second term after the discovery, on which a summons shall issue, as on other complaints, requiring the adverse party to appear and answer it on or before the first day of the next term. The application shall stand for hearing at the term to which the summons is returned executed, and shall be summarily decided by the court. The evidence may either be by depositions or by witnesses examined in court. But no such application shall be made more than three years after the final judgment was rendered."

Many of the cases cited by the petitioner have no application to the present proceeding. They relate to suits in equity and to the power of the court of original jurisdiction in an equity suit to prevent or stay the execution of the mandate of an appellate court. There can be no doubt as to what is the rule recognized in cases of that kind in the courts of the United States or in courts established by its authority.

In *Southard* v. *Russell*, 16 How. 547, p. 570, it was said: " Nor will a bill of review lie in the case of newly discovered evidence after the publication, or decree below, where a decision has taken place on an appeal, unless the right is reserved in the decree of the appellate court, or permission be given on an application to that court directly for the purpose. This appears to be the practice of the Court of Chancery and House of Lords, in England, and we think it founded in principles essential to the proper administration of the law, and to a reasonable termination of litigation between the parties in chancery suits." So in *United States* v. *Knight*, 1 Black, 488, 489: "The defeated party, upon the discovery of new evidence, may, after a final decree in this court, obtain leave here to file a bill of review in the court below to review the judgment which this court had rendered." In *Sandford Fork & Tool Co., Petitioner*, 160 U. S. 247, 255, the court said: "When a case has been once decided by this court on appeal, and remanded to the Circuit Court, whatever was before this court, and disposed of by its decree, is considered as finally settled. The Circuit Court is bound by the decree as the law of the case; and must carry it into execution, according to the mandate. That court cannot vary it, or examine it for any other purpose than execution; or give any other or further relief; or review it, even for apparent error, upon any matter decided on appeal; or intermeddle with it, further than to settle so much as has been remanded. *Sibbald* v. *United States*, 12 Pet. 488, 492; *Texas & Pacific Railway* v. *Anderson*, 149 U. S. 237. If the Circuit Court mistakes or misconstrues the decree of this court, and does not give full effect to the mandate, its action may be controlled, either upon a new appeal (if involving a sufficient amount) or by a writ of mandamus to execute the mandate of this court." See also *In re Potts, Petitioner*, 166 U. S. 263, 267–8.

The action against the railway company was one at law, and whether the court in the Indian Territory had authority to grant the new trial of which complaint is made depends upon the Arkansas statute which, by the act of Congress, was made a law of the Indian Territory. Sections 5153 and 5154 evidently refer to the ordinary motion or application for a new trial made during the term at which the verdict of the jury or the decision of the court is rendered. Section 5155 relates to

new trials for grounds disclosed after the term, and requires
such grounds to be set forth in a petition, summons upon which
shall issue against the adverse party.   A proceeding under that
section is, in form, a new, independent suit, although the statute requires the application to be summarily decided by the
court.

These statutory provisions apply to actions at law, not suits
in equity.   This view is supported by the decision of the Supreme Court of Arkansas in *Jacks* v. *Adair*, 33 Arkansas, 161, 167
(1878).   Referring to the statute giving authority to grant a
new trial after the term upon the ground of newly discovered
evidence, that court said : " The correct view of the statute in
question seems to be this : that it extends to cases at law a new
remedy, without taking away any which existed in equity, but
as to the latter being cumulative, where any difference might
exist.   It is noticeable that the word 'decrees' is not used,
which is the apt and ordinary designation of final orders in
equity ; and there are other indications in the language and
context of the provisions in question, that they were primarily
intended for cases at law, and for new trials of facts found by
a jury, or a court sitting as such." [1]

We perceive no reason to doubt that the action of the court
of original jurisdiction was justified by the statute.   So that
the only question remaining is whether it was competent for
Congress to confer upon such court, established under the authority of the United States, the power to grant a new trial in
an action at law upon grounds discovered after the expiration
of the term at which the verdict or decision was rendered.
Some light is thrown upon this question by the cases in this
and in other courts.

Ex parte Russell, 13 Wall. 664, 668, was an action in the
Court of Claims to recover compensation for the seizure and
use by the United States military authorities of certain steamers
belonging to the claimant.   The case involved the construction
and effect of the second section of the act of June 25, 1868,

---

[1] That case was based upon sections 4688, 4690, 4691 and 4692, Gantt's Digest (1874), which are the same as the above sections in Mansfield's Digest.

which provided that the Court of Claims, "at any time while any suit or claim is pending before or on appeal from said court, or within two years next after the final disposition of any suit or claim, may, on motion on behalf of the United States, grant a new trial in any such suit or claim, and stay the payment of any judgment therein, upon such evidence (although the same may be cumulative or other) as shall reasonably satisfy said court that any fraud, wrong or injustice in the premises has been done to the United States; but until an order is made staying the payment of a judgment, the same shall be payable and paid as now provided by law." In the Court of Claims an application for a new trial was made by the United States when the case was pending in this court. The former court dismissed the application for want of jurisdiction, on the ground, in part, that after it was made the mandate of this court affirming the original judgment against the United States was filed in the Court of Claims. From that order an appeal was allowed to this court, and one of the questions presented was whether the Court of Claims should have dismissed the application for a new trial for want of jurisdiction. This court observed that the Court of Claims erred in dismissing the application, and after referring to the causes which probably induced the passage of the act of June 25, 1868, said: "But whatever reasons Congress may have had for passing the act, of its right to pass it there is no question. The erection of the Court of Claims itself, and the giving to parties the privilege of suing the Government therein, though dictated by a sense of justice and good faith, was purely voluntary on the part of Congress; and it has the right to impose such conditions and regulations in reference to the proceedings in that court as it sees fit. The section in question was undoubtedly intended to give the Government an advantage, which, in respect to its form, is quite unusual, if not unprecedented, but which Congress undoubtedly saw sufficient reason to confer. It authorizes the Court of Claims, on behalf of the United States, at any time while a suit is pending before, or on appeal from, said court, or within two years next after the final disposition of such suit, to grant a new trial upon such evidence as shall satisfy the court that the Government has been defrauded or wronged. . . . It has been objected that the granting of a new trial after a decision by this court is, in effect, an appeal from the decision of this court. This would be so if it were granted upon the same case presented to us. But it is not. A new case must be made; a case involving fraud or other wrong practised upon the Government. It is analogous to the case of a bill of review in chancery to set aside a former decree, or a bill impeaching a decree for fraud. We are of opinion, therefore, that the Court of Claims had jurisdiction to grant a new trial, notwithstanding the filing of the mandate of this court." Chief Justice Chase and Mr. Justice Clifford dissented from the opinion because, in their judgment, "the act of Congress did not warrant the granting of a new trial on a petition filed subsequent to an appeal and the return of the mandate of this court." In *Ex parte United States*, 16 Wall. 699, 703, the above case was again before this court, and a peremptory mandamus was awarded requiring the Court of Claims to hear and determine the application for a new trial.

In *United States* v. *Young*, 94 U. S. 258, 260, it appeared that a new trial was granted by the Court of Claims, in a suit at law, while an appeal was pending here from the original judgment. This court said: "The Court of Claims, by granting a new trial, has resumed control of the cause and the parties. This it had the right to do. Such a power may be somewhat anomalous, but it is expressly given, and every person when he submits himself to the jurisdiction of that court for the prosecution of his claim submits himself to its operation. The proceedings under which the new trial was obtained are now a part of the record below, and, after judgment is finally rendered, may be brought here by appeal for review."

In *Belknap* v. *United States*, 150 U. S. 588, 590, 591, the court observed that while ordinarily the Court of Claims would be without power to grant a new trial at a term subsequent to that at which the original judgment was rendered, it had such power under section 1088 of the Revised Statutes—which is the same in substance as the second section of the above act of June 25, 1868. The court said: "In order to give full effect

to this statute the Court of Claims must have power to grant a
new trial at a term subsequent to that at which the judgment
was rendered, for it explicitly provides that it may be exercised
at any time within two years."

In Iowa there is a statute giving the court power to grant a
new trial on grounds discovered after a verdict or decision is rendered—the petition for the new trial to be filed not later than
one year after final judgment, and the case made by it tried as
other cases. In *Cook* v. *Smith,* 58 Iowa, 607, 608, the Supreme
Court of that State said: "The right to apply for, and the
power of the court to entertain, jurisdiction of the application
during the time limited in the statute are absolute and unconditional. There is no such inconsistency between the two proceedings as to require the one to be abated because the other is
pending. It may be both should not be actively prosecuted at
the same time for the determination of one, may render a decision
in the other unnecessary. Upon application this would no doubt
be controlled by the courts. Suppose the ground upon which
a new trial was asked was not discovered until after the appeal
was taken, on the last day allowed therefor, would such appeal
deprive the court of the power to entertain jurisdiction of a
petition for a new trial? Clearly not, we think, for during the
time limited in the statute the power of the court and the right
of the party are unconditional. There are cases where neither
party is satisfied with the judgment below. Would an appeal
by one party oust the court of the power to entertain and grant
a new trial on the application of the other party? We think
not."

In a case arising under a statute similar to the one in Arkansas, the Supreme Court of California said: "The appeal from
the judgment did not divest the trial court of jurisdiction to hear
and determine the motion for a new trial." *Naglee* v. *Spencer,*
60 California, 10. In *Rayner* v. *Jones,* 90 California, 78, 81, the
same court said: "A notice of motion for a new trial was served
and filed in due season, and upon the hearing of the motion the
trial court dismissed it, upon the theory, evidently, that as the
judgment made and entered had been appealed from when the
motion for a new trial came on for hearing, the court below

had lost jurisdiction to determine it. This view of the matter is untenable, and the court should have heard the motion, and either granted or denied it, upon the bill of exceptions presented, which is a part of the record here on the appeal from the order of dismissal." See also *Carpentier* v. *Williamson*, 25 California, 154, 167; *McDonald* v. *McConkey*, 57 California, 325; *Chase* v. *Evoy*, 58 California, 348; *Scott* v. *Scott's Ex'r*, 82 Kentucky, 328; *Duffitt* v. *Crozier*, 30 Kansas, 150; *Hines* v. *Driver*, 89 Indiana, 339; *Railroad Co.* v. *O'Donnell*, 24 Nebraska, 753.

The same principles have been recognized in criminal cases. In *State ex rel. Turner* v. *Circuit Court for Ozaukee County*, 71 Wisconsin, 595, which was a criminal case, an application was made for a new trial after the affirmance of the original judgment. In that State it was provided by statute that " the Circuit Court may at the term in which the trial of any indictment or information shall be had, or within one year thereafter, and in either case before or after judgment, on the petition or motion in writing of the defendant, grant a new trial for any cause for which, by law, a new trial may be granted, or when it shall appear to the court that justice has not been done, and on such terms as the court may direct." The Supreme Court of Wisconsin said: " It appears that a proper motion was made within one year from the judgment, upon the grounds addressed to the discretion of the Circuit Court, and a new trial was undoubtedly granted under the special authority conferred by the above statute; and the question now is, Had the court power to grant it? We can only consider the question of the power or jurisdiction of the court in the matter, not whether it exercised that power wisely or granted the motion on insufficient grounds, for the court may have erred, but error does not affect its jurisdiction. This statute was probably borrowed from Massachusetts. See Pub. Stat. Mass. 1882, c. 114, § 128; *Com'th* v. *Peck*, 1 Met. 428; *Com'th* v. *McElhaney*, 111 Mass. 439; *Com'th* v. *Scott*, 123 Mass. 418. Also Terr. Stat. Wis. 1839, p. 377, § 6; Rev. Stat. 1849, c. 149, § 6; Rev. Stat. 1858, c. 180, § 6. We do not well see upon what grounds the power of the court to grant the new trial can be denied if the provision is valid. The fact that the judgment has been affirmed by this court furnishes no suffi cient reason for denying that power. It is said by the affirmance of the judgment it became a finality, a final determination of·the cause and sentence of the law. That view would certainly be correct had not the legislature conferred this special authority to grant a new trial upon a proper cause shown. On affirmance of a judgment in a civil case no new trial could be granted unless the statute authorized it. Only where the statute does authorize it can a new trial after affirmance be granted, either in a civil or criminal cause. In actions of ejectment the Circuit Court can grant a new trial even after affirmance by this court, and this by virtue of a statute upon the subject. *Haseltine* v. *Simpson*, 61 Wisconsin, 427. Consequently we can perceive no sufficient grounds or reasons for denying the validity of the statute to grant a new trial after judgment has been affirmed in this court, any more in a criminal than in a civil cause."

In *Commonwealth* v. *McElhaney*, 111 Mass. 439, 441, 443, which was an indictment for murder, an application was made by petition for a new trial on the ground of·newly discovered evidence. The question was raised by the Commonwealth whether the application could be entertained after the accused had been sentenced to death and the executive warrant for execution thereof issued. The question depended upon section 7, of the General Statutes of Massachusetts, c. 173, providing that "the Supreme Judicial Court and Superior Court may, at the term in which the trial of any indictment is had, or within one year thereafter, on the petition or motion in writing of the defendant, grant a new trial for any cause for which by law a new trial may be granted, or when it appears to the court that justice has not been done, and on such terms or conditions as the court shall direct." The Supreme Judicial Court of Massachusetts said: "At the time of the passage of the General Statutes, therefore, this court had no original criminal jurisdiction, except of capital cases; and in these cases sentence has always been passed within a very short time after the trial and conviction, and a copy of the record of the conviction and sentence forthwith transmitted to the Governor, in accordance with the Revised Statutes, c. 139, § 11, and the General Statutes, c. 174,

§ 24; and yet the General Statutes, c. 173, § 7, in terms authorize a petition for a new trial to be presented to this court at any time within one year after the trial. The unavoidable conclusion is that so long as that year has not elapsed, and the sentence has not been carried into execution, the court is authorized to entertain a petition for a new trial."

In no one of the above cases, nor indeed in any case to which our attention has been called, was there any suggestion of the want of power in the legislature to authorize the granting of a new trial in an action at law upon evidence discovered after the term at which the verdict or decision was rendered. So far as the power of Congress is concerned, we cannot conceive that legislation of that character in respect of cases at law, as distinguished from cases in equity, infringes upon any right secured by the Constitution of the United States.

In the case now before us it appears that the operation of the original judgment was suspended by a supersedeas. But the statute, reasonably construed, does not declare that the right to apply for a new trial upon newly discovered evidence after the term shall be any the less when the original judgment is superseded. Nor does it declare that a new trial of an action at law shall not be applied for or granted while the case is pending in the appellate court. It is true that, in the absence of legislation to the contrary, neither the filing of a petition for new trial nor the granting of a new trial by the court of original jurisdiction, after the term and upon newly discovered evidence, interferes with the power of the appellate court to proceed with the hearing and determination of the case upon the record before it. But the operation and effect of its final judgment may be ultimately controlled by the disposition made by the court of original jurisdiction of an application for a new trial made in conformity with a statute. If this be regarded as an anomalous rule of procedure in actions at law, it is sufficient to say that Congress, in its wisdom and in order to promote the ends of justice, saw proper to prescribe it, and we know of no reason to question the authority it has exercised upon this subject. All embarrassment in the present case was avoided by the fact that the new trial of which petitioner complains was not granted while the original case was in this court nor until after our mandate had been filed in the court of original jurisdiction.

Our conclusions are: 1. That the statute of Arkansas in question, which was made by Congress the law of the Indian Territory, is to be held applicable only to actions and proceedings at law in the courts of that Territory, as distinguished from suits or proceedings in equity; 2. That an application under that statute, within the time prescribed, for a new trial in an action at law, upon grounds discovered after the term at which the verdict or decision was rendered, was a matter of right, and did not require the leave of any court—the application constituting, on appeal, a new action, in which summons or process would regularly issue against the adverse party, and which must be heard and determined by the court upon evidence adduced by the parties.

It results that the court of original jurisdiction acted within the authority conferred upon it, and the rule for a mandamus compelling it to set aside the order granting a new trial must be discharged, and it is so ordered.

DISTRICT OF COLUMBIA *v.* MOULTON.

ERROR TO THE COURT OF APPEALS OF THE DISTRICT OF COLUMBIA.

No. 234.   Argued April 9, 10, 1901.—Decided May 27, 1901.

Park street is a public highway in the northwest section of the city of Washington. For some days before the accident which was the ground of this action, a steam roller had been used in connection with the work of resurfacing the street with macadam. This roller became disabled, and was placed close to the south curb of the street, a canvas cover was placed over it, and it was left there for two days. On the second day the horse of the plaintiff in error, being driven along the street, became restive from the flapping of the canvas cover, reared, and upset the vehicle, and threw out the plaintiff, injuring him. *Held* that the District of Columbia was not liable for the injuries which the plaintiff so suffered.

THIS action was begun by the defendant in error in the Supreme Court of the District of Columbia. In substance he asserted in his declaration a right to recover from the District of Columbia a specified sum, upon the ground that by its negligence, on November 26, 1896, he had sustained serious personal injury. The negligence averred consisted in this, that for a space of two days prior to and including the date named the District had negligently and knowingly left upon a public highway known as Park street a large steam roller, which was calculated to frighten horses of ordinary gentleness; and while plaintiff was driving along said street, with due care, in a carriage drawn by a horse of that disposition, the animal was frightened and rendered unmanageable by the steam roller, and in the struggles of the horse one of the wheels of the carriage was broken, plaintiff was thrown out upon the ground with great force, and he sustained the injuries for which recovery was asked. Defendant filed a plea of the general issue.

The evidence most favorable to the contention of the plaintiff tended to show the following: Park street is a public highway in the northwest section of the city of Washington, commencing at Fourteenth street and running westwardly. For several days prior to the accident in question a steam roller had been used in connection with the work of resurfacing Park street with macadam. This roller was of the kind usually employed in constructing macadamized gravel roads. It had three wheels, the tread of the rear wheel being about eight feet, which was its extreme width. The machine was about eight feet long and about five or six feet high. The smokestack was a little higher than the other part of the machine. While the roller was in use, on the forenoon of the day before the accident hereinafter referred to, it "broke down." The nature of the injury to the roller does not appear, otherwise than as it may be inferred from the fact that the roller was subsequently removed by horse power, that the machinery was simply disabled. On becoming out of order, the roller was placed close to the south curb of Park street, from twenty to fifty feet west of Pine street—a street fifty feet in width—and distant about nine hundred feet westwardly from Fourteenth street. Over the roller was placed a canvas cover. The roadway proper, at the point where the roller was stationed, was about twenty-eight feet wide, and there was ample room for the passage of vehicles between the roller and the northerly side of Park street.

About 3 o'clock on the afternoon of November 26, 1896, (Thanksgiving Day,) plaintiff drove into Park street from Fourteenth street, and as he did so saw the steam roller. The horse he was driving was one which the plaintiff had owned for several years, was regarded as of an ordinary gentle disposition and had several times been driven safely past steam rollers when they were in actual operation. Plaintiff guided his horse, intending to pass by the roller in the space to the right thereof, but on approaching Pine street the horse became restive—from the flapping of the canvas cover on the roller or from some other cause—and when about opposite the middle of Pine street became unmanageable, reared and upset the vehicle, throwing out and injuring the plaintiff. The evidence also tended to show that other horses in passing the roller had exhibited fear.

The case was tried to a jury, and resulted in a verdict for the plaintiff. On appeal the judgment was affirmed by the Court of Appeals of the District. 15 App. D. C. 363.

Mr. Andrew B. Duvall for plaintiff in error. *Mr. Clarence A. Brandenburg* was on his brief.

Mr. A. S. Worthington for defendant in error. *Mr. Charles L. Frailey* was on his brief.

Mr. Justice White, after making the foregoing statement, delivered the opinion of the court.

That the District of Columbia is not an insurer of the safety of travelers upon its streets is, of course, unquestioned. This being so, we think the lower courts erred in upholding the liability of the District for the injuries sustained by the plaintiff, under the circumstances disclosed in the record.

The steam roller in question had been brought to the place where the accident occurred for a lawful purpose, viz., that of performing a duty enjoined upon the District to keep in repair the streets subject to its control. The use of an appliance such as a steam roller was a necessary means to a lawful end—a means essential to the performance of a duty imposed by law. It must, therefore, follow that if in the legitimate and proper use of such machine, with reasonable notice to the public of such use, an injury is occasioned to one of the public, such injury is *damnum absque injuria. Lane v. Lewiston,* 91 Maine, 292, 294; *Morton v. Frankfort,* 55 Maine, 46; *Cairncross v. Pewaukee,* 78 Wisconsin, 66, commenting upon and explaining *Hughes v. Fond du Lac,* 73 Wisconsin, 380. Conceding that the roller was an object calculated to frighten horses of ordinary gentleness, yet, at the most, the liability of the municipality for negligently permitting such objects to remain within the limits of a highway, if it exists, must primarily be dependent upon the fact that they are unlawfully upon the highway.

The sole negligence complained of in the declaration was averred to consist in keeping the steam roller in question on Park street for the space of two days so as to be a public nuisance and dangerous to travelers passing along said street with their carriages and horses. There was no allegation that the roller in consequence of its being disabled presented such a changed appearance that the danger of its frightening an animal was enhanced. Nor was there any averment that the negligence was committed in the use of the canvas covering, and no proof was offered on the trial tending to show that such a cover was not the means usually employed to protect steam rollers from the weather when they were lawfully on the street and for the time being not in use.

Where but one inference can reasonably be drawn from the evidence the question of negligence or no negligence is one of law for the court. *Northern Pacific Railroad Co. v. Freeman,* 174 U. S. 379, 384; *Metropolitan Railway Co. v. Jackson,* 3 App. Cas. 193. It is only where the evidence is such that reasonable men may fairly differ as to the deductions to be drawn therefrom, that the determination of the fact of negligence should be submitted to a jury. *Warner v. Baltimore & Ohio Railroad,* 168 U. S. 339, 348. The question which here arises then is,

Did the evidence justify the trial court in permitting the jury to determine whether or not in allowing the disabled roller to remain at the place referred to, under the circumstances stated, the District negligently and unlawfully obstructed the highway?

We shall assume that the period when the steam roller became unserviceable while in use on Park street was the forenoon of the day prior to the accident, as claimed by the plaintiff. The right, however, to use a steam roller upon a public street for the purpose of the repair of such street we think necessarily includes the right to retain the roller upon the street until a reasonable time after the necessity for the use of the machine has terminated, in the meantime exercising due care in the deposit of the machine when not in use and giving due notice and warning to the public of the presence of such machine if travel upon the street is permitted. We can perceive no difference in principle between using and keeping a steam roller on the streets until the completion of a particular work and the maintaining a lawful excavation, such as for the construction of a sewer or of an underground road and the use of an engine, derrick, etc., in connection with the hoisting of earth from an excavation. The appliances used in connection with such excavations, even though calculated to frighten horses of ordinary gentleness not familiar with such objects, undoubtedly may be retained at the place where needed until the necessity therefor has ceased, and the circumstance that such appliances become temporarily disabled cannot, in reason, be held to affect the right of the municipal authorities to keep such machinery on the works until in the ordinary course of events and in the exercise of a reasonable discretion it is found convenient either to there make the needed repairs or to remove the appliances elsewhere for that purpose. Now, the only inference warranted by the record is that when the steam roller in question got out of order it was being used upon the street, and the necessity for its further use continued to exist. Had the machine not broken down, or had needed repairs been made to it at the place where the roller was deposited, it might lawfully have been allowed to remain upon the street while its further use was required, and until it was reasonably convenient to remove it. Under such a state of facts as has been detailed there was nothing either in the circumstance of the disabling of the machine, or in the detention, warranting the inference that the right to leave the roller upon the street over a legal holiday did not exist, and that an illegal use of the highway had originated. It follows that the facts in evidence respecting the keeping of the roller on Park street during the period referred to did not justify the submission to the jury of the question whether the District was negligent in so keeping the machine, as it could not reasonably have been inferred that the employés of the District were negligent in failing to remove the machine before the occurrence of the accident.

As respects the notice owing to the plaintiff of the presence of the roller, we agree with the opinion of the Supreme Judicial Court of Maine in *Lane* v. *Lewiston, supra,* that where a steam roller is allowed to remain upon a municipal highway it is requisite that the municipality causing the obstruction should give reasonable notice to the traveling public of its presence, but that a view of its obstruction itself in time to avoid it without injury amounts to notice. In other words, as stated by the Maine court, " No one needs notice of what he already knows," and " Knowledge of the danger is equivalent to prior notice." 91 Maine, 296. That the plaintiff had notice of the presence of the roller on Park street in ample time to have avoided it, is undisputed. When he turned from Fourteenth street into Park street it was broad daylight, there was nothing to obstruct his view westward, and in fact he testified that the roller was in plain sight. He was not induced or directed by the agents of the District to proceed past the roller. He knew that such objects sometimes frightened horses, but from his acquaintance with the disposition of his horse he believed that he could control the animal and drive safely past the roller, and he voluntarily undertook to do so. Now, it seems clear—particularly as the danger was neither hidden nor concealed—that the District was under no obligation to restrain the plaintiff from attempting to pass, either by closing Park street or by other means. The District was not bound to presume that it would be *necessarily* hazardous to attempt to drive past the roller, stationary and quiet as it was, and familiar as horses in a large city usually are to the sight and sounds of electric and cable cars and horseless motors. The District, at best, was only chargeable with notice that the roller was an object which might frighten some horses of ordinary gentleness, not that it would inevitably do so. It was bound to give sufficient warning to drivers of the presence of the roller in time to enable them to avoid passing it, if desired. The District, however, had a right to assume that a driver of mature age was familiar with the habits and disposition of his horse, and was possessed of the common knowledge respecting the tendency of steam rollers to occasionally frighten such animals. The roller being lawfully on the street, the District was not bound to guard against the consequences of a voluntary attempt to drive by this roller. Certainly, if a driver believed that it would not be the natural and proper consequence of such an attempt that his safety would be endangered, the District ought not to be charged with notice that the attempt would be dangerous either to life or to limb.

The foregoing observations sufficiently indicate the errors committed by the trial court in the instructions given to the jury and in the refusal to give requested instructions, to which exceptions were noted. It suffices to say in conclusion that the trial court erred in refusing to instruct the jury, as requested, that upon the whole evidence in the case their verdict should be for the District. As said by this court, speaking through Mr. Justice Blatchford, in *Schofield* v. *Chicago, Milwaukee & St. Paul Railway Co.*, 114 U. S. 615, 618:

"It is the settled law of this court that, when the evidence given at the trial, with all the inferences which the jury could justifiably draw from it, is insufficient to support a verdict for the plaintiff, so that such a verdict, if returned, must be set aside, the court is not bound to submit the case to the jury, but may direct a verdict for the defendant. *Improvement Co.* v. *Munson*, 14 Wall. 442; *Pleasants* v. *Fant*, 22 Wall. 116; *Herbert* v. *Butler*, 97 U. S. 319; *Bowditch* v. *Boston*, 101 U. S. 16; *Griggs* v. *Houston*, 104 U. S. 553; *Randall* v. *Baltimore & Ohio Railroad Co.*, 109 U. S. 478; *Anderson County Comrs.* v.

Beal, 113 U. S. 227; *Baylis* v. *Travelers' Insurance Co.,* 113 U. S. 316."

The judgment of the Court of Appeals of the District of Columbia is reversed, with instructions to that court to reverse the judgment of the Supreme Court of the District of Columbia, and to grant a new trial.

JACOBS v. MARKS.

ERROR TO THE SUPREME COURT OF THE STATE OF ILLINOIS.

No. 410. Submitted January 7, 1901.—Decided May 27, 1901.

The question whether the record and judicial proceedings in the Michigan court received full faith and credit, in the courts of Illinois is one for this court to consider and determine; and it holds that, upon the facts disclosed in the record, the courts of Illinois did give to the judgment and judicial proceedings of the state court of Michigan full faith and credit, within the meaning of the Constitution.

The judgment in question in this case did not necessarily import that the plaintiff had received satisfaction of her claim.

The distinction between *Halderman* v. *United States,* 91 U. S. 584, and *United States* v. *Parker,* 120 U. S. 89, shown.

In June, 1896, Dora Marks brought an action in the circuit court of Cook County, Illinois, against Lewis Jacobs, for false representations and deceit whereby the plaintiff had been induced to become a member of a corporation known as the Chicago Furniture and Lumber Company of Escanaba, Michigan, composed of said Jacobs and one Nathan Neufeldt, and to pay into such concern the sum of $5000. The plaintiff sought to recover in this action the money so expended by her, alleging that the shares of stock so taken by her in said company were worthless.

The defendant filed a demurrer to the declaration, which was overruled, and thereupon he filed a plea of not guilty; and also several special pleas, in which he set up, in substance, that the

plaintiff, on or about December 4, 1893, instituted an action in
the circuit court of Delta County, Michigan, against the Chicago
Furniture and Lumber Company, to recover the sum claimed
in the present suit; that service was duly had upon said company, which entered its appearance, and said court acquired
jurisdiction of the parties to said cause and the subject-matter
thereof; that, afterwards, the said parties came to a settlement
of said cause; that, on July 25, 1894, the said court entered the
following order : " This cause having been settled, it is hereby
discontinued by consent of both parties, without cost to either
party ; " and that the said plaintiff had, therefore, received full
satisfaction of the claim upon which the present suit is based.
These special pleas were traversed, and the trial resulted in a
verdict in favor of the plaintiff for $4000.　At the trial of the
present case the plaintiff put in evidence a written agreement
between the Chicago Furniture and Lumber Company and Dora
Marks, in the following terms :

" Articles of agreement made and entered into this 14th of
July, A. D. 1894, by and between the Chicago Furniture and
Lumber Company, a corporation, of the city of Escanaba, Delta
County, Michigan, parties of the first part, and Dora Marks of
Denver, Colorado, party of the second part.　Party of the first
part agrees to purchase the twenty thousand dollar ($20,000)
worth of stock of the said Chicago Furniture and Lumber Company, which the party of the second part holds, for the sum of
$4000, to be paid for as follows : $1000 to Mead and Jennings,
attorneys for said party of the second part, as soon as the parties of the first part dispose of their treasury stock to the amount
of $1000 or interest other capital in said company to the
amount of $1000 and $3000 to said party of the second part,
on the day that the plant now occupied by the parties of the
first part in said city of Escanaba is turned over to them, and a
clear title to the property earned by them.　Parties of the first
part further agree to discontinue the damage suit now pending
against the party of the second part without cost.　Said parties of the first part further agree to release said party of the
second part from all liability of said second party for the balance due on unpaid stock.　Party of the second part agrees to

sell her said stock of $20,000 to the parties of the first part and accept payment as aforesaid mentioned. Party of the second part also agrees to discontinue the suit now pending under attachment proceedings against party of the first part, without cost. Said stock to be transferred as paid for.

" In witness whereof the parties have hereunto set their hands and seals the day and year first above written."

Thereupon, over the objections of the defendant Jacobs, the plaintiff was permitted to testify that the company never carried out the agreement under which the suit was brought, and that she never recovered a single dollar in satisfaction of her claim. The defendant requested the court to instruct the jury that the settlement of the Michigan case constituted a bar to this action. These instructions were refused, and the trial resulted in a verdict and judgment in favor of the plaintiff in the sum of $4000.

The cause was taken to the Appellate Court of Illinois, which first reversed, and then, on rehearing, affirmed, the judgment of the trial court; and, afterwards, to the Supreme Court of Illinois, which, on December—, 1899, affirmed the judgment of the Appellate Court. A writ of error was thereupon allowed by this court.

Mr. Louis J. Blum and *Mr. Edgar C. Blum* for plaintiff in error.

Mr. John F. Dillon, Mr. Andrew J. Hirschl and *Mr. John W. Byam* for defendant in error.

MR. JUSTICE SHIRAS after stating the case, delivered the opinion of the court.

The plaintiff in error alleges error in the action of the Illinois courts in failing to give full faith and credit to the judicial record and proceedings of the circuit court of Delta County, Michigan.

A contention is made on behalf of the defendant in error that the decision of the state Supreme Court did not rest on a

Federal question, and that, hence, under the doctrine of *Seeberger* v. *McCormick*, 175 U. S. 274, and cases therein cited, we have no jurisdiction to review it.

But the record discloses that, at the trial in the circuit court of Cook County, the defendant, after having put in evidence the record of proceedings in the circuit court of Delta County, Michigan, wherein Dora Marks was plaintiff and the Chicago Furniture and Lumber Company was defendant, asked the court to give the following instruction:

"You are instructed that if you find from the evidence that the plaintiff herein instituted a suit in the circuit court of Delta County, Michigan, against the Chicago Furniture and Lumber Company for the purpose of recovering the $4000 involved in this suit now before you, and that she made a settlement of this cause with the defendant therein or any one else, that the plaintiff is barred from the further prosecution of this suit, and the verdict of the jury must be for the defendant." And, in support of the motion for a new trial, it appears that the defendant alleged that "the verdict and action of the court fail to give full faith and credit to the judgment of the circuit court of Delta County, Michigan, in the case of *Dora Marks* v. *The Chicago Furniture and Lumber Company*, contrary to art. 4, sec. 1, of the Constitution of the United States, which provides: 'Full faith and credit shall be given in every State to the public acts, records and judicial proceedings of every other State.'"

It also appears that, in the tenth assignment of error filed in the Appellate Court, it was alleged that the circuit court had erred in failing to give full faith and credit to the judgment, records and judicial proceedings of the circuit court of Delta County, Michigan, as required by the Constitution of the United States.

It further appears that, in the assignment of errors filed in the Supreme Court of Illinois to the judgment and action of the Appellate Court, it was alleged that the Appellate Court erred in "not reversing said judgment by reason of the error of the circuit court in failing to give full faith and credit to the judgment and judicial proceedings of the circuit court of Delta County, Mich.," and also error was alleged in that "the

Appellate Court erred, as did the circuit court, in failing to give full faith and credit to the judgment of the circuit court of Delta County, Michigan, rendered in the case of *Dora Marks* v. *The Chicago Furniture and Lumber Company*, and introduced in evidence in this cause, which judgment is as follows: 'This cause having been settled, it is hereby discontinued by consent of both parties without cost to either party,' as required by article four, section one, of the Constitution of the United States."

And it is assigned for error in this court that the courts below failed to give full faith and credit to the judicial records and proceedings of the circuit court of Delta County, Michigan, in the case of *Dora Marks* v. *The Chicago Furniture and Lumber Company*, and thus deprived the plaintiff in error of his rights and privileges under said article 4, section 1, of the Constitution of the United States; and, indeed, this is the sole error relied on here by the plaintiff in error.

We think, therefore, that the question whether the record and judicial proceedings in the Michigan court received full faith and credit in the courts of Illinois is one for us to consider and determine, and we hence decline to dismiss the writ of error. *Green* v. *Van Buskirk*, 5 Wall. 307, 314; *Carpenter* v. *Strange*, 141 U. S. 87, 103; *Huntington* v. *Attrill*, 146 U. S. 657, 684.

We come, then, to the question whether, upon the facts disclosed in this record, the courts of Illinois gave full faith and credit, within the meaning of the Constitution of the United States, to the judgment and judicial proceedings of the state court of Michigan.

And, first, what was the case made by the pleadings?

The declaration was in action on the case, and alleged that the defendant induced the plaintiff, by false and fraudulent representations, to join him and one Neufeldt in a scheme to form a corporation for the purpose of carrying on the business of the manufacture and sale of furniture in the town of Escanaba, in the State of Michigan, and to furnish and pay to the defendant the sum of $5000, for which the plaintiff was to receive shares of stock in the proposed company; that, relying on the said false and fraudulent representations, (the nature of which were stated in the declaration,) the plaintiff paid over the said sum of $5000, and became a member of the corporation known as the Chicago Furniture and Lumber Company, composed of the plaintiff, the defendant and said Neufeldt; that, owing to the fact that the said representations as to the defendant and Neufeldt putting in large sums of money into the enterprise proved to be false and untrue, as the defendant well knew, the shares of stock taken by plaintiff were valueless, and so the defendant falsely deceived and defrauded the plaintiff, to her damage in the sum of ten thousand dollars.

To this declaration the defendant pleaded the general issue of not guilty, and several special pleas, setting forth, in several phases, that after the making of the said alleged false representations by the defendant, and after the plaintiff had parted with her money on the strength thereof, as set out in the declaration, the plaintiff, on or about the 4th of December, 1893, instituted an action in the circuit court of Delta County, Michigan, against the Chicago Furniture and Lumber Company, whereby she sought to recover from said company the sum of four thousand dollars, which she asserted the said company owed her as having been fraudulently contracted and procured; that the company was served and appeared; that afterwards the plaintiff and the defendant company came to a settlement of the said cause of action, and an order was duly entered on July 25, 1894, in said circuit court of Delta County, Michigan, in the following terms:

" This cause having been settled, it is hereby discontinued by consent of both parties without cost to either party;" that the said cause of action set forth in the declaration in this cause is brought upon the same claim upon which the said action was brought by the said Dora Marks against the said Chicago Furniture and Lumber Company; that thus " the plaintiff has received satisfaction and payment of her said claim; and this the defendant is ready to verify."

To these special pleas the plaintiff filed a replication, alleging that the cause of action set forth in her said declaration was not the same claim as that sued on by the plaintiff against the Chicago Furniture and Lumber Company in the circuit court of

Delta County, Michigan, and that she, the plaintiff, did not, nor has she at any time received satisfaction of her said claim sued on herein, and of this put herself upon the country.

In the trial of the issues thus made up the defendant put in evidence a certified copy of the proceedings in the Michigan court, and the plaintiff, in connection therewith, put in evidence an agreement between the Chicago Furniture and Lumber Company and Dora Marks, in the following terms:

" Articles of agreement made and entered into this 14th day of July, A. D. 1894, by and between the Chicago Furniture and Lumber Company, a corporation, of the city of Escanaba, Delta County, Michigan, parties of the first part, and Dora Marks, Denver, Colorado, party of the second part. Party of the first part agrees to purchase the twenty thousand dollars' worth of stock of the said Chicago Furniture and Lumber Company, which the party of the second part holds, for the sum of $4000, to be paid as follows: $1000 to Mead & Jennings, attorneys for said party of the second part, as soon as the said parties of the first part dispose of their treasury stock to the amount of $1000 or interest other capital in said company to the amount of $1000 and $3000 to said party of the second part, on the day that the plant now occupied by the parties of the first part in said city of Escanaba is turned over to them and a clear title to the property earned by them. Parties of the first part further agree to discontinue the damage suit now pending against the party of the second part without cost. Said parties of the first part further agree to release said party of the second part from all liability of said second party for the balance due on unpaid stock. Party of the second part agrees to sell her said stock of $20,000 to the parties of the first part and accept payment as aforesaid mentioned. Party of the second part also agrees to discontinue the suit now pending under attachment proceedings against party of the first part without cost. Said stock to be transferred as paid for."

On July 21, 1897, the jury found the defendant guilty, and assessed the plaintiff's damages at four thousand dollars, and on November 29, 1897, after a motion for a new trial had been made and overruled, a final judgment was entered according to the verdict.

As already stated, the judgment of the circuit court was affirmed by the Appellate Court, whose judgment was affirmed by the Supreme Court of Illinois.

It is, of course, obvious that none of the errors assigned to the rulings of the trial court, in the admission or rejection of evidence, or to its instructions to the jury, nor those assigned to the judgments of the Appellate and Supreme Courts, can be considered by us except as they affect the question of the legal import of the Michigan judgment, as concluding the controversy between the parties in the Illinois courts.

The trial court did not reject the record of the proceedings in the Michigan court as evidence entitled to be considered by the court and jury in the Illinois court. Did those proceedings disclose that the cause of action in the Michigan court was, in legal contemplation, the same with that asserted in the Illinois court? Did they disclose that the plaintiff, by making the settlement therein, had received satisfaction of her claim against Jacobs asserted in the present action? Did they disclose that the plaintiff, by bringing and discontinuing an action against the furniture company, accept the agreement of July 14, 1894, as a satisfaction of her alleged claim, and did such conduct on her part operate as a release of that company, and, if so, did the release operate in favor of the defendant in the present suit?

So far as these questions involve matters of fact they are concluded by the verdict of the jury. That verdict imports, under the issues formed by the pleadings, that the claim asserted against the corporation in the Michigan court was not the same with that asserted against Jacobs in the circuit court of Illinois, and that, whether or not the claims were the same, the plaintiff never received payment or satisfaction of her claim.

The plaintiff in error, therefore, is bound to maintain that, as a necessary implication of law, regardless of the verdict of the jury, the two actions asserted the same claim, and that the judgment and proceedings in the Michigan court precluded the plaintiff from maintaining a subsequent suit against the defendant in the Illinois court.

It is, no doubt, true that the object of the plaintiff was the same in both suits, namely, to be indemnified for the loss incurred by putting her money into the venture, but it does not follow that the causes of action were the same. Apparently, the theory of her action against the company was to treat the money advanced as a loan made to the company and induced by false representations. But if she found herself mistaken in her choice of a remedy, she was not thereby deprived of a right of redress against the person who had deceived her. It is, however, contended that she was entitled to but one satisfaction, and that the legal import of the judgment in the Michigan court is that she had received a satisfaction of her claim in that suit. But we think that the judgment in question did not necessarily import that the plaintiff had received satisfaction of her claim. The recital that the cause had been settled was not an adjudication by the court. It evidently had reference to the agreement of July 14, 1894, which was matter *dehors* the record, and with which the court had nothing to do. The entry that the "cause is hereby discontinued by consent of both parties, without cost to either party," although entered as a judgment of the court, does not of itself import an agreement to terminate the controversy, nor imply an intention to merge the cause of action in the judgment. The case of *Halderman* v. *United States*, 91 U. S. 584, is quite in point. There a judgment entry in the words "dismissed agreed" was pleaded, in a subsequent action, as a former recovery, but it was held by the Circuit Court and by this court that such an entry did not sustain the plea. It was said by Mr. Justice Davis, delivering the opinion:

"It is a general rule that a plea of former recovery, whether it be by confession, verdict, or demurrer, is a bar to any new action of the same or the like nature for the same cause. This rule conforms to the policy of the law, which requires an end to the litigation after its merits have been determined. But there must be at least one decision on the *right* between the parties before there can be said to be a termination of the controversy, and before a judgment can avail as a bar to a subsequent suit. Conceding that this action is between the same parties as well as for the same subject-matter as the former one, are the United

States barred from a recovery by reason of anything alleged in the pleas? The first, second and fourth pleas are not essentially different. In each the judgment relied on is 'that the said suit is not prosecuted and be dismissed.' This entry is nothing more than the record of a nonsuit, although the customary technical language is not used. But the plaintiffs in error deny that this is the effect of the order, and insist that the pleas present a case of *retraxit*, by which the United States forever lost their action, because they voluntarily announced to the court that, on the defendants paying the costs, the suit be dismissed. Such an announcement does not imply that they had no cause of action, or, if they had, that they intended to renounce it, or that it was adjusted. Nonsuits are frequently taken on payment of costs by the adverse party, in order that the controversy may be arranged out of court; but they do not preclude the institution and maintenance of subsequent suits in case of failure to settle the matters in dispute. . . . Whatever may be the effect given by the courts of Kentucky to a judgment entry 'dismissed agreed,' it is manifest that the words do not of themselves import an agreement to terminate the controversy, nor imply an intention to merge the cause of action in the judgment. Suits are often dismissed by the parties; and a general entry is made to that effect, without incorporating in the record, or even placing on file, the agreement. It may settle nothing, or it may settle the entire dispute. If the latter, there must be a proper statement to that effect to render it available as a bar. But the general entry of the dismissal of a suit by agreement is evidence of an intention, not to abandon the claim on which it is founded, but to preserve the right to bring a new suit thereon, if it becomes necessary. It is a withdrawal of a suit on terms, which may be more or less important. They may refer to costs, or they may embrace a full settlement of the contested points; but, if they are sufficient to bar the plaintiff, the plea must show it."

Such views apply still more strongly in the present case, because, as we have seen, the parties in the two suits were not the same, and because the agreement which led to the discontinuance of the suit in the Michigan court proved, when produced at the trial of the present suit, to have been executory in its terms, and not, in any sense, a renunciation of the plaintiff's claim. It was also shown, to the satisfaction of the jury, that this agreement was never fulfilled by the company, and that the plaintiff had never received the money therein promised.

The case above cited also answers the contention of the plaintiff in error that it was not competent for the plaintiff to show that the discontinuance of the suit in Michigan was induced by an executory agreement on the part of the defendant company, and that such an agreement had not been fulfilled. If the defendant, instead of going to trial on the plaintiff's replication that she had never satisfaction of the claim sued on, had demurred thereto on the ground that it was not competent to contradict the legal import of the Michigan judgment by the evidence offered, upon the principle of the case cited the demurrer must have been overruled.

We are of opinion that the trial court did not err in permitting the plaintiff to show that the entry of discontinuance in the Michigan case was not intended by the parties as a release and satisfaction of the cause of action, but was the result of a promissory agreement on the part of the defendant company which was never complied with. Such evidence was competent to support the plaintiff's replication to the defendant's plea in the present suit, that the plaintiff had received full satisfaction and payment of her said claim. In admitting such evidence the court did not refuse to give full faith and credit to the Michigan judgment, but properly allowed evidence, not to contradict the necessary legal import of that judgment, but to show the real meaning of the parties to that suit in agreeing upon its discontinuance.

As against the case of *Halderman* v. *United States*, the counsel for the plaintiff in error cite the subsequent case of *United States* v. *Parker*, 120 U. S. 89, 96, which they contend must be understood as overruling the prior case. In this view of the two cases we do not agree.

In the latter case the question arose whether a former judgment in a suit by the United States against Parker as principal and Stuart as surety, upon an official bond, was a judgment of

nonsuit, which would have permitted the United States to bring
another action, or whether it was equivalent to a *retraxit*, by
which the United States forever lost their action, and the latter
was held.

But this court did not thereby disapprove of the doctrine of
the *Halderman* case, or depart from its reasoning, as is seen in
the fact that that case was cited, with others, as establishing
the principle that a nonsuit is not conclusive as an estoppel, because it does not determine the right of the parties. This
court in discussing the facts of the case, (after quoting the text
of the Practice Act of Nevada, in which State the action had
been tried,) said:

"It thus appears that there are five instances in which the
dismissal of an action has the force only of a judgment of nonsuit; 'in every other case,' the statute provides, 'the judgment
shall be rendered on the merits.' If the case at bar is not included among the enumerated cases in which a dismissal is equivalent to a nonsuit, it must therefore be a judgment on the merits.
In the present case the suit was not dismissed by the plaintiff
himself before trial, nor by one party upon the written consent
of the other, nor by the court for the plaintiff's failure to appear
on the trial, nor by the court at the trial for an abandonment
by the plaintiff of his cause; neither was a dismissal by the court
upon motion of the defendant, on the ground that the plaintiff
had failed to prove a sufficient case for the jury at the trial.
The judgment was rendered upon the evidence offered by the
defendants, which could only have been after the plaintiff had
made out a *prima facie* case. That evidence was passed upon
judicially by the court, who determined its effect to be a bar to
the cause of action. This was confirmed by the consent of the
attorney representing the United States. The judgment of dismissal was based on the ground of the finding of the court, as
matter of fact and matter of law, that the subject-matter of the
suit had been so adjusted and settled by the parties that there
was no cause of action then existing. This was an ascertainment judicially that the defence relied on was valid and sufficient, and consequently was a judgment upon the merits, finding
the issue for the defendants. Being, as already found, for the

same cause of action as now sued upon, it operates as a bar to the present suit by way of estoppel."

This statement of the facts and law in that case clearly shows that the decision is not inconsistent with that announced in the case of *Halderman* v. *United States*, and also that it is not applicable to the case in hand.

These views dispose of the only question which our jurisdiction enables us to review.

Finding, as we do, that the courts of Illinois gave all that faith and credit to the judgment and judicial proceedings in the Michigan court to which they were entitled under the Constitution of the United States, the other errors assigned we cannot consider, and the judgment of the Supreme Court of Illinois is

Affirmed.

GLAVEY *v.* UNITED STATES.

APPEAL FROM THE COURT OF CLAIMS.

No. 235. Argued April 11, 12, 1901.—Decided May 27, 1901.

When an office with a fixed salary has been created by statute, and a person duly appointed to it has qualified and entered upon the discharge of his duties, he is entitled, during his incumbency, to be paid the salary prescribed by statute.

Such an appointment is complete when duly made by the President and confirmed by the Senate, and the giving of a bond required by law is a mere ministerial act for the security of the Government, and not a condition precedent to his authority to act in performance of the duties of the office.

As the act of 1882 created a distinct, separate office, with a fixed annual salary for the incumbent, to be paid by the Secretary of the Treasury; as the plaintiff was legally appointed thereto, by the Secretary under and by virtue alone of that act; and as he entered upon the discharge of the duties appertaining to that position, he was entitled to demand the salary attached by Congress to the office.

THE case is stated in the opinion of the court.

Mr. Robert D. Benedict for appellant. *Mr. E. S. Mussey* was on his brief.

Mr. Assistant Attorney General Pradt for the United States. *Mr. Felix Brannigan* was on his brief.

MR. JUSTICE HARLAN delivered the opinion of the court.

This action was brought May 22, 1897, to recover from the United States the sum of $6011.98, which amount the plaintiff Glavey, who was formerly a local inspector of vessels at New Orleans, alleged that he was entitled to receive for services performed by him as a special inspector of foreign steam vessels at the same city, at the rate of two thousand dollars per annum from May 25, 1891, to May 27, 1894.

The Court of Claims dismissed the petition. The majority of that court were of opinion that under the terms of his appointment the plaintiff was precluded from demanding compensation for any services performed by him as special inspector of foreign steam vessels. The minority were of opinion that the statute having fixed the **salary of** a special inspector of foreign steam vessels, it was beyond the power of the Secretary, in whom was vested the power of appointment, to prescribe as a condition of the plaintiff's appointment that he should serve as such special inspector without compensation beyond that received by him as a local inspector. 35 C. Cl. 242.

By section 4400 of the Revised Statutes of the United States, Title "Regulation of Steam Vessels," as the revision stood prior to August 7, 1882, it was provided: "All steam vessels navigating any waters of the United States which are common highways of commerce, or open to general or competitive navigation, excepting public vessels of the United States, vessels of other countries, and boats propelled in whole or in part by steam for navigating canals, shall be subject to the provisions of this Title."

Section 4415 of the same title relates to local boards of inspectors and the appointment of local inspectors.

Section 4400 was amended and enlarged by the act of Congress approved August 7, 1882, c. 441, by adding at the end of that section these words: "And all private foreign steam vessels carrying passengers from any port of the United States to any other place or country shall be subject to the provisions of sections 4417, 4418, 4421, 4422, 4423, 4424, 4470, 4471, 4472, 4473, 4479, 4482, 4488, 4489, 4496, 4497, 4499 and 4500 of this Title, and ·shall be liable to visitation and inspection by the proper officer, in any of the ports of the United States, respecting any of the provisions of the sections aforesaid." 22 Stat. 346.

By that act it was further provided that for the purpose of carrying into effect its provisions "the Secretary of the Treasury shall appoint officers to be designated as special inspectors of foreign steam vessels, at a salary of two thousand dollars per annum each, and there shall be appointed of such officers at the port of New York, six; at the port of Boston, two; at the port of New Orleans, two; and at the port of San Francisco, two," § 2; that "the special inspectors of foreign steam vessels shall perform the duties of their office and make reports thereof to the Supervising Inspector General of Steam Vessels, · under such regulations as shall be prescribed by the Secretary of the Treasury," § 3; that "each special inspector of foreign steam vessels shall execute a proper bond, to be approved by the Secretary of the Treasury, in such form and upon such conditions as the Secretary may prescribe, for the faithful performance of the duties of his office," § 4; that "the Secretary of the Treasury shall procure for the several inspectors heretofore referred to such instruments, stationery, printing, and other things necessary, including clerical help, where he shall deem the same necessary for the use of their respective offices, as may be required therefor," § 5; and that "the salaries of the special inspectors of foreign steam vessels and clerks provided for, together with their traveling and other expenses, when ·on official duty, and all instruments, books, blanks, stationery, furniture, and other things necessary to carry into effect the provisions of this act, shall be paid for by the Secretary of the Treasury, out of any moneys in the Treasury not otherwise appropriated," § 6. 22 Stat. 346.

The judgment of the Court of Claims was based upon a finding of facts which is here given in full:

"I. The claimant, a citizen of the United States, residing at New Orleans, La., was, on the 17th day of April, 1891, duly appointed, pursuant to Revised Statutes, section 4415, to the office of local inspector of hulls of steam vessels, for the district of New Orleans, La., and on April 21, 1891, he accepted said appointment and duly qualified by taking the prescribed oath of office and by forwarding the same together with the official bond prescribed by law therefor to the Treasury Department. He then and there entered upon the discharge of his duties and continued to discharge the same until May 27, 1894. During the claimant's incumbency of said office he claimed each month the salary thereof by rendering his accounts therefor, which were promptly paid by the defendants.

"II. The report of the supervising inspector general for the fiscal year ending June 30, 1889, recommended:

"'That sections 2 to 6, inclusive, of the amendment to section 4400, Revised Statutes, which provides a separate set of officers and clerks for the inspection of foreign steam vessels, be repealed, the reasons for the creation of such offices having ceased to exist upon the passage of the act approved June 19, 1886, which abolished the fees formerly collected from domestic steam vessels and their licensed officers, which fees were permanently appropriated previously for the support of the domestic inspection service and which could not legally be diverted therefrom for the support of officers and clerks inspecting foreign steam vessels, from whom no fees could legally be collected for such support. The action of Congress in the matter of creating the separate offices was based on the reasons given in the following extract from the special report of the supervising inspector general, dated January 21, 1882: "... Authority should be given the Secretary of the Treasury to appoint these special inspectors and to pay their salaries, ... per annum, and necessary traveling expenses, from funds appropriated from moneys in the Treasury not otherwise appropriated, as it would seem obviously improper that such special officers should be paid from the appropriation for the salaries and expenses of steamboat inspection from funds collected by a tax on American steamboat owners and the licensed officers of such vessels." As the officers and clerks of both services are now paid from funds in the general Treasury, the advantage of uniting the two services must be clearly obvious, both as to public interests and economy in conducting the service. In the latter respect a saving can be made of all the salaries now being paid, except at the port of New York, where two of the officers and the clerk might be retained by transfer to the domestic service, dispensing with the services of the other two now employed. The inspectors at San Francisco, Boston, Philadelphia, Baltimore and New Orleans could be dispensed with altogether, thereby saving to the Government the sum of $14,000 annually, the total of salaries now paid those officers. The additional work that would fall upon the domestic service by such dispensation would be as follows: At New York, 138 steamers; San Francisco, 11; Boston, 18; Portland, Me., 7; Philadelphia, 8; Baltimore, 10; Port Huron, 3; Marquette, 11; Buffalo, 8; Oswego, 22; Burlington, Vt., 3; Detroit, 2; New Orleans, 16. Total steamers, 257.'

" III. By the finance report of the Secretary of the Treasury to the Speaker of the House of Representatives, first session Fifty-first Congress (1889), it was recommended 'that all laws be repealed which provide a separate establishment for the inspection of foreign steam vessels, and that the inspectors of domestic steam vessels be authorized and required to perform all necessary services in connection with the inspection of foreign steamships. The offices proposed for abolition are virtually sinecures, and until they are abolished the Executive will remain subjected to importunity to fill them. The services of three of these officers have been dispensed with.' The three offices disposed of were those at San Francisco, Cal., New Orleans, La., and Philadelphia, Pa.

" IV. While the claimant was holding the office aforesaid, to wit, May 25, 1891, he received from the Secretary of the Treasury a communication, of which the following is a true copy, viz: 'Treasury Department, Office of the Secretary, Washington, D. C., May 15, 1891. Mr. John Glavey, New

Orleans, Louisiana. Sir: Under the provisions of an act of Congress, approved August 7, 1882, entitled 'An act to amend section 4400 of title LII of the Revised Statutes of the United States, concerning the regulation of steam vessels,' you are hereby appointed to serve in connection with your appointment as local inspector of hulls of steam vessels, as a special inspector of foreign steam vessels, without additional compensation, for the port of New Orleans, Louisiana, the appointment to take effect from date of oath. Respectfully yours, Charles Foster, Secretary.'

"V. May 25, 1891, the claimant took the oath therein referred to, which was in the usual form of an oath of office, and transmitted the same to the Secretary of the Treasury on that date. He was not required to and did not give or offer to give the bond prescribed by statute for the office of special inspector of foreign steam vessels. From the time of taking the oath aforesaid until May 27, 1894, the claimant performed the duties of a special inspector of foreign steam vessels at said port.

"VI. During the time the claimant was performing the duties of special inspector of foreign steam vessels, as aforesaid, he made no request or demand upon the Secretary of the Treasury or any other officer of the defendants, to be paid the salary prescribed by law for the incumbent of the office of special inspector of foreign steam vessels at said port, nor did he when he subscribed the oath as aforesaid; nor did he at any time thereafter while he held said office of local inspector of hulls of steam vessels, for which he was paid as aforesaid, make to the Secretary of the Treasury or to any other officer of the Government any protest or objection whatever to the performance of the duties of special inspector of foreign steam vessels in connection with his appointment as local inspector of hulls of steam vessels at said port without additional compensation.

"VII. Prior to the time the claimant ceased to perform the services aforesaid he received from the acting Secretary of the Treasury a communication of which the following is a true copy: 'Treasury Department, Office of the Secretary, Washington, D. C., December 15, 1893. Mr. John Glavey, inspector of hulls of steam vessels, New Orleans, La. Sir: Department letter of the 7th instant requesting you to tender your resignation as inspector of hulls of steam vessels for the tenth district is hereby revoked, and you are requested to tender your resignation as inspector of hulls of steam vessels for the district of New Orleans, La., also as special inspector of foreign steam vessels for the port of New Orleans, La., to take effect upon the appointment and qualification of your successor. Respectfully yours, W. E. Curtis, Acting Secretary.' Thereafter he received from the acting Secretary another communication, of which the following is a copy: 'Treasury Department, Office of the Secretary, Washington, D. C., April 14, 1894. Mr. John Glavey, inspector of hulls of steam vessels, New Orleans, La. Sir: Your services as inspector of hulls of steam vessels for the district of New Orleans, La., are hereby discontinued, to take effect upon the appointment and qualification of your successor. Respectfully yours, S. Wike, Acting Secretary.' And, thereafter, May 28, 1894, the claimant's duly appointed and qualified successor as local inspector of hulls of steam vessels entered upon the discharge of the duties of said office, after which the claimant ceased to perform the duties of said office. The claimant performed the duties of said office as special inspector of foreign steam vessels until said May 26, 1894, a period of three years and two days."

The learned Assistant Attorney General admits it to be a general principle that when an office with a fixed salary has been created by statute and a person duly appointed to it has qualified and entered upon the discharge of his duties, he is entitled during his incumbency to be paid the salary prescribed by statute. He insists, however, that this principle is not applicable in the present case because, he contends, the Secretary of the Treasury did not mean, by his letter or communication of May 15, 1891, to appoint Glavey to the office of special inspector of foreign steam vessels at the port of New Orleans.

We cannot sustain this contention. Section 4400 of the Revised Statutes was so amended by the act of August 7, 1882, as to bring foreign steam vessels within the provisions of certain other specified sections; and by the same act, and for the purpose of carrying its provisions into effect, the Secretary of the

Treasury was directed to appoint special inspectors of foreign steam vessels at designated ports, one of which was the port of New Orleans. In view of the express words of the act, his failure or refusal to appoint might have been regarded as a failure or refusal to discharge a duty distinctly imposed upon him by statute. And that seems to have been the view of that officer, for although he had officially declared to Congress that the office of special inspector of foreign steam vessels was virtually a "sinecure," he shows by his communication of May 15, 1891, that he regarded the act of August 7, 1882, as mandatory, and that he appointed Glavey in obedience to its provisions. As he had no authority to appoint Glavey except in virtue of that act, we cannot assume that he proceeded or intended to proceed outside of its provisions. We must take it that he meant just what he plainly and expressly declared, and consequently that he intended, in virtue of the authority given by the act of 1882, to appoint Glavey to the office of special inspector of foreign steam vessels at New Orleans.

The next contention of the Government is that if the communication of May 15, 1891, is to be taken as showing a valid appointment to the office in question, Glavey did not legally qualify as special inspector in that he did not give or tender the bond prescribed by section 4 of the act of 1882; consequently, it is argued, he was at most only an officer *de facto*.

Is it true that the execution of the required bond was necessary in order that Glavey could lawfully proceed in the discharge of the duties of the office to which he was appointed?

Some light is thrown upon this question by *United States* v. *Bradley*, 10 Pet. 343, 357, 364. That was an action upon a bond of one who acted as paymaster in the army. The act under which the bond was taken provided that "all officers of the pay, commissary and quartermaster's department, shall, previous to entering on the duties of their offices, give good and sufficient bonds to the United States, fully to account for all moneys and public property which they may receive, in such sum as the Secretary of War shall direct." 3 Stat. 298, c. 69, § 6. This court, speaking by Mr. Justice Story, after observing that the proper officers of a department to which the disbursement of public moneys was entrusted could take a valid bond to secure the Government in respect of such moneys, said: "Before concluding this opinion, it may be proper to take notice of another objéction raised by the third plea, and pressed at the argument. It is that Hall was not entitled to act as paymaster until he had given the bond required by the act of 1816, in the form therein prescribed; and that not having given any such bond, he is not accountable as paymaster for any moneys received by him from the Government. We are of a different opinion. Hall's appointment as paymaster was complete when his appointment was duly made by the President and confirmed by the Senate. The giving of the bond was a mere ministerial act for the security of the Government, and not a condition precedent to his authority to act as paymaster. Having received the public moneys as paymaster, he must account for them as paymaster."

The doctrine announced in that case was reaffirmed in *United States* v. *Linn*, 15 Pet. 290, 313, which was an action upon a writing obligatory given by a receiver of public moneys in a certain land office. The case came before this court upon questions in respect of which the judges of the Circuit Court were divided. Those questions were: 1. Whether the obligation of the receiver and his sureties, being without seal, was a bond within the act of Congress of May 10, 1800, which provided that a receiver of public moneys for lands of the United States "shall, before he enters upon the duties of his office, give bond, with approved security . . . for the faithful discharge of his trust." 2 Stat. 73, 75, c. 55, § 6. 2. Whether such an instrument was good at common law. The court, speaking by Mr. Justice Thompson, and referring to the emoluments which the receiver was entitled to have, said: "These emoluments were the considerations allowed him for the execution of the duties of his office; and his appointment and commission entitled him to receive this compensation, whether he gave any security or not. His official rights and duties attached upon his appointment. This was so held by this court in the case of *United States* v. *Bradley*, 10 Pet. 364." After stating what had been decided in that case, the court proceeded: "According to this doctrine, which is undoubtedly sound, Linn was a receiver *de jure*, as well as *de facto*, when the instrument in question was given."

In *United States* v. *Le Baron*, 19 How. 73, 78, the question was as to the time when a person nominated and confirmed as a deputy postmaster, and whose commission was put into the hands of the Postmaster General for delivery to the appointee, was to be deemed to have been invested with such office. This court, speaking by Mr. Justice Curtis, said : " When a person has been nominated to an office by the President, confirmed by the Senate, and his commission has been signed by the President, and the seal of the United States affixed thereto, his appointment to that office is complete. Congress may provide, as it has done in this case, that certain acts shall be done by the appointee before he shall enter on the possession of the office under his appointment. These acts then become conditions precedent to the complete investiture of the office ; but they are to be performed by the appointee, not by the Executive ; all that the Executive can do to invest the person with his office has been completed when the commission has been signed and sealed ; and when the person has performed the required conditions, his title to enter on the possession of the office is also complete."

It may be here observed that the above cases are stronger than the present case in that the act of 1882 contained no provision requiring a special inspector of foreign steam vessels to execute a bond before entering on the duties of his office. We observe also that the principles announced in the *Bradley* and *Linn* cases were recognized in *United States* v. *Eaton*, 169 U. S. 331.

In view of the former decisions of this court, it cannot be held that the execution by Glavey of the bond required by the act of 1882 was a condition precedent to his right to exercise the functions of the office to which he was appointed by the Secretary of the Treasury. Congress did not so direct. His appointment was complete, at least, when he took the required oath and transmitted evidence of that fact to the Secretary. After taking the oath, evidencing thereby his acceptance of the appointment, he was entitled to proceed in the execution of the duties of his office and became liable for any failure to properly discharge them.

It remains to inquire whether, by reason of the statement in the Secretary's letter of communication of May 15, 1891, that the appointment in question was "without additional compensation" beyond that received by the appointee as local inspector of hulls of steam vessels, Glavey was estopped to demand the salary fixed by the act of 1882 for special inspectors of foreign steam vessels.

In *United States* v. *Symonds*, 120 U. S. 46, 49, the question was whether certain services were performed "at sea" within the meaning of section 1556 of the Revised Statutes fixing the pay of lieutenants in the navy when at sea, or when on shore duty, or when on leave or waiting orders. Symonds claimed that the services for which he sued were performed "at sea," and that he was entitled to the compensation fixed by the statute for services of that kind. This court said: "If the regulations of 1876 had not recognized services 'on board a practice ship at sea' as sea services, the argument on behalf of the Government would imply that they could not be regarded by the courts, or by the proper accounting officers, as sea services; in other words that the Secretary of the Navy could fix, by order and conclusively, what was and was not sea service. But Congress certainly did not intend to confer authority upon the Secretary of the Navy to diminish an officer's compensation, as established by law, by declaring that to be shore service which was in fact sea service , or to increase his compensation by declaring that to be sea service which was in fact shore service. The authority of the Secretary to issue orders, regulations and instructions, with the approval of the President, in reference to matters connected with the naval establishment, is subject to the condition, necessarily implied, that they must be consistent with the statutes which have been enacted by Congress in reference to the navy. He may, with the approval of the President, establish regulations in execution of, or supplemental to, but not in conflict with, the statutes defining his powers or conferring rights upon others. The contrary has

never been held by this court. What we now say is entirely
consistent with *Gratiot* v. *United States*, 4 How. 80, and *Ex
parte Reed*, 100 U. S. 13, upon which the Government relies.
Referring in the first case to certain army regulations, and in
the other to certain navy regulations, which had been approved
by Congress, the court observed that they had the force of law.
See also *Smith* v. *Whitney*, 116 U. S. 181. In neither case
however, was it held that such regulations, when in conflict
with the acts of Congress, could be upheld. If the services of
Symonds were, in the meaning of the statute, performed 'at
sea,' his right to the compensation established by law for sea
service is as absolute as the right of any other officer to his salary as established by law." To the same effect was *United
States* v. *Barnette*, 165 U. S. 174, 179.

In *People ex rel. Satterlee* v. *Board of Police*, 75 N. Y. 38, 42,
the question was whether the compensation of a police surgeon
was that fixed by statute or that named in a resolution of a
board of police under which he was appointed. He accepted
the appointment and performed the duties of the office for more
than two years, drawing only the salary fixed by the resolution
and which was less than that fixed by statute. The Court of
Appeals of New York, speaking by Judge Miller—all the members of the court who voted in the case concurring—said: " As
the statute gave the salary, I think fixing the amount at a less
rate, by resolution, could not make it less than the statute declared. There is no principle upon which an individual, appointed or elected to an official position, can be compelled to
take less than the salary fixed by law. The acceptance and discharge of the duties of the office, after appointment, is not a
waiver of the statutory provision fixing the salary therefor, and
does not constitute a binding contract to perform the duties of
the office for the sum named. The law does not recognize the
principle that a board of officers can reduce the amount fixed
by law for a salaried officer, and procure officials to act, at a
less sum than the statute provides, or that such official can make
a binding contract to that effect. The doctrine of waiver has
no application to any such case, and cannot be invoked to aid
the respondent."

The ruling in that case was reaffirmed in *Kehn* v. *State*, 93 N. Y. 291, 294, which involved the claim of a fireman whose compensation had been reduced by his superior officer below that fixed by law.   The court, speaking by Judge Rapallo, re-affirmed the principles of the *Satterlee* case, and approved the decision in *Goldsborough* v. *United States*, Taney's Decisions, 80, 88, saying: " The present case, however, is stronger than either of those cases cited.   At the time the appellant entered into the service his pay was fixed by law, and there is no evi-dence that he ever consented to a change.   It was reduced by the superintendent, and for a portion of the time the appellant took the reduced pay, but that does not estop him from claiming his full pay if he was legally entitled to it."

In the *Goldsborough* case referred to, Chief Justice Taney said : " Where an act of Congress declares that an officer of the Government or public agent, shall receive a certain compensa-tion for his services, which is specified in the law, undoubtedly that compensation can neither be enlarged or diminished, by any regulation or order of the President, or of a Department, unless the power to do so is given by act of Congress."

In *Adams* v. *United States*, 20 C. Cl. 115, which involved the compensation due to one who had performed the duties of an inspector and also of deputy collector of customs, the court said : " The law creates the office, prescribes its duties, and fixes the compensation.   The selection of the officer is left to the col-lector and Secretary.   The appointing power has no control, beyond the limits of the statute, over the compensation, either to increase or diminish it."   In the same case it was also said : " Monthly vouchers were drawn up, reciting the number of days the claimant was employed during the month and the amount of compensation allowed by the collector and Secretary, ending with a receipt ' in full for compensation for the period above stated,' which the claimant signed.   We do not think he thereby relinquished his right to claim the further compensation al-lowed by law.   If the appointing officer has no power to change the compensation of an inspector, certainly the paying officer has not.   He had no right to exact such a receipt and the claimant lost nothing by signing it. *Fisher's Case*, 15 C. Cl. 323; *Bostwick* v. *United States*, 94 U. S. 53."

We are of opinion that as the act of 1882 created a distinct, separate office—special inspector of foreign steam vessels—with a fixed annual salary for the incumbent, to be paid by the Secretary of the Treasury out of any moneys in the Treasury not otherwise appropriated; as the plaintiff was legally appointed by the Secretary a special inspector under and by virtue alone of that act; and as he entered upon the discharge of the duties appertaining to that position, he was entitled to demand the salary attached by Congress to the office in question.

It is said that the Secretary, before appointing the plaintiff, had reached the conclusion that the office of special inspector of foreign steam vessels was unnecessary and that all laws providing a separate establishment for the inspection of foreign steam vessels should be repealed. Such undoubtedly was the opinion expressed by the Secretary in his report to the Speaker of the House of Representatives at the first session, 1889, of the Forty-first Congress. But Congress did not immediately heed his recommendation on that subject, and there was no repeal of the act of 1882 until the passage of the statute of March 1, 1895, 28 Stat. 699, c. 146, § 1. During the entire term of his service as special inspector the act of 1882 was in force. If the Secretary, having become convinced that the special inspectors of foreign steam vessels were not needed and the public interests did not require the appointment of such officers, could properly, for such reasons, have withheld any action under the statute of 1882 until he again communicated his views to Congress, it does not follow that he could make an appointment under that statute conditioned that the appointee should accept a less salary than Congress prescribed. Whether a local inspector should be required to inspect foreign steam vessels without additional compensation, or whether the visitation and inspection of such vessels should be done by an officer acting under an appointment for that particular purpose, was a matter for the determination of Congress. The purpose of Congress, as indicated by the act of 1882, was to compensate the services of a special inspector of foreign steam vessels by an annual salary of a specified amount. It was not competent for the Secretary of the Treasury, having the power of appointment, to defeat that purpose by what was, in effect, a bargain or agreement between him and his appointee that the latter should not demand the compensation fixed by statute. Judge Lacombe, speaking for the Circuit Court of the United States for the Southern District of New York in *Miller* v. *United States,* 103 Fed. Rep. 413, 415, well said: "Any bargain whereby, in advance of his appointment to an office with a salary fixed by legislative authority, the appointee attempts to agree with the individual making the appointment that he will waive all salary or accept something less than the statutory sum, is contrary to public policy, and should not be tolerated by the courts. It is to be assumed that Congress fixes the salary with due regard to the work to be performed, and the grade of man that such salary may secure. It would lead to the grossest abuses if a candidate and the executive officer who selects him may combine together so as entirely to exclude from consideration the whole class of men who are willing to take the office on the salary Congress has fixed but will not come for less. And, if public policy prohibit such a bargain in advance, it would seem that a court should be astute not to give effect to such illegal contract by indirection, as by spelling out a waiver or estoppel." If it were held otherwise, the result would be that the Heads of Executive Departments could provide, in respect of all offices with fixed salaries attached and which they could fill by appointments, that the incumbents should not have the compensation established by Congress, but should perform the service connected with their respective positions for such compensation as the Head of a Department, under all the circumstances, deemed to be fair and adequate. In this way the subject of salaries for public officers would be under the control of the Executive Department of the Government. Public policy forbids the recognition of any such power as belonging to the Head of an Executive Department. The distribution of officers upon such a basis suggests evils in the administration of public affairs which it cannot be supposed Congress intended to produce by its legislation. Congress may control the whole subject of sal-

aries for public officers; and when it declared that for the purpose of carrying into effect the provisions of the act of 1882 the
Secretary of the Treasury "*shall* appoint officers to be designated as special inspectors of foreign steam vessels, *at a salary
of two thousand dollars per annum each*," it was not for the
Secretary to make the required appointments under a stipulation
with the appointee that he would take any less salary than that
prescribed by Congress.   The stipulation that Glavey, who was
local inspector, should exercise the functions of his office of
special inspector of foreign steam vessels "without additional
compensation" was invalid under the statute prescribing the
salary he should receive, was against public policy, and imposed
no legal obligation upon him.   And the mere failure of the appointee to demand his salary as such officer until after he had
ceased to be local inspector, was not in law a waiver of his right
to the compensation fixed by the statute.

*The judgment of the Court of Claims is reversed, and the cause
is remanded for further proceedings consistent with this
opinion.*

The CHIEF JUSTICE, MR. JUSTICE BROWN, MR. JUSTICE PECK-
HAM and MR. JUSTICE McKENNA dissented.

DECISIONS ANNOUNCED WITHOUT OPINIONS DURING THE TIME COVERED BY THIS VOLUME.

That "time" was only two days, May 27 and May 28, and the "decisions announced" are all stated in Volume 181, viz.: No. 389 on page 616, and Nos. 671, 674, 665, 678, 683, 581, 635, 672, 680, 694, 697 and 703, on pages 621, 622 and 623.

SUMMARY STATEMENT OF BUSINESS OF THE SUPREME COURT OF THE UNITED STATES FOR OCTOBER TERM, 1900.

Original Docket.

Number of cases, ...	19
Number of cases disposed of, ...	9
Leaving undisposed of	10

Appellate Docket.

Number of cases at close of October Term, 1899, . .	303
Number of cases docketed at October Term, 1900, .	401
Total, ...	704
Number of cases disposed of at October Term, 1900, .	368
Number of cases remaining undisposed of . . .	336
Showing an increase of	33

INDEX.

ADMIRALTY.

1. Vessels engaged in trade between Porto Rican ports and ports of the United States are engaged in the coasting trade in the sense in which those words are used in the New York pilotage statutes; and steam vessels engaged in such trade are coastwise steam vessels under Revised Statutes, section 4444. *Huus* v. *New York & Porto Rico Steamship Co.*, 392.

2. The statutes of New York impose compulsory pilotage on foreign vessels inward and outward bound to and from the port of New York by way of Sandy Hook. *Homer Ramsdell Transportation Co.* v. *La Compagnie Générale Transatlantique*, 406.

3. In an action at common law the ship owner is not liable for injuries inflicted exclusively by negligence of a pilot accepted by a vessel compulsorily. *Ib.*

BANKRUPT.

1. Frank Brothers were adjudged bankrupts in February, 1899. For a long time prior to that Pirie & Co. had dealt with them, selling them merchandise. Within four months prior to the adjudication of bankruptcy Pirie & Co. received from them $1336.79, leaving a balance still due and unpaid of $3093.98. When this payment was made Frank Brothers were hopelessly insolvent to the knowledge of Frank Brothers, but Pirie & Co. and their agents had no knowledge of it, and had no reasonable cause to believe that the bankrupts by such payment intended to give a preference, nor did they intend to do so. Pirie & Co. proved their claim against the estate, and received a dividend thereon, which they still hold. *Pirie* v. *Chicago Title & Trust Co.*, 438.

2. The provisions in the Bankrupt Act of July 1, 1898, c. 541, § 60, that "a person shall be deemed to have given a preference if, being insolvent, he has procured or suffered a judgment to be entered against himself in favor of any person, or made a transfer of any of his property, and the effect of the enforcement of such judgment or transfer will be to enable any one of his creditors to obtain a greater percentage of his debt than any other of such creditors, of the same class," means that a transfer of property includes the giving or conveying anything of value, anything which has debt paying or debt securing power; and money is property. If the person receiving such preference did not have cause to believe that it was intended, he may keep the property transferred, but, if it be only a partial discharge of his debt, cannot prove the balance. *Ib.*

(613)

3. When the purpose of a prior law is continued, its words usually are, and an omission of the words implies an omission of the purpose. *Ib.*

4. The object of a bankrupt act is, so far as creditors are concerned, to secure equality of distribution among all, of the property of the bankrupt. *Ib.*

5. Subdivision *c* of section 60 of the bankrupt act is applicable to the cases arising under subdivision *b*, and allows a set-off, which might not be otherwise allowed. *Ib.*

CASES AFFIRMED OR FOLLOWED.

De Lima v. *Bidwell, ante,* 1, followed by reversing the action of the general appraisers. *Goetze* v. *United States,* 221.

Dooley v. *United States, ante,* 222, followed. *Armstrong* v. *United States,* 243.

Lantry v. *Wallace, ante,* 536, followed. *Hood* v. *Wallace,* 555.

CASES DISTINGUISHED.

The distinction between *Halderman* v. *United States,* 91 U. S. 584, and *United States* v. *Parker,* 120 U. S. 89, shown. *Jacobs* v. *Marks,* 583.

CONCURRENCE IN JUDGMENT, BUT NOT IN OPINION.

In announcing the conclusion and judgment of the court in *Downes* v. *Bidwell,* MR. JUSTICE BROWN delivered an opinion. MR. JUSTICE WHITE delivered a concurring opinion which was also concurred in by MR. JUSTICE SHIRAS and MR. JUSTICE MCKENNA. MR. JUSTICE GRAY also delivered a concurring opinion. The Chief Justice, MR. JUSTICE HARLAN, MR. JUSTICE BREWER, and MR. JUSTICE PECKHAM dissented. Thus it is seen that there is no opinion in which a majority of the court concurred. Under these circumstances the reporter made headnotes of each of the sustaining opinions, and placed before each the names of the justices or justice who concurred in it, as follows:

I. By MR. JUSTICE BROWN, in announcing the conclusion and judgment of the court.

1. The Circuit Courts have jurisdiction, regardless of amount, of actions against a collector of customs for duties exacted and paid under protest upon merchandise alleged not to have been imported. *Downes* v. *Bidwell,* 244.

2. The island of Porto Rico is not a part of the United States within that provision of the Constitution which declares that "all duties, imposts, and excises shall be uniform throughout the United States." *Ib.*

3. There is a clear distinction between such prohibitions of the Constitution as go to the very root of the power of Congress to act at all, irrespective of time or place, and such as are operative only throughout the United States, or among the several States. *Ib.*

4. A long continued and uniform interpretation, put by the executive and legislative departments of the Government, upon a clause in the Constitution should be followed by the judicial department, unless such interpretation be manifestly contrary to its letter or spirit. *Ib.*

II. By MR. JUSTICE WHITE, with whom MR. JUSTICE SHIRAS and MR. JUS-
TICE MCKENNA concurred.

1. The government of the United States was born of the Constitution, and
all powers which it enjoys or may exercise must be either derived expressly or by implication from that instrument. Even then, when an
act of any department is challenged, because not warranted by the
Constitution, the existence of the authority is to be ascertained by determining whether the power has been conferred by the Constitution,
either in express terms or by lawful implication, to be drawn from the
express authority conferred or deduced as an attribute which legitimately inheres in the nature of the powers given, and which flows from
the character of the government established by the Constitution. In
other words, whilst confined to its constitutional orbit, the government of the United States is supreme within its lawful sphere. *Ib.*

2. Every function of the government being thus derived from the Constitution, it follows that that instrument is everywhere and at all times
potential in so far as its provisions are applicable. *Ib.*

3. Hence it is that wherever a power is given by the Constitution and there
is a limitation imposed on the authority, such restriction operates upon
and confines every action on the subject within its constitutional limits. *Ib.*

4. Consequently it is impossible to conceive that where conditions are
brought about to which any particular provision of the Constitution
applies its controlling influence may be frustrated by the action of any
or all of the departments of the government. Those departments,
when discharging, within the limits of their constitutional power, the
duties which rest on them, may of course deal with the subject committed to them in such a way as to cause the matter dealt with to
come under the control of provisions of the Constitution which may
not have been previously applicable. But this does not conflict with
the doctrine just stated, or presuppose that the Constitution may or
may not be applicable at the election of any agency of the government. *Ib.*

5. The Constitution has undoubtedly conferred on Congress the right to
create such municipal organizations as it may deem best for all the
territories of the United States whether they have been incorporated
or not, to give to the inhabitants as respects the local governments
such degree of representation as may be conducive to the public well-
being, to deprive such territory of representative government if it is
considered just to do so, and to change such local governments at discretion. *Ib.*

6. As Congress in governing the territories is subject to the Constitution,
it results that all the limitations of the Constitution which are applicable to Congress in exercising this authority necessarily limit its
power on this subject. It follows also that every provision of the Constitution which is applicable to the territories is also controlling therein.
To justify a departure from this elementary principle by a criticism
of the opinion of Mr. Chief Justice Taney in *Scott v. Sandford*, 19
How. 300, is unwarranted. Whatever may be the view entertained of

the correctness of the opinion of the court in that case, in so far as it interpreted a particular provision of the Constitution concerning slavery and decided that as so construed it was in force in the territories, this in no way affects the principle which that decision announced, that the applicable provisions of the Constitution were operative. *Ib.*

7. In the case of the territories, as in every other instance, when a provision of the Constitution is invoked, the question which arises is, not whether the Constitution is operative, for that is self-evident, but whether the provision relied on is applicable. *Ib.*

8. As Congress derives its authority to levy local taxes for local purposes within the territories, not from the general grant of power to tax as expressed in the Constitution, it follows that its right to locally tax is not to be measured by the provision empowering Congress "To lay and collect Taxes, Duties, Imposts, and Excises," and is not restrained by the requirement of uniformity throughout the United States. But the power just referred to, as well as the qualification of uniformity, restrains Congress from imposing an impost duty on goods coming into the United States from a territory which has been incorporated into and forms a part of the United States. This results because the clause of the Constitution in question does not confer upon Congress power to impose such an impost duty on goods coming from one part of the United States to another part thereof, and such duty besides would be repugnant to the requirement of uniformity throughout the United States. *Ib.*

III. By MR. JUSTICE GRAY.

1. The civil government of the United States cannot extend immediately, and of its own force over territory acquired by war. Such territory must necessarily, in the first instance, be governed by the military power under the control of the President as commander in chief. Civil government cannot take effect at once, as soon as possession is acquired under military authority, or even as soon as that possession is confirmed by treaty. It can only be put in operation by the action of the appropriate political department of the Government at such time and in such degree as that department may determine. *Ib.*

2. In a conquered territory, civil government must take effect, either by the action of the treaty-making power, or by that of the Congress of the United States. The office of a treaty of cession ordinarily is to put an end to all authority of the foreign government over the territory; and to subject the territory to the disposition of the Government of the United States. *Ib.*

3. The government and disposition of territory so acquired belong to the Government of the United States, consisting of the President, the Senate, elected by the States, and the House of Representatives, chosen by and immediately representing the people of the United States. *Ib.*

4. So long as Congress has not incorporated the territory into the United States, neither military occupation nor cession by treaty makes the conquered territory domestic territory, in the sense of the revenue laws.

But those laws concerning "foreign countries" remain applicable to the conquered territory, until changed by Congress. *Ib.*

5. If Congress is not ready to construct a complete government for the conquered territory, it may establish a temporary government, which is not subject to all the restrictions of the Constitution. *Ib.*

CONSTITUTIONAL LAW.

1. By the Customs Administrative Act of 1890 an appeal is given from the decision of the collector "as to the rate and amount of the duties chargeable upon imported merchandise," to the Board of General Appraisers, who are authorized to decide "as to the construction of the law and the facts respecting the classification of such merchandise, and the rate of duties imposed thereon under such classification;" but where the merchandise is alleged not to have been imported at all, but to have been brought from one domestic port to another, the Board of General Appraisers has no jurisdiction of the case, and an action for money had and received will lie against the collector to recover back duties assessed by him upon such property, and paid under protest. *De Lima* v. *Bidwell*, 1.

2. With the ratification of the treaty of peace between the United States and Spain, April 11, 1899, the island of Porto Rico ceased to be a "foreign country" within the meaning of the tariff laws. *Ib.*

3. Whatever effect be given to the act of March 24, 1900, applying for the benefit of Porto Rico the duties received on importations from that island after the evacuation by the Spanish forces, it has no application to an action brought before the act was passed. *Ib.*

See Concurrence in Judgment,   Customs Duties;
     but not in Opinion;   Jurisdiction, 1;
       Due Process of Law.

CONTRACT.

See Trust.

CUSTOMS DUTIES.

1. Duties upon imports from the United States to Porto Rico, collected by the military commander and by the President as Commander-in-Chief, from the time possession was taken of the island until the ratification of the treaty of peace, were legally exacted under the war power. *Dooley* v. *United States*, 222.

2. As the right to exact duties upon importations from Porto Rico to New York ceased with the ratification of the treaty of peace, the correlative right to exact duties upon imports from New York to Porto Rico also ceased at the same time. *Ib.*

See Concurrence in Judgment, but not in Opinion;
     Constitutional Law, 1, 2, 3;
     Jurisdiction.

DISTRICT OF COLUMBIA.

Park street is a public highway in the northwest section of the city of

Washington. For some days before the accident which was the ground of this action, a steam roller had been used in connection with the work of resurfacing the street with macadam. This roller became disabled, and was placed close to the south curb of the street, a canvas cover was placed over it, and it was left there for two days. On the second day the horse of the plaintiff in error, being driven along the street, became restive from the flapping of the canvas cover, reared, and upset the vehicle, and threw out the plaintiff, injuring him. *Held*, that the District of Columbia was not liable for the injuries which the plaintiff so suffered. *District of Columbia v. Moulton*, 576.

DUE PROCESS OF LAW.

1. The essential elements of due process of law are notice and opportunity to defend, and in determining whether such rights are denied, the court is governed by the substance of things and not by mere form. *Simon v. Craft*, 427.

2. A person charged with being of unsound mind is not denied due process of law by being refused an opportunity to defend, when, in fact, actual notice was served upon him of the proceedings, and when, if he had chosen to do so, he was at liberty to make such defences as he deemed advisable. *Ib.*

3. The due process clause in the Fourteenth Amendment to the Constitution does not necessitate that the proceedings in a state court should be by a particular mode, but only that there shall be a regular course of proceedings, in which notice is given of the claim asserted, and an opportunity afforded to defend against it. *Ib.*

4. This court accepts as conclusive the ruling of the supreme court of Alabama that the jury which passed upon the lunacy proceeding considered in this case was a lawful jury, that the petition was in compliance with the statute, and that the asserted omissions in the recitals in the verdict and order thereon were at best but mere irregularities which did not render void the order of the state court, appointing a guardian. *Ib.*

EVIDENCE.

See PRACTICE.

INDIAN TERRITORY COMMISSIONERS.

1. In 1896, commissioners, appointed by Judges of the United States Court in the Indian Territory were inferior officers, not holding their offices for life, or by any fixed tenure, but subject to removal by the appointing power. *Reagan v. United States*, 419.

2. Commissioners appointed by that court prior to the act of March 1, 1895, were entitled to reappointment under that act, but were removable at pleasure unless at that date, or at the date of removal, causes for removal were prescribed by law. *Ib.*

3. As no causes for removal had been prescribed by law at the date of removal of claimant in 1896, he was subject to removal by the judge of his district, and the action of that judge in removing him was not open to review in an action for salary. *Ib.*

JUDGMENT.

1. The question whether the record and judicial proceedings in the Michigan court received full faith and credit in the courts of Illinois is one for this court to consider and determine; and it holds that, upon the facts disclosed in the record, the courts of Illinois did give to the judgment and judicial proceedings of the state court of Michigan full faith and credit, within the meaning of the Constitution. *Jacobs* v. *Marks*, 583.

2. The judgment in question in this case did not necessarily import that the plaintiff had received satisfaction of her claim. *Ib.*

JURISDICTION.

1. The Court of Claims, and the Circuit Courts, acting as such, have jurisdiction of actions for the recovery of duties illegally exacted upon merchandise, alleged not to have been imported from a foreign country. *Dooley* v. *United States*, 222.

2. Under the circumstances set forth in its opinion this court thinks that the rule respecting appeals to the Court of Appeals of the District of Columbia must receive the interpretation here which was given to it by the Court of Appeals. *United States* v. *Alvey*, 456.

MANDATE OF THIS COURT.

The action of the Supreme Court of Illinois in this case on April 17, 1901, was a full compliance with the mandate of this court in this case, reported 177 U. S. 51. *Lake Street Elevated Railroad Co.* v. *Farmers' Loan and Trust Co.*, 417.

MINERAL LANDS.

1. The rights conferred upon the locators of mining locations by Rev. Stat. section 2322, are not subject to the right of way expressed in § 2323, and are not limited by § 2336. *Calhoun Gold Mining Co.* v. *Ajax, Gold Mining Co.*, 499.

2. As to § 2336, by giving to the oldest or prior location, where veins unite, all ore or mineral within the space of intersection, and the vein below the point of union, the prior location takes no more, notwithstanding that § 2322 gives to such prior location the exclusive right of possession and enjoyment of all the surface included within the limits of the location, and of all veins, lodes and ledges throughout their entire depth, the top or apex of which lies inside of such surface lines extended downward, vertically. *Held*, that § 2336 does not conflict with § 2332, but supplements it. *Ib.*

3. A locator is not confined to the vein upon which he based his location, and upon which the discovery was made. *Ib.*

4. A patent is not simply a grant for the vein, but a location gives to the locator something more than the right to the vein which is the subject of the location. *Ib.*

5. Patents are proof of the discovery. They relate back to the location of the claims, and cannot be collaterally attacked. *Ib.*

NATIONAL BANK.

1. This was an action, brought by the receiver of a national bank under Rev. Stat. § 5151, providing that shareholders of every such association shall be held individually responsible, equally and ratably, and not one for another, for all contracts, debts and engagements of such association to the amount of their stock therein, at the par value thereof, in addition to the amount invested in such share. *Lantry* v. *Wallace*, 536.

2. Assuming that the defendant became a shareholder in a national bank in consequence of fraudulent representations of the bank's officers, two questions are presented for determination: 1, Whether such representations, relied upon by defendant, constituted a defence in this action, brought by the receiver only for the purpose of enforcing the individual liability imposed by § 5151, Rev. Stat., upon shareholders of national banking associations? which question is answered in the negative; and, 2, Can the defendant, because of frauds of the bank whereby he was induced to become a purchaser of its stock, have a judgment against the receiver, on a counterclaim for money paid by him for stock, to be satisfied out of the bank's assets and funds in his control and possession? which question is also answered in the negative. *Ib.*

3. The present action is at law, its object being to enforce a liability created by statute for the benefit of creditors who have demands against the bank of which the plaintiff is receiver. If the defendant was entitled, under the facts stated, to a rescission of his contract of purchase, and to a cancellation of his stock certificate, and to be relieved from responsibility as a shareholder of the bank, he could obtain such relief only by a suit in equity, to which the bank and the receiver were parties. *Ib.*

4. If the defendant was entitled, under the facts stated, to a rescission of his contract of purchase, and to a cancellation of his stock certificate, and consequently to be relieved from all responsibility as a shareholder of the bank, he could obtain such a relief only by a suit in equity, to which the bank and the receiver were parties. *Ib.*

5. Whether a decree based upon the facts set forth in the answer, even if established in a suit in equity, would be consistent with sound principles, or with the statutes regulating the affairs of national banks, and securing the rights of creditors, is a question upon which the court does not express an opinion. *Ib.*

6. The purchase of this stock by the bank under the circumstances was *ultra vires*, but that did not render the purchase void. *Ib.*

7. As the constitution of Utah distinguished between stock and credits in determining the amount of property of a national bank subject to taxation, shares of stock were not credits, and resident and non-resident shareholders were not entitled to deduct *bona fide* indebtedness from their shares of stock. *Commercial Bank* v. *Chambers*, 556.

8. The assessed value of real estate owned by a bank in other States than that in which the bank is located, is not to be deducted in determining the amount of assessable property of the bank, unless authorized by the laws of the State in which the bank is situated. *Ib.*

NEW TRIAL.

1. The court below, of original jurisdiction in this case, had authority, upon newly discovered evidence, to grant to the railway company a new trial, after the final decision of this case, in an action at law in that court. *Fuller* v. *United States*, 562.

2. It was competent for Congress to confer upon such court, established under the authority of the United States, the power to grant a new trial in an action at law upon grounds discovered after the expiration of the term at which the verdict or decision was rendered. *Ib.*

3. The statute does not declare that the right to apply for a new trial upon newly discovered evidence after the term shall be any the less when the original term is superseded; nor that a new trial of an action at law shall not be applied for or granted, while the case is pending in the appellate court. *Ib.*

4. The statute of Arkansas in question is applicable only to actions and proceedings at law in the courts of that State, as distinguished from suits or proceedings in equity; and as application under that statute, within the time prescribed, for a new trial in an action at law, upon grounds discovered after the term at which the verdict or decision was rendered, was a matter of right, it did not require the leave of any court. *Ib.*

PATENT FOR INVENTION.

This was an action at law against the United States upon an alleged implied contract to pay for the use of. a patented invention belonging to the plaintiffs in error, in rifles used by the Government which had been purchased under contract from a Norwegian Company. It was conceded that a contract must be established in order to entitle appellants to recover, as the Court of Claims has no jurisdiction of demand against the United States founded on torts. *Held*, that on the facts proved in this case no such contract was proved against the United States, and that if the petitioners have suffered injury, it has been through the infringement of their patent, and not by a breach of contract. *Russell* v. *United States*, 516.

PRACTICE.

In this case this court holds, (1) that it was not error in the court below to try the case on the amended petition; (2) that the report to the Government of a person employed by the Attorney General in this case was properly rejected as evidence; (3) that there was no error in the rulings of the court below. *District of Columbia* v. *Talty*, 510.

PUBLIC SEWER.

1. Whether the construction of a public sewer by assessments upon adjoining property entitles the owners of such property to the free use of such sewer, or only to the right to a free entrance to their particular sewers, is a question of local policy. *Carson* v. *Brockton Sewerage Commission*, 398.

2. Notwithstanding that such sewer was built by assessments upon the property benefited, it is competent for the legislature to require persons making use of it to pay a reasonable sum for such use. *Ib.*

3. Where an ordinance fixes the charges that shall be paid for the use of a common sewer, no notice is required to be given to the property owners of an assessment for that purpose. *Ib.*

SALARIED OFFICES.

1. When an office with a fixed salary has been created by statute, and a person duly appointed to it has qualified and entered upon the discharge of his duties, he is entitled, during his incumbency, to be paid the salary prescribed by statute. *Glavey* v. *United States*, 595.

2. Such an appointment is complete when duly made by the President and confirmed by the Senate, and the giving of a bond required by law is a mere ministerial act for the security of the Government, and not a condition precedent to his authority to act in performance of the duties of the office. *Ib.*

3. As the act of 1882 created a distinct, separate office, with a fixed annual salary for the incumbent, to be paid by the Secretary of the Treasury; as the plaintiff was legally appointed thereto, by the Secretary under and by virtue alone of that act; and as he entered upon the discharge of the duties appertaining to that position, he was entitled to demand the salary attached by Congress to the office. *Ib.*

STATUTES.

A. STATUTES OF THE UNITED STATES.

B. STATUTES OF THE STATES AND TERRITORIES.

TREATIES.

TAX AND TAXATION.

TRUST.

1. As the governing committee of the stock exchange had no personal interest to the fund in question in this suit, which was placed in its possession in the trust and confidence that it would see that the purposes of the deposit were fulfilled, and that the moneys were paid out only in accordance with the terms of the trust under which it was deposited,

Lightning Source UK Ltd.
Milton Keynes UK
UKHW041348270119
336190UK00019B/435/P